Soil Biology

Volume 26

Series Editor
Ajit Varma, Amity Institute of Microbial Technology, Amity University Uttar Pradesh, Noida, UP, India

For further volumes:
http://www.springer.com/series/5138

Else K. Bünemann • Astrid Oberson •
Emmanuel Frossard
Editors

Phosphorus in Action

Biological Processes in Soil Phosphorus Cycling

Springer

Editors
Dr. Else K. Bünemann
ETH Zurich
Group of Plant Nutrition, Institute of Plant, Animal and Agro-ecosystem Sciences
Eschikon 33
8315 Lindau
Switzerland
else.buenemann@ipw.agrl.ethz.ch

Dr. Astrid Oberson
ETH Zurich
Group of Plant Nutrition, Institute of Plant, Animal and Agro-ecosystem Sciences
Eschikon 33
8315 Lindau
Switzerland
astrid.oberson@ipw.agrl.ethz.ch

Prof. Dr. Emmanuel Frossard
ETH Zurich
Group of Plant Nutrition, Institute of Plant, Animal and Agro-ecosystem Sciences
Eschikon 33
8315 Lindau
Switzerland
emmanuel.frossard@ipw.agrl.ethz.ch

ISSN 1613-3382
ISBN 978-3-642-15270-2 e-ISBN 978-3-642-15271-9
DOI 10.1007/978-3-642-15271-9
Springer Heidelberg Dordrecht London New York

Cover design: SPi Publisher Services

Printed on acid-free paper

Springer is part of Springer Science+Business Media (www.springer.com)

Preface

Phosphorus (P) is an essential element for all living organisms. In terrestrial ecosystems, P is often the most limiting nutrient. Due to human alteration of the P cycle, pollution with excess P is prevalent in many aquatic ecosystems and can also affect biodiversity in terrestrial ecosystems. Because economically mineable P deposits are finite, a better management of the P cycle is mandatory. This requires a good understanding of the processes of soil P dynamics.

Soil P occurs in inorganic and organic forms. Forms of P in the mineral phase, the physicochemical processes of sorption and desorption, and their effects on P cycling and availability are well known and predictable. Organic P forms are less well characterized and the biological processes (e.g., mineralization and immobilization) affect P dynamics to varying and often unknown degrees. In addition, the interactions of P dynamics with carbon and nitrogen are not well understood.

The chemical characterization of organic P and its dynamics in terrestrial as well as aquatic environments have been reviewed in a book quite recently (Turner et al. 2005). One conclusion from this book was that information on the mechanisms and rates of organic P transformations is urgently needed in order to better manage soil organic P. Given the great amount of information included in this "Organic P Bible", we hesitated when we were asked by the series editor Ajit Varma to publish a book on P in the Springer Series on Soil Biology. However, we realized that significant progress has been made in the last 5 years, and that it was indeed timely to collect the existing information on the biological processes in soil P cycling and present the state of the art. Measuring the rates of any reaction in soil is a step towards understanding the "action", and therefore we decided that our book should focus on 'phosphorus in action'.

We planned three sections: one on methods, one on processes, and one on ecosystems and management. We asked the authors to present case studies rather than a complete review of the topic, wherever appropriate. We also asked the authors to consider two aspects in particular: (1) that the role of biological processes in soil P cycling has to be examined against the availability of inorganic P resulting from physicochemical processes and (2) that interactions between carbon, nitrogen, and P have to be examined to fully understand similarities and differences of P cycling to the cycling of the other elements. Lastly, authors were asked to point out gaps in methods and understanding.

We are extremely grateful to all authors for contributing such an impressive number of high quality chapters. Some chapters were written by scientists who have been working together before. In other chapters, authors published together for the first time and found the exchange of ideas during the preparation stimulating. Some chapters present the state-of-the-art of topics that have been reviewed previously. Other topics have never been summarized before and the authors thus compiled truly original chapters.

As a teaser for the book, we would like to point out some of its highlights (Table 1). To our minds, this table is an attempt to assess the progress that has been made in the field, but of course you will find much more in-depth and significant information within each chapter. At the end of the book, we have tried to draw some general conclusions and to summarize the main research needs.

Table 1 Editors' choice of highlights in each chapter

Chapter	Authors	Highlight
1	Doolette and Smernik	Useful summary of the relative merits of solution and solid state ^{31}P nuclear magnetic resonance (NMR), and X-ray absorption near edge structure (XANES) in terms of sample preparation, sensitivity, resolution and quantification
2	Bünemann et al.	Novel data on P forms in microbial cells extracted from soil, i.e. cells that grew in situ, including non-culturable soil microorganisms
3	Frossard et al.	First combined review of radioactive and stable isotopes in the study of P cycling, also summarizing recent progress in the measurement of gross and net organic P mineralization rates
4	Wasaki and Maruyama	Informative table listing the molecular tools to study soil P cycling and applications in the pioneering studies done in the past 10 years
5	Schnepf et al.	Three modeling case studies as excellent examples of how the most important processes of P cycling in soil-plant systems can be identified and quantified on different temporal and spatial scales
6	Jansa et al.	Stimulating information on the accessibility of different soil P forms by arbuscular mycorrhizal and ectomycorrhizal fungi
7	Jones and Oburger	Thorough summary of mechanisms of P solubilization by soil microorganisms
8	Chapuis-Lardy et al.	Interesting presentation of the role of earthworms' surface casts and termite mounds, respectively, in P transfer and erosion
9	Nannipieri et al.	Critical examination of the limitations of conventional enzyme assays
10	George et al.	Eye-opening section on the co-ordination of plant responses to variations in P supply
11	Jouany et al.	Noteworthy figure illustrating that a grassland sward is more efficient in converting nitrogen into biomass under non-limiting than under limiting P supply
12	Weintraub	Comprehensive section on P limitation in arctic and alpine soils, including results from fertilization studies

(*continued*)

Table 1 (continued)

Chapter	Authors	Highlight
13	Fox et al.	Concise summary of P cycling and management in temperate forest plantations
14	Reed et al.	Original section on the implications of P limitation on tropical forests under climate and land use change
15	Belnap	Novel case study on the interaction between exotic annual grasses and soil P availability that exemplifies effects of invasive plants on P availability
16	Dao and Schwartz	Detailed description of P transformations in semi-solid and liquid manures
17	Oberson et al.	Comprehensive table comparing the P status and various measurements of biological P cycling in a field experiment under organic or conventional management
18	Tiessen et al.	Eye-opening description of how global change affects P cycling and vice versa, which in essence is a plea for much more efficient P use everywhere in the world

Finally, we would like to acknowledge everyone who has contributed to this book. All 63 authors deserve a very warm thank you for their willingness to contribute, the many hours of work that went into each chapter, and for the excellent communication with us. We had been warned that we would have to write many reminders, but in fact it was a very positive experience for us to work with these 18 leading authors.

Each chapter was reviewed by one or two external reviewers. All of them readily accepted the task and their efforts helped to improve the quality of the book in a very significant way. Thus, we would like to thank all of them for their dedication:

Mark Bakker, INRA Bordeaux
David E. Bignell, Queen Mary, University of London
Ruth E. Blake, Yale University
Barbara J. Cade-Menun, Agriculture and Agri-Food Canada
Cory Cleveland, University of Montana
Peter Cornish, University of Western Sydney
Tim Crews, Prescott College, Arizona
Warwick J. Dougherty, Primary Industries, Industry and Investment NSW
Reiner Giesler, Umeå University
Guy Kirk, Cranfield University
Ellen Kandeler, Hohenheim University
Petra Marschner, University of Adelaide
Roel Merckx, Katholieke Universiteit Leuven
Mike McLaughlin, CSIRO Land and Water, Adelaide
Claude Plassard, INRA Montpellier
Alan Richardson, CSIRO Plant Industry, Canberra
Peter Schweiger, University of Natural Resources and Applied Life Sciences, Vienna
Robert Sinsabaugh, University of New Mexico
Harry Olde Venterink, ETH Zurich
Peter Vitousek, Stanford University

Jutta Lindenborn and the entire staff at Springer (including Mr. Balamurugan Elumalai and his team at SPi Technologies India Private Ltd.) were extremely helpful during the publishing process. Cornelia Bühlmann and Kathrin von Arx thoroughly checked the formal aspects and reference lists of each chapter. Last but not least, we thank Ajit Varma for initiating this book.

Lindau, Switzerland

Else K. Bünemann
Astrid Oberson
Emmanuel Frossard

Reference

Turner BL, Frossard E, Baldwin DS (2005) Organic phosphorus in the environment. CABI, Wallingford

Contents

Contributors

Achat, David L. ENITA de Bordeaux, UMR 1220 TCEM (INRA-ENITAB), 1 cours du Général de Gaulle, 33175 Gradignan, France

Albaugh, Timothy J. Department of Forestry and Environmental Resources, North Carolina State University, 3108 Jordan Hall, Raleigh, NC 27695, USA

Ballester, Maria Victoria Centro de Energia Nuclear na Agricultura, Universidade de São Paulo, Av. Centenário 303, 13416-000 Piracicaba, SP, Brasil

Belnap, Jayne US Geological Survey, Canyonlands Research Station, 2290 S. West Resource Boulevard, Moab, UT 84532, USA

Bernasconi, Stefano M. Geological Institute, ETH Zurich, Sonneggstrasse 5, 8092 Zürich, Switzerland

Blanchart, Eric Institut de Recherche pour le Développement (IRD), UMR 210 Eco&Sols, Campus SupAgro, 2 place Viala, 34 060, Montpellier CEDEX 2, France

Brossard, Michel Institut de Recherche pour le Développement (IRD), UMR 210 Eco&Sols, Campus SupAgro, 2 place Viala, 34 060, Montpellier CEDEX 2, France

Bünemann, Else K. Group of Plant Nutrition, Institute of Plant, Animal and Agro-ecosystem Sciences, ETH Zurich, Eschikon 33, 8315 Lindau, Switzerland

Chapuis-Lardy, Lydie Institut de Recherche pour le Développement (IRD), UMR 210 Eco&Sols, BP 1386, CP 18524, Dakar, Senegal

Cleveland, Cory C. Department of Ecosystem and Conservation Sciences, University of Montana, Missoula MT 59812, USA

Cruz, Pablo INRA-UMR AGIR, Chemin de Borde Rouge, BP 52627, 31326 Castanet-Tolosan Cedex, France

Dao, Thanh H. US Department of Agriculture, Agricultural Research Service, Beltsville, MD, USA

Daufresne, Tanguy INRA-CEFS, Chemin de Borde Rouge, BP 52627, 31326 Castanet-Tolosan Cedex, France

Doolette, Ashlea L. School of Agriculture, Food and Wine, University of Adelaide, Waite Campus, Urrbrae, SA 5064, Australia

Duru, Michel INRA-UMR AGIR, Chemin de Borde Rouge, BP 52627, 31326 Castanet-Tolosan Cedex, France

Ehlers, Knut Group of Plant Nutrition, Institute of Plant, Animal and Agro-ecosystem Sciences, ETH Zurich, Eschikon 33, 8315 Lindau, Switzerland

Fardeau, Jean-Claude 117 allée du Petit Colombier, 04100 Manosque, France

Finlay, Roger Department of Forest Mycology and Pathology, Uppsala BioCenter, Swedish University of Agricultural Sciences, 75007 Uppsala, Sweden

Fox, Thomas R. Department of Forest Resources and Environmental Conservation, Virginia Polytechnic Institute and State University, 228 Cheatham Hall, Blacksburg VA 24061, USA

Fransson, Ann-Mari Plant Ecology and Systematics, Lund University, Ecology Building, 2236 Lund, Sweden

Frossard, Emmanuel Group of Plant Nutrition, Institute of Plant, Animal and Agro-ecosystem Sciences, ETH Zurich, Eschikon 33, 8315 Lindau, Switzerland

George, Tim S. SCRI, Invergowrie, Dundee DD25DA, UK

Giagnoni, L. Department of Plant, Soil and Environmental Sciences, University of Firenze, Piazzale delle Cascine 18, 50144 Firenze, Italy

Hammond, John P. Warwick HRI, University of Warwick, Wellesbourne, Warwick CV359EF, UK

Jansa, Jan Group of Plant Nutrition, Institute of Plant, Animal and Agro-ecosystem Sciences, ETH Zurich, Eschikon 33, 8315 Lindau, Switzerland

Jones, David L. Environment Centre Wales, Bangor University, Gwynedd LL57 2UW, UK

Jouany, Claire INRA-UMR AGIR, Chemin de Borde Rouge, BP 52627, 31326 Castanet-Tolosan Cedex, France

Klepsch, Sabine Department of Forest and Soil Sciences, Institute of Soil Science, BOKU – University of Natural Resources and Applied Life Sciences Vienna, Peter Jordan Strasse 82, 1190 Vienna, Austria; Department of Health and Environment Austrian Institute of Technology (AIT), 2444 Seibersdorf, Austria

Landi, L. Department of Plant, Soil and Environmental Sciences, University of Firenze, Piazzale delle Cascine 18, 50144 Firenze, Italy

Bayon, Renée-Claire Le Soil and Vegetation Laboratory, Institute of Biology, University of Neuchâtel, Rue Emile-Argand 11, CP 158, 2009 Neuchâtel, Switzerland

Leitner, Daniel Department of Forest and Soil Sciences, Institute of Soil Science, BOKU – University of Natural Resources and Applied Life Sciences Vienna, Peter Jordan Strasse 82, 1190 Vienna, Austria

López-Hernández, Danilo Laboratorio de Estudios Ambientales, Instituto de Zoología y Ecología Tropical, Facultad de Ciencias, Universidad Central de Venezuela, Caracas, Venezuela

Maruyama, Hayato Graduate School of Biosphere Science, Hiroshima University, Kagamiyama 1-7-1, Higashi-Hiroshima 739-8521, Japan

Miller, Bradley W. U. S. Environmental Protection Agency, National Risk Management Research Laboratory, Land Remediation and Pollution Control Division, 5995 Center Hill Avenue, Cincinnati, OH 45224-1702, USA

Mollier, Alain Department for the Environment and Agronomy, Joint Research Unit for Soil–Plant Transfer and the Cycle of Nutrients and Trace Elements (TCEM), National Institute for Agricultural Research (INRA), 71, Avenue Edouard Bourlaux, BP 81, 33883 Villenave d'Ornon Cedex, France

Morel, Christian Institut National de la Recherche Agronomique (INRA), UMR 1220 TCEM (INRA-ENITAB), 89 Av. E Bourleau, BP 81, 33883 Villenave d'Ornon Cedex, France

Nannipieri, P. Department of Plant, Soil and Environmental Sciences, University of Firenze, Piazzale delle Cascine 18, 50144 Firenze, Italy

Oberson, Astrid Group of Plant Nutrition, Institute of Plant, Animal and Agroecosystem Sciences, ETH Zurich, Eschikon 33, 8315 Lindau, Switzerland

Oburger, Eva Department of Forest and Soil Sciences, BOKU – University of Natural Resources and Applied Life Sciences, Peter Jordan Strasse 82, 1190 Vienna, Austria

Pellerin, Sylvain Department for the Environment and Agronomy, Joint Research Unit for Soil–Plant Transfer and the Cycle of Nutrients and Trace Elements (TCEM), National Institute for Agricultural Research (INRA), 71, Avenue Edouard Bourlaux, BP 81, 33883 Villenave d'Ornon Cedex, France

Prusisz, Bartlomiej Group of Plant Nutrition, Institute of Plant, Animal and Agro-ecosystem Sciences, ETH Zurich, Eschikon 33, 8315 Lindau, Switzerland

Pypers, Pieter Tropical Soil Biology and Fertility Institute of the International Centre for Tropical Agriculture (TSBF-CIAT), Nairobi, Kenya

Rabeharisoa, Lilia Laboratoire des Radio-Isotopes, Route d'Andraisoro 101, Antananarivo, Madagascar

Randriamanantsoa, Lalajaona Institute of Plant, Animal and Agro-ecosystem Sciences, ETH Zurich, Eschikon 33, 8315 Lindau, Switzerland

Reed, Sasha C. U.S. Geological Survey, Southwest Biological Science Center, Moab UT 84532, USA

Renella, G. Department of Plant, Soil and Environmental Sciences, University of Firenze, Piazzale delle Cascine 18, 50144 Firenze, Italy

Rubilar, Rafael Departamento de Silvicultura, Facultad de Ciencias Forestales, Universidad de Concepción, Concepción, Chile

Salcedo, Ignacio Departamento de Energia Nuclear, Ignacio Salcedo. Universidade Federal de Pernambuco, Centro de Tecnologia, Av. Prof. Luiz Freire 1000, Cidade Universitária, 50740-540 Recife, PE, Brasil

Schnepf, Andrea Department of Forest and Soil Sciences, Institute of Soil Science, BOKU – University of Natural Resources and Applied Life Sciences Vienna, Peter Jordan Strasse 82, 1190 Vienna, Austria

Schwartz, Robert C. US Department of Agriculture, Agricultural Research Service, Bushland, TX, USA

Sinaj, Sokrat Station de recherche Agroscope Changins-Wädenswil ACW, CP 1202, 1260 Nyon, Switzerland

Smernik, Ronald J. School of Agriculture, Food and Wine, University of Adelaide, Waite Campus, Urrbrae, SA 5064, Australia

Smith, F. Andrew School of Agriculture, Food and Wine, University of Adelaide, Waite Campus DX 650 636, Adelaide, SA 5005, Australia

Smith, Sally E. School of Agriculture, Food and Wine, University of Adelaide, Waite Campus DX 650 636, Adelaide, SA 5005, Australia

Stape, Jose L. Department of Forestry and Environmental Resources, North Carolina State University, 3108 Jordan Hall, Raleigh, NC 27695, USA

Tamburini, Federica Group of Plant Nutrition, Institute of Plant, Animal and Agro-ecosystem Sciences, ETH Zurich, Eschikon 33, 8315 Lindau, Switzerland

Taylor, Philip G. Department of Ecology and Evolutionary Biology and the Institute of Arctic and Alpine Research, University of Colorado, Boulder, CO 80303, USA

Tiessen, Holm Inter-American Institute for Global Change Research, Av. dos Astronautas 1758, 12227-010 São José dos Campos SP, Brasil

Townsend, Alan R. Department of Ecology and Evolutionary Biology and the Institute of Arctic and Alpine Research, University of Colorado, Boulder, CO 80303, USA

Wallander, Håkan Microbial Ecology, Department of Ecology, Lund University, 22362 Lund, Sweden

Wasaki, Jun Graduate School of Biosphere Science, Hiroshima University, Kagamiyama 1-7-1, Higashi-Hiroshima 739-8521, Japan

Weintraub, Michael N. Department of Environmental Sciences, University of Toledo, 2801 W. Bancroft Street, Mail Stop 604, Toledo, OH 43606, USA

White, Philip J. SCRI, Invergowrie, Dundee DD25DA, UK

Part I
Methods

Chapter 1
Soil Organic Phosphorus Speciation Using Spectroscopic Techniques

Ashlea L. Doolette and Ronald J. Smernik

1.1 Introduction

Phosphorus (P) is present in many different forms in soil. At any one time, only a small fraction of total soil P is in a form directly available for plant or microbial uptake. It is generally accepted that this directly available form closely equates to orthophosphate ($H_2PO_4^-$ and HPO_4^{2-}) in soil solution. However, this is not to say that all of the remaining soil P is forever inaccessible to biota. On the contrary, much of it can be converted to the directly available form. For some forms of P, this conversion can be rapid or practically instantaneous – weakly sorbed orthophosphate is constantly coming into and out of soil solution and is in rapid equilibrium with solution orthophosphate. For other forms of P, this conversion can be very slow and these forms of P can remain unavailable for centuries or longer. Therefore, the likelihood that any given P atom in soil is going to be taken up by a plant or a microbe is highly dependent on the form or chemical speciation of that P atom. Consequently, the ability of the soil to provide P to biota depends on what forms of P are present and their relative amounts.

The most commonly used differentiation of soil P is between inorganic and organic forms. This is partly due to the importance of this distinction. Transformations of inorganic forms of soil P are controlled by the processes of precipitation, dissolution and sorption. At the molecular level, nearly all inorganic P in soil is orthophosphate, and its chemistry is determined by the strength of ionic bonds to surrounding atoms. It is the relative strength of these ionic bonds that explains why phosphate behaves so differently to the common mineral forms of other nutrients, e.g. nitrate and sulfate. Whereas phosphate generally has low solubility, is easily fixed and is relatively immobile, nitrate and sulfate have much higher solubility, are not easily fixed and are mobile in soils. What distinguishes organic forms of soil P from inorganic forms is that they contain at least one covalent bond to a carbon

A.L. Doolette and R.J. Smernik (✉)
School of Agriculture, Food and Wine, University of Adelaide, Waite Campus, Urrbrae, SA 5064, Australia
e-mail: ronald.smernik@adelaide.edu.au

E.K. Bünemann et al. (eds.), *Phosphorus in Action*, Soil Biology 26,
DOI 10.1007/978-3-642-15271-9_1, © Springer-Verlag Berlin Heidelberg 2011

atom, generally via an ester linkage (i.e. through an oxygen atom). Most transformations of organic P, and in particular their conversion to inorganic P, require the breaking of this covalent bond. Precipitation, dissolution and sorption also affect organic forms of P (Berg and Joern 2006).

The other reason that the differentiation between inorganic and organic forms is so fundamental to P speciation is that this distinction has been easy to make using long-established techniques. Inorganic P is traditionally detected spectrophotometrically as a blue-coloured phosphomolybdenum complex formed when free phosphate reacts with an acidified molybdate reagent. Organic P does not form a coloured complex with this reagent, and so can be determined as the difference between total P (usually measured as inorganic P after digestion of the soil extract) and inorganic P. However, there are drawbacks to this method (Turner et al. 2003c, 2006). Organic P is overestimated when inorganic polyphosphates are present because they do not react with the molybdate reagent, and therefore are included in the organic P fraction. Additionally, high organic matter concentrations in alkaline extracts can interfere with the colorimetry, but can be minimised by acidifying the extracts to precipitate organic matter (Tiessen and Moir 1993). There are also several other species that interfere with the formation of the phosphomolybdenum complex. These include silica, arsenic, chromium, nitrite, nitrate and sulfide, though these are more of a concern when determining phosphorus species in water samples (Neal et al. 2000). These interferences are usually negligible in soils (Turner et al. 2006).

The differentiation of inorganic and organic P is only the beginning of soil P speciation, because each of these broad classes encompasses a huge range of chemical forms. The subject of this chapter is speciation of organic P forms. There are three general approaches to detailed soil organic P characterisation: sequential extraction, enzymatic hydrolysis and spectroscopic analysis. Organic P speciation using spectroscopic techniques is the subject of this chapter.

1.1.1 Why Spectroscopy?

Chemical spectroscopy involves the differentiation of species based on their differential absorption or irradiation of electromagnetic radiation. The absorption of radiation by matter results in an increase in its energy. At the atomic and molecular scale, energy levels are quantised, i.e. a material can only absorb radiation if it allows the matter to go from one defined level (usually the "ground state") to another defined level (an "excited state"). Electromagnetic radiation consists of photons, which are tiny "packets" of energy. The amount of energy in any packet is a function of the frequency of the radiation. For example, blue light has a higher energy than red light. Photons with energy less than that of visible light are in the infrared, microwave and radiowave regions (decreasing in that order), whereas photons with energy greater than that of visible light are in the ultraviolet, X-ray and gamma-ray regions (increasing in that order).

Some spectroscopic techniques are based on measuring the decrease in intensity (i.e. the number of photons) observed when radiation impacts on matter, whereas others measure the electromagnetic radiation produced as the excited state returns to a lower energy state (often the ground state). Once again, since energy states are quantised, this involves the release of a photon of a specific energy. The key to chemical spectroscopy is that energy levels of an atom or group of atoms depend on their chemical environment. Therefore, the specific energy absorbed or irradiated conveys information about chemical speciation. There are many types of spectroscopy that differ in the type of radiation involved and in the aspect of chemical structure they probe. For organic P speciation of soils, the main types of spectroscopic techniques that have been used are NMR spectroscopy and X-ray absorption spectroscopy (XAS). The specifics of each of these techniques are considered in the following sections. However, first we will discuss some general aspects of P speciation and spectroscopic techniques.

The history of soil organic P speciation using spectroscopic techniques is as yet a short one, dating back to the first use of NMR spectroscopy to characterise soil organic P (Newman and Tate 1980). Speciation of soil organic P before this (and there is a long history of such studies dating back to at least 1940) relied on "wet" chemical analyses. There are a number of limitations of these methods. For a start, separate analyses are required for each class of organic P compound, the main ones being inositol phosphates, phospholipids and nucleic acids, and these analyses generally involve multiple steps and are time-consuming. The development of these techniques and the understanding they enabled are well documented in the reviews by Dalal (1977) and Anderson (1980).

Another major drawback to these wet chemical techniques is that a large proportion of soil organic P remains unidentified, around 50% according to Dalal (1977). Furthermore, these methods generally require fairly harsh extraction conditions to solubilise the organic P – to "break the links between the [phosphate] esters and other soil components", as described by Anderson (1980) – and consequently they run the risk of "breakdown or alteration of the esters themselves".

Spectroscopic techniques can, to a large extent, overcome these problems. Importantly, the spectroscopic techniques used for soil organic P analysis can identify several organic P species simultaneously, though not always quantitatively. However, some problems are as relevant to spectroscopic techniques as they are to wet chemical techniques, but spectroscopic techniques also have their own unique problems.

The two main measures of the capability and performance of any spectroscopic technique are *resolution* and *sensitivity*. Resolution refers to the ability of the technique to distinguish between species, and depends on both the difference in frequency or energy of radiation absorbed or transmitted between different species and also the line-width or signal broadness. Sensitivity refers to the ability of the technique to detect signal against background signal or noise. The better the sensitivity, the lower the detection limit and the greater the potential for detecting minor components.

For the analysis of complex mixtures such as soils, there is a third key measure of spectroscopic performance, *quantitation,* which is often overlooked. Quantitation refers to the relative amount of signal generated by different species. Clearly, a technique in which the amount of signal produced is the same (or at least predictable) for all species is superior for quantifying the relative amounts of these species.

Another consideration that is especially important for soil analysis is the need for pretreatment. In particular, the problems associated with solubilisation discussed above in relation to wet chemical techniques are equally relevant to solution-based spectroscopic techniques. Some spectroscopic methods require more specific pretreatments, e.g. NMR may require removal of paramagnetic species.

Finally, there are practical considerations, including the cost of analysis and the availability of equipment. A downside of the increasing sophistication of spectroscopic techniques is that the equipment is generally expensive to purchase and therefore is not available in most soil laboratories. For example, most NMR analysis of soils is carried out in a dozen or so specialised facilities worldwide, even though NMR spectrometers are widely available in the chemistry departments of most research institutions. Synchrotron-based analyses require even more specialised and expensive equipment. However, there is an up-side to this specialisation: the expensive and sophisticated instruments usually come with operators who are specialists in the technique and are (or at least should be) keen to collaborate with end-users.

We have organised this chapter primarily by spectroscopic technique and secondarily under the subheadings of "sample preparation", "sensitivity", "resolution" and "quantitation". The intention is to enable end-users of these spectroscopic techniques to compare techniques on the basis of each of these primary measures of analytical performance. Thus, we have approached this subject from a spectroscopist's viewpoint. We have tried to avoid spectroscopists' jargon and unnecessary technical details of the techniques as much as possible but, as pointed out, the trend towards more sophisticated methods of speciation will increasingly mean that end-users will need to interact with specialised spectroscopists to get the best analyses possible.

1.2 Solution ^{31}P NMR Spectroscopy

Solution ^{31}P nuclear magnetic resonance (NMR) spectroscopy is by far the most widely used spectroscopic technique for the speciation of soil organic P. This is partly because it was the first spectroscopic technique used for this purpose (Newman and Tate 1980). However, the main reason is that, of currently available techniques, it provides the most detailed and accurate information (or in spectroscopic terms, the best resolution and quantitation) in most circumstances. Nonetheless, it is has several limitations and these need to be considered when using the technique or interpreting solution ^{31}P NMR data. It is not possible to cover all aspects of the use of P NMR spectroscopy within this chapter. Further information can be found in numerous helpful reviews (Preston 1996; Condron et al. 1997; Cade-Menun 2005a, b).

Of all the spectroscopic techniques, NMR is the hardest to fully understand. It has a fairly obscure physical basis (a nuclear property called "spin") and can only properly be described using quantum mechanics. Simplified descriptions can be found elsewhere (Veeman 1997; Cade-Menun 2005a) and more complex descriptions in NMR textbooks (Derome 1987; Yoder and Schaeffer 1987). Fortunately, one does not need to completely understand the physical basis of NMR in order to use NMR spectroscopy or to interpret NMR spectra. In fact, one of the great advantages of NMR spectroscopy is that "ordinary" NMR spectra (and most NMR used for the speciation of organic P in soils come under this heading) are very easily interpreted.

Figure 1.1 shows a typical solution ^{31}P NMR spectrum of a soil extract. In general, each peak represents P in a different chemical environment. Some peaks are well separated from all others (e.g. pyrophosphate, polyphosphate and phosphonates) and can easily be assigned without ambiguity. However, other peaks are very close or overlap, as shown in the expanded orthophosphate monoester region, and assigning and quantifying these is more difficult, as discussed in detail in the

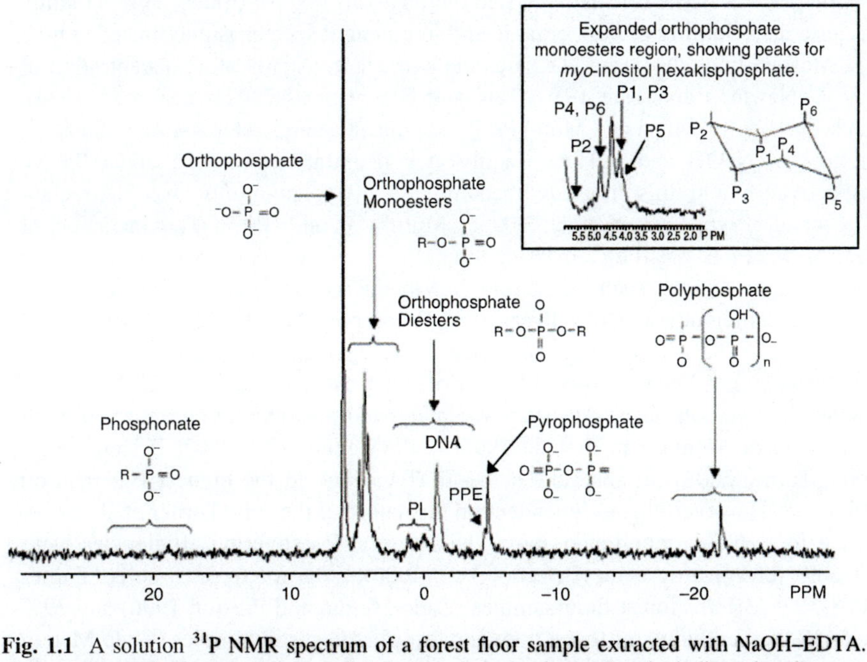

Fig. 1.1 A solution ^{31}P NMR spectrum of a forest floor sample extracted with NaOH–EDTA. This spectrum shows the diversity of P species in natural samples, including phosphonates, orthophosphate, orthophosphate monoesters, orthophosphate diesters such as phospholipids (*PL*) and deoxyribonucleic acid (*DNA*), pyrophosphates and polyphosphate, with the terminal P in the polyphosphate chain indicated by *PPE*. The *inset* shows the expanded orthophosphate monoester region, indicating the peaks and structure for *myo*-inositol hexakisphosphate (phytate). Reprinted from Cade-Menun (2005b), with permission from Elsevier

rest of this section. In general, similar species give rise to peaks in similar parts of the spectrum. As a result, whole regions of the ^{31}P NMR spectrum can be assigned to classes of compound, as shown in Fig. 1.1. Thus NMR can provide both broad and detailed speciation of P types.

1.2.1 Sample Preparation

Being a solution technique, solution ^{31}P NMR spectroscopy has the disadvantage of requiring an extraction step prior to analysis. The aim of such an extraction is to maximise solubilisation of P while minimising alteration of P speciation and optimising the conditions for subsequent NMR analysis. Often these are competing goals, and choices have to be made based on the purpose of the analysis and the nature of the soil. A comprehensive comparison of the performance of different extractants and extraction conditions, and recommendations for optimising these can be found elsewhere (Cade-Menun 2005a, b; Turner et al. 2005). Here, we provide a brief overview of these issues.

Solution ^{31}P NMR is usually carried out on alkaline soil extracts. This is mainly because the solubility of both organic and inorganic P species is maximised at high pH. Most early studies used NaOH as the extractant, usually at a concentration of 0.5 M (Newman and Tate 1980; Tate and Newman 1982; Hawkes et al. 1984). Subsequently, Bowman and Moir (1993) developed a single-step extraction using a mixture of NaOH and EDTA (usually at concentrations of 0.25 and 0.05 M, respectively), and this has now become the most commonly used extractant (Dai et al. 1996; Turner et al. 2003c; Murphy et al. 2009). The inclusion of EDTA, which is a strong chelating ligand, serves two purposes: it complexes paramagnetic cations such as Fe and Mn in the extract and it increases soil P extraction efficiency and the diversity of P compounds extracted (Bowman and Moir 1993). The effectiveness of NaOH–EDTA as an extractant for ^{31}P NMR spectroscopy has been compared to that of other extractants, e.g. 0.25 M NaOH, Chelex (a chelating resin) plus 0.25 M NaOH, and post-extraction treatment with Chelex (Cade-Menun and Preston 1996; Cade-Menun et al. 2002; Briceño et al. 2006; Turner 2008). In general, NaOH–EDTA achieved the highest P extraction efficiency. However, this is dependent on the nature of the soil (Turner et al. 2005). In the four studies mentioned above, NaOH total P extraction efficiencies were 30–60% for volcanic soils (Briceño et al. 2006), 27% for tropical soils (Turner 2008), 22–43% for forest floor samples (Cade-Menun and Preston 1996) and 79% and 43%, respectively, for forest floor and low pH forest soil samples (Cade-Menun et al. 2002). Corresponding extraction efficiencies using NaOH–EDTA were 37–60% for volcanic soils (Briceño et al. 2006), 37% for tropical soils (Turner 2008), 71–91% for forest floor samples (Cade-Menun and Preston 1996; Cade-Menun et al. 2002) and 34% for a forest soil sample (Cade-Menun et al. 2002).

Comparisons cannot be made between Chelex treatments as the techniques differ slightly amongst these studies.

Interestingly, Cade-Menun and Preston (1996) originally advised against the use of NaOH–EDTA because the paramagnetic ions remain in solution (unlike when Chelex is used) and this causes line broadening and overlap of resonances. Their original recommendation was that NaOH–EDTA was only suitable for samples with high P concentrations and low levels of paramagnetic species, unless the metal complexes could be removed prior to ^{31}P NMR analysis. However, their results also showed that NaOH–EDTA extracted a higher concentration and diversity of P compounds. Some P compounds such as polyphosphates were not detected when other extractants were used. Finally, there was less hydrolysis of P compounds when NaOH–EDTA was used (Cade-Menun and Preston 1996). These findings were corroborated by Cade-Menun et al. (2002), Briceño et al. (2006) and Turner (2008) for their soil samples. The fact that paramagnetic ions remain in solution when EDTA is used can also be advantageous because they induce rapid relaxation and this can improve sensitivity (discussed in Sect. 1.2.2). For these reasons, NaOH–EDTA has been generally accepted as the preferred extractant for solution ^{31}P NMR analysis.

Obviously, the nature of the soil P that cannot be extracted cannot be determined using solution ^{31}P NMR spectroscopy and this is one of the major limitations of the technique. There are several reasons why P might not be extractable, and therefore the composition of this non-extractable P fraction can be quite variable: some P may be present in complex polymeric molecules that are alkaline-insoluble (i.e. the "humin" fraction of the classical humic fractionation scheme), some may be strongly complexed to minerals, and some may just be water-insoluble. For example, phospholipids, given their hydrophobic nature, will not be extracted into aqueous extractants. This raises serious doubt on the widespread assignment of signals in ^{31}P NMR spectra of soil extracts to phospholipids, as pointed out by Doolette et al. (2009).

Besides the problem of non-extractable P, the greatest problem with solution ^{31}P NMR analysis of organic P is the potential for hydrolysis. A number of papers have reported the instability of some organic P compounds, particularly orthophosphate diesters, in alkaline solution. Turner et al. (2003b) tested the stability of numerous organic P compounds added to NaOH–EDTA extracts. Most were found to be stable for several days at room temperature. However, ribonucleic acid (RNA) was shown to be very unstable, completely degrading in 24 h to orthophosphate monoesters. This is consistent with other reports (Anderson 1967; Makarov et al. 2002b). By contrast, deoxyribonucleic acid (DNA) is more stable (Turner et al. 2003b) and can remain intact in alkaline solutions for at least 24 h (Makarov et al. 2002a).

Phospholipids are also susceptible to hydrolysis. Turner et al. (2003b) tested the stability of three common phospholipids: phosphatidyl choline, phosphatidyl serine and phosphatidyl ethanolamine. Phosphatidyl ethanolamine showed no degradation, phosphatidyl serine showed minimal degradation after 24 h and partial degradation after 19 days, but phosphatidyl choline was completely degraded within 24 h to two orthophosphate monoester compounds. These compounds have been identified as α- and β-glycerophosphate (Folch 1942; Baer et al. 1953; Doolette et al. 2009).

Since orthophosphate diesters are susceptible to hydrolysis, solution ^{31}P NMR is likely to underestimate the true concentrations of orthophosphate diesters and overestimate the concentrations of orthophosphate monoesters. The pH of the extract can also influence the rate of degradation. Doolette et al. (2009) reported that although phosphatidyl choline degraded over 8 days in 0.25 M NaOH, it completely degraded in just 90 min in 1.16 M NaOH. Although such high NaOH concentrations are not used for soil extractions, freeze-drying and re-dissolving of soil extracts results in these concentrations during solution ^{31}P NMR analysis.

Hydrolysis and modification of native P compounds can be avoided by acquiring NMR spectra at lower pH. McDowell and Stewart (2005a) analysed several water extracts of soil and dung and detected a variety of well-resolved peaks (orthophosphate, orthophosphate monoesters and diesters, pyrophosphate, polyphosphate and phosphonates). Increasing the pH to >13 resulted in a decrease in many P species, and this was attributed to either hydrolysis or precipitation. Adams (1990) analysed neutral soil extracts and found the diester-P concentrations to be greater than those of monoester-P. This technique may have minimised the hydrolysis of some diester-P, but Adams (1990) noted that the extractant conditions would have undoubtedly favoured the soluble P fractions. So, although strong alkaline extractions introduce the risk of hydrolysis they also maximise P extractability. To accurately and quantitatively assess soil organic P using solution ^{31}P NMR, maximum recovery is vital and alkaline reagents are the most effective for this.

1.2.2 *Sensitivity*

Sensitivity is often the limiting factor in solution ^{31}P NMR analysis of soil organic P. NMR is an inherently insensitive technique. The main reason for this is that the difference in energy levels between the ground and excited spin states is very small and at room temperature the energy levels are nearly equally populated. NMR signal is only generated by the difference in these populations, which is typically around 1 in 10,000.

There are several factors that contribute to NMR sensitivity. Each different type of nucleus has an inherent sensitivity related to a property called its "magnetogyric ratio". Fortunately, the ^{31}P nucleus is one of the more sensitive nuclei; it is less sensitive than ^{1}H, but more sensitive than ^{13}C or ^{15}N. Sensitivity is also linearly related to the number of nuclei in the sample. This depends on the isotopic abundance of the NMR-active nucleus, and the concentration of the element. Again it is fortunate that ^{31}P has 100% isotopic abundance. By comparison, ^{13}C and ^{15}N are 1% and 0.4% abundant, respectively. Against this, P usually has a much lower abundance than C or N in soil. Sensitivity also increases more than linearly with magnetic field strength, so there is a clear benefit in acquiring spectra on high field instruments.

There are two main ways to combat the inherently low sensitivity of ^{31}P NMR. One is to maximise the amount of P in the sample analysed and the other is to acquire and average a large number of scans. Soil extracts are usually concentrated prior to NMR analysis in order to improve sensitivity. This can be achieved by lyophilisation (freeze drying), rotary evaporation or evaporating under a stream of nitrogen at 40°C (Cade-Menun 2005b). Lyophilisation is the most widely used technique because it avoids an increase in temperature that might degrade the sample (Cade-Menun et al. 2002; Turner et al. 2003b). Dried extracts are re-dissolved immediately prior to ^{31}P NMR analysis in order to minimise hydrolysis. The amount of sample re-dissolved needs to be sufficient to obtain an optimum P concentration but not so great as to increase the viscosity of the sample, which can cause line broadening.

Compared with other spectroscopic techniques, NMR generally requires the collection of many more (often orders of magnitude more) scans. Sensitivity improves as the square root of the number of scans. Solution ^{31}P NMR spectra of soil extracts are acquired using as few as 500 or as many as 110,000 scans (Cade-Menun 2005b) but typically <10,000 scans. However, collecting a large number of scans is not always practical due to the cost involved and the increased risk of hydrolysis. Sensitivity can also be improved by using a 10 mm rather than a 5 mm probe, or by using a higher field spectrometer. Of more importance than the number of scans is the acquisition time, which is the product of the number of scans times the recycle time, i.e. the time between consecutive scans. Acquisition of a ^{31}P NMR spectrum of a soil extract typically takes only a few tenths of a second. However, it may take much longer than this for the nuclei to regain their equilibrium magnetisation. If insufficient time is allowed for this to occur, the result is signal saturation, or a decrease in the amount of signal obtained per scan. The process by which equilibrium magnetisation is regained is called relaxation, and the parameter that describes it is the spin-lattice relaxation time constant (T_1). Recycle times of five times T_1 are required to ensure saturation losses are <1% (Yoder and Schaeffer 1987). Shorter recycle delays can be used to give higher sensitivity (per unit time), by trading off some loss of signal per scan through saturation for running more scans per unit time. However, this can be at the expense of quantitation, as discussed in Sect. 1.2.4.

A wide range of recycle times have been used to acquire solution ^{31}P NMR spectra of alkaline extracts: 10–30 s (Doolette et al. 2009), 15–20 s (Smernik and Dougherty 2007), 1–20 s (Newman and Tate 1980; Tate and Newman 1982), 4.32 s (He et al. 2007a; McDowell et al. 2007), 2 s (Turner 2008), 1.5 s (Dai et al. 1996), 1 s (Koopmans et al. 2007), 0.2 s (Bedrock et al. 1994; Guggenberger et al. 1996) and 0.808 s (Turner et al. 2003a). Given that T_1 varies between P species (Newman and Tate 1980; Cade-Menun et al. 2002; McDowell et al. 2006) and with sample temperature (Crouse et al. 2000; Turner et al. 2003b; Puppato et al. 2007) and concentration of paramagnetic ions (McDowell et al. 2006), it makes sense to measure T_1 for every sample. This can be achieved with either saturation-recovery or inversion-recovery experiments, as first discussed by Newman and Tate (1980) and later strongly recommended by Cade-Menun et al. (2002). However, few researchers explicitly state that they implement this experiment prior to running their samples. As an alternative to measuring T_1 values, McDowell et al. (2006) have suggested that these can be estimated from the P/(Fe + Mn) ratio. This may have some merit, but more research is warranted before this technique can be used with certainty.

1.2.3 Resolution

Solution ^{31}P NMR spectroscopy provides better resolution of organic P species than any other technique currently available. The resolving power of solution ^{31}P NMR spectroscopy resides in the simple Gaussian–Lorentzian peak shape and the narrow peak width compared to peak spread. Despite this, there is inevitably some signal overlap, especially in the orthophosphate monoester region. However, a further advantage of solution ^{31}P NMR spectroscopy is that similar species appear in similar parts of the spectrum, so even when individual organic P compounds cannot be resolved, groups of peaks can still be assigned to the broad compound classes.

The resolution of solution ^{31}P NMR spectra of soil extracts varies considerably. A major influence is the concentration of paramagnetic ions, the presence of which decreases resolution by increasing line broadening. For tropical soils, Turner (2008) found that maintaining pH >13 can prevent the extraction of excess paramagnetic ions. Poor resolution was apparent when a lower concentration of NaOH was used for the extraction. Resolution was most strongly influenced by Mn. Iron had relatively little influence on line broadening and even at low Mn concentration, Mn was the main paramagnetic ion responsible for reduced resolution. Cade-Menun et al. (2002) reported that the removal of paramagnetic ions from soil extracts reduced line widths and improved resolution in the monoester region. The effect of paramagnetic ions is also evident when pretreatment techniques are used to specifically remove paramagnetic ions. McDowell and Stewart (2005b) pretreated soil with Ca–EDTA–dithionite before extracting with NaOH–EDTA. The pretreatment successfully diminished Fe and Mn concentrations and this decreased the line widths of NMR signals by up to 46%, while having little effect on the forms of P detected.

Resolution is also affected by solution pH. Crouse et al. (2000) evaluated the effects of pH ranging from 4.0 to 13.2 on the solution ^{31}P NMR spectrum of a turkey litter extract composed mainly of orthophosphate and phytate. The appearance and position of the orthophosphate peak in particular was very variable. This peak was sharpest at the highest pH (13.2), and was also clear of all but one of the phytate peaks at this pH. This provides a further reason to acquire solution ^{31}P NMR spectra at pH >13. Resolution can also be affected by the presence of suspended particles. Solution-state NMR spectroscopy will only detect those P nuclei in solution, but often small amounts of dried extract fail to redissolve. Thus it is often beneficial to filter or centrifuge the sample prior to analysis.

1.2.3.1 Identification of P Species

Good spectral resolution is a necessary but not sufficient condition for achieving detailed speciation. The other necessary condition is accurate assignment of peaks.

Many studies rely on comparison to reported literature values to assign peaks to P compounds (Dai et al. 1996; Cade-Menun et al. 2002; Makarov et al. 2002b; Briceño et al. 2006; McDowell and Stewart 2006). These chemical shifts, however, can vary with parameters such as pH, temperature, the concentration of paramagnetic ions and ionic strength (Costello et al. 1976; Derome 1987; Crouse et al. 2000; McDowell and Stewart 2005a; Puppato et al. 2007; Smernik and Dougherty 2007). Some species, such as pyrophosphate and polyphosphate, resonate at very distinct chemical shifts (Turner et al. 2003b). These species can be identified without fear of misassignment, since they do not overlap with other resonances. Other compounds, particularly the orthophosphate monoesters, appear in crowded regions of the NMR spectrum and are much more difficult to identify.

Turner et al. (2003b) approached the problem of extract composition affecting peak positions by acquiring spectra of standard P compounds added to an NaOH–EDTA soil extract. These chemical shifts are likely to be more accurate than those determined in NaOH–EDTA alone. However, the concentrations of many species vary from soil extract to soil extract, and so these should still be taken as a guide only. Smernik and Dougherty (2007) introduced a spiking procedure using low concentrations of added model compounds. Even this resulted in slight changes in the chemical shift of peaks, but because the peaks of the native organic P compounds were still visible, the identity of the spiked species could be determined with high precision.

Spiking soils to assign P species has only been undertaken in a small number of studies (e.g. Adams and Byrne 1989; McDowell and Stewart 2005a; McDowell et al. 2007; Smernik and Dougherty 2007; Doolette et al. 2009). One reason for this is a misplaced belief that maintaining a pH >13 will not only ensure optimal resolution but also consistent chemical shifts. Although the largest changes in chemical shift occur below pH 13 (Costello et al. 1976; Crouse et al. 2000; Puppato et al. 2007), there can still be small variations (up to 0.3 ppm) in the chemical shifts of P compounds between different soils when pH >13, these variations generally being largest for the orthophosphate resonance (Turner et al. 2003c; Doolette et al. 2009). This can become problematic in both the orthophosphate and orthophosphate monoester regions, due to the number and close separation of peaks. Even spiking might not enable definitive identification of some peaks. For example, Doolette et al. (2009) noted that the chemical shifts of ethanolamine phosphate and β-glycerophosphate in NaOH–EDTA extracts can be indistinguishable.

Identification of P species, particularly in the orthophosphate monoester region, is further complicated by the fact that, besides specific small P-containing molecules, soils also contain much larger "humic" molecules. These also contain P, but since the P is in a variety of slightly different chemical environments, these do not produce sharp resonances, but rather a broad signal that is often overlooked. Figure 1.2 shows the ^{31}P NMR spectra of NaOH–EDTA soil extracts of a calcareous soil (Hart), an acidic soil (Wagga) and of acid and calcareous model soils (mixtures of pure clay + sand). All soils were incubated with the addition of cellulose for 25 weeks. Whilst all spectra contain sharp resonances that can be attributed to specific small P-containing molecules, resulting from microbial

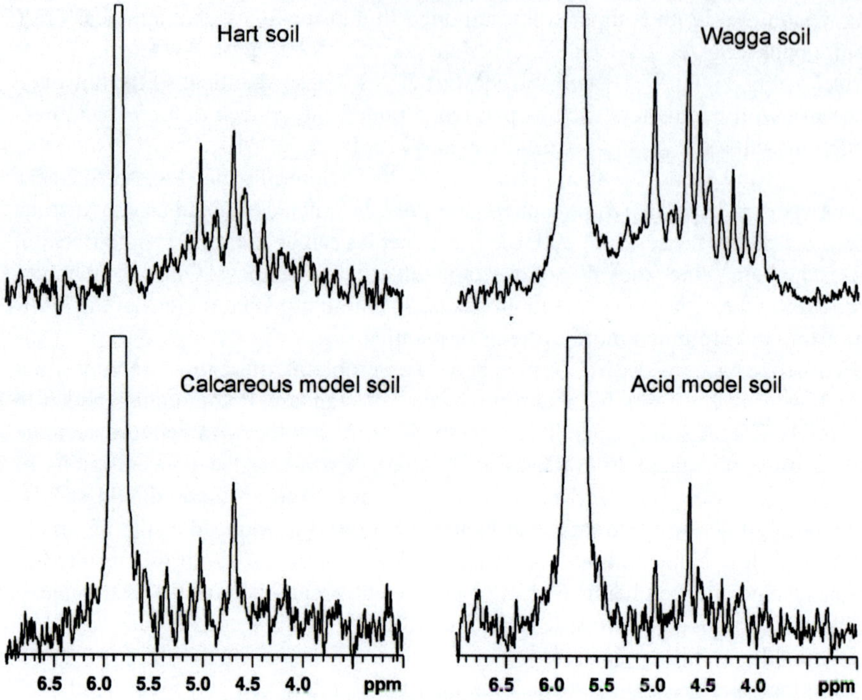

Fig. 1.2 Expansion of the monoester region of ^{31}P solution NMR spectra of NaOH–EDTA extracts of a calcareous soil (Hart), an acidic soil (Wagga) and acidic and calcareous model soils following a 25-week incubation with cellulose addition. Reprinted from Bünemann et al. (2008), with permission from Elsevier

P immobilisation after carbon addition, there is clearly an underlying broad signal for the "real" soils that is absent for the model soils.

Support for this interpretation comes from humic fractionations that show that P associated with humic and fulvic acid in volcanic soils can account for 32–75% and 51–68% of organic P, respectively (Borie et al. 1989; Escudey et al. 2001; Borie and Rubio 2003). Using ^{31}P NMR, He et al. (2006) examined the spectral characteristics of P in humic substances and concluded that the organic compounds had molecular weights greater than 3,000. Furthermore, the ^{31}P NMR spectra they presented of the humic acid fraction contained a broad resonance in the orthophosphate monoester region. Similar spectra were presented by Makarov et al. (1997), who also found that orthophosphate monoesters were the dominant alkali-extractable P species associated with humic acid fractions. However, the poor resolution of their spectra could also be due to the use of NaOH alone as the extractant.

The presence of a broad underlying signal in the monoester region of solution ^{31}P NMR spectra of soil extracts warrants further consideration when attempting to identify and quantify P compounds. Although most of the focus has been on accurately identifying the sharp resonances in the monoester region, the identification and characterisation of the broader resonance has been limited to a small number of samples (Bünemann et al. 2008).

1.2.4 Quantitation

In most cases where speciation of organic P is being sought, it is not just identification of the species that is important, but also their quantification. This is a considerably more difficult task for a spectroscopic technique. Many studies equate quantification of spectral signal with quantification of the species that give rise to them. Although this may be true for solution ^{31}P NMR, it is by no means assured. Quantification of P species from ^{31}P NMR spectra is usually carried out by multiplying peak areas by the total P concentration of the extract. This assumes that all the P in the reconstituted NaOH–EDTA extract is soluble and is observed with equal sensitivity. This may not always be the case, and more attention should be paid to this issue.

As discussed in Sect. 1.2.2, when recycle times are insufficient to ensure complete relaxation between scans, signal saturation occurs, i.e. less signal is produced by the affected nuclei. This decreases sensitivity, but more importantly it compromises quantitation if not all P species relax at the same rate. Therefore, there can be a trade-off between sensitivity and quantitation. For samples where identification is more important than quantification, shorter relaxation times may be appropriate; but, for any sample where quantification is sought, sufficient time for complete relaxation of *all* species must be allowed for or a bias will result.

A useful method for dealing with quantitation issues is spin counting, which involves the use of a signal intensity standard to gauge the overall NMR observability of a sample. This has been used extensively in solid-state NMR analysis of soil C (Smernik and Oades 2000a, b), N (Smernik and Baldock 2005) and P (Dougherty et al. 2005; McBeath et al. 2006). Since most quantitation problems in NMR involve the under-detection of nuclei (e.g. through saturation), showing that NMR observability for a sample is close to 100% is usually sufficient to prove that the NMR spectrum is quantitative.

The use of an internal standard, although not commonly implemented, will help to overcome the problems associated with reduced and variable observability of different P species. For the successful use of an internal standard, the standard should be chemically inert, soluble in the extract, and produce peaks that do not overlap with those of the sample (Metz and Dunphy 1996; Al Deen et al. 2002). Methylenediphosphonic acid (MDP) appears to satisfy these criteria and has been used in a couple of soil studies (Bedrock et al. 1994; Turner 2008). Other compounds, such as trimethyl phosphate, sodium phosphate (Al Deen et al. 2002) and triphenyl phosphate (Maniara et al. 1998), have been used for the quantitative analysis of P-containing agricultural chemicals, but their suitability for soil studies is yet to be determined.

The other major limitation to accurate quantitation is poor resolution, especially in regions of ^{31}P NMR spectra where there is considerable overlap of resonances. This problem is usually addressed using spectral deconvolution, which can be used to quantify signal in overlapping resonances. Spectral deconvolution involves a numeric least-squares fit of the spectrum as the sum of multiple peaks of standard shape (Lorentzian or Gaussian). Spectral deconvolution has been used to identify the complex signals associated with *myo*-inositol hexakisphosphate (phytic acid), as shown in Fig. 1.3 (Turner et al. 2003a; McDowell and Stewart 2006) and with orthophosphate monoesters (McDowell and Stewart 2005a).

Although the mathematical process of spectral deconvolution is quite straightforward (and is included in standard NMR software), special care and consideration is required to produce accurate and reliable results. To date, these problems appear to have not been fully appreciated and understood within the soil science community. By contrast, researchers in the biomedical sciences community, who experience similar problems with solution ^{31}P NMR analyses, have studied the problems with spectral deconvolution in quite some detail, and their findings are worthy of discussion here. They have adopted different approaches, including the generation of model test and simulated spectra (Corbett 1993) and the incorporation of prior knowledge, which takes into account what is known about the sample and the spectrum (Changani et al. 1999). The aim of these approaches is to account for the overlap of peaks and improve quantitation of spectra with poor resolution and sensitivity that occur due to the complexity and heterogeneity of samples and limited time available to obtain a sufficient number of scans (Šárka and Mika 2001).

The incorporation of prior knowledge takes into account baseline adjustment (essential for estimation of accurate peak areas), as well as the selection of the

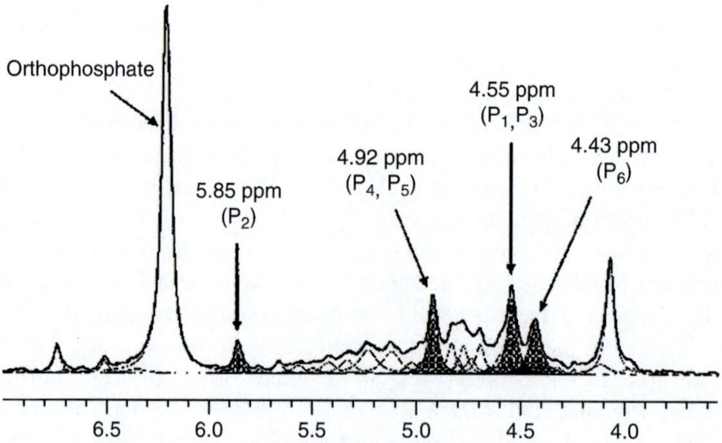

Fig. 1.3 ^{31}P NMR spectrum of a NaOH–EDTA extract of a lowland permanent pasture soil. The *shaded peaks* are phytate resonances that have been identified using spectral deconvolution. Reprinted from Turner et al. (2003a) with permission from Soil Science

correct line shape (Gaussian or Lorentzian) and line width (Corbett 1993; Šárka and Mika 2001). The researcher inputs or fixes these values using what they know about the sample and the peaks they can identify. Inclusion of this prior knowledge in the analysis enables the generation of more information with increased accuracy and reliability (Changani et al. 1999). When analysing soil extracts, for example, this can include fixing line widths of the "sharp" and "broad" peaks where this is appropriate (Dougherty et al. 2007; Smernik and Dougherty 2007).

Corbett (1993) showed that as signal-to-noise decreased, deconvolution produced larger linewidths as it attempted to fit the local noise in the vicinity of the peak. As a consequence, deconvolution tended to overestimate the true area of small or noisy peaks. In other spectra, where there was a sharp noise peak located at the same chemical shift as a true signal, deconvolution attempted to fit the peak shape to the noise. This usually resulted in the true peak area being underestimated. Their analysis also showed that when deconvolution is applied to regions with substantial overlap of peaks, the predicted areas for adjacent peaks are interrelated (i.e. when one is overestimated the other is underestimated) and this error increases in proportion to the degree of overlap. Both of these errors can be minimised by fixing line widths, provided that information about the true line widths is known.

Although these more sophisticated methods of deconvolution and curve fitting are likely to be time-consuming, the adoption of these techniques will result in more accurate and reliable quantification of ^{31}P NMR soil spectra.

An alternative approach to quantification using spectral deconvolution is to use a spiking procedure, but to date this method has only been used by Smernik and Dougherty (2007). By directly spiking a known concentration of phytate into soil extracts immediately prior to analysis, they confirmed the identification of the peaks and were also able to quantify the native phytate concentration.

1.3 Solid-State ^{31}P NMR Spectroscopy

Although most soil P NMR analysis is done in solution mode, NMR analysis can also be carried out on solid samples. Acquiring NMR spectra on solid-state samples introduces some additional problems and requires the use of substantially different and specialised equipment and techniques. The differences between solution and solid-state NMR modes have important consequences for sample preparation, sensitivity, resolution and quantitation, as detailed below.

A complete description of the differences between solution and solid-state NMR can be found in NMR textbooks (e.g. Fyfe 1983). In essence, the two modes differ mainly because molecules in solution are in constant, rapid motion. This constant movement averages out all aspects of the chemical environment of a given nucleus (e.g. a ^{31}P nucleus) except those associated with chemical bonding. Furthermore, even though at any instant a molecule will have a particular orientation with respect to the applied magnetic field, this orientation is constantly changing, and on average all orientations are equally likely. As a consequence, the frequency or chemical shift of a solution NMR peak is the isotropic value averaged over all orientations. On the other hand, in solid samples, the orientations of molecules with respect to the magnetic field are fixed. In this case, chemically equivalent nuclei can have different chemical shifts depending on their orientation and so signals are much broader. The extent of this broadening depends on the chemical shift anisotropy (CSA), i.e. how much the chemical shift varies with orientation, and this varies substantially among molecules.

In general, for ^{31}P NMR of soils, CSA is much greater (~10–200 ppm) than differences in chemical shift between species (generally <10 ppm) and so unless steps are taken to counter CSA, solid-state ^{31}P NMR analysis of soils would result in one broad lump and no speciation information could be gained. Fortunately, there is a way to overcome CSA broadening, known as magic angle spinning (MAS). MAS involves rapidly spinning (in the kHz range) a sample at 54.7° (the "magic angle") to the applied magnetic field. A description of the physical basis for the technique can be found elsewhere (Fyfe 1983). When MAS is used, nuclei produce relatively sharp signals, mainly at their isotropic chemical shift (i.e. equivalent to the solution NMR shifts). However, solid-state NMR signals are still nearly always 1–2 orders of magnitude broader than for solution NMR, even when MAS is used. Furthermore, if MAS is not fast enough, artefacts called "spinning sidebands" (SSBs) appear. The intensity of SSBs increases with the CSA of the nucleus, increases with increasing magnetic field strength and decreases with increasing MAS rate.

One advantage of acquiring NMR spectra in the solid-state is that one can take advantage of cross polarisation (CP), a technique in which magnetisation is transferred from ^{1}H to ^{31}P nuclei (Fyfe 1983). In solid-state NMR, the "normal" mode of signal acquisition in which ^{31}P nuclei are directly irradiated is usually called direct polarisation (DP). The CP technique relies on strong dipolar coupling between ^{31}P and ^{1}H nuclei that only occurs in the solid-state. CP provides a direct signal enhancement and also allows for a shorter recycle time that is controlled by relaxation of ^{1}H nuclei rather than ^{31}P nuclei (the former generally being 1–2 orders of magnitude faster). Thus, CP offers a substantial improvement in sensitivity over DP. Unfortunately, this comes at a cost of poorer quantitation, because the transfer of magnetisation to ^{31}P nuclei can be adversely affected by a number of factors including a lack of nearby ^{1}H nuclei, the presence of paramagnetic species, and molecular motion (McDowell et al. 2003a; Benitez-Nelson et al. 2004; Dougherty et al. 2005; He et al. 2007b). These issues are discussed in detail in Sect. 1.3.4.

1.3.1 Sample Preparation

Minimal sample preparation is required for solid-state ^{31}P NMR spectroscopy and this is its main benefit over solution ^{31}P NMR. It is best to grind soil samples to a fine powder, as this improves homogeneity and also facilitates sample spinning. Samples should be as dry as possible for acquisition of CP spectra because molecular motion affects this technique. Only a small amount of soil is required (approximately 0.2 g) for analysis.

1.3.2 Sensitivity

Solid-state ^{31}P NMR suffers the same sensitivity problem as solution ^{31}P NMR, and generally a large number of scans needs to be acquired to produce a good quality spectrum. One advantage of the solid-state technique is that the amount of sample that can be analysed is not restricted by solubility. Sensitivity is thus influenced by (a) the P content of the soil, (b) the number of scans that can be acquired and (c) the relative "NMR observability" of P species present. Condron et al. (2005) suggested that concentrations of around 1 mg P/gram of soil are needed to produce high-quality ^{31}P NMR spectra and that the low natural abundance of P in some soils may result in the signal being below detection limits.

As is the case for solution ^{31}P NMR, sensitivity per unit time is strongly influenced by the time required between scans and, as discussed in Sect. 1.3, there are large differences between CP and DP techniques in this respect. For example, for the same samples, Frossard et al. (2002) used a 3 s relaxation delay for CP spectra and 20 s for DP, whilst McBeath et al. (2006) used relaxation delays of 4 and 100 s for CP and DP, respectively. A range of recycle times have been used for the acquisition of solid-state ^{31}P NMR spectra: e.g. 0.2 s (Shand et al. 1999) and 10 s (Williams et al. 1981; Frossard et al. 1994) for CP; and 1 s (Benitez-Nelson et al. 2004), 1.5 s (He et al. 2009) and 10 s (Conte et al. 2008) for DP. However, in most cases the reasons for these choices of recycle times are not made clear. Dougherty et al. (2005) used saturation-recovery to show that, for their samples, a recycle delay of at least 20 s was required to detect at least 80% of potential signal intensity in DP spectra and that the use of 5 s recycle times resulted in the detection of around 50% of signal intensity.

NMR observability in solid-state ^{31}P NMR spectra of soils is greatly influenced by the presence of paramagnetic species. Manganese and Fe phytate salts are reported as being undetectable by solid-state NMR (He et al. 2007c), as are other Fe-associated P species (Hinedi and Chang 1989; McDowell et al. 2002b). Using spin counting, Dougherty et al. (2005) found that the majority of potential NMR signal for their soils was undetected, and attributed this to the effect of paramagnetic species.

Dougherty et al. (2005) also found that the ^{31}P NMR observability of P in DP spectra was 2–2.7 times greater than for CP spectra. Similarly, Benitez-Nelson et al. (2004) found that the CP spectra of marine particulates required over four times more scans to achieve the same signal-to-noise ratio as corresponding DP spectra. McDowell et al. (2002b) reported similar findings for cropping soils. Consequently, the use of DP alone and CP with DP is becoming more common due to the low sensitivity of CP alone (McDowell et al. 2002b; Benitez-Nelson et al. 2004; Conte et al. 2008).

1.3.3 Resolution

The greatest disadvantage of solid-state compared to solution ^{31}P NMR is the poorer resolution of the former. Rarely have solid-state ^{31}P NMR studies focused on determining organic P concentrations or identifying individual P compounds in whole soils (Newman and Condron 1995; Condron et al. 1997; McBeath et al. 2006; Conte et al. 2008). Instead, much research has dealt with the application of solid-state NMR in differentiating between inorganic metal–P species and determining the relative amount of whole-soil inorganic and organic P (Hinedi and Chang 1989; Frossard et al. 1994, 2002; McDowell et al. 2002a, b).

Identifying Na, K, Ca and Mn phytate salts (He et al. 2007c) and their corresponding inorganic phosphate salts (Turner et al. 1986) in reference materials can be undertaken with relative ease because a linear correlation exists between the chemical shift and electronegativity of the cation. That is, the chemical shift decreases with increasing metal ion valence. The major peaks for phytate and other reference organic P compounds have been shown to be broader than those of the corresponding inorganic orthophosphate compounds (He et al. 2007b).

In addition to differences in the average (isotropic) chemical shift, strong SSBs, can be used to aid P characterisation. Variations in the intensity of SSBs have been noted by Frossard et al. (1994), Williams et al. (1981) and Hinedi and Chang (1989), but they were not used to identify specific P pools or compounds in soils. Dougherty et al. (2005) used a combination of CP and DP with selective extractions, and reported that organic P could be identified by its broad resonance and prominent SSBs, in comparison with the considerably sharper resonances and smaller SSBs of inorganic P. It has been proposed that SSBs can also be used to distinguish between phytate and inorganic orthophosphate species when the major resonances of both compounds are similar (He et al. 2007c). In particular, the intensity of the SSBs was greatest in the spectra of phytate compounds, minor SSBs were present in the spectra of inorganic hydrogen phosphate compounds, and SSBs were absent in tribasic phosphate compounds (He et al. 2007b).

The relative sizes of SSBs of inorganic and organic P species can be explained by differences in their electronic structures (Fig. 1.4). As discussed earlier, the size of SSBs is proportional to the CSA of the species, which is a measure of how symmetric the chemical environment is around the P nucleus. In both inorganic and the vast majority of organic P species, the P atom is surrounded by four O atoms arranged in a symmetric tetrahedral fashion. However, these O atoms are not equivalent, in that one has a double-bond to P, whereas the rest have a single bond to P. In orthophosphate itself, the remaining three O atoms each carry a negative charge, resulting in a polar (and hence non-symmetric) environment around the P nucleus. However, in the case of orthophosphate, the distinction between these two types of O environments is illusory, because there are four equivalent resonant electronic structures that in fact make these O atoms chemically equivalent. In contrast, for organic P molecules, at least one of the oxygen atoms is distinguished by being bound through a covalent bond to an organic residue

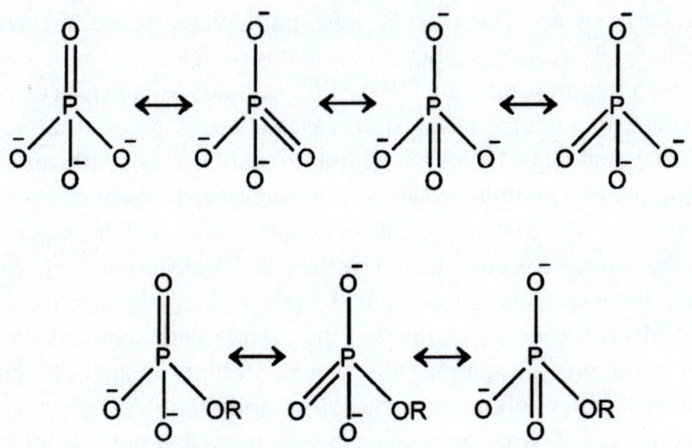

Fig. 1.4 The four resonance structures of orthophosphate (*above*) and three resonance structures of organic phosphate molecules (*below*). The organic residue is denoted as *R*

(denoted R in Fig. 1.4). Therefore, the P in organic P species is always in an unsymmetrical environment and would be expected to have larger SSBs than orthophosphate.

As is the case for solution ^{31}P NMR spectra, poor spectral resolution can also be caused by the presence of paramagnetic ions, but this does not always appear to be the case. Increased line broadening (McDowell et al. 2003a) and high intensity SSBs (Hinedi and Chang 1989) have been ascribed to the presence of paramagnetic species (Fe and Mn) in soil samples. However, Shand et al. (1999) decreased the Fe and Mn concentrations in soil humic acid samples and found there to be no improvement in spectral resolution. Instead, the authors attributed the poor spectral resolution to the various ways in which the P species were associated with the organic phase.

Other factors that may affect resolution have not been widely examined. He et al. (2007b) and Conte et al. (2008) both recommend analysis of only completely dry samples. They reported that moisture in the soil sample at the time of analysis altered the intensity and position of peaks and the size of SSBs.

1.3.4 Quantitation

Quantification of P species using solid-state ^{31}P NMR spectra is hampered by low observability and poor resolution. The low P observabilities reported by Dougherty et al. (2005) for whole soils (an average of 9% for CP and 22% for DP) give a pessimistic outlook for using solid-state ^{31}P NMR for quantitative P speciation in soils. On the other hand, much higher values of 71–88% for DP spectra have been reported for low-Fe calcareous soils (McBeath et al. 2006). At the very least, there is a clear need to address the issue of selective observation of P species and the bias this is likely to create, and spin counting appears to be best way to do this. It may be that the Dougherty et al. (2005) soils were particularly prone to paramagnetic interferences, but only further studies will make this clear.

The poor resolution in solid-state ^{31}P NMR spectra of soils makes deconvolution necessary to quantify signals in almost all cases (McDowell et al. 2002b; Hunger et al. 2004; Conte et al. 2008). One exception is where species with very distinctive chemical shifts, such as pyrophosphate, are of interest (McBeath et al. 2006). As is the case for solution ^{31}P NMR, the results of deconvolution will be dependent on the way in which the resonances are selected by the user. Peak assignments can be made using an automatic peak-picking routine in the processing software (He et al. 2009), through visual identification by comparing the sample with standard P compounds acquired prior to running the sample (Shand et al. 1999; Frossard et al. 2002; Conte et al. 2008) or by directly spiking into the soil sample (Benitez-Nelson et al. 2004). Usually, assignments are based on previous literature assignments (McDowell et al. 2002b, 2003b). Directly spiking the soil sample would provide the most accurate results, but Benitez-Nelson et al. (2004) reported that the peaks broadened when the standards were mixed into the sample matrix. Therefore, for reliable quantitation, careful consideration of peak selection is required when fitting peaks.

The main issues of deconvolution that affect solution NMR analysis also affect solid-state NMR analysis. However, the greater degree of overlap in solid-state spectra makes these problems greater. For example, Fig. 1.5 shows solid-state ^{31}P NMR spectra of two soils and the fits of these spectra to seven components using deconvolution. Although a good fit is obtained in each case, quantifying individual species from these fits would appear optimistic, as acknowledged in the paper (McDowell et al. 2003b). The incorporation of prior knowledge has been attempted by Hunger et al. (2004), who used a minimum set number of peaks, a Lorentzian lineshape for the narrower peaks and a Gaussian lineshape for the broad peaks, and in doing so improved the quality of the fit. There is quite some way to go before deconvolution of solid-state ^{31}P NMR spectra of soils can be considered a reliable technique for quantification of P species.

1.4 P XANES Spectroscopy

XANES (X-ray absorption near-edge structure) spectroscopy, which is synonymous with NEXAFS (near-edge X-ray fine structure) spectroscopy, is a technique that can be used for speciation of numerous elements in soils. XANES requires an energy-tunable source of X-rays that is currently only possible with a synchrotron. The use of synchrotron-based techniques for soil analyses is a recent development, coinciding with the rapid and continuing increase in the number of synchrotron facilities worldwide since the 1990s (Lombi and Susini 2009).

XANES is a type of XAS and is based on the photoelectric effect. When materials are irradiated with high-energy electromagnetic radiation, electrons (photoelectrons) are released. In XAS, the energy of the incident radiation is varied close to the threshold required to excite strongly bound (core) electrons of an atom to produce

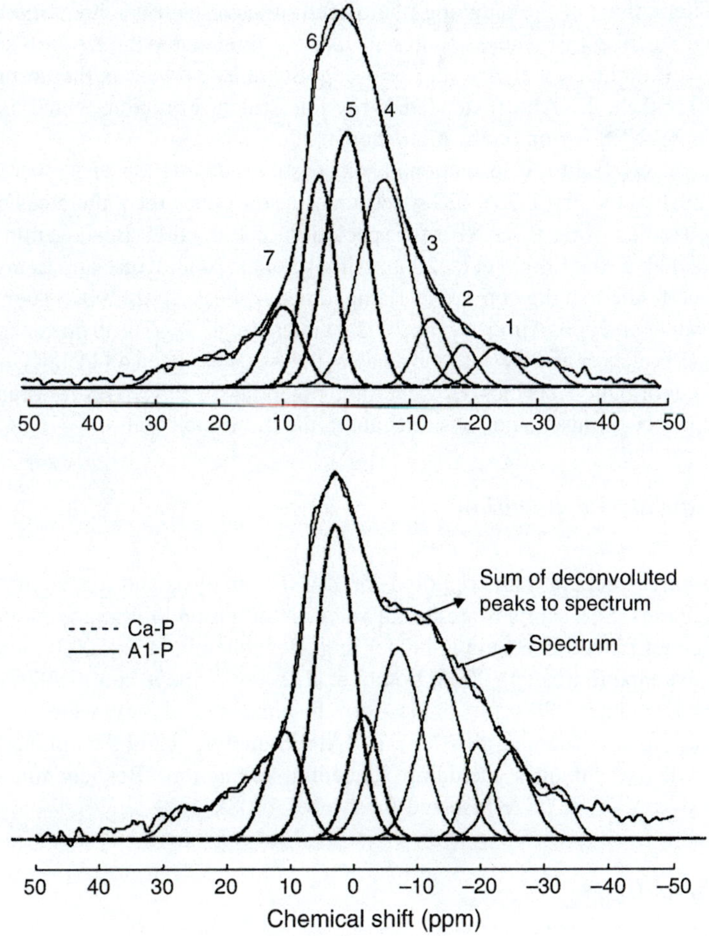

Fig. 1.5 ^{31}P HP/Dec MAS NMR spectra of soils with pH of 5.91 (*top*) and 4.48 (*bottom*) showing the deconvolution of the NMR signal into peaks assigned to Al–P and Ca–P species. The *numbers* in the upper spectrum refer to assignments made to individual P species. Reprinted from McDowell et al. (2003b), with permission from Elsevier

photoelectrons. This energy is quite specific for each different element, ensuring that in most cases a spectrum can be generated for a given element free of interferences from other elements that may be present. Below the threshold energy for photoelectron production there may be some absorption that corresponds to excitation of core electrons into high-energy excited states (Ajiboye et al. 2008). The main feature of the XAS spectrum is the "absorption edge", which basically corresponds to the threshold energy of photoelectron release. The position (in terms of frequency or energy) of the absorption edge is strongly affected by the oxidation state of the atom, since the removal of electrons from an atom results in the remaining electrons being held more strongly. The electronegativity of neighbouring atoms will have a similar, but smaller effect. At energies beyond the absorption edge there are features that relate to interactions of the outgoing photoelectron wave with neighbouring atoms (Lombi and Susini 2009). These features are usually divided into the XANES spectra region (which includes pre-edge and post-edge features as well as the absorption edge itself) and the EXAFS (extended X-ray fine structure) region, which extends beyond the XANES region on the high-energy side.

The use of XANES for P speciation in soils is still in its infancy. Hesterberg et al. (1999) published the first P XANES spectrum of a soil. Since then, there has been a handful of studies using P XANES for speciation of P in soils (Beauchemin et al. 2003; Sato et al. 2005; Lombi et al. 2006; Ajiboye et al. 2008; Kruse and Leinweber 2008), some related studies on organic amendments (Peak et al. 2002; Toor et al. 2005; Shober et al. 2006; Ajiboye et al. 2007a; Gungor et al. 2007) and one on marine sediments (Brandes et al. 2007). In most cases, the focus has been on identifying and quantifying inorganic P species. However, the potential for P XANES for organic P speciation has been raised and so is relevant to the discussion here.

1.4.1 Sample Preparation

Minimal sample preparation is required for XANES analysis and this is one of its great advantages. Soil samples are usually sieved and ground to improve homogeneity and either placed in a sample holder (e.g. Beauchemin et al. 2003) or spread onto adhesive tape (Sato et al. 2005; Lombi et al. 2006; Ajiboye et al. 2007a, 2008; Kruse and Leinweber 2008). Samples can be analysed at any water content, including as pastes (Khare et al. 2004). P XANES analysis can be combined with other analytical techniques including sequential extraction (Beauchemin et al. 2003; Ajiboye et al. 2007a; Kruse and Leinweber 2008).

1.4.2 Sensitivity

The sensitivity of P XANES analysis is generally very high. This can be attributed to the high energy of the transitions involved (especially compared to NMR) and also to the high intensity radiation produced by synchrotrons. Most published P XANES spectra were acquired in two to ten scans and have good signal-to-noise ratios, although Ajiboye et al. (2008) have commented on the low signal-to-noise ratios for low P soils.

1.4.3 Resolution

The ability to differentiate species is the main limitation of P XANES analysis. This is basically a resolution issue in spectroscopic terms. The nature of a XANES signal (e.g. Fig. 1.6), which is complex and extends across most of the spectral range for every species, is very different to that of an NMR signal, which is a simple Gaussian–Lorentzian "peak". This complicates the issue of resolution, and greatly

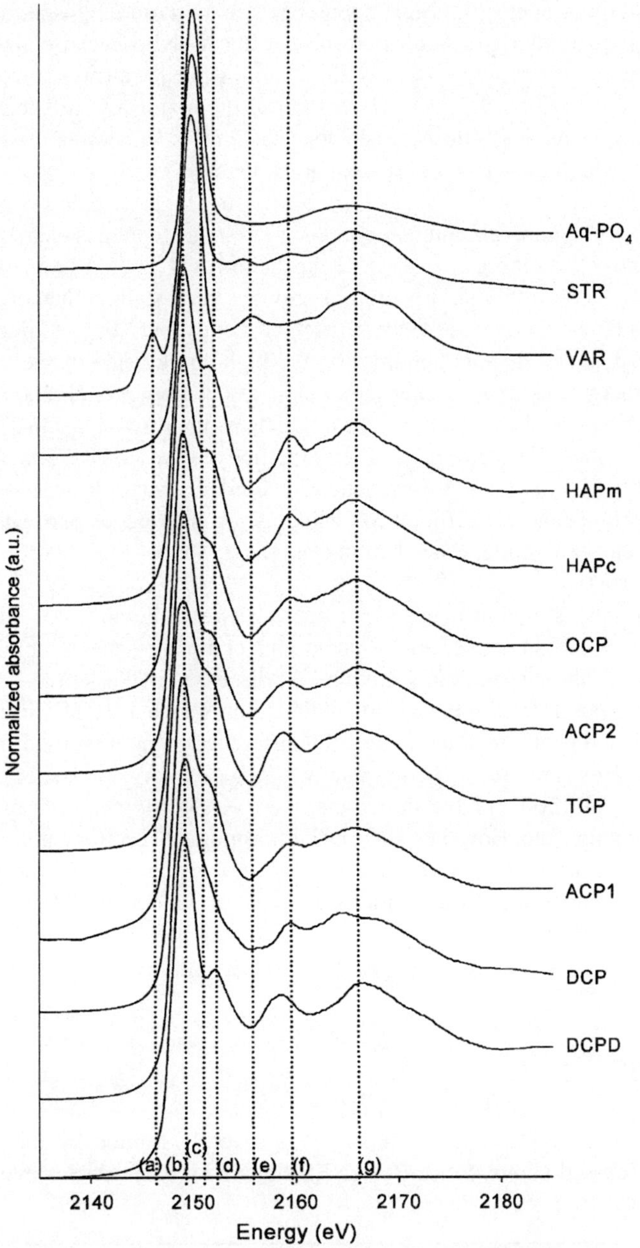

Fig. 1.6 Phosphorus *K*-edge XANES spectra for different inorganic phosphate standard species: $H_2PO_4^-$ in 0.1 M NaCl solution (*Aq-PO₄*); strengite (*STR*; $FePO_4{\cdot}2H_2O$); variscite (*VAR*; $AlPO_4{\cdot}2H_2O$); synthesised hydroxyapatite [*HAPc*; $Ca_5(PO_4)_3OH$]; hydroxyapatite in mineral form (*HAPm*); octacalcium phosphate [*OCP*; $Ca_4H(PO_4)_3{\cdot}2.5H_2O$]; β-tricalcium calcium phosphate [*TCP*; $Ca_3(PO_4)_2$]; amorphous calcium phosphate phase 1 (*ACP1*) and amorphous calcium

reduces the ability to detect multiple components in a mixture. Phosphorus speciation in soils is particularly difficult to resolve using XANES because all P species have the same oxidation state (+V) and the P atom is nearly always surrounded by four O atoms. Therefore, the spectral position of the major XANES feature – the absorption edge – varies little between species. This is in contrast to sulfur and many metals, which exist in multiple oxidation states.

Some features of P XANES spectra are diagnostic. Figure 1.6 shows P XANES spectra of several P minerals and compounds. A pre-edge feature is characteristic of iron phosphates (Hesterberg et al. 1999; Beauchemin et al. 2003; Lombi et al. 2006; Ajiboye et al. 2008), although it is more prominent in crystalline than non-crystalline minerals (Hesterberg et al. 1999; Kruse and Leinweber 2008). Calcium phosphates usually have a distinct shoulder on the high-energy side of the absorption edge (Hesterberg et al. 1999; Beauchemin et al. 2003; Lombi et al. 2006; Ajiboye et al. 2008; Kruse and Leinweber 2008), which is again more distinct for some minerals than others (Hesterberg et al. 1999; Ajiboye et al. 2008). The P XANES spectra of organic P compounds are generally similar, contain little in the way of diagnostic features and are difficult to distinguish from aqueous or weakly bound phosphate (Peak et al. 2002; Beauchemin et al. 2003; Shober et al. 2006; Kruse and Leinweber 2008).

An important distinction needs to be made between the ability to distinguish between the XANES spectra of pure P-containing compounds and minerals and the ability to resolve these species in a mixture. As discussed by Peak et al. (2002), the former may be possible on the basis of subtle features such as small differences in peak position and peak broadness, but these differences cannot be used for resolving mixtures. This is particularly important with regard to organic P speciation. For example, He et al. (2007c) have shown that the P XANES spectra of pure phytate salts are distinguishable. However, their lack of prominent spectral features, mutual similarity and similarity to P XANES spectra of soluble and sorbed orthophosphate would make it impossible to distinguish different phytate salts in heterogeneous matrices such as soils.

1.4.4 Quantitation

Quantitation in XANES analysis of soil P is severely limited by the issues of resolution discussed above. Since P XANES signals for each species overlap over

---

Fig. 1.6 (Continued) phosphate phase 2 (*ACP2*) [both $Ca_3(PO_4)_2{\cdot}xH_2O$, where *x* depends on the phase]; dibasic calcium phosphate [*DCP*; ($CaHPO_4$)]; and dibasic calcium phosphate dihydrate [*DCPD*; ($CaHPO_4{\cdot}2H_2O$)]. The *dashed lines* show energy levels of importance to indicate unique spectra features for different species: (**a**) STR; (**b**) absorption edge; (**c, d**) CaP species related to their solubility; (**e**) STR and VAR; (**f**) CaP species; (**g**) oxygen oscillation. Reprinted from Sato et al. (2005) with permission. Copyright 2005 American Chemical Society virtually the whole spectral width, quantitative analysis of P XANES spectra of soils is usually achieved by fitting the experimental spectrum to a linear combination of XANES spectra of model materials. This is the equivalent of the deconvolution process used in NMR spectroscopy. There are inherent problems with this process. The main one is that even if a good fit is achieved, one cannot determine whether or not an equally good or even better fit could be achieved with a different combination of input spectra. This fundamental problem appears to be unappreciated in some studies, where the results of "best-fits" to XANES spectra are reported as actual compositions.

A key step in any linear combination fit (LCF) to a XANES spectrum is choosing the number and identity of component spectra. Beauchemin et al. (2003) tackled the first issue, choosing the number of components for the LCF by using principle component analysis (PCA). They followed a similar approach to one they employed to analyse sulfur XANES spectra of soils (Beauchemin et al. 2002), which involves using PCA to identify how many independent (orthogonal) components are required to explain the variance of P XANES spectra of a set of soils. In their case, they found that just two components were required for their set of five soils. The second step of this approach is "target transformation" (TT), which identifies the likely nature of these components through comparison of spectra generated for each component with those of model materials. Beauchemin et al. (2003) found that of their 11 model materials, eight were identified as likely component species. They then carried out LCFs for each soil using two to three component spectra (Fig. 1.7). Importantly, similar quality fits were achieved using different input spectra. This highlights the non-uniqueness of the LCF procedure. However, when components were considered as belonging to two general classes (calcium phosphate and phosphate sorbed to iron or aluminium oxide), rather than as specific minerals, there was good consistency in the fits.

It should be noted that the P compositions of soils derived from P XANES spectra reported by Beauchemin et al. (2003) and in subsequent studies (Sato et al. 2005; Lombi et al. 2006; Gungor et al. 2007) that used the same method identify no organic P forms. This is clearly not reasonable, and highlights the limitations of the method. Other studies (Toor et al. 2005; Shober et al. 2006; Ajiboye et al. 2007a, 2008), again using the same method for analysis of P XANES spectra, do include an organic P component (invariably phytate). However, given the lack of features in P XANES spectra of organic P compounds and their similarity to some inorganic P forms, this method of quantifying organic P, let alone organic P speciation does not appear promising.

An important aspect of verifying any quantification method is testing on simple mixes. Ajiboye et al. (2007b) have done this for binary mixes of mineral phosphates. The results were mixed at best, with the quantification of minor components (25%) causing particular difficulties. Khare et al. (2004) successfully used XANES to quantify phosphate sorbed to binary mixtures of iron and aluminium hydroxides; however, it should be emphasised that this is a very simple system compared with soils. Another important prerequisite for quantitation is that the signal produced by all P species is independent of the matrix they are in. This requires validation

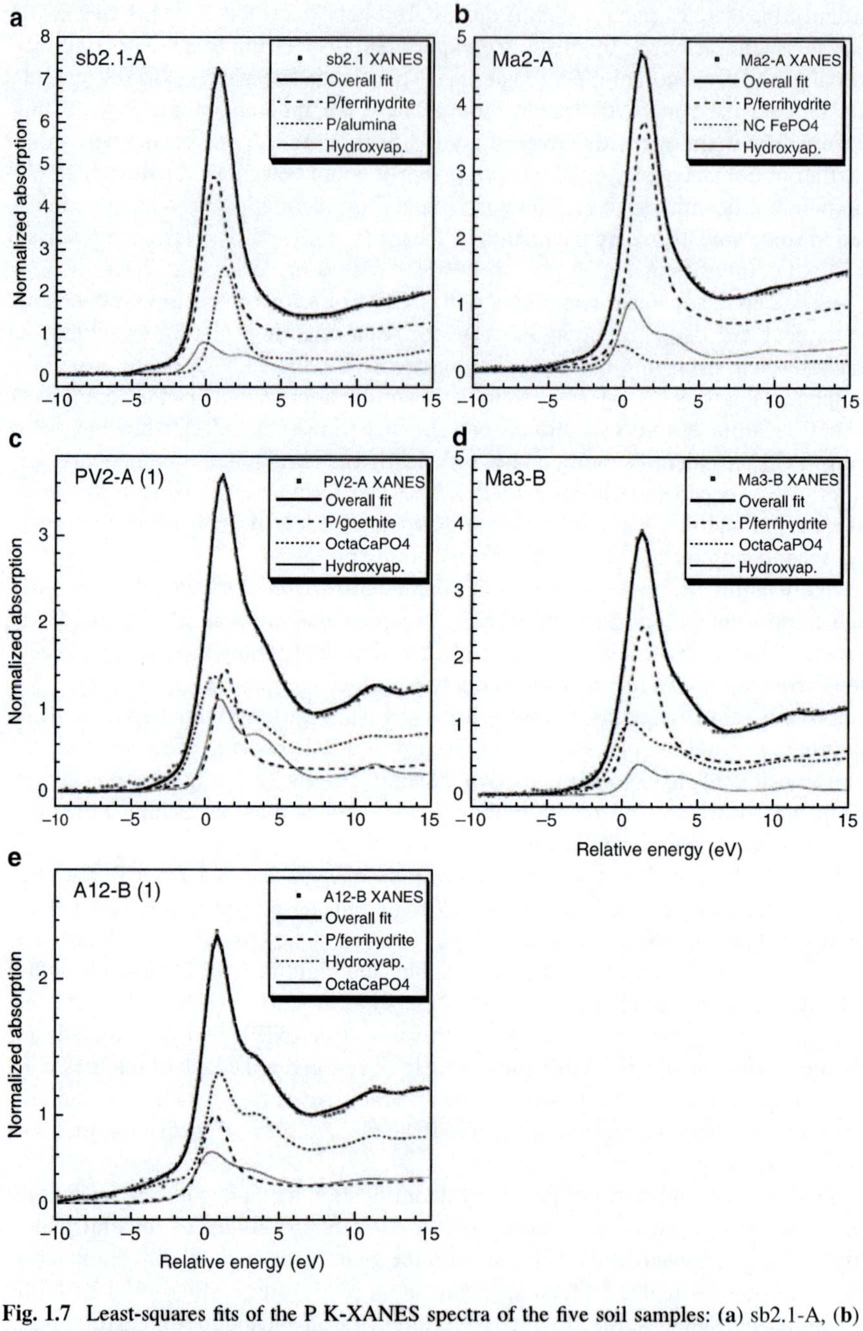

Fig. 1.7 Least-squares fits of the P K-XANES spectra of the five soil samples: (a) sb2.1-A, **(b)** Ma2-A, **(c)** PV2-A, **(d)** Ma3-B and **(e)** AI2-B. *P/ferrihydrite*, *P/goethite* and *P/alumina* refer to PO_4 adsorbed on ferrihydrite, goethite or alumina; *octaCaPO$_4$* octacalcium phosphate; *hydroxyap.* hydroxyapatite; *NC FePO$_4$* non-crystalline $FePO_4$. Reprinted from Beauchemin et al. (2003), with permission from Journal of Environmental Quality

through spiking experiments analogous to those described above for NMR analysis. Quantitative analysis is certainly made easier if all P species in a sample produce equivalent amounts of signal. Kruse and Leinweber (2008) have tested this for a set of peat soils, with encouraging results.

Finally, it must be remembered that P XANES analysis of soils is still in its infancy and further developments are likely. The identification of only two to four principle components in studies so far (Beauchemin et al. 2003; Sato et al. 2005; Toor et al. 2005; Lombi et al. 2006) partly reflects the small number of spectra analysed. It is likely that the analysis of larger sets of soils would result in the identification of more principle components. P XANES also offers the added ability of mapping distributions of P species at the submicron scale, the utility of which is only just being explored (Lombi et al. 2006; Brandes et al. 2007). Another promising direction involves the use of $L_{2,3}$-edge P XANES spectra. Until now, all P XANES spectra of soils have been acquired at the so-called *K*-edge, which involves excitation of electrons from 1s orbitals. Kruse et al. (2009) have recently shown that $L_{2,3}$-edge P XANES spectra, which involve excitation from p orbitals, are much more information-rich, especially for organic P species. This will not completely overcome the limitations of LCF, but may enable the reliable identification and quantification of more than two to three components in P XANES analysis of soils, possibly including multiple organic P components.

1.5 Conclusions

Spectroscopic techniques offer the best potential for determining the speciation of organic P in soils. The three main spectroscopic techniques that have been used for this purpose are solution ^{31}P NMR spectroscopy, solid-state ^{31}P NMR spectroscopy and P XANES spectroscopy. All techniques have their advantages and their limitations and these need to be understood in order to choose the best technique for any given purpose. All techniques are also highly technical and are best carried out with assistance from experts in the field. To help in deciding between techniques, we summarise the relative merits of each on four criteria (sample preparation, sensitivity, resolution and quantitation) in Table 1.1, on a scale of one to five stars. This is of course a subjective assessment, but we believe it is a useful starting point. It is also based on the techniques as they currently exist and does not take into account the potential for future development.

Table 1.1 indicates that where the main goal is quantification of organic P species, solution ^{31}P NMR is likely to be the best alternative. However, this comes with the proviso that species not soluble in alkaline extracts will be missed, and that hydrolysis is likely to cause some species to be misidentified. Furthermore, solution ^{31}P NMR cannot differentiate between organic P associated with different cations or minerals. The solid-state techniques (solid-state ^{31}P NMR and P XANES) may help here, but detailed identification and quantification is unlikely. Solid-state ^{31}P NMR is particularly unsuited to speciation of Fe-associated organic

Table 1.1 Summary of the relative merits of each spectroscopic technique for organic P speciation

	Solution ^{31}P NMR	Solid-state ^{31}P NMR	P XANES
Sample preparation	⋆⋆⋆ Requires extraction into alkaline solution. Not all organic P is extracted and some organic P compounds may be hydrolysed under these conditions	⋆⋆⋆⋆ Virtually no sample preparation required. CP spectra can be affected if samples are not dry	⋆⋆⋆⋆⋆ Virtually no sample preparation required. Spectra can be acquired at any water content
Sensitivity	⋆⋆⋆ Sensitivity of NMR is generally poor. Can be overcome by acquiring many scans	⋆⋆⋆ Sensitivity of NMR is generally poor. Can be overcome by acquiring many scans. Not limited by solubility	⋆⋆⋆⋆ Sensitivity very good
Resolution	⋆⋆⋆⋆ Resolution is very good compared to other techniques. However, not every individual organic P molecule can be resolved	⋆⋆ Resolution is quite poor due to broadness of signals. Overlap inevitable	⋆⋆ Resolution (of mixtures) is poor due to signal for each species being similar and covering whole spectral range. Overlap inevitable
Quantitation	⋆⋆⋆⋆ Best prospects for quantification of organic P species. Somewhat limited by overlap of signals in crowded regions. Cannot quantify non-extractable species and hydrolysis may cause bias	⋆ Severely limited by poor resolution and also interferences by paramagnetic species	⋆ Severely limited by poor ability to resolve individual organic P signals

P (due to low observability), whereas P XANES is particularly suited (due to a diagnostic pre-edge feature).

1.6 Future Directions

All three spectroscopic techniques can be improved. A particular focus should be the quantitative aspects of each technique, as this is what end-users mostly want from spectroscopic analysis of organic P in soils, and there currently appears to be a lack of understanding of the limitations in this area.

Some issues are common to several techniques, and foremost amongst these is deconvolution and fitting of spectra. There needs to be a wider appreciation amongst both spectroscopists and end-users that a good fit to a spectrum doesn't equate to a definitive composition, and much more can be done to test the validity and identify the limitations of these procedures.

Other issues are specific to each technique. Key areas for improvement for solution ^{31}P NMR include accurate and reliable identification of as many peaks as possible, a quantitative understanding of hydrolysis, and quantification of the apparent broad resonance of large P-containing molecules. A key area for improvement in solid-state ^{31}P NMR analysis would be routine gauging of NMR observability, as this would identify under what circumstances a quantitative assessment is possible. For P XANES analysis, further assessment of the potential of $L_{2,3}$-edge spectra seems warranted, along with development of submicron mapping of P speciation. Understanding how P speciation varies in space would be invaluable, even if the detail of P speciation is limited.

The use of spectroscopic techniques for organic P speciation is still in its infancy. There have been many important advances already and there are surely more to come. However, we believe the time has come for research in this area to mature to a new phase of development, from being dominated by "proof-of-concept" studies, to being dominated by hard-nosed and rigorous assessment of capabilities and limitations. It is only through the feedback of such assessment into further development that accurate, reliable and quantitative methods will emerge that can be applied to all of the areas where organic P speciation is important.

References

Adams MA (1990) ^{31}P-NMR identification of phosphorus compounds in neutral extracts of Mountain Ash (*Eucalyptus regnans* F. Muell.) soils. Soil Biol Biochem 22:419–421

Adams MA, Byrne LT (1989) ^{31}P-NMR analysis of phosphorus compounds in extracts of surface soils from selected karri (*Eucalyptus diversicolor* F. Muell.) forests. Soil Biol Biochem 21:523–528

Ajiboye B, Akinremi OO, Hu Y, Flaten DN (2007a) Phosphorus speciation of sequential extracts of organic amendments using nuclear magnetic resonance and X-ray absorption near-edge structure spectroscopies. J Environ Qual 36:1563–1576

Ajiboye B, Akinremi OO, Jurgensen A (2007b) Experimental validation of quantitative XANES analysis for phosphorus speciation. Soil Sci Soc Am J 71:1288–1291

Ajiboye B, Akinremi OO, Hu Y, Jurgensen A (2008) XANES speciation of phosphorus in organically amended and fertilized Vertisol and Mollisol. Soil Sci Soc Am J 72:1256–1262

Al Deen TS, Hibbert DB, Hook JM, Wells RJ (2002) Quantitative nuclear magnetic resonance spectrometry – II. Purity of phosphorus-based agrochemicals glyphosate (*N*-(phosphonomethyl)-glycine) and profenofos (*O*-(4-bromo-2-chlorophenyl) *O*-ethyl *S*-propyl phosphorothioate) measured by ^{1}H and ^{31}P QNMR spectrometry. Anal Chim Acta 474:125–135

Anderson G (1967) Nucleic acids, derivatives, and organic phosphates. In: McLaren AD, Peterson GH (eds) Soil biochemistry. Edward Arnold, London, pp 67–90

Anderson G (1980) Assessing organic phosphorus in soils. In: Khasawneh FE, Sample EC, Kamprath EJ (eds) Role of phosphorus in agriculture. ASA, Madison, pp 411–428

Baer E, Stancer HC, Korman IA (1953) Migration during hydrolysis of esters of glycerophosphoric acid. III. Cephalin and glycerylphosphorylethanolamine. J Biol Chem 200:251–255
Beauchemin S, Hesterberg D, Beauchemin M (2002) Principle component analysis approach for modeling sulfur K-XANES spectra of humic acids. Soil Sci Soc Am J 66:83–91
Beauchemin S, Hesterberg D, Chou J, Beauchemin M, Simard RR, Sayers DE (2003) Speciation of phosphorus in phosphorus-enriched agricultural soils using X-ray near-edge structure spectroscopy and chemical fractionation. J Environ Qual 32:1809–1819
Bedrock CN, Cheshire MV, Chudek JA, Goodman BA, Shand CA (1994) Use of ^{31}P-NMR to study the forms of phosphorus in peat soils. Sci Total Environ 152:1–8
Benitez-Nelson CR, O'Neill L, Kolowith LC, Pellechia P, Thunell R (2004) Phosphonates and particulate organic phosphorus cycling in an anoxic marine basin. Limnol Oceanogr 49:1593–1604
Berg AS, Joern BC (2006) Sorption dynamics of organic and inorganic phosphorus compounds in soil. J Environ Qual 35:1855–1862
Borie F, Rubio R (2003) Total and organic phosphorus in Chilean volcanic soils. Gayana Bot 60:69–78
Borie F, Zunino H, Martãnez L (1989) Macromolecule-P associations and inositol phosphates in some Chilean volcanic soils of temperate regions. Commun Soil Sci Plant Anal 20:1881–1894
Bowman RA, Moir JO (1993) Basic EDTA as an extractant for soil organic phosphorus. Soil Sci Soc Am J 57:1516–1518
Brandes JA, Ingall E, Paterson D (2007) Characterization of minerals and organic phosphorus species in marine sediments using soft X-ray fluorescence spectromicroscopy. Mar Chem 103:250–265
Briceño M, Escudey M, Galindo G, Borchardt D, Chang AC (2006) Comparison of extraction procedures used in determination of phosphorus species by ^{31}P-NMR in Chilean volcanic soils. Commun Soil Sci Plant Anal 37:1553–1569
Bünemann EK, Smernik RJ, Marschner P, McNeill AM (2008) Microbial synthesis of organic and condensed forms of phosphorus in acid and calcareous soils. Soil Biol Biochem 40:932–946
Cade-Menun B (2005a) Using phosphorus-31 nuclear magnetic resonance spectroscopy to characterise organic phosphorus in environmental samples. In: Turner BL, Frossard E, Baldwin DS (eds) Organic phosphorus in the environment. CABI, Wallingford, pp 21–44
Cade-Menun BJ (2005b) Characterizing phosphorus in environmental and agricultural samples by P-31 nuclear magnetic resonance spectroscopy. Talanta 66:359–371
Cade-Menun BJ, Preston CM (1996) A comparison of soil extraction procedures for ^{31}P NMR spectroscopy. Soil Sci 161:770–785
Cade-Menun BJ, Liu CW, Nunlist R, McColl JG (2002) Soil and litter phosphorus-31 nuclear magnetic resonance spectroscopy: extractants, metals, and phosphorus relaxation times. J Environ Qual 31:457–465
Changani KK, Ala-Korpela M, Fuller BJ, Mierisova S, Bryant DJ, Taylor-Robinson SD, Davidson BR, Bell JD (1999) Incorporation of metabolite prior knowledge for data analysis: biochemical implications of dynamic ^{31}P NMR ex vivo pig liver studies. NMR Biomed 12:197–204
Condron LM, Frossard E, Newman RH, Tekely P, Morel J-L (1997) Use of ^{31}P NMR in the study of soils and the environment. In: Nanny MA, Minear RA, Leenheer JA (eds) Nuclear magnetic resonance in environmental chemistry. Oxford University Press, New York, pp 247–271
Condron LM, Turner BL, Cade-Menun BJ (2005) Chemistry and dynamics of soil organic phosphorus. In: Sims JT, Sharpley AN (eds) Phosphorus: agriculture and the environment. ASA, CSSA and SSSA, Madison, pp 87–122
Conte P, Šmejkalová D, Piccolo A, Spaccini R (2008) Evaluation of the factors affecting direct polarization solid state ^{31}P-NMR spectroscopy of bulk soils. Eur J Soil Sci 59:584–591
Corbett RJT (1993) How to perform automated curve fitting to in vivo ^{31}P magnetic resonance spectroscopic data. Magn Reson Mater Phys Biol Med 1:65–76
Costello AJR, Glonek T, Myers TC (1976) ^{31}P nuclear magnetic resonance-pH titrations of *myo*-inositol hexaphosphate. Carbohydr Res 46:159–171

Crouse DA, Sierzputowska-Gracz H, Mikkelsen RL (2000) Optimization of sample pH and temperature from phosphorus-31 nuclear magnetic resonance spectroscopy of poultry manure extracts. Commun Soil Sci Plant Anal 31:229–240

Dai K, David MB, Vance GF, Krzyszowska AJ (1996) Characterization of phosphorus in a spruce-fir spodosol by phosphorus-31 nuclear magnetic resonance spectroscopy. Soil Sci Soc Am J 60:1943–1950

Dalal RC (1977) Soil organic phosphorus. Adv Agron 29:83–117

Derome AE (1987) Modern NMR techniques for chemistry research. Pergamon, Singapore

Doolette AL, Smernik RJ, Dougherty WJ (2009) Spiking improved solution phosphorus-31 nuclear magnetic resonance identification of soil phosphorus compounds. Soil Sci Soc Am J 73:919–927

Dougherty WJ, Smernik RJ, Chittleborough DJ (2005) Application of spin counting to the solid-state ^{31}P NMR analysis of pasture soils with varying phosphorus content. Soil Sci Soc Am J 69:1–13

Dougherty WJ, Smernik RJ, Bünemann EK, Chittleborough DJ (2007) On the use of HF pre-treatment of soils for ^{31}P NMR analyses. Soil Sci Soc Am J 71:1111–1118

Escudey M, Galindo G, Forster JE, Briceño M, Diaz P, Chang A (2001) Chemical forms of phosphorus of volcanic ash-derived soils in Chile. Commun Soil Sci Plant Anal 32:601–616

Folch J (1942) The nature of the glycerophosphoric acid present in phosphatides. J Biol Chem 146:31–33

Frossard E, Tekely P, Grimal JY (1994) Characterization of phosphate species in urban sewage sludges by high-resolution solid-state ^{31}P NMR. Eur J Soil Sci 45:403–408

Frossard E, Skrabal P, Sinaj S, Bangerter F, Traore O (2002) Forms and exchangeability of inorganic phosphate in composted solid organic wastes. Nutr Cycl Agroecosyst 62:103–113

Fyfe CA (1983) Solid state NMR for chemists. CFC, Guelph

Guggenberger G, Christensen BT, Rubæk G, Zech W (1996) Land-use and fertilization effects on P forms in two European soils: resin extraction and ^{31}P-NMR analysis. Eur J Soil Sci 47:605–614

Gungor K, Jurgensen A, Karthikeyan KG (2007) Determination of phosphorus speciation in dairy manure using XRD and XANES spectroscopy. J Environ Qual 36:1856–1863

Hawkes GE, Powlson DS, Randall EW, Tate KR (1984) A P-31 nuclear magnetic-resonance study of the phosphorus species in alkali extracts of soils from long-term field experiments. J Soil Sci 35:35–45

He Z, Ohno T, Cade-Menun BJ, Erich MS, Honeycutt CW (2006) Spectral and chemical characterization of phosphates associated with humic substances. Soil Sci Soc Am J 70:1741–1751

He Z, Cade-Menun B, Toor G, Fortuna A, Honeycutt C, Sims J (2007a) Comparison of phosphorus forms in wet and dried animal manures by solution phosphorus-31 nuclear magnetic resonance spectroscopy and enzymatic hydrolysis. J Environ Qual 36:1086–1095

He Z, Honeycutt C, Xing B, McDowell R, Pellechia P, Zhang T (2007b) Solid-state Fourier transform infrared and P-31 nuclear magnetic resonance spectral features of phosphate compounds. Soil Sci 172:501–515

He Z, Honeycutt CW, Zhang T, Pellechia PJ, Caliebe WA (2007c) Distinction of metal species of phytate by solid-state spectroscopic techniques. Soil Sci Soc Am J 71:940–943

He Z, Honeycutt CW, Griffin TS, Cade-Menun BJ, Pellechia PJ, Dou Z (2009) Phosphorus forms in conventional and organic dairy manure identified by solution and solid state ^{31}P NMR spectroscopy. J Environ Qual 38:1909–1918

Hesterberg D, Zhou W, Hutchinson KJ, Beauchemin S, Sayers DE (1999) XAFS study of adsorbed and mineral forms of phosphate. J Synchrotron Radiat 6:636–638

Hinedi ZR, Chang AC (1989) Solubility and phosphorus-31 magic angle spinning nuclear magnetic resonance of phosphorus in sludge-amended soils. Soil Sci Soc Am J 53:1057–1061

Hunger S, Cho H, Sims JT, Sparks DL (2004) Direct speciation of phosphorus in alum-amended poultry litter: solid-state ^{31}P NMR investigation. Environ Sci Technol 38:674–681

Khare N, Hesterberg D, Beauchemin S, Wang S-L (2004) XANES determination of adsorbed phosphate distribution between ferrihydrite and boehmite in mixtures. Soil Sci Soc Am J 68:460–469

Koopmans GF, Chardon WJ, McDowell RW (2007) Phosphorus movement and speciation in a sandy soil profile after long-term animal manure applications. J Environ Qual 36:305–315

Kruse J, Leinweber P (2008) Phosphorus in sequentially extracted fen peat soils: a K-edge X-ray absorption near-edge structure (XANES) spectroscopy study. J Plant Nutr 171:613–620

Kruse J, Leinweber P, Eckhardt K-U, Godlinski F, Hu Y, Zuin L (2009) Phosphorus L2, 3-edge XANES: overview of reference compounds. J Synchrotron Radiat 16:247–259

Lombi E, Susini J (2009) Synchrotron-based techniques for plant and soil science: opportunities challenges and future perspectives. Plant Soil 320:1–35

Lombi E, Scheckel KG, Armstrong RD, Forrester S, Cutler JN, Paterson D (2006) Speciation and distribution of phosphorus in a fertilized soil: a synchrotron-based investigation. Soil Sci Soc Am J 70:2038–2048

Makarov MI, Malysheva TI, Haumaier L, Alt HG, Zech W (1997) The forms of phosphorus in humic and fulvic acids of a toposequence of alpine soils in the northern Caucasus. Geoderma 80:61–73

Makarov MI, Haumaier L, Zech W (2002a) The nature and origins of diester phosphates in soils: a ^{31}P-NMR study. Biol Fertil Soils 35:136–146

Makarov MI, Haumaier L, Zech W (2002b) Nature of soil organic phosphorus: an assessment of peak assignments in the diester region of ^{31}P NMR spectra. Soil Biol Biochem 34:1467–1477

Maniara G, Rajamoorthi K, Rajan S, Stockton GW (1998) Method performance and validation for quantitative analysis by ^{1}H and ^{31}P NMR spectroscopy. Applications to analytical standards and agricultural chemicals. Anal Chem 70:4921–4928

McBeath TM, Smernik RJ, Lombi E, McLaughlin MJ (2006) Hydrolysis of pyrophosphate in a highly calcareous soil: a solid-state phosphorus-31 NMR study. Soil Sci Soc Am J 70:856–862

McDowell RW, Stewart I (2005a) Peak assignments for phosphorus-31 nuclear magnetic resonance spectroscopy in pH range 5–13 and their application in environmental samples. Chem Ecol 21:211–226

McDowell RW, Stewart I (2005b) An improved technique for the determination of organic phosphorus in sediments and soils by ^{31}P nuclear magnetic resonance spectroscopy. Chem Ecol 21:11–22

McDowell RW, Stewart I (2006) The phosphorus composition of contrasting soils in pastoral, native and forest management in Otago, New Zealand: sequential extraction and ^{31}P NMR. Geoderma 130:176–189

McDowell RW, Brookes PC, Mahieu N, Poulton PR, Johnston AE, Sharpley AN (2002a) The effect of soil acidity on potentially mobile phosphorus in a grassland soil. J Agric Sci 139:27–36

McDowell RW, Condron LM, Mahieu N, Brookes PC, Poulton PR, Sharpley AN (2002b) Analysis of potentially mobile phosphorus in arable soils using solid state nuclear magnetic resonance. J Environ Qual 31:450–456

McDowell RW, Condron LM, Mahieu N (2003a) Analysis of phosphorus in sequentially extracted grassland soils using solid state NMR. Commun Soil Sci Plant Anal 34:1623–1636

McDowell RW, Mahieu N, Brookes PC, Poulton PR (2003b) Mechanisms of phosphorus solubilisation in a limed soil as a function of pH. Chemosphere 51:685–692

McDowell RW, Stewart I, Cade-Menun BJ (2006) An examination of spin-lattice relaxation times for analysis of soil and manure extracts by liquid state phosphorus-31 nuclear magnetic resonance spectroscopy. J Environ Qual 35:293–302

McDowell RW, Cade-Menun B, Stewart I (2007) Organic phosphorus speciation and pedogenesis: analysis by solution ^{31}P nuclear magnetic resonance spectroscopy. Eur J Soil Sci 58:1348–1357

Metz K, Dunphy L (1996) Absolute quantitation of tissue phospholipids using ^{31}P NMR spectroscopy. J Lipid Res 37:2251–2265

Murphy PNC, Bell A, Turner BL (2009) Phosphorus speciation in temperate basaltic grassland soils by solution P-31 NMR spectroscopy. Eur J Soil Sci 60:638–651

Neal C, Neal M, Wickham H (2000) Phosphate measurement in natural waters: two examples of analytical problems associated with silica interference using phosphomolybdic acid methodologies. Sci Total Environ 251–252:511–522

Newman RH, Condron LM (1995) Separating subspectra from cross-polarization magic-angle spinning nuclear magnetic resonance spectra by proton spin relaxation editing. Solid State Nucl Magn Reson 4:259–266

Newman RH, Tate KR (1980) Soil phosphorus characterization by 31P nuclear magnetic resonance. Commun Soil Sci Plant Anal 11:835–842

Peak D, Sims JT, Sparks DL (2002) Solid-state speciation of natural and alum-amended poultry litter using XANES spectroscopy. Environ Sci Technol 36:4253–4261

Preston CM (1996) Applications of NMR to soil organic matter analysis: history and prospects. Soil Sci 161:144–166

Puppato A, DuPré DB, Stolowich N, Yappert MC (2007) Effect of temperature and pH on ^{31}P nuclear magnetic resonances of phospholipids in cholate micelles. Chem Phys Lipids 150:176–185

Šárka M, Mika A-K (2001) MR spectroscopy quantitation: a review of frequency domain methods. NMR Biomed 14:247–259

Sato S, Solomon D, Hyland C, Ketterings QM, Lehmann J (2005) Phosphorus speciation in manure and manure-amended soils using XANES spectroscopy. Environ Sci Technol 39:7485–7491

Shand C, Cheshire M, Bedrock C, Chapman P, Fraser A, Chudek J (1999) Solid-phase ^{31}P NMR spectra of peat and mineral soils, humic acids and soil solution components: influence of iron and manganese. Plant Soil 214:153–163

Shober AL, Hesterberg DL, Sims JT, Gardner S (2006) Characterization of phosphorus species in biosolids and manures using XANES spectroscopy. J Environ Qual 35:1983–1993

Smernik RJ, Baldock JA (2005) Does solid-state ^{15}N NMR spectroscopy detect all soil organic nitrogen? Biogeochemistry 75:507–528

Smernik RJ, Dougherty WJ (2007) Identification of phytate in phosphorus-31 nuclear magnetic resonance spectra – the need for spiking. Soil Sci Soc Am J 71:1045–1050

Smernik RJ, Oades JM (2000a) The use of spin counting for determining quantitation in solid state ^{13}C NMR spectra of natural organic matter 1. Model systems and the effects of paramagnetic impurities. Geoderma 96:101–129

Smernik RJ, Oades JM (2000b) The use of spin counting for determining quantition in solid state ^{13}C NMR spectra of natural organic matter 2. HF-treated soil fractions. Geoderma 96:159–171

Tate KR, Newman RH (1982) Phosphorus fractions of a climosequence of soils in New Zealand tussock grassland. Soil Biol Biochem 14:191–196

Tiessen H, Moir JO (1993) Characterization of available P by sequential extraction. In: Carter MR (ed) Soil sampling and methods of analysis. Lewis, Boca Raton, pp 75–86

Toor GS, Peak JD, Sims JT (2005) Phosphorus speciation in broiler litter and turkey manure produced from modified diets. J Environ Qual 34:687–697

Turner B (2008) Soil organic phosphorus in tropical forests: an assessment of the NaOH-EDTA extraction procedure for quantitative analysis by solution ^{31}P NMR. Eur J Soil Sci 59:453–466

Turner GL, Smith KA, Kirkpatrick RJ, Oldfieldt E (1986) Structure and cation effects on phosphorus-31 NMR chemical shifts and chemical-shift anisotropies of orthophosphates. J Magn Reson 70:408–415

Turner BL, Mahieu N, Condron LM (2003a) Quantification of *myo*-inositol hexakisphosphate in alkaline soil extracts by solution ^{31}P NMR spectroscopy and spectral deconvolution. Soil Sci 168:469–478

Turner BL, Mahieu N, Condron LM (2003b) Phosphorus-31 nuclear magnetic resonance spectral assignments of phosphorus compounds in soil NaOH-EDTA extracts. Soil Sci Soc Am J 67:497–510

Turner BL, Mahieu N, Condron LM (2003c) The phosphorus composition of temperate pasture soils determined by NaOH-EDTA extraction and solution ^{31}P NMR spectroscopy. Org Geochem 34:1199–1210

Turner BL, Cade-Menun BJ, Condron LM, Newman S (2005) Extraction of soil organic phosphorus. Talanta 66:294–306

Turner BL, Newman S, Reddy KR (2006) Overestimation of organic phosphorus in wetland soils by alkaline extraction and molybdate colorimetry. Environ Sci Technol 40:3349–3354

Veeman WS (1997) Nuclear magnetic resonance, a simple introduction to the principles and applications. Geoderma 80:225–242

Williams RJP, Giles RGF, Posner AM (1981) Solid state phosphorus N.M.R. spectroscopy of minerals and soils. J Chem Soc Chem Comm:1051–1052

Yoder CH, Schaeffer CD (1987) Introduction to multinuclear NMR. Benjamin Cummings, Menlo Park

Chapter 2
Characterization of Phosphorus Forms in Soil Microorganisms

Else K. Bünemann, Bartlomiej Prusisz, and Knut Ehlers

2.1 Microorganisms as a Pool of Phosphorus in Soil

Soil microorganisms act as sink and source of phosphorus (P) and mediate key processes in the soil P cycle, e.g., P mineralization and immobilization (Oberson and Joner 2005). The role of mycorrhizal fungi in P absorption and transport to plant roots is well documented (Jakobsen et al. 2002; Jansa et al. 2011), but the role of the soil microbial biomass (i.e., soil bacteria, archaea, and mostly saprophytic fungi) in soil P cycling is less well defined. Microbial P immobilization can affect P availability by removing inorganic P from the soil solution, especially when soluble carbon is available for microbial growth (Bünemann et al. 2004a; Olander and Vitousek 2004). In the absence of recent carbon inputs, gross organic P mineralization rates of between 1.4 and 2.5 mg P kg^{-1} day^{-1} have been measured for arable soils (Oehl et al. 2004; Frossard et al. 2011). Microbial activity in a grassland soil was also shown to be required to replenish organic P in the soil solution (Seeling and Zasoski 1993). Thus, although the microbial biomass contains only 0.4–2.5% of total P in cropped soils and up to 7.5% in grassland soils, it can play a fundamental role in the soil P cycle, especially when its turnover time is only a few months (Oberson and Joner 2005).

The main forms of P in microorganisms are shown in Table 2.1. Based on values for P-limited aquatic bacteria (Vadstein 2000), the dominant forms of microbial P are nucleic acids and phospholipids (together 60%), cytoplasmic inorganic P (10%), cytoplasmic organic P (10%), and polyphosphate (20%). However, the distribution of P forms in microbial cells changes with environmental conditions, in particular with carbon and nutrient availability (Herbert 1961). Changes may be either qualitative (i.e., some cell components may occur in cells only under certain environmental conditions), or quantitative (i.e., the concentration of cell components that are always present may increase or decrease). This could ultimately affect

E.K. Bünemann (✉), B. Prusisz, and K. Ehlers
Group of Plant Nutrition, Institute of Plant, Animal and Agro-ecosystem Sciences, ETH Zurich, Eschikon 33, 8315 Lindau, Switzerland
e-mail: else.buenemann@ipw.agrl.ethz.ch

E.K. Bünemann et al. (eds.), *Phosphorus in Action*, Soil Biology 26,
DOI 10.1007/978-3-642-15271-9_2, © Springer-Verlag Berlin Heidelberg 2011

Table 2.1 Quantitatively main forms of P found in microorganisms: chemical groups, NMR regions, examples, location and function

Chemical group	NMR region	Examples	Description
Nucleic acids	Diester	DNA, RNA	Biopolymers present in all living cells; storage of genetic information. About 9% P
Phospholipids	Diester	Phosphatidylglycerol, phosphatidylethanolamine	Major components of cell membranes; ensure semipermeability. Usually one phospholipid molecule consists of glycerol, two saturated or unsaturated fatty acids, and one P atom
Teichoic acids	Diester	Glycerol teichoic acid	P-containing polysaccharides present in cell walls of Gram-positive bacteria
Metabolites	Mostly monoester	AMP, glucose-6-phosphate, phosphoenol pyruvate, ATP	Intermediates and products of all metabolic processes
Phosphonates	Phosphonate	2-Aminoethyl-phosphonic acid	Compounds with direct bonds between carbon and P atoms. Biological function not well known
Organic polyphosphates	Pyro- and polyphosphate	ATP	Produced by ATP synthase. Transfers chemical energy when converted back to AMP or ADP and orthophosphate
Inorganic pyrophosphate	Pyrophosphate	Pyrophosphate	P storage compound consisting of two condensed phosphate groups
Inorganic polyphosphates	Polyphosphate	Polyphosphate	P storage compounds consisting of three or more condensed phosphate groups
Orthophosphates	Orthophosphate	HPO_4^{2-}; $H_2PO_4^-$	Inorganic anions of phosphoric acid. Used in metabolic processes and for synthesis of complex molecules

DNA deoxyribonucleic acid, *RNA* ribonucleic acid, *AMP* adenosine monophosphate, *ADP* adenosine diphosphate, *ATP* adenosine triphosphate the availability of microbial P forms to plants. For example, organic P, but not polyphosphate, contained in filtered extracts of a cultured soil bacterium was shown to be hydrolyzed and taken up by plants growing in a nutrient solution to which the filtrates had been added (Macklon et al. 1997).

Several recent studies have investigated carbon:nitrogen:phosphorus (C:N:P) stoichiometry in aquatic bacteria. Under low P conditions, the P concentration, growth rate, and RNA content were positively related in *Escherichia coli* as well as in lake bacteria, and P in RNA represented 82% of the total bacterial P pool in an oligotrophic lake (Elser et al. 2003). When P was abundant, a lower and more variable proportion of total P was contained in RNA. At constant growth rate of *E. coli* in a chemostat (see Sect. 2.2.1), the proportion of P in RNA varied between 40% and 70% of total P, depending on the temperature (Cotner et al. 2006). From experiments with *E. coli* under controlled conditions and from a literature survey on heterotrophic prokaryotes in freshwater lakes, Makino et al. (2003) concluded that each bacterial species regulates its elemental composition homeostatically within a relatively narrow range of a characteristic biomass C:P ratio, depending mainly on its RNA content, and that shifts in the dominance of different bacterial species in the environment are responsible for the large variation in community C:P ratios.

These examples suggest that the concentration and forms of microbial P are affected by the composition of the microbial community, P availability, and growth stage. To date, however, our knowledge of the variation in microbial P forms has been obtained in aquatic systems or with pure cultures, and methodological constraints have limited our ability to investigate these relationships in soil. The classical approach of measuring the black box "microbial P" by fumigation-extraction methods (Oberson and Joner 2005) has yielded important information, especially in combination with isotopic labeling (Frossard et al. 2011). For example, 2 days after labeling soil with ^{33}P, Oberson et al. (2001) found up to 25% of ^{33}P in microbial P, without a change in the size of the microbial P pool. This recovery of ^{33}P in microbial P was about five times greater than that in available inorganic P and suggests a P flux of about 0.5 mg P kg^{-1} soil day^{-1} through the soil microbial biomass. However, hydrolysis of microbial P during fumigation is an inherent part of the method and thus it provides no information on the chemical forms of soil microbial P present. Therefore, Oberson and Joner (2005) suggested that microorganisms be extracted from soil and their P forms analyzed.

Here, we argue that a more detailed picture of the P forms within the soil microbial biomass is an important next step in our efforts to understand the role of microorganisms as sink and source of P. In addition, a better understanding of the fate of microbial P forms released into the soil could provide insight into the origin of soil organic P, its availability to mineralization processes and its environmental behavior, e.g., with respect to losses to water courses. Ultimately, such an improved knowledge will contribute to more efficient P use in agroecosystems, which is mandatory because of dwindling mineable P deposits as well as increasing P eutrophication worldwide (Tiessen et al. 2011).

In this chapter, we summarize the current knowledge on P forms in soil microorganisms. We begin with an overview of methods for culturing microorganisms and for direct extraction of cells from soil. Next, we describe a suite of analytical methods for determining the forms of P in soil microorganisms. Two case studies are presented: one on the use of NMR spectroscopy to analyze P forms in pure cultures of soil microorganisms, and one on the forms of P in microbial cells extracted from soil. We discuss the potential microbial origin of soil organic P, and methodological limitations and future applications of the approach of combining the extraction of microbial cells from soil with chemical analysis.

2.2 Cultivation and Extraction Methods

2.2.1 Culturing of Microorganisms

Cultivation of soil microorganisms is important for detailed taxonomic and physiological studies. However, only a small fraction of microorganisms can be cultured (1–5%), even though this percentage can probably be increased with improved cultivation techniques (Janssen et al. 2002).

Batch cultures represent the simplest approach for culturing microorganisms, where microorganisms grow from a small inoculum introduced into a given amount of mineral or complex medium. Typically, after an initial stationary and lag phase, exponential growth at a constant growth rate occurs, which eventually slows down until the maximum stationary phase is reached, followed by a decrease in cell numbers (Herbert 1961). Thus, biomass production stops quickly after depletion of the growth-limiting nutrient, although the number of cells can still increase if initiated cell division is completed. Unlike for carbon, biomass production can continue due to relocation of the nutrient. For example, teichoic acids can be replaced by P-free teichuronic acids in Gram-positive bacteria under P limitation, while in other species the degradation of RNA and polyphosphate can deliver P upon which the cells can continue to grow (Wanner and Egli 1990).

The main disadvantage of batch culture is the highly dynamic nature of the system, which prevents identification of the precise factors that affect the composition of cells (Tempest and Wouters 1981). This is overcome by continuous culture in a chemostat, which is at steady state when the exponential growth rate μ is equal to the dilution rate D (ratio of flow rate to the volume of the culture). Thus, very low growth rates can be achieved at low dilution rates, and low growth rates are likely to be the typical situation in soils. So far, chemostat studies on the chemical forms of P have been done with aquatic microbial communities (Makino and Cotner 2004) but not with soil microbial communities.

Harvesting of cultured cells is done by centrifugation or filtration. For subsequent nutrient analysis, washing steps may be required to obtain cells without contamination by the growth medium. For filtration, various pore sizes may have to be tested to avoid losses of very small cells (Wang et al. 2007). However, there is a trade-off between pore size and duration of filtration.

2.2.2 *Extraction of Microorganisms from Soil*

Ideally, a method for extracting microorganisms from soil in order to analyze microbial P forms would meet the following (often conflicting) conditions:

1. A high percentage of the soil microbial biomass is recovered (high quantitative efficiency).
2. The fraction of the soil microbial biomass that is recovered either represents the soil microbial community well (high qualitative efficiency) or is a proportion that can be regarded as an operational pool (i.e., a proportion that is defined by the extraction procedure and is highly reproducible).
3. The cells are free from adhering soil colloids.
4. The extracted cells are physically intact and their metabolic state is not altered. In particular, microbial P forms are unaffected.

The approach used most often to extract microbial cells from soil is density gradient centrifugation (Bakken and Lindahl 1995). This method consists of mechanical soil dispersion using a blender to destroy aggregates and to detach microbial cells from soil particles, followed by separation of structurally intact microbial cells from soil particles using the nonionic density gradient medium Nycodenz. The quantitative efficiency is assessed by microscopic cell counts after staining with acridin orange or similar DNA stains (Lindahl 1996). The qualitative efficiency is evaluated by comparing the microbial community composition in the soil using either biochemical methods such as phospholipid fatty acid (PLFA) analysis or molecular methods such as denaturing gradient gel electrophoresis (DGGE). The purity of the cells from soil particles can be assessed by determining the loss of cell weight after ignition (Bakken and Lindahl 1995) or by measuring the content of typical soil elements such as iron and aluminum in the extracted fraction (Ehlers et al. 2008). Whether the cells are physically intact or not is seen under the microscope, while changes in the metabolic state can be assessed by measuring adenosine triphosphate (ATP) in the cells (Lindahl and Bakken 1995). Potential changes in the biochemical composition during extraction have to be tested using pure cultures.

The recovery of structurally intact bacterial cells using density gradient centrifugation can be as high as 70–80% when the soil is extracted repeatedly (Lindahl 1996). However, repeated extractions are often impractical and in most studies the recovery is much lower. For example, the recovery was around 20% free from adhering soil colloids (Sitaula et al. 1999). For acid clayey tropical soils, cell yields were as low as 0.5% (Maron et al. 2006). In such instances, soil-specific adaptations might be required, e.g., to obtain enough material for chemical analyses. Working with a tropical Ferralsol, Ehlers et al. (2008) increased cell yield from 4.9% to 10.6% by adjusting the pH to 7.5, but in both cases the extracted fraction was highly contaminated with soil material. Although the addition of 0.8% NaCl solution eliminated this contamination, it also decreased cell yield to 4.6%. However, in the absence of NaCl the extracted microbial communities were more similar to the soil microbial communities as assessed both with PLFA and DGGE.

Using molecular-based methods, Courtois et al. (2001) showed that apart from the preferential extraction of γ-proteobacteria, bacterial diversity was similar for extracted cells and for extraction of DNA from soil after cell lysis. However, there might be a bias for specific groups, as demonstrated for methane-oxidizing (Priemé et al. 1996) and ammonia-oxidizing bacteria (Aakra et al. 2000).

How well the extracted cells represent the soil microbial community always poses a problem because the method of Bakken and Lindahl (1995) was designed to extract bacteria and not fungi. In fact, fungal hyphae are destroyed during mechanical dispersion of soil (Faegri et al. 1977). In addition, they are probably only partly detached from soil particles or become pelleted with soil particles during centrifugation. Nevertheless, the fungal PLFA 18:2ω6 was found in the extracted fraction at a similar abundance as in soil (Ehlers et al. 2008). The authors argued that single-cell fungal structures such as yeasts and spores were recovered together with bacteria. In a follow-up study, however, addition of carbon and nitrogen sources led to a great increase in the fungal PLFA in soil, but not in the extracted fraction, suggesting discrimination against fungi during extraction (Ehlers et al. 2010). So far, DNA from the extracted fraction has only been analyzed using primers for prokaryotes. Analysis with eukaryotic primers could yield more information about the composition of the extracted community.

Due to the difficulties of extracting intact and pure fungal hyphae from soil, methods designed to extract mycelium for elemental analysis have used mesh compartments filled with sand (Wallander et al. 2003) or with a mixture of glass beads and soil of <40 μm particle size (Neumann and George 2005). In the first case (Wallander et al. 2003), mesh bags were buried in spruce forests and recovered after 12–18 months for extraction of ectomycorrhizal mycelia. The C:N ratios in the hyphae were much wider (20) than those found using fumigation-extraction methods (6–14), although the microbial biomass in this soil was dominated by fungi. Presumably, fumigation-extraction methods extract more of the cytoplasm, while cell wall material is likely to remain in the soil. In the second case (Neumann and George 2005), P concentrations in arbuscular mycorrhizal mycelium recovered after 45 days from pots planted with potatoes were lower than values reported for hyphae grown in vitro, suggesting either lack of polyphosphate accumulation due to P limitation or efficient transfer of P to the host plant. Although such approaches deliver useful results, a method to extract fungal hyphae directly from soil so that they can be chemically analyzed is yet to be devised.

2.3 Methods for Analysis of Chemical Forms of Phosphorus

2.3.1 Total P

Most methods for elemental P analysis require liquid samples and, thus, digestion procedures prior to measurements. Acidic or alkaline persulfate digestion, usually in combination with autoclaving (e.g., Ebina et al. 1983), has been applied, as well as mineralization with single acids such as nitric acid under heating (Danku et al. 2009).

After sample mineralization, total P is often measured photometrically (e.g., Lovdal et al. 2008). The method requires only simple instrumentation, but suffers from severe matrix effects and a short dynamic range. In addition, complete conversion of all P compounds into orthophosphate has to be assured. Alternatively, methods that are based on the atomization and excitation/ionization in an inductively coupled plasma (ICP) torch, with detection by optical emission spectrometry (ICP-OES; e.g., Duboc et al. 1995) or mass spectrometry (ICP-MS; e.g., Danku et al. 2009) can be used. These methods are more resistant to matrix effects and offer broader dynamic ranges, multi-elemental capabilities, and higher sample throughput.

An analytical technique that does not require liquid samples and has been successfully applied to the analysis of total P in single bacteria is X-ray fluorescence (XRF) in combination with scanning electron microscopy (Heldal et al. 1985). In fact, X-ray microanalysis allows measurement of total amounts of all major elements except hydrogen in single microbial cells (Norland et al. 1995). Comparison of native and cultured aquatic bacteria by X-ray microanalysis revealed major differences in size and nutrient contents (Fagerbakke et al. 1996). In particular, the content of P was greatest in cultured cells in the growth phase (17–31 fg P $cell^{-1}$), intermediate in cultured cells in the stationary phase (3.4–6.7 fg P $cell^{-1}$) and yet lower in native cells (0.5–1.1 fg P $cell^{-1}$). Thus, it appears that analyses of the elemental composition of single cells can give information on the physiological status and nutrient limitations. To understand the processes in the soil, such analyses have to be done on extracted native bacteria.

2.3.2 P Speciation Using ^{31}P NMR

Speciation analysis of P forms by means of ^{31}P NMR is the most promising analytical tool to date (see Doolette and Smernik 2011). However, further developments in instrumentation, spectra analysis, and sample preparation are necessary. The sensitivity is still unsatisfactory for analysis of microorganisms. Up to now, it has been necessary to acquire a large amount of sample, i.e., on the order of 0.1–1 mg P, which is feasible for pure cultures but not when extracting microbial cells from soil. For example, using the cell P content listed in Table 2.2 for the treatment designated H_2O (i.e., without addition of carbon, nitrogen, or P), the total cell count of 4.7×10^9 in this sample and an extraction yield of 5% gives 0.0075 mg P extracted from 10 g of soil, which is 100 times lower than would be required for ^{31}P NMR. Other problems include correct identification of resonances and hydrolysis of P compounds during extraction. Nevertheless, the ability to measure the ratio between different P forms (e.g., orthophosphate, monoester, and diester P) can provide very useful information about P species, even at the current stage of method development. For example, solution ^{31}P NMR has been applied to the analysis of P forms in pure cultures of soil bacteria and fungi (see Sect. 2.4.1).

The application of ^{31}P NMR to living systems is called in vivo ^{31}P NMR spectroscopy and has the advantages of being non-invasive and suitable for the study of time-dependent phenomena (Rasmussen et al. 2000). A study of the actinomycete *Corynebacterium glutamicum* used 45 mg dry weight per ml (total volume in the NMR tube was 1.8 ml) to study fluctuations of polyphosphate after additions of glucose, acetate, and phosphate and under changing oxygen supply (Lambert et al. 2002). A study of P forms in arbuscular mycorrhizal hyphae or cucumber roots by in vivo ^{31}P NMR distinguished orthophosphate, orthophosphate monoesters, nucleic acid triphosphates, and polyphosphate (Viereck et al. 2004). The fresh weight of hyphae or roots in the NMR tube ranged between 0.03 and 0.14 g.

These studies suggest that in vivo ^{31}P NMR has the potential to overcome the degradation problems that occur in the analysis of microbial P forms using solution ^{31}P NMR. However, sufficient material must be available, which again is problematic when studying microorganisms extracted from soil.

2.3.3 P Speciation by Chromatographic, Spectrometric, Staining, and Enzymatic Techniques

2.3.3.1 P in DNA and RNA

Before P analysis, nucleic acids have to be released from the cells. Most often, a combination of physical, chemical, and enzymatic treatment is used (Bakken and Frostegård 2006). Among physical treatments, bead-beating with glass beads or sterilized sand in mini-beadbeaters and grinding in liquid nitrogen are most commonly used. Chemical agents such as EDTA, SDS, and Triton make membranes more permeable, enhancing the effect of the other treatments. Enzymes can destroy bacterial cell walls, e.g., by hydrolyzing glycoside bonds (lysozyme) or peptide bonds (achromopeptidase) in the peptidoglycan layer. However, all treatments are somewhat selective in releasing nucleic acids from different species at different growth stages. For example, growing cells lyse more easily than cells at the stationary stage (Bakken and Frostegård 2006).

Since the concentration of P in DNA and RNA is constant at about 9%, the amount of P in DNA and RNA is usually calculated after quantification of nucleic acids (Makino and Cotner 2004). Methods of DNA and RNA quantification use either spectrophotometry or fluorometry. Photometric methods are based either on the absorbance of the analytes at 260 nm (Sambrook and Russel 2006) or on color development after reaction with orcinol (Endo 1970) or diphenylamine (Burton

1956). Absorbance at 260 nm requires the sample to be protein-free. The purity can be judged from the 260/280 nm absorbance ratio, with a low ratio indicating contamination by protein. If potentially interfering substances have to be removed, every cleaning step introduces the possibility of analyte loss and extends the time of analysis. In contrast, fluorometric measurements based on specific or unspecific fluorescent dyes can be performed on rough cellular extracts, omitting cleaning steps. Their sensitivity is usually higher than spectrophotometric measurements and the time of analysis is considerably shorter. After binding of the dye to the nucleic acids (e.g., of ethidium bromide, Hoechst 33258 or PicoGreen to double-stranded DNA, and of RiboGreen to RNA and DNA), fluorescent yield increases substantially. An example for the application of an unspecific fluorescent dye (RiboGreen) is the analysis of RNA and DNA in rough extracts of Daphnia (Gorokhova and Kyle 2002): after addition of the dye, RNase and DNase were subsequently applied and the decrease in the measured fluorescent signals was assigned to the amount of RNA and DNA, respectively.

Introduction of fluorescent dyes has shortened and simplified the quantitative analysis of nucleic acids. However, there are still analytical pitfalls. For example, when enzymes are added, correction has to be made for the increase in fluorescent signal due to the added protein. Some enzymes need specific cofactors, appropriate temperatures, and assay times. Finally, species-specific DNA and RNA structures can produce different spectro- or fluorometric responses. Thus, standard curves using DNA and RNA from the species being assayed assure best quantification (De Mey et al. 2006).

2.3.3.2 P in Lipids

Before analysis of phospholipids, they have to be extracted from the sample. The classical lipid extraction method uses a mixture of chloroform, methanol, and water in the ratio of 1:2:0.8 by volume (Bligh and Dyer 1959). Subsequently, a mixture of chloroform and methanol (2:1 by volume) has been recommended for the quantitative extraction of total phospholipids (Van Der Meeren et al. 1992).

The first analysis of phospholipid classes in microorganisms utilized thin layer chromatography (TLC) with subsequent elution of the phospholipids and analysis of total P in each fraction (Hossack and Rose 1976). This simple method can still provide reliable and relatively sensitive data regarding P distribution among various phospholipid classes. More recently, normal-phase high-performance liquid chromatography (HPLC) with a flame ionization detector (FID) has been applied successfully for analysis of phospholipids in bacteria (Moreau et al. 1995). Alternatively, evaporative light scattering detector (ELSD) and MS are used for analysis of lipids and phospholipids (Arnoldsson and Kaufmann 1994; Valeur et al. 1993).

Knowing the molar content of P in a given phospholipid, the quantity of P in it can be calculated from the quantitative analysis of the phospholipid. Similarly, P in phospholipids can be calculated from the quantitative analysis of PLFAs. This method usually consists of phospholipid separation on silica columns, followed by methylation and gas chromatography (GC) analysis with detection by FID or MS (Frostegård et al. 1993). Calculation of the P in phospholipids is based on an assumed molar ratio between fatty acids and P of 2:1. Although this assumption is not always valid (e.g., cytidine diphosphate diacylglycerol and pyrophosphatidic acid have a molar ratio of 1:1), it is legitimate for most fatty acids and, importantly, for the most abundant ones, i.e., phosphatidylglycerol, phosphatidylethanolamine, cardiolipin, phosphatidylcholine, and phosphatidylinositol. Other potential sources of error are losses of phospholipids during the separation step and during methylation.

There are also a few reports on phospholipid analysis without chromatographic separation. Electrospray ionization mass spectrometry was applied to the analysis of crude lipid extracts of two marine bacteria (Mazzella et al. 2005), with proposed fragmentation pathways for the two most abundant phospholipids, i.e., phosphatidylglycerol and phosphatidylethanolamine. Fast atom bombardment with mass spectrometric detection was used for the characterization of bacterial phospholipids (Heller et al. 1988).

2.3.3.3 P in Metabolites

Many metabolites in microbial cells contain P. Examples are P monoesters such as adenosine monophosphate (AMP), glucose-6-phosphate, phosphoenol pyruvate and 3-phosphoglycerate, and ATP as a condensed phosphate.

Extraction of metabolites from bacterial cells is usually done with hot ethanol, pure water, or water solutions containing various additives, commonly acids or bases (Mashego et al. 2007). Because fungal cells are much less susceptible to lysis, extraction with organic solvents is employed, often in several steps (Frisvad and Thrane 1987) or with assistance of sonication (Smedsgaard 1997).

Analysis of ATP extracted from living cells has been used for more than four decades to estimate active biomass in aquatic systems (Holmhans and Booth 1966) and also in soils (Jenkinson and Oades 1979). Many other metabolites have been analyzed, usually by means of enzymatic assays. Analysis of total metabolites (metabolomics) has been gaining importance in recent years but requires highly sophisticated instrumentation, i.e., one or several mass spectrometers coupled with liquid (LC-MS) or gas chromatography (GC-MS) or capillary electrophoresis (CE-MS) (Mashego et al. 2007).

Fortunately, since all P-containing metabolites are ionic molecules their analysis is a much simpler task than analysis of all intracellular metabolites. Thus, there is a broad range of possible extractants to choose from (boiling ethanol, KOH, perchloric acid, hot water, hydrochloric acid, and acetic acid) and ion chromatography can be used, usually with conductometric and UV detection (Bhattacharya et al. 1995) or with pulse amperometric detection for analysis of sugar phosphates (Smits et al. 1998).

2.3.3.4 P in Teichoic Acids and Other Compounds

Teichoic acids can be a significant P pool in Gram-positive bacteria. For example, Ellwood and Tempest (1972) found that the purified cell walls of *Bacillus subtilis* contributed 17–25% of cell dry matter and that the cell walls contained 2.4–6.2% P, mostly in teichoic acids. Due to its complexity, however, quantitative analysis of teichoic acids is rarely performed. The procedure consists of separation of bacterial cell walls by centrifugation, followed by cleaning with chemicals (Ellwood and Tempest 1972) or application of enzymes (Bhavsar et al. 2004), and subsequent analysis of total P. Spectrophotometric and fluorometric titration have also been employed in the analysis of teichoic acids (Pal et al. 1989).

Another P compound in microorganisms is inorganic polyphosphate. Its analysis comprises separation of the inorganic polyphosphate chains on a solid sorbent, application of a polyphosphatase, and subsequent spectrophotometric determination of orthophosphate (Werner et al. 2005).

Analysis of phospho-proteins can be expected to be developed quickly for proteomic studies. A hyphenated quantitative method for the analysis of phosphoproteins has already been developed (Krueger et al. 2009).

Finally, myo-inositol hexakisphosphate (phytate) has been reported to be synthesized by soil microorganisms (Caldwell and Black 1958), although the majority of phytate in soils is thought to be derived from plants (Turner et al. 2002).

2.4 Phosphorus Forms in Cultured and Extracted Soil Microorganisms

2.4.1 Analysis of P Forms in Pure Cultures Using ^{31}P NMR

Bacteria and fungi isolated from soil and cultivated in batch cultures were analyzed by solution ^{31}P NMR in two studies (Bünemann et al. 2008a; Makarov et al. 2005). In both cases, the concentration of P was about twice as high in bacteria (mean values of 18 and 23 mg g^{-1} dry weight) than in fungi (mean value of 9.5 mg g^{-1} in both studies). For the fungi, both studies showed quite a similar pattern of P forms (Fig. 2.1): most of the P was more or less equally distributed over the orthophosphate, monoester P, and the pyro- and polyphosphate regions, even though the proportions of orthophosphate and condensed P were highly variable. Very small proportions of P were found in the diester region (mean values of 2% and 6%). In contrast, the diester pool was more significant in bacteria (18% and 38%), whereas very little condensed P was found (1% and 3%). Comprising on average 66% of P, the monoester region was more dominant in the study by Bünemann et al. (2008a) than in that by Makarov et al. (2005) where the average was 34%. This could be due to the difference in bacterial species but more likely due to hydrolysis of diesters during sample preparation and even during NMR analysis, as shown by Turner et al. (2003).

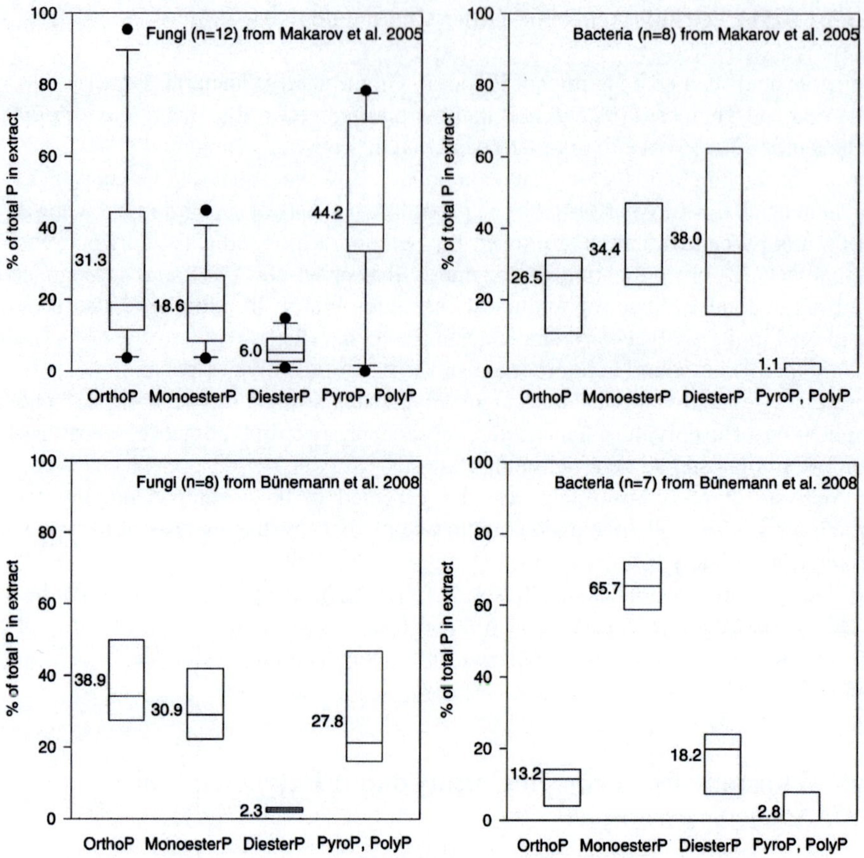

Fig. 2.1 Distribution of P species in pure cultures of fungi and bacteria extracted with 0.5 M NaOH (Makarov et al. 2005) (*upper figures*) or 0.25 M NaOH plus 0.05 M EDTA (Bünemann et al. 2008a) (*lower figures*) as determined by solution ^{31}P NMR. The *lower boundary of each box* indicates the 25th percentile, the *full line within the box* the median, the *dotted line* the mean (with the value given *left of the box*), and the *upper boundary* the 75th percentile. *Error bars* indicate the 90th and 10th percentiles (only if $n > 8$). Outliers are shown as *dots*

Both NMR studies (Bünemann et al. 2008a; Makarov et al. 2005) aimed to elucidate the origin of soil organic P. The spectra from bacteria and plants in the study by Makarov et al. (2005) were rather similar, suggesting that it will be difficult to distinguish between microbial and plant origin of organic P in soils. In their study, the proportions of different diesters in two acid soils corresponded to those in the spectra of bacteria and plants, whereas in a calcareous soil the pattern was more similar to that of fungi. However, fungal growth would be expected to dominate in acid soils (Rousk et al. 2009). In addition, the contribution of bacteria and fungi to soil organic P depends not only on the relative biomasses of the two groups but also on the P concentration, turnover time, and selective stabilization. Thus, the origin of soil organic P cannot be deduced from the broad P forms in different organisms, with the exception of condensed phosphates. As shown in Fig. 2.1, pyro- and polyphosphates occur in much higher proportions in fungi than in bacteria, and a positive correlation of pyrophosphate and fungal abundance was indeed seen by Bünemann et al. (2008b).

The possibility that specific compounds in the monoester region could be characteristic for either bacteria or fungi was tested by Bünemann et al. (2008a). Of the 15 peaks found by deconvolution of the monoester region, two were unique to but not present in all fungi. For this limited dataset, bacteria and fungi were significantly separated, based on the signal distributions in the monoester region. However, a much more extensive dataset would have to be obtained before any conclusions with respect to the bacterial or fungal origin of soil organic P could be drawn using such an approach.

In contrast to the P forms discussed so far, phosphonates have not been detected in plants. Using ^{31}P NMR, Koukol et al. (2008) found phosphonates in basidiocarps and vegetative mycelia of basidiomycetes. These data support the microbial origin of phosphonates in soils, even though the concentrations (14–140 mg kg^{-1} dry matter, equal to 0.1–4.4% of extracted P) were lower than those of all other P forms, and even though phosphonate production of associated bacteria cannot be excluded because the samples were taken from the field.

2.4.2 *Analysis of P Forms in Microbial Cells Extracted from a Ferralsol*

Highly weathered, acidic tropical Ferralsols are usually characterized by strong P sorption capacities, leading to low amounts of available P. Under these conditions, the role of microbial P as a highly dynamic P pool could be of particular importance.

An incubation experiment was conducted to study the effect of carbon and nutrient additions on microbial activity and cell internal P pools in a Ferralsol (Ehlers et al. 2010). Four treatments were chosen:

1. Without C, N, or P addition (H_2O)
2. With P addition (P)
3. With C and N addition (CN)
4. With C, N, and P addition (CNP)

Microbial cells were extracted from soil by density gradient centrifugation (Bakken and Lindahl 1995), as adapted for highly weathered tropical soils (Ehlers et al. 2008). Extracted cells were analyzed for total P content and P contained in PLFA by using extraction and total P determination, and for P in DNA and RNA by using fluorescent dye. Microbial communities were analyzed in their most active state, as given by the peak of the respiration rates (Fig. 2.2). Since the H_2O and P treatments did not show a peak in the respiration curves, the microbial community was analyzed at the same time as in the CN treatment.

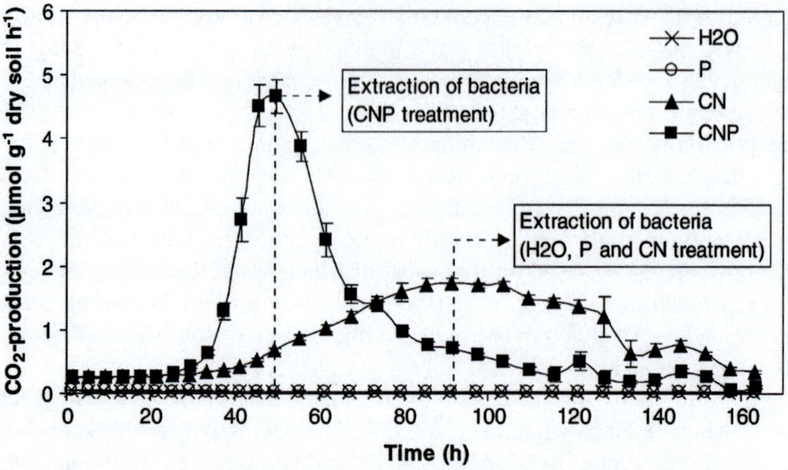

Fig. 2.2 Respiration rates in a Ferralsol from Kenya as affected by addition of P alone (*P*), carbon and nitrogen (*CN*), or carbon, nitrogen, and P (*CNP*) during the first 163 h of incubation (Ehlers et al. 2010). H_2O signifies no additions of C, N or P. *Error bars* show the standard deviation derived from three independent replicates. *Arrows* indicate the time of extraction of bacteria from soil for different treatments. Reprinted from Ehlers et al. (2010) with permission from Elsevier

Table 2.2 P content and P pools in soil microorganisms extracted from a Ferralsol[a]

Treatment[a]	Total P		DNA-P		RNA-P		PLFA-P	
	fg P cell^{-1}		fg P cell^{-1}		fg P cell^{-1}		fg P cell^{-1}	
H_2O	3.22 (0.83)	ab	0.52 (0.08)	a	0.06 (0.01)	b	0.15 (0.07)	a
P	2.70 (0.60)	b	0.56 (0.08)	a	0.09 (0.03)	b	0.05 (0.04)	a
CN	4.60 (1.13)	ab	0.74 (0.17)	a	0.28 (0.01)	a	0.15 (0.11)	a
CNP	6.53 (2.46)	a	0.57 (0.16)	a	0.09 (0.03)	b	0.13 (0.08)	a

H_2O no additions; *P* added P; *CN* added carbon and nitrogen; *CNP* added carbon, nitrogen, and P
All values are means of three replicates (standard deviations in brackets). Values within a column followed by the same letter do not differ significantly ($P < 0.05$) according to Tukey's test. Modified from Ehlers et al. (2010). Reprinted with permission from Elsevier
[a]See text for treatment details

Total P contents in extracted cells ranged from 2.1 to 8.9 fg P cell^{-1}, with a tendency for higher contents in treatments CN and CNP (Table 2.2). These observed total P contents (Ehlers et al. 2010) seem to be within a reasonable range compared to 3.4–31 fg P cell^{-1} for cultured aquatic bacteria (Fagerbakke et al. 1996), or values between 0.01 and 10 fg P cell^{-1} for in situ analysis of marine bacteria (Gundersen et al. 2002). We found a positive correlation between maximum respiration rates and total P content per cell ($r = 0.59$, $P = 0.003$). This is in accordance with the growth rate hypothesis, which states that biomass P content increases with growth rate due to increased P allocation to P-rich ribosomal RNA (Elser et al. 2000, 2003).

However, the proportion of RNA-P to total P in our study ranged only between 1% and 6%. There are several potential explanations, including:

1. Incomplete cell lysis and RNA extraction
2. Degradation of RNA during extraction and filtration
3. Low growth rates
4. Extraction of a mixture of active and dormant cells from soil with high and low proportions of RNA, respectively
5. Absence of P limitation

These points are discussed in more detail in the following paragraphs.

We used cells of *E. coli* to test the protocol for cell lysis and RNA extraction. Because we found a proportion of RNA-P in *E. coli* of 53%, the protocol was deemed to be efficient. Nevertheless, cell lysis might still have been incomplete for the microorganisms extracted from soil. Degradation of RNA during extraction and filtration was tested by subjecting a pure culture of *Arthrobacter* sp. to the same procedure as soil bacteria, i.e., centrifugation over Nycodenz, sampling of the bacterial layer, and filtration. Compared to samples that were only filtered, centrifugation over Nycodenz decreased the proportion of RNA-P to total cell P by about 18%, whereas total cell P and P in DNA were unaffected (Prusisz and Bünemann unpublished results). However, RNA losses during filtration may have occurred in both cases. Since the filtration of cells extracted from soil usually takes longer than the filtration of pure cultures, this could be an important factor contributing to low RNA recovery.

Apart from methodological problems, the low proportions of RNA-P in cells extracted from soil could point to low growth rates. For example, Makino et al. (2003) found that RNA-P in *E. coli* grown in a chemostat contributed about 40% of total P at low growth rates and 80% at high growth rates. In carbon-amended soils, growth rates were rather high (Fig. 2.2). However, the extracted cells were probably a mixture of active and dormant cells. In that case, the higher RNA-P content in activated cells would have been diluted by the low RNA-P content in dormant cells.

Finally, P availability can affect the proportion of RNA-P in microbial cells. In a study on a mixed bacterial community from a lake, RNA-P contained 25–43% of total P when grown in P-sufficient culture, compared with 79–93% under P limitation (Makino and Cotner 2004). Since the formation of microbial biomass was similar for the treatments CN and CNP (Ehlers et al. 2010), limitation of the microorganisms by carbon rather than P was indeed indicated. This is in agreement with an earlier study on the same soil type (Bünemann et al. 2004b).

Cell contents of DNA and PLFA were similar to those reported in other studies. The DNA content of 6.6 fg cell^{-1} was higher than the 1.6–2.4 fg cell^{-1} reported by Bakken and Olsen (1989) for indigenous soil bacteria, but well within the range of 2–9 fg DNA cell^{-1} that the same authors found for cultured soil bacteria. Similarly, for soil bacteria Torsvik and Goksoyr (1978) and Sandaa et al. (1998) found values of 8.4 fg DNA cell^{-1} and 8.8–11.5 fg DNA cell^{-1}, respectively. For PLFA, the overall average of 0.77×10^{-17} mol PLFA cell^{-1} for all treatments is similar to PLFA values of 0.62×10^{-17}–2.35×10^{-17} mol cell^{-1} for 15 Swedish soils covering a wide range of pH and organic matter contents (Frostegård and Bååth 1996).

Together, the investigated P pools explained between 10% (CNP) and 25% (P) of the total P content per cell. Due to the low recovery of RNA discussed above, these numbers are low compared to the results of, e.g., Vadstein (1998) who found that about 60% of the total P in heterotrophic planktonic bacteria cultured under P limitation was bound in DNA, RNA, and phospholipids. Other P compounds that were not measured in this study include metabolites and teichoic acids. For a complete understanding of carbon and P effects it would be important to analyze all P pools in the extracted cells and to overcome the problem of degradation of compounds during the procedure.

2.5 Conclusions and Outlook

This chapter has provided some evidence that characterization of P forms in microorganisms can contribute to an improved understanding of the role of bacteria and fungi in soil P cycling. Importantly, this approach delivers for the first time data on P forms in soil microorganisms that actually grew in the soil and not in batch culture or chemostat. Thus, unculturable microorganisms are likely to be included in the data. The presence of dormant as well as active cells might partly explain the low proportions of RNA in extracted cells, although incomplete extraction or degradation during the procedure are likely to contribute as well.

In addition, a predominantly fungal origin of condensed phosphates as well as phosphonates in soils was shown. To identify the origin of monoester and diester P in soils, specific compounds of unequivocal microbial origin would have to be found. For microbial communities extracted from a Ferralsol, carbon addition rather than changes in P supply affected the chemical composition of the microbial P pool. This points to the lack of P limitation of microorganisms in high-P-sorbing tropical soils.

However, there are two main problems with this approach of characterizing P forms in soil microorganisms growing in situ:

1. Different methods have to be applied for extraction of bacteria and fungi, and they are not exclusive, i.e., fungal structures can be extracted together with bacterial cells and vice versa. In addition, available methods for extraction of hyphae are not suitable for most soils except possibly those with a high sand content. Future work should therefore focus on improving methods for extraction of fungal hyphae.
2. Changes in cell internal P forms during extraction and sample preparation can occur, and the sensitivity of comprehensive analytical techniques such as ^{31}P NMR is generally too low. When different methods are applied to quantify different P species, there is potential for overlap between P pools, and many different methods have to be applied for a complete speciation analysis, including a range of extraction protocols. Direct comparison of P speciation by ^{31}P

NMR and by a combination of chromatographic, fluorometric, and colorimetric techniques would reveal the scope and limits of each approach.

Recent analytical developments in metabolomics and single-cell ecophysiology (Wagner 2009) hold great promise, and the combination of cell extraction with molecular analysis of microbial communities and functions will eventually give further insight into the role of specific groups of microorganisms in soil P cycling.

Acknowledgements We would like to thank Astrid Oberson for initiating this work and the entire Group of Plant Nutrition for fruitful discussions on the topic. We are grateful to Cory Cleveland, Astrid Oberson, and Emmanuel Frossard for helping to improve the manuscript.

References

Aakra Å, Hesselsoe M, Bakken LR (2000) Surface attachment of ammonia-oxidizing bacteria in soil. Microb Ecol 39:222–235

Arnoldsson KC, Kaufmann P (1994) Lipid class analysis by normal phase high performance liquid chromatography. Development and optimization using multivariate methods. Chromatographia 38:317–324

Bakken LR, Frostegård Å (2006) Nucleic acid extraction from soil. In: Nannipieri P, Smalla K (eds) Nucleic acids and proteins in soil. Springer, Berlin, pp 49–73

Bakken LR, Lindahl V (1995) Recovery of bacterial cells from soil. In: Van Elsas JD, Trevors JT (eds) Nucleic acids in the environment: methods and applications. Springer, Berlin, pp 9–27

Bakken LR, Olsen RA (1989) DNA-content of soil bacteria of different cell size. Soil Biol Biochem 21:789–793

Bhattacharya M, Fuhrman L, Ingram A, Nickerson KW, Conway T (1995) Single-run separation and detection of multiple metabolic intermediates by anion-exchange high-performance liquid chromatography and application to cell pool extracts prepared from *Escherichia coli*. Anal Biochem 232:98–106

Bhavsar AP, Erdman LK, Schertzer JW, Brown ED (2004) Teichoic acid is an essential polymer in *Bacillus subtilis* that is functionally distinct from teichuronic acid. J Bacteriol 186:7865–7873

Bligh EG, Dyer WJ (1959) A rapid method of total lipid extraction and purification. Can J Biochem Physiol 37:911–917

Bünemann EK, Bossio DA, Smithson PC, Frossard E, Oberson A (2004a) Microbial community composition and substrate use in a highly weathered soil as affected by crop rotation and P fertilization. Soil Biol Biochem 36:889–901

Bünemann EK, Smithson PC, Jama B, Frossard E, Oberson A (2004b) Maize productivity and nutrient dynamics in maize-fallow rotations in western Kenya. Plant Soil 264:195–208

Bünemann EK, Smernik RJ, Doolette AL, Marschner P, Stonor R, Wakelin SA, McNeill AM (2008a) Forms of phosphorus in bacteria and fungi isolated from two Australian soils. Soil Biol Biochem 40:1908–1915

Bünemann EK, Smernik RJ, Marschner P, McNeill AM (2008b) Microbial synthesis of organic and condensed forms of phosphorus in acid and calcareous soils. Soil Biol Biochem 40:932–946

Burton K (1956) Study of the conditions and mechanism of the diphenylamine reaction for the colorimetric estimation of deoxyribonucleic acid. Biochem J 62:315–323

Caldwell AG, Black CA (1958) Inositol hexaphosphate. II. Synthesis by soil microorganisms. Soil Sci Soc Am Proc 22:293–296

Cotner JB, Makino W, Biddanda BA (2006) Temperature affects stoichiometry and biochemical composition of *Escherichia coli*. Microb Ecol 52:26–33

Courtois S, Frostegard A, Goransson P, Depret G, Jeannin P, Simonet P (2001) Quantification of bacterial subgroups in soil: comparison of DNA extracted directly from soil or from cells previously released by density gradient centrifugation. Environ Microbiol 3:431–439

Danku JMC, Gumaelius L, Baxter I, Salt DE (2009) A high-throughput method for *Saccharomyces cerevisiae* (yeast) ionomics. J Anal At Spectrom 24:103–107

De Mey M, Lequeux G, Maertens J, De Maeseneire S, Soetaert W, Vandamme E (2006) Comparison of DNA and RNA quantification methods suitable for parameter estimation in metabolic modeling of microorganisms. Anal Biochem 353:198–203

Doolette AL, Smernik RJ (2011) Soil organic phosphorus speciation using spectroscopic techniques. In: Bünemann EK, Oberson A, Frossard E (eds) Phosphorus in action: biological processes in soil phosphorus cycling. Soil biology, vol 26. Springer, Heidelberg. doi:10.1007/978-3-642-15271-9_1

Duboc P, Schill N, Menoud L, Vangulik W, Vonstockar U (1995) Measurements of sulfur, phosphorus and other ions in microbial biomass – influence on correct determination of elemental composition and degree of reduction. J Biotechnol 43:145–158

Ebina J, Tsutsui T, Shirai T (1983) Simultaneous determination of total nitrogen and total phosphorus in water using peroxodisulfate oxidation. Water Res 17:1721–1726

Ehlers K, Bünemann EK, Oberson A, Frossard E, Frostegård Å, Yuejian M, Bakken LR (2008) Extraction of soil bacteria from a Ferralsol. Soil Biol Biochem 40:1940–1946

Ehlers K, Bakken LR, Frostegård Å, Frossard E, Bünemann EK (2010) Phosphorus limitation in a Ferralsol: impact on microbial activity and cell-internal P pools. Soil Biol Biochem 42:558–566

Ellwood DC, Tempest DW (1972) Influence of culture pH on content and composition of teichoic acids in walls of *Bacillus subtilis*. J Gen Microbiol 73:395–402

Elser JJ, Sterner RW, Gorokhova E, Fagan WF, Markow TA, Cotner JB, Harrison JF, Hobbie SE, Odell GM, Weider LJ (2000) Biological stoichiometry from genes to ecosystems. Ecol Lett 3:540–550

Elser JJ, Acharya K, Kyle M, Cotner J, Makino W, Markow T, Watts T, Hobbie S, Fagan W, Schade J, Hood J, Sterner RW (2003) Growth rate-stoichiometry couplings in diverse biota. Ecol Lett 6:936–943

Endo Y (1970) A simultaneous estimation method of DNA and RNA by orcinol reaction and a study on reaction mechanism. J Biochem 67:629–633

Faegri A, Torsvik VL, Goksoyr J (1977) Bacterial and fungal activities in soil – separation of bacteria and fungi by a rapid fractionated centrifugation technique. Soil Biol Biochem 9:105–112

Fagerbakke KM, Heldal M, Norland S (1996) Content of carbon, nitrogen, oxygen, sulfur and phosphorus in native aquatic and cultured bacteria. Aquat Microb Ecol 10:15–27

Frisvad JC, Thrane U (1987) Standardized high-performance liquid chromatography of 182 mycotoxins and other fungal metabolites based on alkylphenone retention indexes and UV-VIS spectra (diode-array detection). J Chromatogr 404:195–214

Frossard E, Achat DL, Bernasconi SM, Bünemann EK, Fardeau J-C, Jansa J, Morel C, Rabeharisoa L, Randriamanantsoa L, Sinaj S, Tamburini F, Oberson A (2011) The use of tracers to investigate phosphate cycling in soil–plant systems. In: Bünemann EK, Oberson A, Frossard E (eds) Phosphorus in action: biological processes in soil phosphorus cycling. Soil biology, vol 26. Springer, Heidelberg. doi:10.1007/978-3-642-15271-9_3

Frostegård Å, Bååth E (1996) The use of phospholipid fatty acid analysis to estimate bacterial and fungal biomass in soil. Biol Fertil Soils 22:59–65

Frostegård Å, Bååth E, Tunlid A (1993) Shifts in the structure of soil microbial communities in limed forests as revealed by phospholipid fatty acid analysis. Soil Biol Biochem 25:723–730

Gorokhova E, Kyle M (2002) Analysis of nucleic acids in *Daphnia*: development of methods and ontogenetic variations in RNA-DNA content. J Plankton Res 24:511–522

Gundersen K, Heldal M, Norland S, Purdie DA, Knap AH (2002) Elemental C, N, and P cell content of individual bacteria collected at the Bermuda Atlantic Time-Series Study (BATS) site. Limnol Oceanogr 47:1525–1530

Heldal M, Norland S, Tumyr O (1985) X-ray microanalytic method for measurement of dry matter and elemental content of individual bacteria. Appl Environ Microbiol 50:1251–1257

Heller DN, Murphy CM, Cotter RJ, Fenselau C, Uy OM (1988) Constant neutral loss scanning for the characterization of bacterial phospholipids desorbed by fast atom bombardment. Anal Chem 60:2787–2791

Herbert D (1961) The chemical composition of microorganisms as a function of their environment. In: Meynell GG, Gooder H (eds) Microbial reaction to environment: eleventh symposium of the Society for General Microbiology. Cambridge University Press, Cambridge, pp 391–416

Holmhans O, Booth CR (1966) Measurement of adenosine triphosphate in the ocean and its ecological significance. Limnol Oceanogr 11:510–519

Hossack JA, Rose AH (1976) Fragility of plasma membranes in *Saccharomyces cerevisiae* enriched with different sterols. J Bacteriol 127:67–75

Jakobsen I, Smith SE, Smith FA (2002) Function and diversity of arbuscular mycorrhizae in carbon and mineral nutrition. In: van der Heijden MGA, Sanders I (eds) Mycorrhizal ecology. Springer, Berlin, pp 75–92

Jansa J, Finlay R, Wallander H, Smith FA, Smith SE (2011) Role of mycorrhizal symbioses in phosphorus cycling. In: Bünemann EK, Oberson A, Frossard E (eds) Phosphorus in action: biological processes in soil phosphorus cycling. Soil biology, vol 26. Springer, Heidelberg. doi:10.1007/978-3-642-15271-9_6

Janssen PH, Yates PS, Grinton BE, Taylor PM, Sait M (2002) Improved culturability of soil bacteria and isolation in pure culture of novel members of the divisions *Acidobacteria*, *Actinobacteria*, *Proteobacteria*, and *Verrucomicrobia*. Appl Environ Microbiol 68:2391–2396

Jenkinson DS, Oades JM (1979) A method for measuring adenosine triphosphate in soil. Soil Biol Biochem 11:193–199

Koukol O, Novak F, Hrabal R (2008) Composition of the organic phosphorus fraction in basidiocarps of saprotrophic and mycorrhizal fungi. Soil Biol Biochem 40:2464–2467

Krueger R, Zinn N, Lehmann WD (2009) Quantification of protein phosphorylation by μLC-ICP-MS. In: de Graauw M (ed) Phospho-proteomics. Humana, Clifton, pp 201–218

Lambert C, Weuster-Botz D, Weichenhain R, Kreutz EW, AAd G, Schoberth SM (2002) Monitoring of inorganic polyphosphate dynamics in *Corynebacterium glutamicum* using a novel oxygen sparger for real time ^{31}P in vivo NMR. Acta Biotechnol 22:245–260

Lindahl V (1996) Improved soil dispersion procedures for total bacterial counts, extraction of indigenous bacteria and cell survival. J Microbiol Methods 25:279–286

Lindahl V, Bakken LR (1995) Evaluation of methods for extraction of bacteria from soil. FEMS Microbiol Ecol 16:135–142

Lovdal T, Skjoldal EF, Heldal M, Norland S, Thingstad TF (2008) Changes in morphology and elemental composition of *Vibrio splendidus* along a gradient from carbon-limited to phosphate-limited growth. Microb Ecol 55:152–161

Macklon AES, Grayston SJ, Shand CA, Sim A, Sellars S, Ord BG (1997) Uptake and transport of phosphorus by Agrostis capillaris seedlings from rapidly hydrolysed organic sources extracted from ^{32}P-labelled bacterial cultures. Plant Soil 190:163–167

Makarov MI, Haumaier L, Zech W, Marfenina OE, Lysak LV (2005) Can ^{31}P NMR spectroscopy be used to indicate the origins of soil organic phosphates? Soil Biol Biochem 37:15–25

Makino W, Cotner JB (2004) Elemental stoichiometry of a heterotrophic bacterial community in a freshwater lake: implications for growth- and resource-dependent variations. Aquat Microb Ecol 34:33–41

Makino W, Cotner JB, Sterner RW, Elser JJ (2003) Are bacteria more like plants or animals? Growth rate and resource dependence of bacterial C:N:P stoichiometry. Funct Ecol 17:121–130

Maron PA, Schimann H, Ranjard L, Brothier E, Domenach AM, Lensi R, Nazaret S (2006) Evaluation of quantitative and qualitative recovery of bacterial communities from different soil types by density gradient centrifugation. Eur J Soil Biol 42:65–73

Mashego MR, Rumbold K, De Mey M, Vandamme E, Soetaert W, Heijnen JJ (2007) Microbial metabolomics: past, present and future methodologies. Biotechnol Lett 29:1–16

Mazzella N, Molinet J, Syakti AD, Dodi A, Bertrand JC, Doumenq P (2005) Use of electrospray ionization mass spectrometry for profiling of crude oil effects on the phospholipid molecular species of two marine bacteria. Rapid Commun Mass Spectrom 19:3579–3588

Moreau RA, Powell MJ, Osman SF, Whitaker BD, Fett WF, Roth L, Obrien DJ (1995) Analysis of intact hopanoids and other lipids from the bacterium *Zymomonas mobilis* by high performance liquid chromatography. Anal Biochem 224:293–301

Neumann E, George E (2005) Extraction of extraradical arbuscular mycorrhizal mycelium from compartments filled with soil and glass beads. Mycorrhiza 15:533–537

Norland S, Fagerbakke KM, Heldal M (1995) Light element analysis of individual bacteria by X-ray microanalysis. Appl Environ Microbiol 61:1357–1362

Oberson A, Joner EJ (2005) Microbial turnover of phosphorus in soil. In: Turner BL, Frossard E, Baldwin DS (eds) Organic phosphorus in the environment. CABI, Wallingford, pp 133–164

Oberson A, Friesen DK, Rao IM, Bühler S, Frossard E (2001) Phosphorus transformations in an Oxisol under contrasting land-use systems: the role of the soil microbial biomass. Plant Soil 237:197–210

Oehl F, Frossard E, Fliessbach A, Dubois D, Oberson A (2004) Basal organic phosphorus mineralization in soils under different farming systems. Soil Biol Biochem 36:667–675

Olander LP, Vitousek PM (2004) Biological and geochemical sinks for phosphorus in soil from a wet tropical forest. Ecosystems 7:404–419

Pal MK, Ghosh JK, Das S (1989) Spectrophotometric and spectrofluorometric titrations of teichoic acid. Indian J Biochem Biophys 26:311–314

Priemé A, Sitaula JIB, Klemedtsson ÅK, Bakken LR (1996) Extraction of methane-oxidizing bacteria from soil particles. FEMS Microbiol Ecol 21:59–68

Rasmussen N, Lloyd D, Ratcliffe R, Hansen P, Jakobsen I (2000) ^{31}P NMR for the study of P metabolism and translocation in arbuscular mycorrhizal fungi. Plant Soil 226:245–253

Rousk J, Brookes PC, Baath E (2009) Contrasting soil pH effects on fungal and bacterial growth suggest functional redundancy in carbon mineralization. Appl Environ Microbiol 75:1589–1596

Sambrook J, Russel DW (2006) Quantitation of nucleic acids. In: Sambrook J, Russel DW (eds) The condensed protocols: from molecular cloning: a laboratory manual. Cold Spring Harbor Laboratory Press, Cold Spring Harbor, pp 739–741

Sandaa RA, Enger O, Torsvik V (1998) Rapid method for fluorometric quantification of DNA in soil. Soil Biol Biochem 30:265–268

Seeling B, Zasoski RJ (1993) Microbial effects in maintaining organic and inorganic solution phosphorus concentrations in a grassland topsoil. Plant Soil 148:277–284

Sitaula BK, Almas A, Bakken LR, Singh BR (1999) Assessment of heavy metals associated with bacteria in soil. Soil Biol Biochem 31:315–316

Smedsgaard J (1997) Micro-scale extraction procedure for standardized screening of fungal metabolite production in cultures. J Chromatogr A 760:264–270

Smits HP, Cohen A, Buttler T, Nielsen J, Olsson L (1998) Cleanup and analysis of sugar phosphates in biological extracts by using solid-phase extraction and anion-exchange chromatography with pulsed amperometric detection. Anal Biochem 261:36–42

Tempest DW, Wouters JTM (1981) Properties and performance of microorganisms in chemostat culture. Enzyme Microb Technol 3:283–290

Tiessen H, Ballester MV, Salcedo I (2011) Phosphorus and global change. In: Bünemann EK, Oberson A, Frossard E (eds) Phosphorus in action: biological processes in soil phosphorus cycling. Soil biology, vol 26. Springer, Heidelberg. doi:10.1007/978-3-642-15271-9_18

Torsvik VL, Goksoyr J (1978) Determination of bacterial DNA in soil. Soil Biol Biochem 10:7–12

Turner BL, Papházy MJ, Haygarth PM, McKelvie ID (2002) Inositol phosphates in the environment. Philos Trans R Soc Lond B Biol Sci 357:449–469

Turner BL, Mahieu N, Condron LM (2003) Phosphorus-31 nuclear magnetic resonance spectral assignments of phosphorus compounds in soil NaOH-EDTA extracts. Soil Sci Soc Am J 67:497–510

Vadstein O (1998) Evaluation of competitive ability of two heterotrophic planktonic bacteria under phosphorus limitation. Aquat Microb Ecol 14:119–127

Vadstein O (2000) Heterotrophic, planktonic bacteria and cycling of phosphorus: phosphorus requirements, competitive ability, and food web interactions. Adv Microb Ecol 16:115–167

Valeur A, Michelsen P, Odham G (1993) Online straight phase liquid chromatography plasmaspray tandem mass spectrometry of glycerolipids. Lipids 28:255–259

Van Der Meeren P, Vanderdeelen J, Baert L (1992) Phospholipid analysis by HPLC. In: Nollet LML (ed) Food analysis by HPLC. Food science and technology, vol 52. Marcel Dekker, New York, pp 241–258

Viereck N, Hansen PE, Jakobsen I (2004) Phosphate pool dynamics in the arbuscular mycorrhizal fungus *Glomus intraradices* studied by in vivo ^{31}P NMR spectroscopy. New Phytol 162:783–794

Wagner M (2009) Single-cell ecophysiology of microbes as revealed by Raman microspectroscopy or secondary ion mass spectrometry imaging. Annu Rev Microbiol 63:411–429

Wallander H, Nilsson LO, Hagerberg D, Rosengren U (2003) Direct estimates of C:N ratios of ectomycorrhizal mycelia collected from Norway spruce forest soils. Soil Biol Biochem 35:997–999

Wang Y, Hammes F, Boon N, Egli T (2007) Quantification of the filterability of freshwater bacteria through 0.45, 0.22, and 0.1 μm pore size filters and shape-dependent enrichment of filterable bacterial communities. Environ Sci Technol 41:7080–7086

Wanner U, Egli T (1990) Dynamics of microbial growth and cell composition in batch culture. FEMS Microbiol Rev 75:19–44

Werner TP, Amrhein N, Freimoser FM (2005) Novel method for the quantification of inorganic polyphosphate (iPoP) in *Saccharomyces cerevisiae* shows dependence of iPoP content on the growth phase. Arch Microbiol 184:129–136

Chapter 3
The Use of Tracers to Investigate Phosphate Cycling in Soil–Plant Systems

Emmanuel Frossard, David L. Achat, Stefano M. Bernasconi, Else K. Bünemann, Jean-Claude Fardeau, Jan Jansa, Christian Morel, Lilia Rabeharisoa, Lalajaona Randriamanantsoa, Sokrat Sinaj, Federica Tamburini, and Astrid Oberson

3.1 Introduction

Phosphorus (P) is indispensable for all living organisms and cannot be replaced in most of its biological functions. Whereas agricultural production is limited in many areas by the lack of available P, excessive P inputs in other agro-ecosystems result in the pollution of surface waters (Frossard et al. 2009). Furthermore, there are indications that the current reserves of rock phosphates that can be mined at a relatively low cost to be processed into fertilizers will be exhausted within the next century (Cordell et al. 2009). P use must therefore become much more efficient in the future. Concepts and management practices for a better crop P use efficiency of P derived from soil or from fertilizer will be based on a better understanding and quantification of soil–plant processes at different spatial and temporal scales.

E. Frossard (✉), E.K. Bünemann, J. Jansa, L. Randriamanantsoa, F. Tamburini, and A. Oberson
Institute of Plant, Animal and Agro-ecosystem Sciences, ETH Zurich, Eschikon 33, 8315 Lindau, Switzerland
e-mail: emmanuel.frossard@ipw.agrl.ethz.ch

D.L. Achat
ENITA de Bordeaux, UMR 1220 TCEM (INRA-ENITAB), 1 cours du Général de Gaulle, 33175 Gradignan, France

S.M. Bernasconi
Geological Institute, ETH Zurich, Sonneggstrasse 5, 8092 Zürich, Switzerland

J.-C. Fardeau
117 allée du Petit Colombier, 04100 Manosque, France

C. Morel
Institut National de la Recherche Agronomique (INRA), UMR 1220 TCEM (INRA-ENITAB), 89 Av. E Bourleau, BP 81, 33883 Villenave d'Ornon Cedex, France

L. Rabeharisoa
Laboratoire des Radio-Isotopes, Route d'Andraisoro 101, Antananarivo, Madagascar

S. Sinaj
Station de recherche Agroscope Changins-Wädenswil ACW, CP 1202, 1260 Nyon, Switzerland

E.K. Bünemann et al. (eds.), *Phosphorus in Action*, Soil Biology 26,
DOI 10.1007/978-3-642-15271-9_3, © Springer-Verlag Berlin Heidelberg 2011

The use of tracers is relevant for studying the release of phosphate ions (Pi is used in this chapter as an abbreviation for phosphate ions, i.e., for H_3PO_4, $H_2PO_4^-$, HPO_4^{2-}, and PO_4^{3-}) into the soil solution, where these are available to plants and microbes (Fig. 3.1) for at least two reasons. Firstly, the total P content of a soil is usually at least one order of magnitude larger than the amount of P that is rapidly cycling in the soil–plant system and two to three orders of magnitude larger than the amount of P present as Pi in the soil solution (Frossard et al. 1995). Secondly, Pi undergoes many abiotic and biotic reactions in the soil (Fig. 3.1), some occurring within a few seconds, others over several years, and is (re)distributed in a large number of pools (Fardeau 1996; Bünemann and Condron 2007).

Phosphorus has one stable isotope (^{31}P), making up virtually 100% of the total P on earth, and seven radioactive isotopes. Only two radioactive isotopes, ^{32}P and ^{33}P, can be used in soil–plant and soil–solution studies. These radioisotopes have a relatively short half-life (14.3 days for ^{32}P and 25.3 days for ^{33}P) and emit β^- radiation with a maximum energy of 1.71 and 0.25 MeV for ^{32}P and ^{33}P, respectively (Endt 1990). These radioisotopes are introduced at extremely low rates in natural systems from the atmosphere (Benitez-Nelson and Buesseler 1999). For research needs, they must be artificially produced either from sulfur in nuclear reactors or by neutron activation of ^{31}P. These radioisotopes have been used for decades to probe P pools in soils, to evaluate reactions in which P is involved, and to

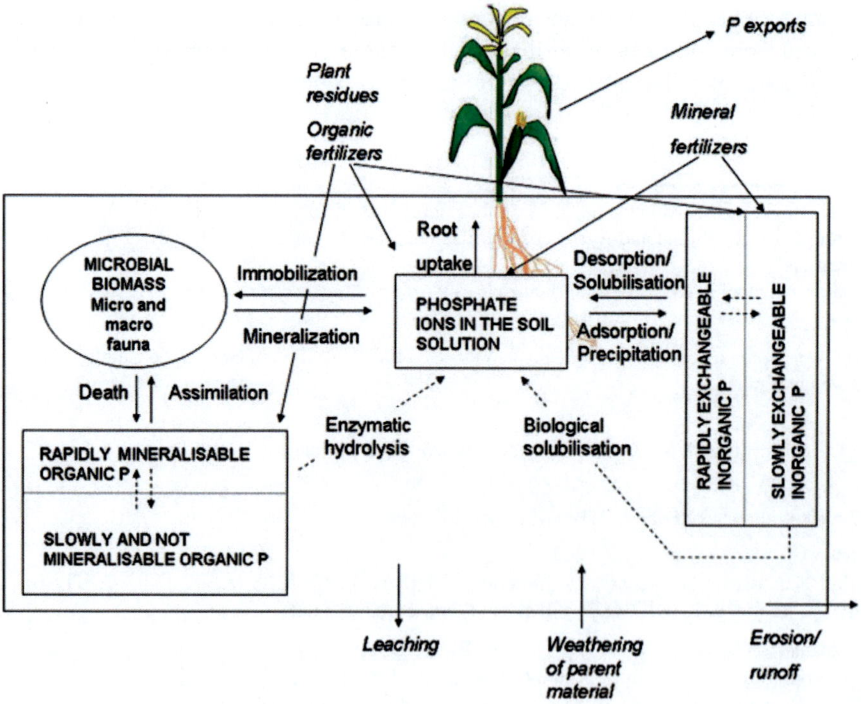

Fig. 3.1 The phosphorus cycle in the soil–plant system

trace the fate of P added as fertilizer from the soil to the plant (Dean et al. 1947; McAuliffe et al. 1947; Nelson et al. 1947; White et al. 1949; Wiklander 1950; Larsen 1952; Talibudeen 1957; Mattingly 1975; Fardeau 1996). But, given their relatively short half-lives, they can only deliver information on reactions for up to a few months, and since they are radioactive they are not easy to use under field conditions.

In soils, the P atom is always accompanied by oxygen (O) atoms in the form of phosphate (PO_4^{3-}, which is the dominant form in soils), phosphonate ($C\text{-}PO_3^{2-}$) or polyphosphate of different lengths [$(PO_3^{-})_n$] (Frossard et al. 2011). Recent publications have shown that variations in the natural abundance of the stable ^{18}O isotope bound to P (noted hereafter as $\delta^{18}O_P$) could deliver relevant information on P transformations and P sources in aquatic systems (Colman et al. 2005; McLaughlin et al. 2006a, b; Elsbury et al. 2009; Young et al. 2009; Jaisi and Blake 2010). Although some of these studies have addressed processes that are relevant to the soil–plant system, such as adsorption–desorption on oxides, precipitation of P-containing minerals, and uptake of phosphate by microorganisms (Jaisi et al. 2010; Liang and Blake 2007; Blake et al. 2005), this approach has not yet been successfully applied to soils. Among the problems slowing down progress in this direction are the lack of methods optimized for quantitatively isolating phosphate from the soil in a form that is amenable to $\delta^{18}O_P$ analysis, and the absence of plant studies.

The aims of this review are to present the use of P radioisotopes: (a) to probe pools and to study P transformations in soils, (b) to trace the fate of fertilizers in soil–plant systems, and (c) to assess the foraging strategies of arbuscular mycorrhizal fungi (AMF). In the last section, we will evaluate the potential of using the stable oxygen isotopes bound to P to study soil P dynamics. In each of these areas we will focus on the biological reactions controlling soil P transformations. These are (a) the uptake of P by soil microorganisms and plants (either through their roots or through mycorrhizal hyphae); (b) the release of Pi from the soil solid phase or from fertilizers, induced by the exudation of phosphatases and organic acids from roots or microorganisms; (c) the release of Pi from plant residues or organic fertilizers to the soil solution; and (d) the release of Pi from microorganisms and organic matter mineralization (Fig. 3.1).

3.2 Use of P Radioisotopes to Study Soil Processes and to Probe Soil P Pools

Many review articles have already been published on the use of P radioisotopes in soil science and plant nutrition (e.g., Wiklander 1950; Larsen 1952; Amer et al. 1969; Fardeau 1996; Di et al. 1997; Frossard and Sinaj 1997). Therefore, we will restrict our review to the principles of the main approaches used, their limits, and examples of their latest applications. We will show how the principles of isotopic dilution can be used to assess the size of different soil P pools and to quantify the rate of transfer of P between these pools, first by the analysis of soil–solution systems and then by the analysis of soil–plant systems. Note that the methods presented in Sects. 3.2–3.4 of this manuscript can be used indifferently with both P radioisotopes (^{32}P or ^{33}P) and will give identical results because we assume that no significant isotopic fractionation occurs between ^{31}P, ^{32}P and ^{33}P (see the explanation in Sect. 3.2.3.2).

3.2.1 *Isotopic Dilution of P in Soil–Solution Systems: The Principles*

More than 60 years ago, McAuliffe et al. (1947) noted that the concentration of Pi labeled with ^{32}P (noted hereafter as ^{32}Pi) added to a soil–solution system decreased with time. They attributed this decrease to an exchange between ^{32}Pi in the solution and ^{31}Pi located on the solid phase of the soil. They interpreted their results by the presence of two P pools, one containing the fraction of readily exchangeable P and the other the fraction of less-exchangeable P. Later, Wiklander (1950) noted that after the addition of ^{32}P to a soil suspension that was at steady-state for Pi, the concentration of ^{32}Pi in solution decreased with time according to a simple power function (3.1). Fardeau (1996) confirmed that this equation was able to describe the decrease of radioactivity in soil–solution systems for many soils for isotopic exchange times as long as 3 months:

$$r_{(t)}/R = \left[r_{(1)}/R\right]t^{-n} \tag{3.1}$$

where R is the total radioactivity introduced in the soil–solution system as ^{32}Pi (or ^{33}Pi) (units are megabecquerel, MBq,), $r_{(t)}$ is the radioactivity remaining in the solution after t minutes of isotopic exchange, $r_{(1)}/R$ is the proportion of the total radioactivity remaining in the solution after 1 min of isotopic exchange, and n is a parameter describing the rate of decrease of radioactivity in the solution. Both $r_{(1)}/R$ and n vary with the concentration of Pi in the soil solution. Soils with a $r_{(1)}/R$ lower than 0.2 are usually considered to have a high sorbing capacity for Pi (Frossard et al. 1993). Parameters $r_{(1)}/R$ and n calculated from (3.1) for selected soils from Switzerland and Madagascar are shown in Table 3.1.

Equation (3.1) shows that Pi labeled with ^{32}P added in the soil–solution system are diluted in a large pool of ^{31}P. This can be interpreted as a homo-ionic exchange between the ^{32}Pi added in the solution and Pi located on the solid phase of the soil that can be desorbed (Fardeau 1996). The fact that this equation is valid for exchange times as long as 3 months shows that a much longer time is needed in most soils to reach a true isotopic equilibrium, which has been confirmed by sorption–desorption experiments (e.g., Torrent 1987). This is why Fardeau (1996), following the ideas of McAuliffe et al. (1947), proposed the

Table 3.1 Examples of results obtained from isotopic exchange kinetic experiments in Swiss (Cadenazzo, Ellighausen, Changins and Vaz) and Malagasy (Betafo, Ivory, Laniera, Lazaina) soils

Soil	Soil type	$r_{(1)}/R$ or m	n	C_P (mg P L^{-1})	T_m (min)	g_m (min^{-1})	F_m (mg P kg^{-1} min^{-1})	$E_{(1\ min)}$ (mg P kg^{-1})	$E_{(1\ day)}$ (mg P kg^{-1})	Total inorganic P (mg P kg^{-1})	Total P (mg P kg^{-1})
Cadenazzo[a]	Fluvisol	0.43	0.32	0.23	0.23	4.3	9.9	5.29	54	910	1,174
Ellighausen[a]	Cambisol	0.26	0.37	0.12	0.07	13.6	16.4	4.56	67	287	789
Changins[a]	Cambisol	0.15	0.41	0.04	0.02	40.9	16.4	2.64	52	199	590
Vaz[a]	Fluvisol	0.71	0.25	0.77	1.04	0.96	7.4	10.8	66	358	1,370
Betafo[b,c]	Andosol	0.002	0.57	0.0008	3.2×10^{-5}	3.1×10^{4}	2.5×10^{2}	3.93	117	218	864
Ivory[b,c]	Ferralsol	0.014	0.36	0.0019	2.0×10^{-5}	5.1×10^{4}	9.7×10^{2}	1.32	13	46	161
Laniera[b]	Ferralsol	0.016	0.37	0.013	3.8×10^{-5}	2.6×10^{4}	3.4×10^{3}	7.18	41	62	220
Lazaina[b]	Ferralsol	0.033	0.43	0.011	8.3×10^{-4}	1.2×10^{3}	1.3×10^{2}	3.05	24	36	165

The mean residence time of Pi in the solution (T_m), the mean turnover rate (g_m) and the mean flux of Pi between the solid phase and the solution (F_m) were calculated for all soils as described in Fardeau et al. (1991)

[a]These results come from Gallet et al. (2003a). The samples were taken in 1998, in the 0–20 cm horizon of a treatment that had not received phosphate fertilizers since 1989. The parameters [$r_{(1)}/R$ and n] describing the kinetics of isotopic exchange were calculated from (3.1)

[b]The parameters (m and n) describing the kinetics of isotopic exchange were calculated from (3.2) and the ratio $r_{(\infty)}/R$ was estimated as the ratio 10 C_P/total inorganic P

[c]The concentration of Pi in the solution was measured in these samples after a concentration of the phosphomolybdate complex in hexanol

characterization of this isotopic exchange through its kinetic parameters [$r_{(1)}/R$ and n] instead of measuring radioactive and stable Pi in the soil solution after a single exchange time. These parameters can be obtained from the isotopic exchange kinetic experiment. In this experiment, a known amount of Pi labeled with ^{32}P or ^{33}P (the tracer) is introduced into a soil–solution system (10 g of soil in 100 mL water) at steady-state equilibrium for Pi (the tracee). After different exchange times of up to 2 h, small volumes of the soil–water suspension are extracted with a syringe, filtered at 0.2 μm, and the tracer and the tracee in solution measured.

As (3.1) does not fit the decrease of radioactivity in the solution in all cases, Fardeau et al. (1985) adapted it as follows:

$$r_{(t)}/R = m\left(t + m^{1/n}\right)^{-n} + r_{(\infty)}/R \tag{3.2}$$

where m and n are parameters determined by nonlinear regression, and $r_{(\infty)}$ the amount of radioactivity that would remain in the solution at isotopic equilibrium. The term $r_{(\infty)}/R$ is operationally estimated as the ratio of the concentration of Pi present in the soil solution expressed in milligrams of P per kilogram of soil to the total inorganic soil P also expressed in milligrams of P per kilogram of soil (Fardeau 1993). The term $r_{(\infty)}/R$ can in some cases be neglected (Achat et al. 2009a, b). Both m and n vary with the concentration of Pi in the soil solution and other factors.

Equations (3.1) and (3.2) can be interpreted as a sum of many exponentials, which demonstrates that the tracer enters in a large number of compartments (Atkins 1969; Probert and Larsen 1972; Diesing et al. 2008). These equations reflect two approaches to the statistical description of the decrease of radioactivity in solution with time (Fardeau 1981). Therefore, especially when working with long-term kinetics (i.e., with isotopic exchange times longer than 1 day), it is necessary to evaluate which of these equations gives the best description of the radioactivity decrease in solution (Bünemann et al. 2007).

Fardeau et al. (1991) derived from (3.2) the mean turnover rate of Pi in the soil solution ($g_{\mathbf{m}}$ in min^{-1}), the mean residence time of Pi in the soil solution ($T_{\mathbf{m}}$ in min^{-1}), and the mean flux of Pi between the solid phase of the soil and the solution ($F_{\mathbf{m}}$ mg P kg^{-1} soil in min^{-1}). The principle underlying these calculations is not explained here, but some results for selected soils are presented in Table 3.1. These results suggest that the mean residence time of Pi in the soil solution is in most cases shorter than 1 min. The isotopic dilution of the tracer introduced in the soil is, in this experiment, mostly caused by its exchange with Pi located on the solid phase of the soil. It is important to note that this exchange refers to the swapping of intact phosphate groups and does not involve any breaking of the P–O bonds. The tracer can, however, also be diluted by a release of non-labeled Pi to the solution (e.g., from an added fertilizer) or as a result of soil organic P mineralization or soil inorganic P solubilization, as discussed below.

If we assume that ^{31}Pi, ^{32}Pi and ^{33}Pi have exactly the same behavior in the soil–solution system, then the specific activity of Pi in the soil solution, i.e., the ratio between radioactivity and mass in the solution, is identical to the specific activity of the Pi that has been isotopically exchanged in the entire system and that is noted as $E_{(t)}$ hereafter (3.3):

$$r_{(t)}/(VC_{\mathrm{P}}) = R/E_{(t)} \tag{3.3}$$

where C_{P} is the concentration of Pi in the soil water extract (in mg P L^{-1}), V is the water to soil ratio (in L kg^{-1}), and $E_{(t)}$ is the amount of isotopically exchanged P within t minutes of exchange (in mg P kg^{-1} soil). This equation is only valid for exchange times shorter than the time that would be necessary to exchange the entire amount of soil inorganic P, including the Pi present in the soil solution.

The $E_{(t)}$ value is the sum of Pi in the solution and of Pi located on the solid phase of the soil that can exchange with Pi in the solution within a time t. It therefore yields the total amount of Pi that can potentially be taken up by a plant or a microbe within this time. In their critical review, Hamon et al. (2002) agree that authors assessing the availability of Pi can indeed use (3.3) to calculate the amount of isotopically exchangeable P. Hamon et al., however, state that researchers interested in measurement of the soil buffering capacity for Pi, or in assessing the effect of soil physicochemical properties on phosphate ion exchangeability, should specifically calculate the fraction of exchangeable phosphate present on the solid phase of the soil, leaving aside the fraction of Pi present in the soil solution. This calculation yields the amount of Pi present on the solid phase that can diffuse into the soil solution with time in response to a decrease in the concentration of Pi in the solution in the absence of microbial activity, a process that is known as "desorption" in environmental chemistry (Tan 1993). This amount of desorbable Pi has been named Pr (expressed in mg P kg^{-1} soil) by Ehlert et al. (2003). It is calculated using the following equation (Hamon et al. 2002; Ehlert et al. 2003; Stroia et al. 2007; Achat et al. 2009b):

$$\mathrm{Pr} = E_{(t)} - VC_{\mathrm{P}} = VC_{\mathrm{P}}\left[R/r_{(t)} - 1\right] \tag{3.4}$$

3.2.2 *Isotopic Dilution of Phosphate Ions in Soil–Solution Systems: Selected Applications*

3.2.2.1 Assessment of Availability of Phosphate Ions for Plants

Results of the isotope exchange kinetics experiment suggest that it is not possible to divide soil total inorganic P into an available P pool and a non-available P pool but, on the contrary, that most of the inorganic P can become available at some point in time. Although some inorganic P (a very small amount) will be very rapidly exchangeable with Pi in solution and therefore be very rapidly available, the vast majority of inorganic P will be very slowly exchangeable (Fardeau

1993). These results agree with those obtained earlier by Barrow (1974, 1983, 1991) and Barrow and Shaw (1975a, b) who, on the basis of long-term sorption experiments, concluded that inorganic P ions located on the solid phase of the soil are distributed along a continuum of solubility, some being in rapid equilibrium with Pi in the solution and some being in very slow equilibrium with Pi in the solution.

The isotopic exchange kinetic experiment provides information on the concentration of P in the soil solution (C_P), on the amount of P that is potentially available to plants [$E_{(t)}$] and on the soil buffer capacity, which expresses the changes in desorbable Pi when the concentration of P in the solution varies (dPr/dC_P; Hamon et al. 2002; Stroia et al. 2007). The concentration of Pi in the solution (C_P) and E values calculated for 1 min and 1 day are shown in Table 3.1 for selected soils from Switzerland and Madagascar. Soils with a $E_{(1\ min)}$ lower than 5 mg P kg^{-1} soil are considered to be phosphate-limiting for crops (Gallet et al. 2003a). The buffer capacity is, however, not straightforward to compute because the parameters used for the calculation of desorbable Pi are C_P-dependent (Morel et al. 1994; Achat et al. 2009a).

Morel et al. (2000) and Stroia et al. (2007) working on soils, and Nemery et al. (2005) working on sediments, succeeded in modeling the changes in desorbable Pi for a wide range of concentrations of Pi in the soil solution and for different exchange times using a kinetic Freundlich equation (3.5):

$$Pr = vC_P{}^w t^p \tag{3.5}$$

where v, w and p are parameters determined by nonlinear regression. This equation is valid when desorbable Pi is lower than the difference between total inorganic soil P and soil solution P. These parameters are obtained by conducting isotopic exchange kinetic experiments over short periods in the same soil with increasing concentrations of Pi in the solution. An example of results obtained with this approach is given in Fig. 3.2.

Using this kinetic Freundlich equation parameterized in batch experiments combined with a mass balance model, Stroia et al. (2007) successfully showed that the removal of 1 kg P ha^{-1} would be buffered by desorbable Pi over a few weeks in a Luvisol and a Brunisol. Their approach showed that the changes in Pi concentration in the solution were larger in the Brunisol than in the Luvisol. In the Luvisol, their modeled final C_P value was almost identical to the final C_P value derived from the field experiment for the removal of 1 kg P ha^{-1}. In the Brunisol, however, the final modeled C_P value was three times higher than the final C_P value derived from the field experiments. This shows that the approach taken by Stroia et al. (2007) is promising for prediction of the changes in C_P for different P balances, but it should be completed by taking into account other mechanisms that can affect the concentration of Pi in the soil solution, such as the transfer of P to deeper horizons or the uptake of Pi by soil microbes.

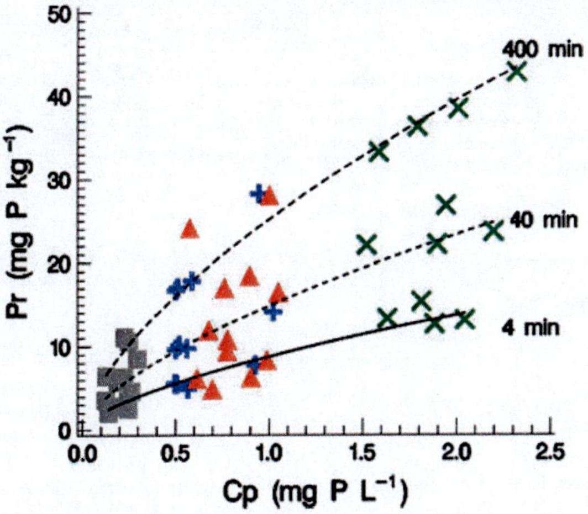

Fig. 3.2 Experimental and modeled values of desorbable Pi (Pr) of soil samples as a function of Pi ion concentration in solution (C_P) and time (Morel 2002). Soils were collected after 17 years of experimentation from the plough layer of a long-term field experiment established on a neutral, non-carbonaceous, sandy loamy soil. It was located close to the village Mant in the south-west of France. Four treatments were repeated four times in blocks. Phosphate was added as supertriple phosphate. The *symbols* represent experimental values: *squares* no P; *plus symbols* on average 27 kg P ha^{-1} applied every year; *times symbols* on average 79 kg P ha^{-1} applied every year; *triangles* on average 52 kg P ha^{-1} applied every 2 years. *Lines* show values after 4, 40, and 400 min of isotopic exchange calculated by the following Freundlich equation: $\text{Pr} = 6.4 \times C_P^{0.65} \times t^{0.23}$ (48 observations, $r^2 = 0.96$)

3.2.2.2 Quantification of Phosphate Mineralized from Soil Organic Matter and from Soil Microorganisms

Isotopic dilution has been used to quantify the amount of P mineralized from soil organic matter and from microbial P (Walbridge and Vitousek 1987; Oehl et al. 2001a; Bünemann et al. 2007; Achat et al. 2009b). The basic principle of the approaches followed by these authors is sketched in Fig. 3.3.

Let us consider a soil in which isotopically exchangeable P has been labeled with ^{32}Pi or ^{33}Pi but in which organic P has not yet been labeled (this applies to the first days after labeling when no fresh organic matter has been introduced). The release of ^{31}Pi induced by the mineralization of organic P will dilute the labeled exchangeable Pi, resulting in an increase in isotopically exchangeable Pi [$E'_{(t)}$]. This increase will be stronger than the increase over time caused solely by physicochemical reactions [$E_{(t)}$]. The difference between these two values [$E'_{(t)} - E_{(t)}$] will deliver the amount of Pi that has been released from soil organic matter through mineralization within a given time t.

The $E'_{(t)}$ value is measured in a biologically active soil labeled with ^{32}Pi or ^{33}Pi during an incubation experiment, whereas the $E_{(t)}$ value reflecting the physicochemical reactions can be estimated from the results of a short-term isotopic exchange kinetics experiment extrapolated to an exchange time equivalent to the

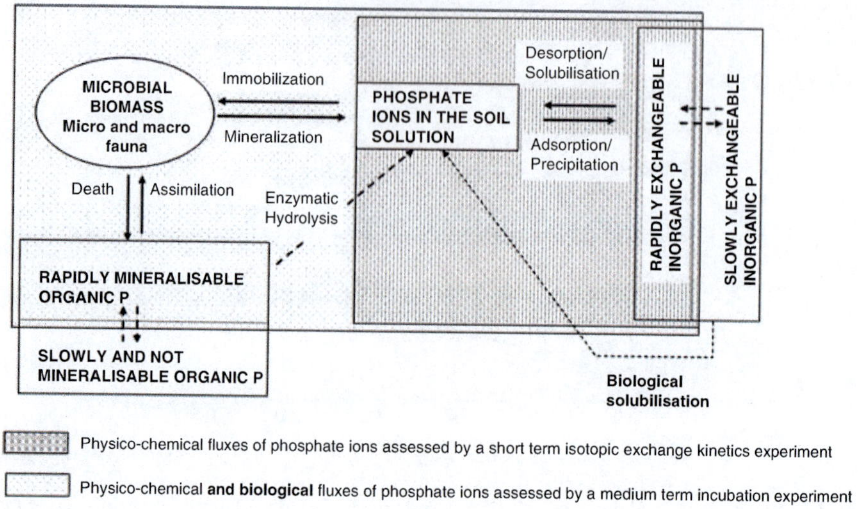

Fig. 3.3 Organic P mineralization is calculated by subtracting the amount of isotopically exchangeable Pi related solely to physicochemical reactions obtained from a short-term batch experiment and extrapolated for a period of, e.g., 14 days, from the amount of isotopically exchangeable Pi measured after a 14-day incubation experiment in which physicochemical and biological reactions were allowed to take place (adapted from Achat et al. 2009b)

duration of the incubation experiment (López-Hernández et al. 1998). Although the principle is simple, its application is more delicate because it requires that all compartments connected to the Pi in the soil solution have a constant isotopic composition. Oehl et al. (2001a) and Bünemann et al. (2007) evaluated each of the steps involved in this approach and made recommendations for the measurement of soil organic P mineralization that should be followed in order to obtain meaningful results. The experiment should be done under steady-state equilibrium for Pi and carbon, i.e., in the absence of rapidly degradable organic matter, at constant and low respiration rate and at constant C_P. They recommend restricting the incubation to 14 days to avoid re-mineralization of organic P compounds that would have been labeled. For each soil, the agreement of measured and extrapolated $E_{(t)}$ values must be checked against the sterilized soil in a batch experiment and in an incubation experiment. The gross P mineralization rate is to be determined by comparing the $E_{(t)}$ values extrapolated from a short-term batch experiment to the $E'_{(t)}$ values measured in a 7–14-day incubation experiment using pre-incubated non-sterile soil. The amount of P immobilized in the microbial biomass during the incubation must then be determined from the amount of microbially held radiolabeled P and ^{31}P, so as to calculate the net P mineralization (gross P mineralization − microbial immobilization). Finally, the detection limit for P mineralization has to be determined for each soil. The work of Bünemann et al. (2007) yielded gross mineralization rates of 0.9–1.2 mg P kg^{-1} soil day^{-1} and net mineralization rates of 0.5–0.9 mg P kg^{-1} soil day^{-1}. The gross mineralization rate measured over 24 h in this study was equivalent to 8–11% of the amount of Pi exchangeable over 24 h [$E_{(24\ h)}$]. Oehl et al. (2004) studied gross mineralization rates of organic P as affected by different cropping systems in a long-term field experiment and showed that gross mineralization reached rates that were equivalent to 5–9% of the amount of Pi exchangeable within 24 h. Although these rates might seem low, they might contribute substantially to plant nutrition in low-input agro-ecosystems or natural ecosystems.

Comparison of the specific activity of soil solution Pi and P held in the microbial biomass was also used to study the kinetics of microbial P uptake and cycling at either constant or changing size of the microbial biomass (Oehl et al. 2001b). This application revealed rapid microbial P turnover that was affected by the cropping system (see Oberson et al. 2011). The addition of glucose and nitrogen resulted in almost complete re-mineralization of microbial P after 70 days of incubation.

Achat et al. (2009b) further adapted the approach proposed by Bünemann et al. (2007) to a sandy forest soil with a very low total P content (31 mg P kg^{-1}), of which 77% was in "dead" soil organic matter, 17% in the microbial biomass, and 6% as inorganic P. The authors measured phosphate fluxes during a 154-day incubation experiment. They estimated the amount of desorbable Pi, the gross mineralization of microbial P, and the gross mineralization of P in dead soil organic matter. The gross mineralization of total organic P (defined by Achat et al. as the sum of P in dead soil organic matter and in the soil microorganisms) was calculated using the difference $E'_{(t)} - E_{(t)}$. The mineralization of microbial P was calculated from the P uptake in the microbial biomass (microbial ^{33}P and P) and from the net decrease in microbial P. The gross phosphate release from dead soil organic matter was calculated as the difference between the gross mineralization of total organic P minus the mineralization of microbial P, or was estimated from the gross carbon mineralization. Using this approach, Achat et al. (2009b) concluded that the gross mineralization rate of total organic P and the increase in Pi concentration in the soil solution were essentially related to the mineralization of microbial P while the release of P from dead soil organic matter and from mineral surfaces remained low. This is to our knowledge the first time that these different fluxes have been quantified in the same experiment. This work shows also that although long-term incubations are difficult to conduct, they can be interesting if one wants to homogeneously label the soil microbial biomass (Achat et al. 2010). This should allow measuring the rate of organic P mineralization from dead organic matter over longer periods (i.e., going beyond the period of 7–14 days proposed above), which would be more realistic for the study of P uptake by plants in agro-ecosystems.

Similar research (assessment of soil organic P and microbial P mineralization and immobilization rates) needs to be done in soils with a high sorbing capacity for Pi since they often have a low to very low P availability. Such analyses could also be considered in order to monitor soil quality, e.g., for soils under different cropping systems or for soils that have received various rates and types of organic or inorganic pollutants.

3.2.3 *Isotopic Dilution of Phosphate Ions in Soil–Solution Systems: Limits and Answers*

Important considerations for correctly interpreting data from isotopic exchange kinetic experiments have been covered in earlier publications (Atkins 1969; Fardeau 1981; Cobelli et al. 2000). These points are the following: (a) the introduction of the tracer (^{32}Pi or ^{33}Pi) should not modify the mass balance of the tracee (^{31}Pi) (i.e., the tracer should add no mass, therefore it should be added carrier-free); (b) the tracer and the tracee must have the same behavior in the soil–solution system, which means that the tracer must be added in the chemical form of the tracee to be studied (here as Pi); (c) the tracer and the tracee must be measured in the same, accessible, compartment (e.g., the soil solution); and (d) the soil–solution system must be at steady-state equilibrium for the tracee when the tracer is introduced. In Sect. 3.2.3.1 and 3.2.3.2, we discuss several potential problems of isotopic dilution techniques and show solutions.

3.2.3.1 Measuring Very Low Concentrations of Phosphate Ions in Solution

Measuring isotopically exchangeable phosphate is extremely difficult in soils that contain very low concentrations of Pi in the soil solution (Salcedo et al. 1991; Hamon and McLaughlin 2002; Bühler et al. 2003; Maertens et al. 2004). This case is often met in highly weathered tropical soils where the very low concentration of available Pi strongly limits plant growth. Different strategies have been proposed to measure these very low concentrations. Phosphate can be concentrated by evaporating the aqueous extract (Bühler et al. 2003), by adding into the aqueous extract a resin that is then eluted in a smaller volume (Salcedo et al. 1991; Hamon and McLaughlin 2002), or by adding a molybdate solution in a high volume of soil water extract and concentrating the phosphomolybdate complex in isobutanol (Pons and Guthrie 1946; Jayachandran et al. 1992a). Two of us (Lalajaona Randriamanantsoa and Christian Morel) recently adapted a standard method (NF EN 1189: Qualité de l'eau. Dosage du phosphore, AFNOR) used to measure low concentrations of Pi in environmental waters for measuring low Pi concentrations in soil water extracts. The method consists in developing the blue phosphomolybdate complex in a high volume of soil extract and subsequently concentrating it in hexanol. When combined with a measurement in a 10-cm cell, this last approach yields a detection limit of 0.3 μg P L^{-1} and a quantification limit of 0.8 μg P L^{-1}, which are lower than those previously reported, e.g., by Bühler et al. (2003). This method needs, however, to be miniaturized because it currently needs large volumes of soil extracts. Another approach to solve this problem is to extract the labeled soil–solution system with a HCO_3^--saturated resin as proposed by Maertens et al. (2004). Because Schneider and Morel (2000) reported that the difference in isotopically exchangeable phosphate measured before and after soil extraction with a HCO_3^--saturated resin was identical to the amount of Pi extracted by the resin, we can assume that the Pi extracted by the HCO_3^--saturated resin and in the soil solution belong to the same compartment. But, before generalizing, this remains to be tested on other soils using isotopic tracers.

3.2.3.2 Are P Radioisotopes Irreversibly Fixed in Soils That Sorb Very High Amounts of Phosphate Ions?

Wolf et al. (1986) suggested that an unknown but significant fraction of the added ^{32}Pi would be irreversibly fixed in soils that sorb very high amounts of Pi, thereby reducing the fraction of radioactivity really participating in the isotopic exchange. They concluded that isotopic dilution would not be a useful tool for assessing phosphate availability in these soils. In other words, they assumed that ^{31}Pi and ^{32}Pi would have a different behavior in the soil–solution system, ^{32}Pi being preferentially sorbed on soil. To our knowledge there is no published information on the sorption of the different P isotopes on soil particles.

We checked the hypothesis that no significant isotopic discrimination would occur between ^{32}Pi and ^{33}Pi during isotopic exchange kinetic experiments in the presence of a strong adsorbant for Pi. This hypothesis sounds reasonable because at a pH close to 5 (e.g., typical for Ferralsols) the mass difference between 1 mol of $H_2{}^{32}PO_4{}^-$ and 1 mol of $H_2{}^{33}PO_4{}^-$ would be 1/98 g. If this hypothesis can be confirmed, then we will deduce that there will be little isotopic discrimination between ^{31}Pi and ^{32}Pi and between ^{31}Pi and ^{33}Pi.

A concentration of 0.25 mM of KH_2PO_4 was added in 100 mL of H_2O or 0.05 M $NaNO_3$ that was bathing 1 g of synthetic goethite prepared according to the protocol of Schwertmann and Cornell (2000). The pH of the solutions was adjusted to 4.5 at the beginning of the experiment. The phosphate–goethite suspensions were shaken overnight. The following day, carrier-free ^{32}Pi and ^{33}Pi were added simultaneously to the suspensions and isotopic exchange kinetics experiments were performed. The experiment was conducted in four replicates. The concentration of Pi in the solution measured after concentrating the phosphomolybdate complex in hexanol was 1.2 μg P L^{-1} in water and 1.7 μg P L^{-1} in $NaNO_3$ (Table 3.2). Figure 3.4 shows that the proportion of radioactivity remaining in the water–phosphate–goethite system decreased slightly more rapidly with ^{33}Pi than with ^{32}Pi, whereas almost identical kinetics were observed in the presence of $NaNO_3$. The $r_{(1)}/R$ parameter was lower with ^{33}Pi than ^{32}Pi in water, whereas the n values were identical. In $NaNO_3$ the n value was lower with ^{32}Pi, whereas the $r_{(1)}/R$ values were identical for both radioisotopes. The $E_{(t)}$ values varied accordingly (Table 3.2). The differences observed in $r_{(1)}/R$ and n could be ascribed to a slight preferential adsorption of the heavier isotope on the goethite. These differences could also be due to the presence of small quantities of ^{32}P-labeled pyro- or polyphosphates in the ^{32}Pi solution, which would not exchange at the same rate as Pi (McBeath et al. 2009). Such chemical contamination has been observed earlier with ^{32}P that had been produced by neutron activation performed at too-high temperatures (Jean-Claude Fardeau, personal communication). Altogether, these values remain very similar. Therefore, although a slight discrimination between the P radioisotopes can occur, especially

Table 3.2 Parameters of the isotopic exchange kinetic experiment [$r_{(1)}/R$ and n], the concentration of Pi (C_P) in the solution measured after pre-concentrating the phosphomolybdate complex in hexanol, and the amount of Pi isotopically exchangeable within 1 min and 1 day [$E_{(1\ min)}$, $E_{(1\ day)}$] calculated for water–phosphate–goethite suspensions and $NaNO_3$–phosphate–goethite suspensions

Suspension	$r_{(1)}/R$	n	C_P (μg P L^{-1})	$E_{(1\ min)}$ (mg P kg^{-1})	$E_{(1\ day)}$ (mg P kg^{-1})
^{32}Pi–water	0.01 ± 0.0002	0.30 ± 0.06	1.2 ± 0.01	10.6 ± 0.9	82.4 ± 9.7
^{33}Pi–water	0.008 ± 0.0001	0.30 ± 0.02	1.2 ± 0.01	12.8 ± 1.2	103 ± 22
^{32}Pi–$NaNO_3$	0.007 ± 0.00004	0.36 ± 0.03	1.7 ± 0.02	25.9 ± 0.6	321 ± 69.9
^{33}Pi–$NaNO_3$	0.007 ± 0.00007	0.40 ± 0.01	1.7 ± 0.02	23.4 ± 0.1	426 ± 38.4

The water–phosphate–goethite suspensions (^{32}Pi–water and ^{33}Pi–water) and $NaNO_3$–phosphate–goethite suspensions (^{32}Pi–$NaNO_3$ and ^{33}Pi–$NaNO_3$) contained 7.75 mg P g^{-1} goethite simultaneously spiked with ^{32}Pi and ^{33}Pi. Means of four replicates are given with their standard deviation. The parameters $r_{(1)}/R$ and n were calculated by fitting the decrease of radioactivity with time shown in Fig. 3.4 with (3.1)

when the concentration of Pi in the solution is very low, we conclude that this is not the most important issue to be solved when conducting isotope exchange kinetics studies in soils that contain low concentrations of Pi in the soil solution. The most important problem is the determination of these low P concentrations (see Sect. 3.2.3.1).

3.2.4 *Application of Isotopic Dilution Techniques to Soil–Plant Systems*

The above-presented principles can be applied to soil–plant systems. In this approach (Larsen 1952), plants are grown on a soil in which exchangeable Pi have been labeled with radioactive Pi. If the fractions of stable and radioactive P taken up by the plants remain small compared to the amounts of stable and radioactive P present in the soil, then the measurement of the specific activity of P in the aerial parts of the plant is equal to the specific activity of the pool of soil exchangeable P in which the added radioactivity has been diluted (3.6), which is called the L (L as in Larsen) value:

$$\mathrm{SA_{plant}} = \mathrm{SA_{soil}} = R/L \tag{3.6}$$

where $\mathrm{SA_{plant}}$ is the ratio between the amount of radioactive P recovered in the plant (in Bq kg^{-1} soil) and the amount of ^{31}P taken up by the plant coming from the soil (in mg P kg^{-1} soil), and $\mathrm{SA_{soil}}$ is the ratio between the total amount of radioactive P (R) added as carrier-free labeled Pi to the soil (in Bq kg^{-1} soil) and L, the amount of Pi that has undergone isotopic exchange within the duration of plant growth (in mg P kg^{-1} soil).

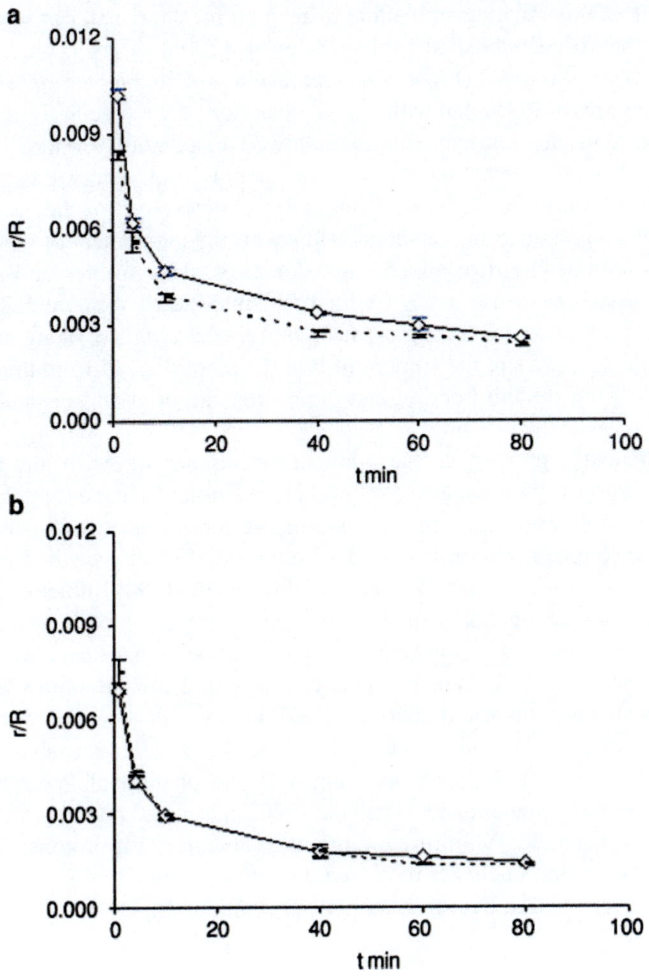

Fig. 3.4 Changes in the proportion of ^{32}P (*continuous line* and *diamonds*) and ^{33}P (*dashed line*) remaining in the solution (r/R) with time (t, expressed in minutes) in (**a**) water–phosphate–goethite suspension and (**b**) $NaNO_3$–phosphate–goethite suspension. *Standard deviations* are shown for each point. Both suspensions had an initial concentration of Pi of 0.25 mmol, were shaken overnight, and were then simultaneously spiked with ^{32}Pi and ^{33}Pi. The $r_{(1)}/R$ and n parameters derived from these curves are shown in Table 3.2

A major problem of this method is to distinguish between the amount of P in the aerial plant parts taken up from the soil or the soil–fertilizer mixture and the amount of P derived from the seed or the planting material (Russell et al. 1957; Truong and Pichot 1976; Brookes 1982). Because seed P or P in the planting material is not labeled, a high transfer of seed P to the aerial plant parts results in an L value that is overestimated. This problem is particularly acute when plants with large seeds (maize, legumes) are grown in soils that contain very little available phosphate

(Pypers et al. 2006). An elegant method to account for the transfer of seed P to the aerial parts has been published by Pypers et al. (2006). They achieved this by growing maize and cowpea (*Vigna unguiculata*) in sand in the presence of increasing concentrations of Pi labeled with ^{32}P. Comparison of the specific activity of P in the plant and in the nutrient solution allowed calculation of the amount of P transferred from the seed to the shoot for any plant P level. Afterwards, regression equations can be used to derive the amount of P derived from the seed for any amount of P accumulated in the shoots. However, the parameters of these regression equations must be determined anew for each plant species and genotype. Another approach is to use a plant with very small seeds (i.e., containing very little P) that can be cut several times, such as *Agrostis tenuis* (Truong and Pichot 1976). With this approach, the amount of P in the plant derived from the seed will be negligible from the third cut on and the results can be used to calculate the L value.

For most plants, growing in the presence or absence of AMF, the measured amount of isotopically exchangeable phosphate is similar to that extrapolated from short-term isotopic exchange kinetic experiments conducted in soil–solution systems for an exchange time equivalent to the duration of plant growth (Fardeau and Jappe 1976; Frossard et al. 1994; Morel and Plenchette 1994; Bühler et al. 2003). This confirms that isotopically exchangeable phosphate is the main source of P for these plants. For other plants, however, L values can be higher than E values. This has been observed, e.g., for rape by Hedley et al. (1982), white lupin (Braum and Helmke 1995), and cowpea (Pypers et al. 2006) in soils containing very little available phosphate. This points to the fact that these plants are able to access P pools other than the isotopically exchangeable phosphate pool. A large body of work reviewed by Neumann and Martinoia (2002) has since confirmed that white lupin, when grown under P-limiting conditions, is able to exude protons, citric acid, and phosphatase from its cluster roots and, therefore, is able to dissolve insoluble calcium phosphate and to hydrolyze organic P compounds.

3.3 Use of P Radioisotopes to Trace the Fate of P Sources in the Soil–Plant System

3.3.1 Labeling the P Source (Direct Labeling Technique)

Labeled P fertilizers have been used in agriculture for over 60 years (Dean et al. 1947; Nelson et al. 1947; Spinks and Barber 1947). At the time, these experiments were called the "Green Cheese Experiments," showing that they were considered with some skepticism. It is interesting to see how famous the majority of these "first" experimenters became!

Phosphate transfer from mineral or organic sources (e.g., mineral P fertilizer, plant residues, microbial bodies, organic compounds) to a plant can easily be studied when the source of P is homogenously labeled with radioactive P. Measuring the amount of radioactive and stable P in a plant allows, knowing the specific activity of the source of P, the calculation of the amount of P in the plant that is derived from the fertilizer. The proportion of P in the plant that is derived from a P fertilizer (%Pdff) can be calculated with (3.7) when using the direct labeling technique:

$$\%Pdff = 100(SA_{+P}/SA_f) \tag{3.7}$$

where %Pdff is the percentage of P in the plant that is derived from the labeled fertilizer, SA_{+P} is the specific activity of P in the aerial parts of the plant that has received the P fertilizer and SA_f the specific activity of P in the fertilizer.

This approach has been used many times. We mention here only a few examples: the mineralization of ^{32}P-labeled organic P compounds and their subsequent use by plants or transfer to other soil pools (Martin and Cartwright 1971; Harrison 1982; Kapoor and Haider 1982; Jayachandran et al. 1992b), the transfer of P added as plant residues to plants and/or to the soil microbial biomass (McLaughlin and Alston 1986; Friesen and Blair 1988; McLaughlin et al. 1988; Thibaud et al. 1988; Armstrong and Helyar 1993; Bünemann et al. 2004), and the transfer of P from ^{32}P-labeled inorganic soluble or insoluble P sources to plants (Boniface et al. 1979; McLaughlin and Alston 1986; Armstrong and Helyar 1993; McBeath et al. 2009). The main limit of this approach is the need for the source to be homogeneously labeled with P. There are, however, many fertilizers such as farmyard manure, compost, and rock phosphate that cannot be readily homogeneously labeled with radioactive P isotopes. In order to do so, it would be necessary to bombard them with neutrons (Kucey and Bole 1984), but this would lead to the production of other radioisotopes (^{24}Na, ^{45}Ca, among others) making the product difficult to handle and to changes in the chemical properties of the substrate. Besides, especially because of the short half-lives of ^{33}P and ^{32}P, this approach is limited to experiments lasting a few months and cannot be used, e.g., to quantify the residual value of P fertilizers for crops. For these types of experiments it is necessary to use the indirect labeling approach, which is described in the next section.

3.3.2 Labeling the Plant-Available Soil Phosphate (Indirect Labeling Technique)

This approach is based on the isotope dilution technique presented earlier in this chapter (Sect. 3.2.1). The release of P from a fertilizer to a plant or to any soil P pool can be studied by introducing a non-labeled fertilizer into a soil in which the isotopically exchangeable phosphate has been labeled (Kucey and Bole 1984; Morel and Fardeau 1989a). The proportion of P in the plant that is derived from a P fertilizer can be calculated with (3.8) when using the indirect labeling technique:

$$\%Pdff = 100(1 - SA_{+P}/SA_{0P}) \tag{3.8}$$

where %Pdff is the percentage of P in the plant that is derived from the fertilizer, SA_{+P} is the specific activity of the plant grown in the presence of the P fertilizer, and SA_{0P} is the specific activity of the plant grown in the absence of the P fertilizer.

This approach has been used, e.g., to assess the transfer of P added as rock phosphate, sewage sludge, compost, and animal manure to plants (Bolan et al. 1987; Hedley et al. 1988; Fardeau et al. 1988; Zapata and Axmann 1995; Kato et al. 1995; Frossard et al. 1996; Sinaj et al. 2002; Oberson et al. 2010). The indirect labeling approach has also been used for calculating the residual value of P fertilizers by comparing a control that was not fertilized to soils that were regularly fertilized (Bowman et al. 1978; Gallet et al. 2003b; Morel and Fardeau 1989b; Oberson et al. 2010).

This indirect approach supposes that the specific activity of the soil exchangeable phosphate is identical in the non-amended and in the amended soil. This basic hypothesis is very difficult to verify. Precautions must be taken when using this indirect approach, especially when the radioactive P is added to the soil at the same time as organic fertilizers. Indeed, these organic fertilizers could cause a substantial microbial immobilization of the radioactive P or could stimulate the mineralization of non-labeled soil organic P, both processes leading to a decrease in the specific activity in the soil solution of the fertilized treatment and therefore to an overestimation of the use of the P derived from the organic fertilizer by the test crop. This phenomenon, which is called "pool substitution," has been recognized for a long time by researchers working on nitrogen (Jenkinson et al. 1985; Hood-Nowotny 2008) but has not yet been addressed explicitly in P studies. Hood-Nowotny (2008) suggests circumventing this problem for nitrogen studies by labeling the soil well in advance with a ^{15}N mineral fertilizer combined with a carbon addition so as to incorporate as rapidly as possible the ^{15}N in the soil microbial biomass, and to add the non-labeled fertilizer only when the ^{15}N excess in the soil solution has reached equilibrium, i.e., several weeks after labeling. The major difference between N and P in this respect is that N dynamics are largely determined by microbial processes, whereas P dynamics are dominated by physicochemical processes in most soils. Oberson et al. (2010) suggest minimizing pool substitution for P by incubating the soil labeled with radioactive P under optimal temperature and humidity conditions for 1–2 weeks before adding the organic P source. Of course, it is very important when using this approach to properly account for the transfer of P from seeds to the aerial plant parts during plant growth (see Sect. 3.2.4). Finally, in principle, this indirect approach can also be used to derive the %Pdff of a non-labeled P source in soil P pools that have been previously labeled with radioactive P.

3.4 Using P Radioisotopes to Assess Foraging Strategies of AMF

Phosphorus radioisotopes have been used for several decades for measuring the acquisition of soil P by mycorrhizal fungi and comparing it to that of roots (Mosse et al. 1973; Powell 1975; Jakobsen et al. 1992). Early studies comparing the specific

P activities in plants colonized by AMF and in non-mycorrhizal plants growing in labeled (and uncompartmented) soils indicated usage of the same P pool by the mycorrhizal and non-mycorrhizal plants (Mosse et al. 1973; Blal et al. 1990; Bolan 1991), in contrast to the ectomycorrhizal and ericoid mycorrhizal associations (Read et al. 2004). Although a few studies scrutinizing the acquisition of P from organic sources by the AMF provided some evidence for differences in the utilization of certain organic P forms between plants colonized by AMF and non-mycorrhizal plants (Jayachandran et al. 1992b; Joner and Jakobsen 1995), this seems to be related to the presence of other soil microbes rather than to the P-mineralization capacity of the AMF themselves (Joner and Jakobsen 1995; Joner et al. 2000; Jansa et al. 2011).

Labeling of soil patches (compartments) inaccessible to roots but accessible to mycorrhizal hyphae through fine mesh-walls (20–35 μm) further contributed to understanding the kinetics of P uptake and the transfer via a mycorrhizal pathway to the plants (Fig. 3.5). Different studies employed systems differing in the construction and mode of labeling. The pros and cons of these different approaches for expanding the knowledge of AMF physiology and functional diversity and their role in plant P nutrition, as well as for studying remaining unanswered questions, are briefly outlined below.

With respect to the construction of the experimental system, important differences can be seen between the studies in terms of (a) the presence of a buffer zone between the root and labeling compartments, (b) the size of the root-free zone as compared to the root zone, and (c) the positioning of the labeling compartment (below or alongside of the root compartment). Large root-free compartments (Jansa et al. 2003; Mikkelsen et al. 2008) allow tracking of P transfer via mycorrhizal hyphae over large and defined distances, but have been criticized for possible bias due to unrealistic constraints imposed on the root development and exaggerated contrasts between AMF and non-mycorrhizal treatments. These studies showed that different AMF species acquire P at distances of several millimeters up to 15 cm from the roots (Jakobsen et al. 1992; Jansa et al. 2003; Smith et al. 2004). Small root-free compartments (vials with mesh lids) containing labeled soil and buried in pots or in the fields (Schweiger and Jakobsen 1999; Smith et al. 2004) are claimed to provide a more realistic comparison of the non-mycorrhizal and mycorrhizal treatments, but are impractical for kinetic studies in which isotopes are either administered by injection or placed at different distances from the roots. These setups do not usually include a buffer zone, which results in some radioactive P transfer to non-mycorrhizal plants due to root hair exploration of the root-free zone (Schweiger et al. 1999).

Two modes of P labeling have been employed – either mixing the isotope with the soil or injecting it into the soil or soilless substrate (Nielsen et al. 2002; Wang et al. 2002; Smith et al. 2004; Jansa et al. 2005). Whereas greater homogeneity of label distribution can be achieved with the first approach, the second (injection) approach allows measuring the efficiency of P uptake by an already established hyphae network (Jansa et al. 2005).

Values between 0% and 30% of the added P radioisotope have been recorded in the literature as proportions of the transported radioactivity via mycorrhizal fungal

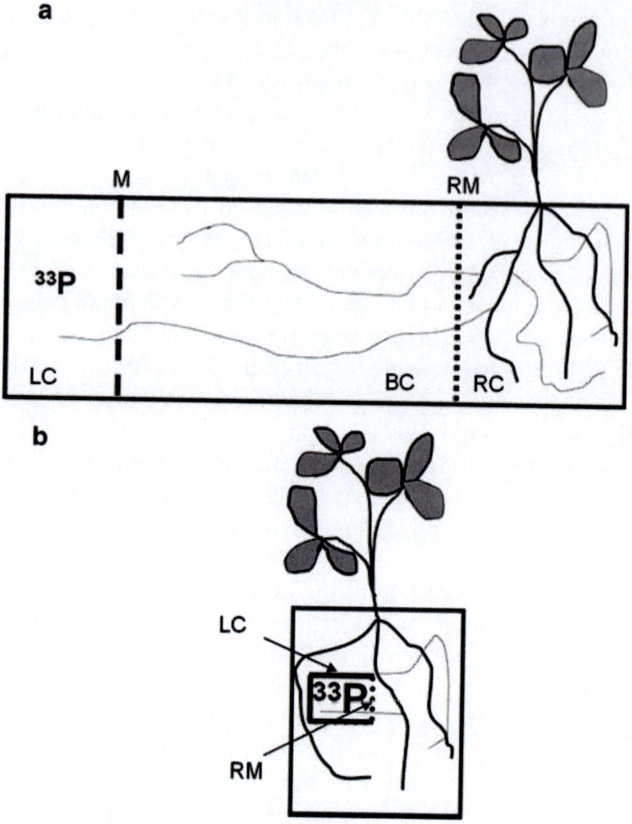

Fig. 3.5 Construction of compartmented cultivation systems for studying acquisition of P via AMF hyphae. The cuvette system (**a**) containing a root compartment (*RC*), and large buffer (*BC*) and labeling (*LC*) compartments (both could vary in size) is suitable for studying mycorrhizal P transfer over a defined distance and for administration of the radioisotope-labeled P by injection. The pot system (**b**) containing a small hyphal compartment, usually without a buffer zone, is claimed to offer less biased determination of the importance of the mycorrhizal acquisition pathway in plant P acquisition from the soil (adapted from Nagy et al. 2005). *RM* root exclusion mesh (20–35 μm), *M* mesh wall (500 μm), *fine lines* AMF hyphae

mycelium from the root-free soil to the roots (Jakobsen et al. 1992; Smith et al. 2000, 2004; Jansa et al. 2003, 2005). Absolute numbers depend on the plant and fungal identities, level of fungal development, substrate properties, time between isotope introduction and harvest, and other features of the experimental system. It has, however, been rather difficult to use this information for quantification of the contribution of the fungus to plant P uptake, although attempts to do this have been made (Smith et al. 2004; Jakobsen et al. 2005). Two major obstacles hinder progress along these lines: (a) quantification of active AMF mycelium has so far only been made by measuring hyphal length density using staining and microscopy, which may not be appropriate due to demonstrated difficulties in distinguishing living and dead hyphal stretches (Gamper et al. 2008), and (b) the specific activity of the soil phosphate pool available to plants and fungi has so far only been assessed in bicarbonate extractions of substrate samples at a single time point (Smith et al. 2004). This could be improved by applying the principles of isotope dilution outlined above, and by quantifying the gene expression of the different P transporters, both in the plant roots and in the AMF hyphae (Maldonado-Mendoza et al. 2001; Grace et al. 2009).

In the future, more attention should be paid to performing functional studies with monoxenic cultures of different AMF, especially those including the whole plant as a host (Voets et al. 2009), to look at realistic source–sink relationships. Little use has been made of the availability of the two P radioisotopes, which could be concomitantly applied to track different P-acquisition pathways in the same system (Cavagnaro et al. 2005). Direct labeling of different organic compounds in combination with monoxenic cultures will provide greater precision in measuring the accessibility of the different organic P forms to the AMF. Isotope labeling studies could in the near future also be coupled to the quantification of expression of fungal gene transporters in a similar manner as done previously for the plant P transporters involved in assimilation of P delivered by the AMF (Nagy et al. 2005).

Finally, this niche labeling can also be used in experiments to study P uptake by the root system, e.g., of specific plant species in mixed stands (with intercropped plants), or to study P uptake from deep soil horizons, as done by Göransson et al. (2006).

3.5 Can the Isotopic Composition of Oxygen Bound to Phosphorus Be Used to Study Biological P Transformations in Soil–Plant Systems?

3.5.1 What Do We Know Already?

Oxygen has three stable isotopes ^{16}O, ^{17}O, and ^{18}O, which are present at 99.759%, 0.037%, and 0.204%, respectively, in the earth's atmosphere. The two most abundant oxygen isotopes ^{16}O and ^{18}O are used in phosphorus studies (Longinelli 1965). The natural abundance of ^{18}O bound to P ($\delta^{18}O_{\mathbf{P\text{-}sample}}$, expressed in parts per thousand, ‰) is calculated according to (3.9):

$$\delta^{18}O_{P-\mathrm{sample}} = 1{,}000\left\{\left[\left({}^{18}O_P/{}^{16}O_P\right)_{\mathrm{sample}}/\left({}^{18}O/{}^{16}O\right)_{\mathrm{reference}}\right] - 1\right\} \qquad (3.9)$$

where $(^{18}O_P/^{16}O_P)_{\mathbf{sample}}$ represents the ratio of ^{18}O to ^{16}O in phosphate in a given sample, and $(^{18}O/^{16}O)_{\mathbf{reference}}$ the ^{18}O to ^{16}O ratio in a reference sample. For oxygen isotopes in phosphate, the reference material used is the Vienna Standard Mean Ocean Water (VSMOW). Several papers have dealt with the oxygen isotope geochemistry of phosphate in adsorption–desorption processes, mineral precipitation, and biological processes, e.g., uptake, intra- and extracellular enzymatic catalysis (Liang and Blake 2007; Blake et al. 2005; Jaisi et al. 2010). In the rest of this section we give a short account of the main findings of these studies, and then in Sect. 3.5.2 we discuss future research to be done to use this tracer to study biological P transformations in soil–plant systems.

In the absence of biological activity and at ambient temperature, isotope exchange between the oxygen present in Pi and water is slow and negligible (Winter et al. 1940; Longinelli 1965; Kolodny et al. 1983; Luz and Kolodny 1985). Studies on precipitation of authigenic P-bearing minerals showed that fractionation between ^{16}O-phosphate and ^{18}O-phosphate linked to this process is small, in the order of about 1‰ (Liang and Blake 2007). Also, sorption of phosphate to Fe oxides does not lead to a significant isotopic fractionation between the sorbed phosphate and the phosphate in solution (Jaisi et al. 2010).

When biological activity is present, however, oxygen exchange between phosphate and water becomes significant and is characterized by both kinetic and equilibrium fractionations, but is dominated by equilibrium isotope fractionation effects (Kolodny et al. 1983; Paytan et al. 2002; Blake et al. 2005). Rapid microbial turnover of phosphate, controlled by reversible intracellular processes such as pyrophosphatase-catalyzed hydrolysis (Blake et al. 2005), is responsible for a temperature-dependent equilibration with ambient water. Such rapid biologically mediated turnover has been demonstrated in natural aquatic systems (Paytan et al. 2002; Colman et al. 2005). The enzyme pyrophosphatase catalyzes the isotopic exchange of oxygen between water and phosphate within a few hours at high enzyme concentrations in laboratory experiments (Blake et al. 2005), and completely overprints the initial $\delta^{18}O_P$ of Pi taken up by microorganisms (Blake et al. 1998). The equation relating the equilibrium between $\delta^{18}O_P$, $\delta^{18}O$ water ($\delta^{18}O_W$), and temperature (Longinelli and Nuti 1973) is shown below:

$$T(°C) = 111.4 - 4.3\left(\delta^{18}O_P - \delta^{18}O_W\right) \tag{3.10}$$

In contrast to intracellular processes, extracellular enzymatic hydrolyses are generally irreversible. Phosphodiesterase catalytic activity results in incorporation of one oxygen atom from ambient water into the phosphate group of a diester P (R–O–PO_2^-–O–R') to produce a monoester P (R–O–PO_3^{2-}). Phosphomonoesterase activity in turn will result in the incorporation of a second oxygen atom from ambient water into the phosphate group released from a phosphomonoester (Blake et al. 2005; Liang and Blake 2009). Furthermore, the effect of the phosphodiesterases on the $\delta^{18}O_P$ of monoester P may depend on the substrate, because ^{16}O is preferentially incorporated into nucleotides released from DNA, whereas ^{18}O is preferentially incorporated into nucleotides released from RNA (Liang and Blake 2009). This substrate and enzyme specificity might be useful to distinguish between different pathways of organic P metabolism in the soil–plant system. The original $\delta^{18}O_P$ signature of organic P compounds might therefore be partially retained when enzymatic catalysis occurs in the extracellular medium and when P is not entirely recycled by microorganisms, i.e., when P is present in high concentrations or when

P is not the primary limiting factor for microbial growth. Finally, studies on *Escherichia coli* showed that this organism preferentially takes up ^{16}O-phosphate into its cells (Blake et al. 2005).

These findings show that the final $\delta^{18}O_P$ signature of Pi derived from monoester P and diester P and of microbially cycled P is a function of the $\delta^{18}O$ of water, the temperature, the speciation and initial $\delta^{18}O_P$ of the phosphates coming into the ecosystem, the abundance of available P, and the biological processes that take place within the ecosystem itself (Blake et al. 2001). Under P-limited conditions, inducing a complete turnover of phosphate (e.g., oligotrophic ocean surface waters), $\delta^{18}O_P$ can be used to track biological processes (e.g., alkaline phosphatase hydrolysis; Paytan et al. 2002; Colman et al. 2005). Under excess P conditions (coastal waters, estuaries, and continental water bodies) $\delta^{18}O_P$ can be used to characterize P sources (McLaughlin et al. 2006a, b; Elsbury et al. 2009). In fact, incomplete turnover of the phosphate pool may result in partial inheritance of the isotope signature of the original P source, which can be tracked in the environment. Elsbury et al. (2009) found $\delta^{18}O_P$ values of dissolved Pi in Lake Erie varying between 10‰ and 17‰, i.e., that were out of the predicted equilibrium with ambient conditions, which would vary between 13‰ and 15‰. By comparing $\delta^{18}O_P$ values of dissolved Pi in lake water to $\delta^{18}O_P$ values in its tributaries, the authors suggest that the rivers supply Pi with a light isotope composition, whereas Pi released from the sediments in the bottom waters following anoxic events have a heavy $\delta^{18}O_P$. Young et al. (2009) showed that the $\delta^{18}O_P$ signature varies widely in materials that could act as sources of P for water. The analyses conducted by Young et al. (2009) yielded a mean $\delta^{18}O_P$ for effluents from wastewater treatment plants of 11‰, a mean $\delta^{18}O_P$ for mineral P fertilizers of 20‰, and a mean $\delta^{18}O_P$ for vegetation and detergents of 17‰.

To date, very few studies have attempted to understand $\delta^{18}O_P$ variations in soils (Ayliffe et al. 1992; Mizota et al. 1992; McLaughlin et al. 2006a). All conclude that anomalous $\delta^{18}O_P$ values in soils compared to parental material might be attributed to biological activity. Because the pioneering study conducted by Mizota et al. (1992) specifically used $\delta^{18}O_P$ to understand soil P dynamics as affected by pedogenesis, we summarize their findings hereafter. Mizota et al. (1992) measured the $\delta^{18}O_P$ of P associated to Ca and Al in volcanic soils from Java and East Africa. They compared $\delta^{18}O_P$ values obtained from soils with those of volcanic ash, animal bones, and local phosphate deposits formed at high temperature. Their results showed a trend of increasing $\delta^{18}O_P$ values from young toward more weathered soils. They interpreted the low $\delta^{18}O_P$ of young soils by the presence of P-bearing apatite produced at high temperatures, and attributed the high $\delta^{18}O_P$ of older soils to intense biogenic recycling.

Two other studies have been conducted with ^{18}O-enriched compounds. Kok and Varner (1967) studied the changes in ^{18}O enrichment in water and showed that no O exchange occurred between H_2O and PO_4 in sterile soils, whereas this exchange was strongly accelerated in biologically active soils or in the presence of micro-organisms. Furthermore, using KH_2PO_4 doubly labeled with ^{32}P and ^{18}O, Larsen et al. (1989) showed that ^{18}O disappeared very rapidly from the Pi when KH_2PO_4

was added to a soil on which ryegrass was growing, whereas ^{18}O remained in Pi when KH_2PO_4 was added to a sterile soil–water suspension. Larsen et al. (1989) concluded that the loss of ^{18}O from Pi could be used as a measure of soil biological activity, which is in line with the conclusions drawn by Kok and Varner (1967) for soils and by Blake et al. (2001) for aquatic systems.

3.5.2 What Should Be Done to Apply This Approach to Soil–Plant Systems?

Many points remain to be clarified before $\delta^{18}O_P$ can be confidently used in soils. The first problem is to isolate phosphate from any other compound containing O and to convert it to silver phosphate for oxygen isotope analysis. This is a challenge in soils, where organic matter is often present in high concentrations and distributed in many forms. Wiedemann-Bidlack et al. (2008) showed that the presence of organic matter strongly affects the measurement of $\delta^{18}O_P$ in apatites and suggested measures to eliminate this contamination. Tamburini et al. (2010) recently adapted a method based on the use of multiple mineral precipitations that does not require extreme pH adjustments of the solutions for the production of pure Ag_3PO_4 amenable to $^{18}O_P$ analysis from HCl extracts of soil or fertilizer. They successfully applied this method to soils rich in organic matter and to fertilizers, showing that the $\delta^{18}O_P$ of HCl-extractable P could be related to the source of P in P-rich soils or to biological activity in low-P soils. Because phosphate in soils is distributed in many pools (e.g., adsorbed on Fe and Al oxides, precipitated on Ca minerals, in organic compounds), another challenge for soil studies is to perform $\delta^{18}O_P$ measurements on different P fractions of a sequential extraction, as was done partially for marine sediments (Jaisi and Blake 2010). Moreover, it is necessary to check for possible exchange of oxygen between water and phosphate for each extraction step by introducing ^{18}O-labeled water into the extraction solutions. This step is particularly important for soils because P forms other than phosphate (e.g., condensed phosphate, pyro- and polyphosphate, phosphate mono- and diester, or phosphonate) can be present in the extracts. In fact, inorganic hydrolysis of such compounds could lead to incorporation of oxygen from the aqueous medium into the newly formed phosphate, thus altering the original isotopic signal. Finally, ultraviolet digestion can be used to determine the $\delta^{18}O_P$ of extracted organic P, as shown by Liang and Blake (2006).

Once these extractions and preparation issues are solved, it will be necessary to assess the oxygen isotope fractionation associated with the most important processes controlling P transformations in the soil–plant system. For instance, whereas a lot is known about partial dissolution of P-bearing minerals like apatite (e.g., rates of dissolution, pH dependency of the process, effects of chelating agents and organic acids), only a few studies have explored the oxygen isotope geochemistry of this process (Lécuyer et al. 1999; Blake et al. 1998). Plants have many types of

P transporters (Bucher 2007) and we do not know whether these transporters preferentially take up ^{16}O-enriched phosphate (like *E. coli*, see Blake et al. 2005), enriching the residual phosphate in the soil solution with ^{18}O.

Another effect to be investigated is the effect of evapotranspiration. Because evapotranspiration leads to an ^{18}O-enrichment in the leaf-water (Helliker and Ehleringer 2000), we can assume that this will also affect the $\delta^{18}O_P$ in the leaves. Finally, it is necessary to investigate how microbial uptake and mineralization processes affect the $\delta^{18}O_P$ of soil available P and soil microbial P.

Information on the $\delta^{18}O_P$ of P sources in P-rich soils will help to trace P fluxes at the ecosystem level (Elsbury et al. 2009; McLaughlin et al. 2006a). In P-limited soil systems, where biological P turnover is high, the dependence of $\delta^{18}O_P$ on water $\delta^{18}O$ and equilibration temperature might be used to study and reconstruct temperature and rainfall gradients over time.

In conclusion, whereas the measure of $\delta^{18}O_P$ in different soil P pools could provide information on the biological transformations of P, and maybe also on the transfer of P in terrestrial ecosystems, there is still much to do to render this approach operational.

3.6 Concluding Remarks and Research Needs

This review shows the importance of using tracers to understand both the physico-chemical and biological transformations of P in soil–plant and soil–solution systems, i.e., to understand "Phosphorus in Action"!

Although P radioisotopes have been used for decades, the last decade has seen some significant improvements in these isotopic approaches, e.g., in the measurement of low P concentration in solution and by taking into account the P derived from the seed in plant P nutrition. Thanks to the use of radioisotopes, significant progress has been made in assessing and modeling relationships between Pi in the solution and Pi bound to soil constituents that can be desorbed to replenish the solution with time under gradients of Pi concentration. Progress has also been made in measuring the mineralization of soil organic P and the immobilization of P in the soil microorganisms, as well as in assessing the foraging strategies of AMF. The most challenging research areas remain (a) to extend the modeling efforts so as to be able to predict the rate of soil solution replenishment with Pi as a function of soil properties and management, (b) to extend the measurement of organic P mineralization to soils that contain very little available P and that have a high sorbing capacity for Pi, and (c) to assess the importance of pool substitution in assessing the fate of P from organic exogenous sources in soil–plant systems.

The measurement of the $\delta^{18}O_P$ in different soil and plant pools bears an interesting potential for studying the biological transformations of P in soil–plant systems and in water bodies, for a better understanding of the origin of the P that triggers their eutrophication. But, a lot of research remains to be done before this tool can become operational.

Finally, other tracers might be of interest, especially for studying the long-term impact of mineral and organic P fertilizers in agro-ecosystems. Rare earth elements, heavy metals, and radionuclides are known to be present in variable concentrations in phosphate deposits and in organic fertilizers, and could in the future be used to record the accumulation of P fertilizers in soils and other parts of the environment (Hu et al. 1998; Otero et al. 2005; de Kok and Schnug 2008). An example of this approach is given by Bertrand et al. (2003), who used the Cd:P ratio of mineral fertilizers to show that most of the P found in the upper horizons of Australian calcarosols was derived from mineral P fertilizer applications.

Acknowledgments We gratefully acknowledge the remarks made by the two reviewers who helped to improve this manuscript.

References

Achat DL, Bakker MR, Augusto L, Saur E, Dousseron L, Morel C (2009a) Evaluation of the phosphorus status of P-deficient podzols in temperate pine stands: combining isotopic dilution and extraction methods. Biogeochemistry 92:183–200

Achat DL, Bakker MR, Morel C (2009b) Process-based assessment of P availability in a low P-sorbing forest soil using isotopic dilution methods. Soil Sci Soc Am J 73:2131–2142

Achat DL, Bakker MR, Saur E, Pellerin S, Augusto L, Morel C (2010) Quantifying gross mineralisation of P in dead soil organic matter: testing an isotopic dilution method. Geoderma 158:163–172

Amer F, Mahdi S, Alradi A (1969) Limitations in isotopic measurement of labile phosphate in soils. J Soil Sci 20:91–100

Armstrong RD, Helyar KR (1993) Utilization of labeled mineral and organic phosphorus sources by grasses common to semiarid mulga shrublands. Aust J Soil Res 31:271–283

Atkins GL (1969) Multicompartmental models for biological systems. Methuen, London

Ayliffe LK, Veeh HH, Chivas AR (1992) The oxygen isotopes of phosphate and the origin of island apatite deposits. Earth Planet Sci Lett 108:119–129

Barrow NJ (1974) The slow reactions between soil and anions. 1. Effects of time, temperature and water content of a soil on the decrease in effectiveness of phosphate for plant growth. Soil Sci 118:380–386

Barrow NJ (1983) A mechanistic model for describing the sorption and desorption of phosphate by soil. J Soil Sci 34:733–750

Barrow NJ (1991) Testing a mechanistic model. XI. The effects of time and of level of application on isotopically exchangeable phosphate. J Soil Sci 42:277–288

Barrow NJ, Shaw TC (1975a) The slow reactions between soil and anions. 5. Effects of period of prior contact on the desorption of phosphate from soils. Soil Sci 119:311–320

Barrow NJ, Shaw TC (1975b) The slow reactions between soil and anions. 3. The effects of time and temperature on the decrease in isotopically exchangeable phosphate. Soil Sci 119:190–197

Benitez-Nelson CR, Buesseler KO (1999) Variability of inorganic and organic phosphorus turnover rates in the coastal ocean. Nature 398(6727):502–505

Bertrand I, Holloway RE, Armstrong RD, McLaughlin MJ (2003) Chemical characteristics of phosphorus in alkaline soils from southern Australia. Aust J Soil Res 41:61–76

Blake RE, O'Neil JR, Garcia GA (1998) Effects of microbial activity on the delta O-18 of dissolved inorganic phosphate and textural features of synthetic apatites. Am Mineral 83:1516–1531

Blake RE, Alt JC, Martini AM (2001) Oxygen isotope ratios of PO4: an inorganic indicator of enzymatic activity and P metabolism and a new biomarker in the search for life. Proc Natl Acad Sci USA 98:2148–2153

Blake RE, O'Neil JR, Surkov AV (2005) Biogeochemical cycling of phosphorus: insights from oxygen isotope effects of phosphoenzymes. Am J Sci 305:596–620

Blal B, Morel C, Gianinazzi-Pearson V, Fardeau JC, Gianinazzi S (1990) Influence of vesicular-arbuscular mycorrhizae on phosphate fertilizer efficiency in two tropical acid soils planted with micropropated oil palm (*Elaeis guineensis* jacq.). Biol Fertil Soils 9:43–48

Bolan NS (1991) A critical review on the role of mycorrhizal fungi in the uptake of phosphorus by plants. Plant Soil 134:189–207

Bolan NS, Hedley MJ, Syers JK, Tillman RW (1987) Single superphosphate-reactive phosphate rock mixtures. 1. Factors affecting chemical composition. Fert Res 13:223–239

Boniface R, Fardeau JC, Guiraud G, Trocme S (1979) Use of phosphorus and nitrogen applied to wheat fertilization – studies using labeled elements. Agrochimica 23:165–178

Bowman RA, Olsen ST, Wanatabe FS (1978) Greenhouse evaluation of residual phosphate by 4 phosphorus methods in neutral and calcareous soils. Soil Sci Soc Am J 42:451–454

Braum SM, Helmke PA (1995) White lupine utilizes soil-phosphorus that is unavailable to soybean. Plant Soil 176:95–100

Brookes PC (1982) Correction for seed-phosphorus effects in L-value determinations. J Sci Food Agric 33:329–335

Bucher M (2007) Functional biology of plant phosphate uptake at root and mycorrhiza interfaces. New Phytol 173:11–26

Bühler S, Oberson A, Sinaj S, Friesen DK, Frossard E (2003) Isotope methods for assessing plant available phosphorus in acid tropical soils. Eur J Soil Sci 54:605–616

Bünemann EK, Condron LM (2007) Phosphorus and sulphur cycling in terrestrial ecosystems. In: Marschner P, Rengel Z (eds) Nutrient cycling in terrestrial ecosystems. Soil biology, vol 10. Springer, Heidelberg, pp 65–92

Bünemann EK, Steinebrunner F, Smithson PC, Frossard E, Oberson A (2004) Phosphorus dynamics in a highly weathered soil as revealed by isotopic labeling techniques. Soil Sci Soc Am J 68:1645–1655

Bünemann EK, Marschner P, McNeill AM, McLaughlin MJ (2007) Measuring rates of gross and net mineralisation of organic phosphorus in soils. Soil Biol Biochem 39:900–913

Cavagnaro TR, Smith FA, Smith SE, Jakobsen I (2005) Functional diversity in arbuscular mycorrhizas: exploitation of soil patches with different phosphate enrichment differs among fungal species. Plant Cell Environ 28:642–650

Cobelli C, Foster D, Toffolo G (2000) Tracer kinetics in biomedical research. From data to model. Kluwer Academic/Plenum, New York

Colman AS, Blake RE, Karl DM, Fogel ML, Turekian KK (2005) Marine phosphate oxygen isotopes and organic matter remineralization in the oceans. Proc Natl Acad Sci USA 102:13023–13028

Cordell D, Drangert JO, White S (2009) The story of phosphorus: global food security and food for thought. Glob Environ Change 19:292–305

de Kok LJ, Schnug E (eds) (2008) Loads and fate of fertilizer-derived uranium. Margraf, Weikersheim

Dean LA, Krantz BA, Colwell WE, Woltz WG, Hawkins A, Nelson WL, MacKenzie AJ, Rubins EJ (1947) Application of radioactive tracer technique to studies of phosphatic fertilizer utilization by crops: greenhouse experiments. Soil Sci Soc Am Proc 12:107–112

Di HJ, Condron LM, Frossard E (1997) Isotope techniques to study phosphorus cycling in agricultural and forest soils: a review. Biol Fertil Soils 27:1–12

Diesing WE, Sinaj S, Sarret G, Manceau A, Flura T, Demaria P, Siegenthaler A, Sappin-Didier V, Frossard E (2008) Zinc speciation and isotopic exchangeability in soils polluted with heavy metals. Eur J Soil Sci 59:716–729

Ehlert P, Morel C, Fotyma M, Destain JP (2003) Potential role of phosphate buffering capacity of soils in fertilizer management strategies fitted to environmental goals. J Plant Nutr Soil Sci 166:409–415

Elsbury KE, Paytan A, Ostrom NE, Kendall C, Young MB, McLaughlin K, Rollog ME, Watson S (2009) Using oxygen isotopes of phosphate to trace phosphorus sources and cycling in Lake Erie. Environ Sci Technol 43:3108–3114

Endt PM (1990) Energy levels of A = 21-44 nuclei (VII). Nucl Phys A 521:1–830

Fardeau JC (1981) Cinétiques de dilution isotopique et phosphore assimilable des sols. Thèse de Doctorat d'Etat, Université Paris 6, Paris

Fardeau JC (1993) Available soil phosphate – its representation by a functional multiple compartmental model. Agronomie 13:317–331

Fardeau JC (1996) Dynamics of phosphate in soils. An isotopic outlook. Fert Res 45:91–100

Fardeau JC, Jappe J (1976) New method for measure of available soil phosphate – extrapolation of isotopic dilution kinetics. C R Acad Sci Hebd Seances Acad Sci D 282(11):1137–1140

Fardeau JC, Morel C, Jappe J (1985) Exchange kinetics of phosphate ions in the soil-solution system – experimental verification of theoretical equations. C R Acad Sci III 300(8):371–376

Fardeau JC, Morel C, Jahiel JM (1988) Does long term contact with the soil improve the efficiency of rock phosphate – results of isotopic studies. Fert Res 17:3–19

Fardeau JC, Morel C, Boniface R (1991) Phosphate transfer from soil to soil solution-kinetic parameters. Agronomie 11:787–797

Friesen DK, Blair GJ (1988) A dual radiotracer study of transformations of organic, inorganic and plant residue phosphorus in soil in presence and absence of plants. Aust J Soil Res 26:355–366

Frossard E, Sinaj S (1997) The isotope exchange kinetic technique: a method to describe the availability of inorganic nutrients. Applications to K, PO_4, SO_4 and Zn. Isotopes Environ Health Stud 33:61–77

Frossard E, Feller C, Tiessen H, Stewart JWB, Fardeau JC, Morel JL (1993) Can an isotopic method allow for the determination of the phosphate fixing capacity of soils? Commun Soil Sci Plant Anal 24:367–377

Frossard E, Fardeau JC, Brossard M, Morel JL (1994) Soil isotopically exchangeable phosphorus: a comparison between E and L values. Soil Sci Soc Am J 58:846–851

Frossard E, Brossard M, Hedley MJ, Metherell A (1995) Reactions controlling the cycling of P in soils. In: Tiessen H (ed) Phosphorus cycling in terrestrial and aquatic ecosystems: a global perspective. SCOPE/Wiley, New York, pp 107–137

Frossard E, Sinaj S, Zhang LM, Morel JL (1996) The fate of sludge phosphorus in soil-plant systems. Soil Sci Soc Am J 60:1248–1253

Frossard E, Bünemann E, Jansa J, Oberson A, Feller C (2009) Concepts and practices of nutrient management in agro-ecosystems: can we draw lessons from history to design future sustainable agricultural production systems? Die Bodenkultur 60:43–60

Frossard E, Bünemann EK, Oberson A, Jansa J, Kertesz MA (2011) Phosphorus and sulfur in soil. In: Huang PM, Sumner ME, Yuncong L (eds) Handbook of soil sciences, 2nd edn. CRC, Boca Raton (in press). ISBN 978-1-4398030-3-5

Gallet A, Flisch R, Ryser J-P, Frossard E, Sinaj S (2003a) Effect of phosphate fertilization on crop yield and soil phosphorus status. J Plant Nutr Soil Sci 166:568–578

Gallet A, Flisch R, Ryser J-P, Nösberger J, Frossard E, Sinaj S (2003b) Uptake of residual phosphate and freshly applied diammonium phosphate by *Lolium perenne* and *Trifolium repens*. J Plant Nutr Soil Sci 166:557–567

Gamper HA, Young JPW, Jones DL, Hodge A (2008) Real-time PCR and microscopy: are the two methods measuring the same unit of arbuscular mycorrhizal fungal abundance? Fungal Genet Biol 45:581–596

Göransson H, Rosengren U, Wallander H, Fransson AM, Thelin G (2006) Nutrient acquisition from different soil depths by pedunculate oak. Trees Struct Funct 20:292–298

Grace EJ, Cotsaftis O, Tester M, Smith FA, Smith SE (2009) Arbuscular mycorrhizal inhibition of growth in barley cannot be attributed to extent of colonization, fungal phosphorus uptake or effects on expression of plant phosphate transporter genes. New Phytol 181:938–949

Hamon RE, McLaughlin MJ (2002) Interferences in the determination of isotopically exchangeable P in soils and a method to minimise them. Aust J Soil Res 40:1383–1397

Hamon RE, Bertrand I, McLaughlin MJ (2002) Use and abuse of isotopic exchange data in soil chemistry. Aust J Soil Res 40:1371–1381

Harrison AF (1982) 32-P method to compare rates of mineralization of labile organic phosphorus in woodland soils. Soil Biol Biochem 14:337–341

Hedley MJ, White RE, Nye PH (1982) Plant-induced changes in the rhizosphere of rape (*Brassica napus* var emerald) seedlings. 3. Changes in L value, soil phosphate fractions and phosphatase activity. New Phytol 91:45–56

Hedley MJ, Bolan NS, Braithwaite AC (1988) Single superphosphate reactive phosphate rock mixtures. 2. The effect of phosphate rock type and denning time on the amounts of acidulated and extractable phosphate. Fert Res 16:179–194

Helliker BR, Ehleringer JR (2000) Establishing a grassland signature in veins: ^{18}O in the leaf water of C3 and C4 grasses. Proc Natl Acad Sci USA 97:7894–7898

Hood-Nowotny R (2008) Isotope labelling methods for evaluating contributions of organic sources to plant nutrition. In: Guidelines on nitrogen management in agricultural systems. Training course series 29. International Atomic Energy Agency, Vienna, pp 182–203

Hu Y, Vanhaecke F, Moens L, Dams R, del Castilho P, Japenga J (1998) Determination of the aqua regia soluble content of rare earth elements in fertilizer, animal fodder phosphate and manure samples using inductively coupled plasma mass spectrometry. Anal Chim Acta 373:95–105

Jaisi DP, Blake RE (2010) Tracing sources and cycling of phosphorus in Peu Margin sediments using oxygen isotopes in authigenic and detrital phosphates. Geochim Cosmochim Acta 74:3199–3212

Jaisi DP, Blake RE, Kukkadapu R (2010) Fractionation of oxygen isotopes in phosphate during its interactions with iron oxides. Geochim Cosmochim Acta 47:1309–1319

Jakobsen I, Abbott LK, Robson AD (1992) External hyphae of vesicular arbuscular mycorrhizal fungi associated with *Trifolium subterraneum* L. 2. Hyphal transport of P-32 over defined distances. New Phytol 120:509–516

Jakobsen I, Chen BD, Munkvold L, Lundsgaard T, Zhu YG (2005) Contrasting phosphate acquisition of mycorrhizal fungi with that of root hairs using the root hairless barley mutant. Plant Cell Environ 28:928–938

Jansa J, Mozafar A, Frossard E (2003) Long-distance transport of P and Zn through the hyphae of an arbuscular mycorrhizal fungus in symbiosis with maize. Agronomie 23:481–488

Jansa J, Mozafar A, Frossard E (2005) Phosphorus acquisition strategies within arbuscular mycorrhizal fungal community of a single field site. Plant Soil 276:163–176

Jansa J, Finlay R, Wallander H, Smith FA, Smith SE (2011) Role of mycorrhizal symbioses in phosphorus cycling. In: Bünemann E, Oberson A, Frossard E (eds) Phosphorus in action: biological processes in soil phosphorus cycling. Soil biology, vol 26. Springer, Heidelberg. doi:10.1007/978-3-642-15271-9_6

Jayachandran K, Schwab AP, Hetrick BAD (1992a) Partitioning dissolved inorganic and organic phosphorus using acidified molybdate and isobutanol. Soil Sci Soc Am J 56:762–765

Jayachandran K, Schwab AP, Hetrick BAD (1992b) Mineralization of organic phosphorus by vesicular-arbuscular mycorrhizal fungi. Soil Biol Biochem 24:897–903

Jenkinson DS, Fox RH, Rayner JH (1985) Interactions between fertilizer nitrogen and soil-nitrogen – the so-called priming effect. J Soil Sci 36:425–444

Joner EJ, Jakobsen I (1995) Uptake of 32P from labelled organic matter by mycorrhizal and non-mycorrhizal subterranean clover (*Trifolium subterraneum* l.). Plant Soil 172:221–227

Joner EJ, van Aarle IM, Vosatka M (2000) Phosphatase activity of extra-radical arbuscular mycorrhizal hyphae: a review. Plant Soil 226:199–210

Kapoor KK, Haider K (1982) Mineralization and plant availability of phosphorus from biomass of hyaline and melanic fungi. Soil Sci Soc Am J 46:953–957

Kato N, Zapata F, Axmann H (1995) Evaluation of the agronomic effectiveness of natural and partially acidulated phosphate rocks in several soils using P-32 isotopic dilution techniques. Fert Res 41:235–242

Kok B, Varner JE (1967) Extraterrestrial life detection based on oxygen isotope exchange reactions. Science 155:1110–1112

Kolodny Y, Luz B, Navon O (1983) Oxygen isotope variations in phosphate of biogenic apatites. 1. Fish bone apatite – rechecking the rules of the game. Earth Planet Sci Lett 64:398–404

Kucey RMN, Bole JB (1984) Availability of phosphorus from 17 rock phosphates in moderately and weakly acidic soils as determined by P-32 dilution, A-value, and total P uptake methods. Soil Sci 138:180–188

Larsen S (1952) The use of P-32 in studies on the uptake of phosphorus by plants. Plant Soil 4:1–10

Larsen S, Middelboe V, Johansen HS (1989) The fate of O-18 labeled phosphate in soil plant systems. Plant Soil 117:143–145

Lécuyer C, Grandjean P, Sheppard SMF (1999) Oxygen isotope exchange between dissolved phosphate and water at temperature ≤135°C: inorganic versus biological fractionations. Geochim Cosmochim Acta 63:855–862

Liang Y, Blake RE (2006) Oxygen isotope signature of Pi regeneration from organic compounds by phosphomonoesterases and photooxidation. Geochim Cosmochim Acta 70:3957–3969

Liang Y, Blake RE (2007) Oxygen isotope fractionation between apatite and aqueous-phase phosphate: 20–45°C. Chem Geol 238:121–133

Liang YH, Blake RE (2009) Compound- and enzyme-specific phosphodiester hydrolysis mechanisms revealed by delta O-18 of dissolved inorganic phosphate: implications for marine P cycling. Geochim Cosmochim Acta 73:3782–3794

Longinelli A (1965) The oxygen isotopic composition of orthophosphate from shells of living marine organisms. Nature 207:716–719

Longinelli A, Nuti S (1973) Revised phosphate-water isotopic temperature scale. Earth Planet Sci Lett 19:373–376

López-Hernández D, Brossard M, Frossard E (1998) P-isotopic exchangeable values in relation to Po mineralization in soils with very low P-sorbing capacities. Soil Biol Biochem 30:1663–1670

Luz B, Kolodny Y (1985) Oxygen isotope varioations in phosphate of biogenic apatites. 4. Mammal teeth and bones. Earth Planet Sci Lett 75:29–36

Maertens E, Thijs A, Smolders E, Degryse F, Cong PT, Merckx R (2004) An anion resin membrane technique to overcome detection limits of isotopically exchanged P in P-sorbing soils. Eur J Soil Sci 55:63–69

Maldonado-Mendoza IE, Dewbre GR, Harrison MJ (2001) A phosphate transporter gene from the extra-radical mycelium of an arbuscular mycorrhizal fungus *Glomus intraradices* is regulated in response to phosphate in the environment. Mol Plant Microbe Interact 14:1140–1148

Martin JK, Cartwright B (1971) The comparative plant availability of phosphorus-32 myo inositol hexaphosphate and phosphorus-32 mono basic potassium phosphate added to soils. Commun Soil Sci Plant Anal 2:375–381

Mattingly GEG (1975) Labile phosphate in soils. Soil Sci 119:369–375

McAuliffe CD, Hall NS, Dean LA, Hendricks SB (1947) Exchange reactions between phosphates and soils: hydroxylic surfaces of soil minerals. Soil Sci Soc Am Proc 12:119–123

McBeath TM, Lombi E, McLaughlin MJ, Bünemann EK (2009) Exchangeability of orthophosphate and pyrophosphate in soils: a double isotopic labelling study. Plant Soil 314:243–252

McLaughlin MJ, Alston AM (1986) The relative contribution of plant residues and fertilizer to the phosphorus-nutrition of wheat in a pasture cereal system. Aust J Soil Res 24:517–526

McLaughlin MJ, Alston AM, Martin JK (1988) Phosphorus cycling in wheat-pasture rotations. I. The source of phosphorus taken up by wheat. Aust J Soil Res 26:323–331

McLaughlin K, Cade-Menun BJ, Paytan A (2006a) The oxygen isotopic composition of phosphate in Elkhorn Slough, California: a tracer for phosphate sources. Estuar Coast Shelf Sci 70:499–506

McLaughlin K, Kendall C, Silva SR, Young M, Paytan A (2006b) Phosphate oxygen isotope ratios as a tracer for sources and cycling of phosphate in North San Francisco Bay. California. J Geophys Res 111:G03003. doi:10.1029/2005JG000079

Mikkelsen BL, Rosendahl S, Jakobsen I (2008) Underground resource allocation between individual networks of mycorrhizal fungi. New Phytol 180:890–898

Mizota C, Domon Y, Yoshida N (1992) Oxygen isotope composition of natural phosphates from volcanic ash soils of the great rift-valley of Africa and East Java, Indonesia. Geoderma 53:111–123

Morel C (2002) Caractérisation de la phytodisponibilité du phosphore du sol par la modélisation du transfert des ions phosphates entre le sol et la solution. Mémoire préparé en vue de l'obtention du diplôme d'habilitation à diriger des recherches. INRA-Bordeaux et Institut National Polytechnique de Lorraine (in French)

Morel C, Fardeau JC (1989a) Native soil and fresh fertilizer phosphorus uptake as affected by rate of applications and P fertilizers. Plant Soil 115:123–128

Morel C, Fardeau JC (1989b) The uptake by crops of fresh and residual phosphatic fertilizers by simultaneous measurements with ^{32}P and ^{33}P. Int J Appl Radiat Isot 40:273–278

Morel C, Plenchette C (1994) Is the isotopically exchangeable phosphate of a loamy soil the plant available P. Plant Soil 158:287–297

Morel C, Tiessen H, Moir JO, Stewart JWB (1994) Phosphorus transformations and availability under cropping and fertilization assessed by isotopic exchange. Soil Sci Soc Am J 58: 1439–1445

Morel C, Tunney P, Plenet D, Pellerin S (2000) Transfer of phosphate ions between soil and solution: perspectives in soil testing. J Environ Qual 29:50–59

Mosse B, Hayman DS, Arnold DJ (1973) Plant growth responses to vesicular-arbuscular mycorrhiza. V. Phosphate uptake by three plant species from P-deficient soils labelled with 32P. New Phytol 72:809–815

Nagy R, Karandashov V, Chague W, Kalinkevich K, Tamasloukht M, Xu GH, Jakobsen I, Levy AA, Amrhein N, Bucher M (2005) The characterization of novel mycorrhiza-specific phosphate transporters from *Lycopersicon esculentum* and *Solanum tuberosum* uncovers functional redundancy in symbiotic phosphate transport in solanaceous species. Plant J 42:236–250

Nelson WL, Krantz BA, Colwell WE, Woltz WG, Hawkins A, Dean LA, MacKenzie AJ, Rubins EJ (1947) Application of radioactive tracer technique to studies of phosphatic fertilizer utilization by crops: field experiments. Soil Sci Soc Am Proc 12:112–118

Nemery J, Garnier J, Morel C (2005) Phosphorus budget in the Marne Watershed (France): urban vs. diffuse sources, dissolved vs. particulate forms. Biochemistry 72:35–66

Neumann G, Martinoia E (2002) Cluster roots – an underground adaptation for survival in extreme environments. Trends Plant Sci 7:162–167

Nielsen JS, Joner EJ, Declerck S, Olsson S, Jakobsen I (2002) Phospho-imaging as a tool for visualization and noninvasive measurement of p transport dynamics in arbuscular mycorrhizas. New Phytol 154:809–819

Oberson A, Tagmann HU, Langmeier M, Dubois D, Mäder P, Frossard E (2010) Fresh and residual phosphorus uptake by ryegrass from soils with different fertilization histories. Plant Soil 334:391–407

Oberson A, Pypers P, Bünemann EK, Frossard E (2011) Management impacts on biological phosphorus cycling in cropped soils. In: Bünemann E, Oberson A, Frossard E (eds) Phosphorus in action: biological processes in soil phosphorus cycling. Soil biology, vol 26. Springer, Heidelberg. doi:10.1007/978-3-642-15271-9_17

Oehl F, Oberson A, Sinaj S, Frossard E (2001a) Organic phosphorus mineralisation studies using isotopic dilution techniques. Soil Sci Soc Am J 65:780–787

Oehl F, Oberson A, Probst M, Fliessbach A, Roth HR, Frossard E (2001b) Kinetics of microbial phosphorus uptake in cultivated soils. Biol Fertil Soils 34:31–41

Oehl F, Frossard E, Fliessbach A, Dubois D, Oberson A (2004) Basal phosphorus mineralisation in soils under different farming systems. Soil Biol Biochem 36:667–675

Otero N, Vitoria L, Soler A, Canals A (2005) Fertiliser characterisation: major, trace and rare earth elements. Appl Geochem 20:1473–1488

Paytan A, Kolodny Y, Neori A, Luz B (2002) Rapid biologically mediated oxygen isotope exchange between water and phosphate. Global Biogeochem Cycles 16(1):1013. doi:10.1029/2001GB001430

Pons WA, Guthrie JD (1946) Determination of inorganic phosphorus in plant materials. Ind Eng Chem Anal Ed 18:184–186

Powell CL (1975) Plant growth responses to vesicular-arbuscular mycorrhiza. VIII. Uptake of p by onion and clover infected with different endogone spore types in 32P labelled soil. New Phytol 75:563–566

Probert ME, Larsen S (1972) The kinetics of heterogeneous isotopic exchange. J Soil Sci 23:76–81

Pypers P, Van Loon L, Diels J, Abaidoo R, Smolders E, Merckx R (2006) Plant-available P for maize and cowpea in P-deficient soils from the Nigerian Northern Guinea Savanna – comparison of E- and L-values. Plant Soil 283:251–264

Read DJ, Leake JR, Perez-Moreno J (2004) Mycorrhizal fungi as drivers of ecosystem processes in heathland and boreal forest biomes. Can J Bot 82:1243–1263

Russell RS, Russell EW, Marais PG (1957) Factors affecting the ability of plants to absorb phosphate from soils. 1. The relationship between labile phosphate and absorption. J Soil Sci 8:248–267

Salcedo IH, Bertino F, Sampaio EVSB (1991) Reactivity of phosphorus in Northeastern Brazilian soils assessed by isotopic dilution. Soil Sci Soc Am J 55:140–145

Schneider A, Morel C (2000) Relationship between the isotopically exchangeable and resin-extractable phosphate of deficient to heavily fertilized soil. Eur J Soil Sci 51:709–715

Schweiger PF, Jakobsen I (1999) Direct measurement of arbuscular mycorrhizal phosphorus uptake into field-grown winter wheat. Agron J 91:998–1002

Schweiger P, Thingstrup I, Jakobsen I (1999) Comparison of two test systems for measuring plant phosphorus uptake via arbuscular mycorrhizal fungi. Mycorrhiza 8:207–213

Schwertmann U, Cornell R (2000) Iron oxides in the laboratory. Preparation and characterization, 2nd edn. Wiley-VCH, Weinheim

Sinaj S, Traore O, Frossard E (2002) Effect of compost and soil properties on the availability of composts phosphate for white clover (*Trifolium repens* L.). Nutr Cycl Agroecosyst 62:89–102

Smith FA, Jakobsen I, Smith SE (2000) Spatial differences in acquisition of soil phosphate between two arbuscular mycorrhizal fungi in symbiosis with *Medicago truncatula*. New Phytol 147:357–366

Smith SE, Smith FA, Jakobsen I (2004) Functional diversity in arbuscular mycorrhizal (am) symbioses: the contribution of the mycorrhizal P uptake pathway is not correlated with mycorrhizal responses in growth or total P uptake. New Phytol 162:511–524

Spinks JWT, Barber SA (1947) Studies of fertilizer uptake using radioactive phosphorus. Sci Agric 27:145–156

Stroia C, Morel C, Jouany C (2007) Dynamics of diffusive soil phosphorus in two grassland experiments determined both in field and laboratory conditions. Agric Ecosyst Environ 119:60–74

Talibudeen O (1957) Isotopically exchangeable phosphorus in soils. II. Factors influencing the estimation of "labile" phosphorus. J Soil Sci 8:86–96

Tamburini F, Bernasconi SM, Angert A, Weiner T, Frossard E (2010) A method for the analysis of the δ18O of inorganic phosphate in soils extracted with HCl. Eur J Soil Sci (in press) doi: 10.1111/j.1365-2389.2010.01290.x

Tan KH (1993) Principles of soil chemistry, 2nd edn. Marcel Dekker, New York

Thibaud MC, Morel C, Fardeau JC (1988) Contribution of phosphorus issued from crop residues to plant nutrition. Soil Sci Plant Nutr 34:481–491

Torrent J (1987) Interactions between phosphate and iron oxide. Adv Geoecol 30:321–344

Truong B, Pichot J (1976) Influence du phosphore des graines de la plante test sur la détermination du phosphore isotopiquement diluable (valeur L). Agron Trop 31:379–385

Voets L, de la Providencia IE, Fernandez K, IJdo M, Cranenbrouck S, Declerck S (2009) Extraradical mycelium network of arbuscular mycorrhizal fungi allows fast colonization of seedlings under in vitro conditions. Mycorrhiza 19:347–356

Walbridge MR, Vitousek PM (1987) Phosphorus mineralization potentials in acid organic soils: processes affecting 32PO4-isotope dilution measurements. Soil Biol Biochem 19:709–717

Wang B, Funakoshi DM, Dalpe Y, Hamel C (2002) Phosphorus-32 absorption and translocation to host plants by arbuscular mycorrhizal fungi at low root-zone temperature. Mycorrhiza 12:93–96

White JL, Fried M, Oherogge AJ (1949) Study of the utlization of phosphorus in green manure crops by the succeeding crop, using radioactive phosphorus. Agron J 41:174–175

Wiedemann-Bidlack FB, Colman AS, Fogel ML (2008) Phosphate oxygen isotope analysis on microsamples of bioapatite: removal of organic contamination and minimization of sample size. Rapid Commun Mass Spectrom 22:1807–1816

Wiklander L (1950) Kinetics of phosphate exchange in soils. Ann R Agric Coll Sweden 17:407–424

Winter ERS, Carlton M, Briscoe HVA (1940) The interchange of heavy oxygen between water and inorganic oxyions. J Chem Soc 32:131–138

Wolf AM, Baker DE, Pionke HB (1986) The measurement of labile phosphorus by the isotopic dilution and anion resin methods. Soil Sci 141:60–70

Young MB, McLaughlin K, Kendall C, Stringfellow W, Rollog M, Elsbury K, Donald E, Paytan A (2009) Characterizing the oxygen isotopic composition of phosphate sources to aquatic ecosystems. Environ Sci Technol 43:5190–5196

Zapata F, Axmann H (1995) P-32 isotopic techniques for evaluating the agronomic effectiveness of rock phosphate materials. Fert Res 41:189–195

Chapter 4
Molecular Approaches to the Study of Biological Phosphorus Cycling

Jun Wasaki and Hayato Maruyama

4.1 Introduction

The organisms involved in phosphorus (P) cycling in soils are highly varied, and microorganisms probably play the most important role. However, more than 99% of soil microorganisms have not been cultured successfully (Torsvik and Øvreås 2002). Therefore, culture-independent methods are required to study the function and ecology of microbes involved in P cycling in soils. Molecular approaches for such culture-independent methods have been developed in the recent past.

Exudates from plant roots have significant effects on soil microbial ecology and P dynamics in the rhizosphere. Therefore, it is important to understand plant–microbe interactions, e.g., symbiosis with mycorrhizal fungi. Plant function and plant–microbe interactions can also be analyzed using molecular approaches.

In this chapter, the applications of molecular tools to study the role of plants and rhizosphere microorganisms in P cycling are discussed. Table 4.1 summarizes the molecular and biochemical tools introduced in this chapter. Since there are some advantages and disadvantages for each method, researchers should select the most appropriate tool(s) for their purpose.

4.2 DNA Extraction

4.2.1 Soil Samples

The starting point for all molecular approaches is the extraction of environmental DNA from the soil. We frequently face difficulties extracting DNA from soils, because soil properties vary strongly in terms of pH and content of mineral and

J. Wasaki (✉) and H. Maruyama
Graduate School of Biosphere Science, Hiroshima University, Kagamiyama 1-7-1, Higashi-Hiroshima 739-8521, Japan
e-mail: junw@hiroshima-u.ac.jp

E.K. Bünemann et al. (eds.), *Phosphorus in Action*, Soil Biology 26,
DOI 10.1007/978-3-642-15271-9_4, © Springer-Verlag Berlin Heidelberg 2011

Table 4.1 Summary of molecular and biochemical tools introduced in this chapter

Molecular tools	Dependency on PCR	Advantage	Disadvantage
DNA-based techniques	–	Ease of use for many approaches	Only the abundant organisms can be detected
RNA-based techniques	–	Active organisms can be detected	Application is difficult because of instability of RNA
Clone library	Yes	Information of DNA sequence for each clone becomes available	Relatively expensive
RFLP (restriction fragment length polymorphism)	Yes	Differences among samples can be detected at genus or higher levels	Sequence information is frequently too short for classification
DGGE (denaturant gradient gel electrophoresis)	Yes	Differences among samples can be detected at species or strain levels	Resolution is frequently not very good, especially in communities with higher diversity
TGGE (temperature gradient gel electrophoresis)	Yes	Differences among samples can be detected at species or strain levels	Resolution is frequently not very good, especially in communities with higher diversity. Special equipment making a temperature gradient is required
SSCP (single-strand conformation polymorphism)	Yes	Differences among samples can be detected at species or strain levels	Resolution is frequently not very good, especially in communities with higher diversity. Temperature-controlled electrophoresis is required
qRT-PCR (quantitative real-time polymerase chain reaction)	Yes	Quantification of specific genes	Specific primer set is required
SIP (stable isotope probing)	Yes (for combination with DGGE)/No	Active organisms can be analyzed with DNA or other stable compounds	Stable isotope of P is absent
Immunocapture using BrdU (5-bromo-2-deoxyuridine)	Sometimes yes	Active organisms can be analyzed. Immunoprecipitation can be applied	Dynamics of BrdU in soil are not well understood
Microarray	No	Whole aspects of specific microbes or mRNAs can be detected	Expensive. Universal microbe arrays are not yet applied
Next generation sequencer	No	Metagenomic and metatranscriptomic data are available	Very expensive
FISH (fluorescence in situ hybridization)	No	Localization of specific microbes or genes can be visualized	Amplification of signals is required for functional genes
PLFA (phospholipid fatty acid)	No	Microbial community structure can be analyzed independently from DNA or RNA	Lack of sequence information
Enzyme activities measured with fluorogenic substrates	No	Applicable for small amounts of soil samples	Dynamic range is different among the substrates dependent on activities
ELF (enzyme-labeled fluorescence)-97 phosphate	No	Localization of phosphatase	Application is specific
Phosphate-reporter bacteria	No	P availability for bacteria can be visualized	Application is specific

PCR polymerase chain reaction

organic compounds. Humic substances in particular, which behave like nucleic acids in some respects, cause problems during DNA extraction, purification, and further experimentation. Thus, the removal of humic substances and their separation from soil DNA are important steps.

Numerous methods for soil DNA extraction have been reported and several kits for soil DNA extraction are commercially provided, e.g., from Epicentre Biotechnologies (Madison, WI, USA), MO Bio Laboratories (Carlsbad, CA, USA), Nippongene (Toyama, Japan), Norgen Biotek (Thoroid, Canada), Omega Bio-Tek (Norcross, GA, USA), and Zymo Research (Orange, CA, USA). Currently, kits that include a bead-beating step are commonly used for soil DNA extraction. Because we mostly work with volcanic ash soils with high affinity to phosphates, we mainly employ the ISOIL kit from Nippongene, which is optimized for humic substance removal from volcanic ash soils (the method for removing the humic substances is protected by patent).

4.2.2 *Plants and Hyphae*

DNA extraction from plants is important for the investigation of the cooperation of plants with symbiotic and associated microbes in P cycling, as well as for the role of plants in these processes. Fungal DNA can be extracted by similar methods to those used for plants. We employ the CTAB (cetyltrimethylammonium bromide; a detergent for solubilizing membranes) method (Rogers and Bendich 1985) or commercially available kits such as the Plant DNA Mini Kit from Qiagen (Hilden, Germany) and Isoplant (Nippongene), for DNA extraction from plants and fungal mycelia. Grinding with pestle and mortar or bead-beating of samples frozen in liquid nitrogen is an effective preparation of tissues prior to DNA extraction. DNA extraction from plants or fungi is relatively easy because of the absence of humic substances.

In the case of arbuscular mycorrhizal (AM) fungi, spores can be collected from soils by sieving. This method enables DNA extraction specifically from spores of AM fungi. The yield of DNA is frequently very low; in such cases, addition of coprecipitating agents such as glycogen is recommended at the ethanol precipitation step.

4.2.3 *Rolling Cycle Amplification*

When rhizosphere soil or spores of AM fungi are targeted, the yield of DNA is sometimes too low (in the nanogram order) for direct analysis. Rolling cycle amplification (RCA) using Phi29 DNA polymerase can provide linearly amplified and purified DNA; commercial kits are available. In our experience, ca. 15 μg DNA can be amplified from 5 ng of DNA extracted from spores of *Gigaspora margarita* by using GenomiPhi v2 (GE Healthcare, Buckinghamshire, UK). Amplified DNA can be used for further analyses, such as polymerase chain reaction (PCR)-dependent molecular tools and metagenomics.

4.3 Molecular Approaches Using the Sequences of SSU rRNA and ITS Regions

4.3.1 Clone Libraries

Small subunit (SSU) rRNA and internal transcribed spacer (ITS) regions are frequently analyzed for the molecular characterization of uncultured microbes. These sequences are easily amplified by PCR. Amplified fragments are compared with respect to their length, restriction fragment patterns, and substantial differences between sequences.

Sequencing of clone libraries is a simple method for the analysis of SSU rRNA or ITS regions. The sequence provides useful information on the phylogenetic position of the microbe. The ratio of overlap of sequences indicates the diversity of the group in the sample. Phylogenetic trees can be prepared based on the sequences. High-throughput analysis is possible using 96- or 384-hole multiplates. The disadvantage of this method is the relatively high expense.

Monteiro et al. (2009) analyzed partial 16S rRNA sequences for 83 bacterial colonies, which were isolated from the rhizosphere of vetiver (*Chrysopogon zizanioides*) as potential biofertilizers. A primer set for PCR-DGGE, which is described in the next section, was used for PCR and sequencing. It was revealed that potentially phosphate solubilizing bacteria coexisted in the vetiver rhizosphere that were closely related to *Acinetobacter*, *Burkholderia*, *Chryseobacterium*, *Dyella*, *Enterobacter*, *Klebsiella*, and *Pseudomonas*.

4.3.2 PCR-DGGE

Denaturant gradient gel electrophoresis (DGGE) is frequently applied for comparing the microbial communities of various environments (Heuer et al. 1997). The work-flow of DGGE is shown in Fig. 4.1. DNA fragments amplified from environmental DNA by PCR are separated on polyacrylamide gels containing gradients of denaturants. Differences of sequences (i.e., GC contents) result in different denaturing points, which show as bands on the denaturant gradient gel. A GC clamp (an artificial sequence rich in guanine and cytosine) that generates a tight bond is often added to the terminus of one primer to increase the resolution of the detected bands (Sheffield et al. 1989). The band patterns obtained by PCR-DGGE are used to

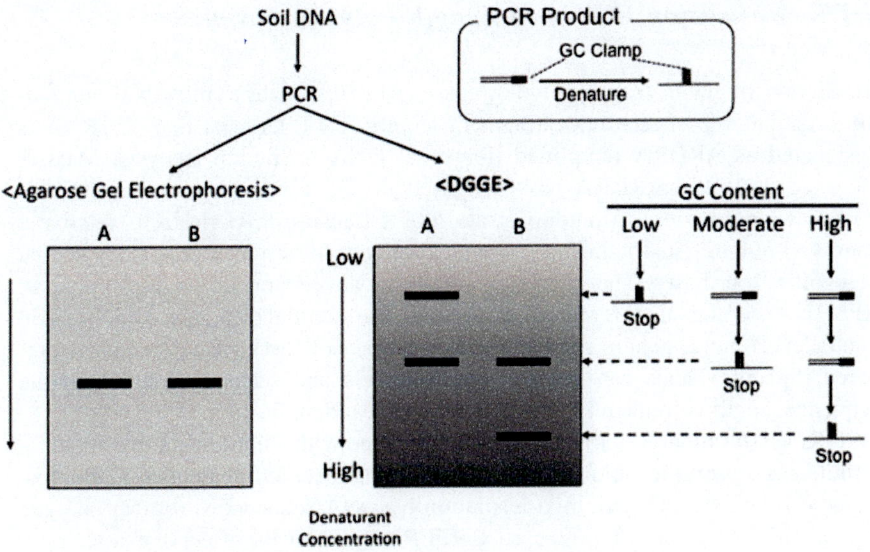

Fig. 4.1 The principle and work-flow of PCR-DGGE

compare samples by multivariate analyses such as principal component analysis or cluster analysis.

The merit of the DGGE method is the ease with which microbial communities of different samples can be compared without DNA sequence determination. The only technical requirement for DGGE is electrophoresis equipment. DGGE can visualize differences at the line or species levels, because bands can be differentiated even if the sequences compared differ merely in one base. By contrast, DGGE is unsuitable for the detection of differences at the genus or any higher levels. When the diversity of the sample is high, the bands separated in the DGGE are poorly resolved.

Marschner et al. (2004) applied PCR-DGGE of 16S rDNA to investigate microbial community structures of the rhizospheres of chickpea, canola, and Sudan grass as affected by P fertilization. Similarly, Wasaki et al. (2005) and Weisskopf et al. (2005) studied the community structures of rhizobacteria associated with white lupin plants, in order to clarify the effects of root exudates on the diversity and community structure of rhizobacteria in cluster roots formed by P-starved white lupin plants. Their results suggested that the diversity and community structure were strongly influenced by root exudates, such as citrate and flavonoids.

Temperature gradient gel electrophoresis (TGGE) resembles DGGE (Henco and Heibey 1990) in principle but differs in having a temperature rather than a denaturant gradient. Single-strand conformation polymorphism electrophoresis (SSCP) allows the separation of single-stranded DNA following the chilling-induced formation of secondary structures in the single-stranded DNA. Both TGGE and SSCP can separate sequences that differ in only one base. These techniques have not yet been applied in studies on soil P cycling.

4.3.3 Restriction Fragment Length Polymorphism

Restriction fragment length polymorphism (RFLP) visualizes microbial communities as patterns of restriction fragment length. RFLP targeted to SSU rRNA is designated as ARDRA (amplified ribosomal DNA restriction analysis; Massol-Deya et al. 1995). ARDRA has been used to characterize cultured isolates of potential biofertilizers (Monteiro et al. 2009). Because SSU rRNA is relatively short (>1,500 bp), its fragments are generally digested by a restriction enzyme that recognizes four bases. The restriction fragments are separated by electrophoresis. RFLP is a useful method for detecting relatively substantial differences (higher than genus level), but it cannot resolve small differences between the sequences compared. Specific bands can be isolated from gels and sequenced, although the sequence length is frequently too short for species identification.

T-RFLP (terminal RFLP) using a primer labeled with a fluorescent dye at the 5′-terminus is a useful technique for the detection of specific fragments. The fluorescence can be detected with high resolution by DNA sequencers or other analytic equipment. George et al. (2009) used T-RFLP to examine the effect of extracellular release of phytase from roots of transgenic plants on microbial community structures in the rhizosphere. They demonstrated that the expression of phytase in transgenic plants had little or no impact on the microbial community structure as compared with control plant lines.

4.4 Methods Targeting Active Microbes and Functional Genes

4.4.1 RNA Extraction

DNA work is relatively easy to apply on environmental samples, but it delivers information not only about active, but also about inactive microbes. Environmental RNA-targeted molecular approaches are required to target the major active players or key genes in the environment. However, RNA extraction from soil is still difficult because of the instability of RNA molecules, and only a few RNA-based studies have succeeded in identifying microorganisms involved in P cycling.

Weisskopf et al. (2005) investigated the bacterial community structures associated with white lupin, a plant that forms "cluster roots" showing an exudative burst under low P conditions, by producing DGGE profiles based on both DNA and RNA. In their study, RNA was isolated using a kit provided by BIO 101 (Vista, CA, USA). They showed that the "present" and "active" populations analyzed by DNA and RNA, respectively, were similar in the root samples but not in the rhizosphere and bulk soils. The effect of root type and age on the structure of bacterial communities living in the root vicinity was more obvious in active communities than in present communities. Thus, monitoring changes in active communities proved to be more informative than dealing only with present communities.

4.4.2 Analysis of Functional Genes by DNA-Based Techniques

Because soil RNA extraction is still problematic in many cases, the identification of functional genes involved in nutrient cycling so far mainly relies on DNA-based techniques.

We developed a PCR-DGGE methodology for the analysis of functional genes. Sakurai et al. (2008) designated specific primer sets for alkaline phosphatase (ALP), which is involved in organic P cycling. The primers were designed on the basis of an alignment of amino acid and nucleotide sequences encoding an ALP-expressing gene derived from various isolates: *Bacillus subtilis* 168, *Nostoc* sp. PCC7120, *Caulobacter crescentus* CB15, *Pseudomonas aeruginosa* PAO1, *Sinorhizobium meliloti* 1021, *Mesorhizobium loti* MAFF303099, and *Corynebacterium glutamicum* ATCC13032. Effects of the application of organic matter and chemical fertilizer on ALP-harboring bacterial communities in the rhizosphere and bulk soil in an experimental lettuce field were analyzed by PCR-DGGE. Numerous *ALP*-expression genes were detected in the DGGE profile, regardless of sampling time, fertilizer treatment, or sampled soil area, which indicated a large diversity in ALP-harboring bacteria in the soil. Several *ALP* gene fragments, which were excised from the DGGE gel and sequenced, were closely related to the ALP-expressing genes of *M. loti* and *Pseudomonas fluorescens*. Using principal component analysis of the DGGE profile it was shown that fertilizer treatment and sampling site significantly affected the community structures of ALP-harboring bacteria.

In the case of functional genes, the data obtained from the DGGE profile is expected to be quite informative. However, missing groups with a lower similarity of the respective functional gene to the sequence used for primer design is an undeniable possibility. For example, the identity of sequences among microorganisms was generally lower for *ALP* genes than for the sequences of SSU rRNA (Sakurai et al. 2008). Therefore, the two primer sets were designed for nested PCR of ALP gene fragments to increase the coverage range.

4.4.3 Quantitative Real-Time PCR: Quantification Technology for SSU rRNA and Transcripts of Functional Genes

Recently, the amount of mRNA in a wide range of samples has been analyzed by PCR-based methods rather than by hybridization-dependent methods. The merits of PCR-based methods are speed, reproducibility, and quantitative capacity. Quantitative RT-PCR (qRT-PCR), in particular, is very effective for microscale samples such as rhizosphere soils.

qRT-PCR for environmental DNA determines the quantity of microorganisms harboring the targeted gene. When genus-specific primers for SSU rRNA are used, the results of the technique can quantify the dominance of the genus. In the case of

RNA-based methods, the amount of mRNA of the target gene can be determined. The method is useful for functional genes that are directly involved in P cycling such as phosphatases, phytases, and nucleases. Melting curves of amplified fragments produced by most of the commercial equipments provide information on the diversity of the amplified fragments.

qRT-PCR was applied to evaluate interactions of arbuscular mycorrhizal fungi. Specific primers for ITS1 and rRNAs of *Glomus mosseae* and *G. intraradices* were designed for qRT-PCR (Alkan et al. 2006). These authors showed some effects of P availability in the soil on the interaction between the two *Glomus* species, and concluded that qRT-PCR was a valuable tool for studying the ecology of AM fungi.

4.4.4 Fluorescence In Situ Hybridization

The distribution of microbial species or functional groups of interest can be visualized by FISH (fluorescence in situ hybridization). A specific primer designed for SSU rRNA or a functional gene, and containing a fluorescent dye, is required for FISH. The dye must not interfere with the autofluorescence of soil particles and plant roots. The sample is hybridized with the dye primer and observed with a fluorescence microscope.

Usually, direct hybridization is possible for SSU rRNA because of the huge amount of molecules. On the other hand, amplification of the signal is required for most functional genes. Several methods have been developed such as MAR- (microautoradiography) (Ito et al. 2002), CARD- (catalyzed reporter deposition) (Pernthaler et al. 2002) and TSA (tyramide signal amplification)-FISH (Schriml et al. 1999). MAR-FISH was applied to study P cycling, even if not in soils. It was applied for the detection of potential polyphosphate-accumulating bacteria in biological phosphorus removal plants (Kong et al. 2005).

4.4.5 Stable Isotope Probing

Stable isotope probing (SIP) is a method for identifying active microbes in the environment. ^{13}C- or ^{15}N-labeled substrates are frequently used for SIP. For example, microbes that respond to root-secreted compounds can be detected in the rhizosphere of $^{13}CO_2$-assimilating plants. Rangel-Castro et al. (2005) determined the effect of liming on the structure of the rhizosphere microbial community metabolizing root exudates by the combination of DGGE and SIP. Phospholipid fatty acid analysis (PLFA) can be applied following SIP, and has been employed, e.g., for the analysis of methanotrophic communities (Chen et al. 2008).

Unfortunately, there are no useful stable isotopes of P. Thus, it seems difficult to apply SIP for studying P cycling. However, as mentioned by Prosser et al. (2006), SIP provides the potential for cultivation-independent characterization of organisms actively assimilating carbon derived from plant root exudates or added to the soil. Since root exudates have impacts on P cycling in the rhizosphere, ^{13}C-SIP could be applied for studying the effects of root exudates on the microbes involved in P cycling.

4.4.6 Bromodeoxyuridine Immunocapture

An analog of thymidine, 5-bromo-2-deoxyuridine (BrdU), can be used for labeling active microorganisms in the environment. BrdU-antibodies are then applied to detect the microorganisms containing BrdU. Artursson et al. (2005) employed BrdU immunocapture in combination with T-RFLP. They found distinct changes in active bacterial community compositions related to *G. mosseae* inoculation, treatment with an antifungal compound, and plant type. The dominant bacterial species that were activated as a result of *G. mosseae* inoculation were found to be mostly uncultured bacteria and *Paenibacillus* species.

4.4.7 Phosphate-Reporter Bacteria

Phosphate-reporter bacteria have been developed and used for studies on P cycling. In particular, de Weger et al. (1994) developed phosphate-reporter strains of *Pseudomonas putida* WCS358 in which the production of β-glucosidase was regulated by the promoter of a gene responsive to P starvation. Kragelund et al. (1997) used two phosphate-reporter strains of *P. fluorescens* DF57 harboring luciferase regulated by the promoter of a gene responsive to P starvation. These phosphate-reporter bacteria were used to assess whether sufficient phosphate was available to the bacteria. It can be a useful tool for studying the plant–microbe interactions involved in P cycling, because the *Pseudomonas* spp. are frequently isolated as potentially plant growth-promoting rhizobacteria.

4.5 Methods for Analysis of Plant Functions

Root-secreted compounds of plants are important for P cycling in soil (Fig. 4.2). Acid phosphatase (APase) secreted from white lupin roots contributes most of the phosphatase activity in rhizosphere soil (Wasaki et al. 2005). Organic acids mobilize sparingly soluble P compounds in the soil. Mobilized inorganic phosphate can then be absorbed by plant roots with the help of phosphate transporters. Secondary plant metabolites are involved in the communication with rhizosphere microbes. Finally, mRNA expression of important genes in plants has been

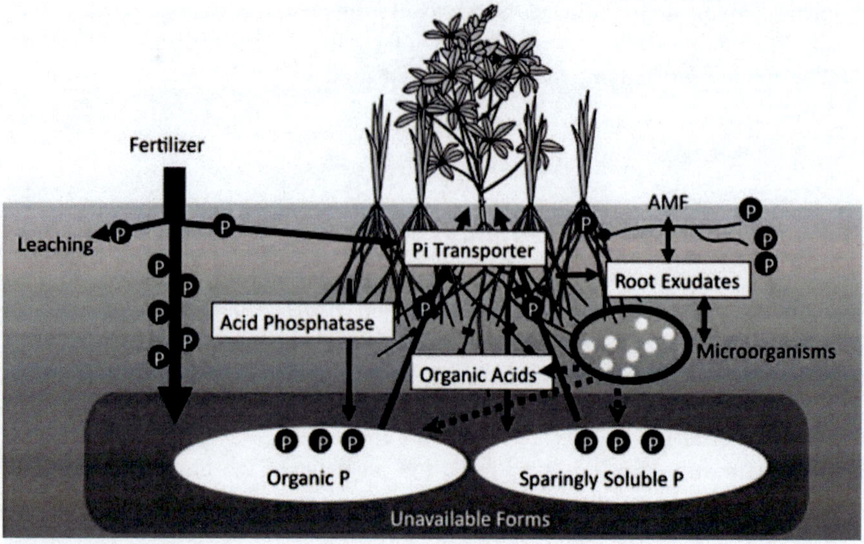

Fig. 4.2 Biological functions in the soil P cycle. *AMF* arbuscular mycorrhizal fungi

frequently analyzed by qRT-PCR. In this section, the methods for analysis of the above-mentioned plant functions involved in P cycling are introduced.

4.5.1 Phosphatase Activities

A method using *p*-nitrophenylphosphate (*p*-NPP) as a substrate is still applied for the measurement of phosphatase activity in soils and root exudates (Ozawa et al. 1995). The degradation product *p*-nitrophenol turns yellow following alkalinization. A fluorogenic substrate, 4-methylumbeliferylphosphate (MUP), is also available for the measurement of phosphatase activities. Many fluorogenic substrates can be used for the determination of soil enzyme activities involved in P, N, and C cycling. For high-throughput analysis of many soil enzyme activities in small samples, 96-well microplates and microplate readers are useful tools that allow detection of the fluorescence of the degradation products (Tscherko et al. 2004; Wasaki et al. 2005). For further detail on these assays see Nannipieri et al. (2011).

Phosphatase activity can be visualized by activity staining of phosphatases in the media or in gels after electrophoresis. For activity staining, the above-mentioned substrates (*p*-NPP and MUP) are often applied, for example in our studies (Wasaki et al. 1997, 1999, 2005). We have shown that the cluster roots of white lupin released a significant amount of APase under P-deficient conditions (Wasaki et al. 1999). Naphtylphosphate/fast-red TR (4-chloro-2-methylbenzenediazonium salt) can also be used for activity staining (Neumann and Martinoia 2002).

Enzyme-labeled fluorescence (ELF)-97 phosphate can be used for histochemical localization of phosphatases. This substrate provides fluorescent precipitates after hydrolysis by phosphatases. We have shown strong APase activity in the epidermal tissues of normal roots and cluster rootlets and in root hairs of cluster rootlets under P deficiency using ELF-97 phosphate as a substrate (Wasaki et al. 2008). Actually, the strong activity in the rhizosphere of cluster roots formed by white lupin under P starvation was derived from root-secreted acid phosphatase (Wasaki et al. 2005). The gene for this root-secreted acid phosphatase was cloned and designated as *LASAP2* (Wasaki et al. 2000). *LASAP2* overexpressing tobacco lines were established and it was tested whether *LASAP2* contributed to P uptake from organic P in the soil (Wasaki et al. 2009). Our results suggest that P uptake by the transgenic plants increased both in aseptic gel media and under soil culture conditions.

4.5.2 Organic Acid Exudation from Plant Roots

Root-secreted organic acids have chelating abilities for metal ions in the soil, and mobilize P compounds bound to the metal ions or adsorbed to soil particles. Knowing the amount and composition of the secreted organic acids is important to better understand P mobilization in the rhizosphere. Rhizobox systems, developed by Dinkelaker and Marschner (1992), are useful for the localization of organic acids in roots. Organic acids trapped with clean filter paper can be measured by HPLC (high-performance liquid chromatography) or enzymatic quantification (Neumann and Römheld 1999; Wasaki et al. 2005). Commercial kits for enzymatic quantification of citrate and malate are available (e.g. Roche Diagnostics, Basel, Switzerland). For more details on P solubilization by organic acids see Jones and Oburger (2011).

4.5.3 Phosphate Transporters

Inorganic phosphate (Pi) is the form of P that is taken up by plants. Plants have two classes of Pi transporters, namely high- and a low-affinity types. Because the Pi concentration in the soil solution is quite low, it is believed that high-affinity Pi transporters play the major role in plant Pi uptake from the rhizosphere. In the case of *Arabidopsis*, there are nine members of high-affinity Pi transporters (Mudge et al. 2002). It was shown that AtPT1 is a high-affinity Pi transporter that cotransports H^+ by overexpression (Mitsukawa et al. 1997). It was indicated that one of the members, Pht1;4 (AtPT2), contributes greatly to phosphate uptake in P-deprived *Arabidopsis* (Misson et al. 2004).

Interestingly, the accumulation of mRNA for a high-affinity Pi transporter is synchronized with the infection of AM fungi (Nagy et al. 2005). Thus, it appears that the Pi transporter plays a role in Pi transport from arbuscules to plant cells. The accumulation of the mRNA for the low-affinity Pi transporter was not regulated by internal P levels of the plants. It appears likely that this type of transporter contributes to internal P translocation rather than P uptake.

4.5.4 Exudation of Antibiotic Compounds and Enzymes

White lupin has a high ability to mobilize P from sparingly soluble P through the secretion of organic acids (Gardner et al. 1983). The major root-secreted organic acid is citrate, which is readily consumed by microbes. To protect the secreted acids from degradation by microbes, white lupin secretes several compounds together with the organic acids (Weisskopf et al. 2006). Net proton release decreases the population of bacteria. Flavonoids and cell-wall-degrading enzymes protect the organic acids from consumption by fungi.

Metabolomic approaches can be useful for analysis of low molecular weight substances in root exudates (see Sect. 4.6.3). Several fluorogenic substrates can be applied for measurement of antibiotic enzyme activities (see Sect. 4.5.1).

It appears that plants and microorganisms compete for P (Unno et al. 2005). Thus, the secretion from roots of substances that protect secreted organic acids from degradation might be of general significance in soils. It seems necessary to consider the function of these antimicrobial substances in the context of plant–microbe interactions involved in P cycling.

4.5.5 Quantitative RT-PCR

qRT-PCR is frequently used for the quantification of plant mRNA, because it is a convenient and straightforward methodology, as discussed above. However, plants sometimes have numerous homologs of a given gene with different functions. Therefore, primers have to be designed carefully to detect a specific gene.

We have tried to analyze the metabolic alterations of P-deficient rice roots by transcriptomic analysis (see Sect. 4.6.1; Wasaki et al. 2003). It was indicated that two Pi transporters were regulated by P deficiency. qRT-PCR revealed that the Pi transporter expression found by cDNA microarray was exactly regulated by P deficiency.

4.6 Novel Technologies: Omics Analyses for P Cycling

4.6.1 Microarrays

The so-called omics analyses have become feasible during the last 10 years. Transcriptomics have been applied prior to other postgenomic strategies such as metabolomics and proteomics. Some transcriptomic studies of low-P adaptation strategies have been carried out using cDNA arrays for plants (Uhde-Stone et al. 2003; Wu et al. 2003; Hammond et al. 2003, 2004; Misson et al. 2005; Wang et al.

2002; Wasaki et al. 2003, 2006). Microarrays on a chip or a glass slide contain thousands of redundant sequences, which can be hybridized with fluorescent dye-labeled samples. The fluorescence is detected using specific scanners, which is the most important requirement. Some new important metabolic changes were suggested in our study of P-deficient rice roots, namely: (1) acceleration of carbon supply for organic acid synthesis through glycolysis; (2) alteration of lipid metabolism; (3) rearrangement of compounds for cell walls; and (4) changes in gene expression related to the response to metallic elements such as Al, Fe, and Zn (Wasaki et al. 2003).

Genomic sequencing or collections of expressed sequence tags are required before microarrays can be designed. Recently, ready-made arrays have become available not only for common model organisms such as *Arabidopsis* and rice, but also for woody plants including poplar, eucalyptus, and grape, and for crop species such as soybean, wheat, and maize.

Microbes involved in P cycling could be detected by the microarray technique if a microarray targeted for soil microbes is developed. He et al. (2007) developed a microarray designated the "GeoChip," containing 24,243 oligonucleotide probes and covering >10,000 genes in >150 functional groups involved in N, C, S, and P cycling, metal reduction and resistance, and organic contaminant degradation. Their array was successfully used for tracking the dynamics of metal-reducing bacteria and associated communities.

4.6.2 Metagenomics and the Next Generation of DNA Sequencers

To study uncultured microorganisms, environmental DNA can be recovered and sequenced. This approach is called "metagenomics." It is expected that metagenomics will provide the information required for understanding the whole microflora in the soil, the functions of the organisms involved (including P cycling), and the isolation of beneficial genes from uncultured microorganisms.

Recently, next-generation DNA sequencers have been developed (Mardis 2008). This development will contribute to progress not only in metagenomics but will also enable metatranscriptomic studies on P cycling. Furthermore, the method will facilitate transcriptomic analyses of not-yet sequenced organisms.

Recently, Badri et al. (2009) reported the first study on the microbial community in the rhizosphere using a pyrosequencer, one of the next-generation DNA sequencers. A metagenomic approach revealed that exudates from an *Arabidopsis* mutant of the ATP-binding cassette (ABC) transporter involved in root secretion of phytochemicals cultivated a microbial community with a relatively greater abundance of potentially beneficial bacteria. The next generation of DNA sequencers could support metagenomic and metatranscriptomic studies on P cycling.

Table 4.2 Application of molecular tools for studies on P cycling

Question	Target molecules	Molecular tools	Examples of application for studying P cycling
Structure of microbial communities involved in P cycling	SSU rRNA	Clone library	Monteiro et al. (2009)
		DGGE/TGGE/SSCP	Marschner et al. (2004), Wasaki et al. (2005), Weisskopf et al. (2007)
		RFLP	George et al. (2009), Monteiro et al. (2009)
		Microarray	–
	Phospholipid fatty acids	PLFA	Tscherko et al. (2004)
Active microbes involved in P cycling	Functional genes	Clone library	–
		DGGE/TGGE/SSCP	Sakurai et al. (2008)
		RFLP	–
	Metagenomics	Next generation sequencer	Badri et al. (2009)
Quantitative analysis of specific microbes	SSU rRNA/functional genes	qRT-PCR	Alkan et al. (1995)
Localization of specific microbes	SSU rRNA/functional genes	FISH	–
P availability in the rhizosphere	P starvation responsive gene	Phosphate-reporter bacteria	de Weger et al. (1994), Kragelund et al. (1997)
Enzyme activities involved in P cycling	Enzyme activities	Fluorogenic substrates	Tscherko et al. (2004), Wasaki et al. (2005)
Localization of phosphatase activities	Phosphatase	ELF-97 phosphate	Wasaki et al. (2008)
Response of microbes to root exudates	SSU rRNA/functional genes	SIP-DGGE	–
	Phospholipid fatty acids	SIP-PLFA	–
Active microbes involved in P cycling	SSU rRNA/functional genes	BrdU	Artursson et al. (2005)
Plant responses to P deficiency	Transcriptomics	Microarray	Hammond et al. (2003, 2004), Uhde-Stone et al. (2003), Wu et al. (2003), Wang et al. (2002), Misson et al. (2005), Wasaki et al. (2003, 2006)
		Next generation sequencer	–
Effects of root-secreted enzymes	Phosphatase/phytase	Overexpression	Wasaki et al. (2008)

4.6.3 *Proteomics and Metabolomics*

Methods for proteomics and metabolomics became available due to the development of mass spectrometric techniques such as TOF (time of flight)-MS (mass spectrometry), GC (gas chromatography)-MS/MS, and LC (liquid chromatography)-MS/MS. These will also contribute to the understanding of P cycling. A proteomic study on P- and Al-stressed rice plants has already been reported (Fukuda et al. 2007). Crude proteins of control and stressed roots were separated on two-dimensional polyacrylamide gel electrophoresis (2D-PAGE), and the unique spots found in samples from stressed plants were determined by peptide mapping using TOF-MS. It was shown that the proteomic alterations under P deficiency and Al stress conditions were similar, indicating that a common metabolic system is responsive to both P deficiency and Al stress.

Suzuki et al. (2009) reported the metabolomic analysis of root exudates of rice seedlings. GC-MS was used for the analysis of primary metabolites in the study. Their results suggested that the physiological change during seedling development was large and that the response to the environment was rather small. We expect that further improvement of these methods and the combination of molecular approaches and metabolomics will provide crucial information.

4.7 Conclusions

In Table 4.2, we have summarized the potential applications of molecular tools for studying P cycling, with examples from the literature. The relationships between tools, samples, and targets are illustrated in Fig. 4.3. We would be happy if this chapter helps readers' studies on P cycling.

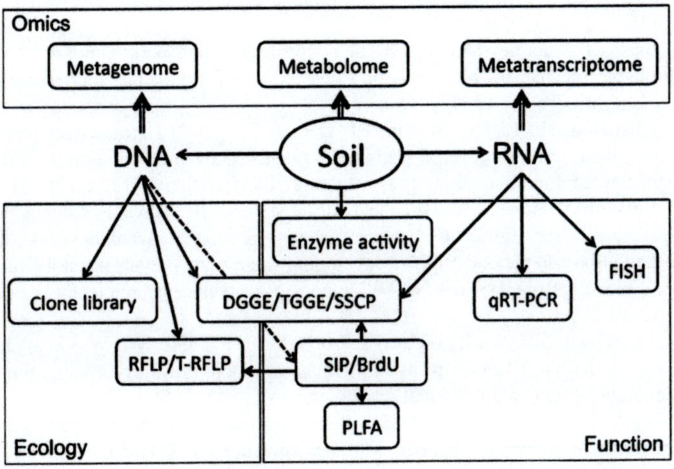

Fig. 4.3 Molecular approaches for analyzing the ecology and function of soil microorganisms

Molecular approaches and recent omics analyses have proven powerful tools for clarifying processes and actors involved in P cycling. On the other hand, we must not forget the advantages of conventional, culture-dependent methods. For example, metagenomic data cannot identify intracellular symbioses. The "living" cell is the central focus of microbial ecology. Adequate molecular approaches have to be selected that provide meaningful complements to conventional approaches for future investigations into P cycling.

References

Alkan N, Gadkar V, Yarden O, Kapulnik Y (2006) Analysis of quantitative interactions between two species of arbuscular mycorrhiz fungi, *Glomus mosseae* and *G. intraradices*, by real-time PCR. Appl Environ Microbiol 72:4192–4199

Artursson V, Finlay RD, Jansson JK (2005) Combined bromodeoxyuridine immunocapture and terminal-restriction fragment length polymorphism analysis highlights differences in the active soil bacterial metagenome due to *Glomus mosseae* inoculation or plant species. Environ Microbiol 7:1952–1966

Badri DV, Quintana N, El Kassis EG, Kim HK, Choi YH, Sugiyama A, Verpoorte R, Martinoia E, Manter DK, Vivanco JM (2009) An ABC transporter mutation alters root exudation of phytochemicals that provoke an overhaul of natural soil microbiota. Plant Physiol 151:2006–2017

Chen Y, Dumont MG, McNamara NP, Chamberlain PM, Bodrossy L, Stralis-Pavese N, Murrell JC (2008) Diversity of the active methanotrophic community in acidic peatlands as assessed by mRNA and SIP-PLFA analyses. Environ Microbiol 10:446–459

de Weger LA, Dekkers LC, van der Bij AJ, Lugtenberg BJJ (1994) Use of phosphate-reporter bacteria to study phosphate limitation in the rhizosphere and in bulk soil. Mol Plant Microb Interact 7:32–38

Dinkelaker B, Marschner H (1992) In vivo demonstration of acid-phosphatase-activity in the rhizosphere of soil-grown plants. Plant Soil 144:199–205

Fukuda T, Saito A, Wasaki J, Shinano T, Osaki M (2007) Metabolic alterations proposed by proteome in rice roots grown under low P and high Al concentration under low pH. Plant Sci 172:1157–1165

Gardner WK, Barber DA, Parbey DG (1983) The acquisition of phosphorus by *Lupinus albus* L. III. The probable mechanism by which phosphorus movement in the soil/root interface is enhanced. Plant Soil 70:107–124

George TS, Richardson AE, Li SS, Gregory PJ, Daniell TJ (2009) Extracellular rerelase of a heterologous phytase from roots of transgenic plants: does manipulation of rhizosphere biochemistry impact microbial community structure? FEMS Microbiol Ecol 70:433–445

Hammond JP, Bennett MJ, Bowen HC, Broadley MR, Eastwood DC, May ST, Rahn C, Swarup R, Woolaway KE, White PJ (2003) Changes in gene expression in Arabidopsis shoots during phosphate starvation and the potential for developing smart plants. Plant Physiol 132:578–596

Hammond JP, White PJ, Broadley MR (2004) Diagnosing phosphorus deficiency in plants. Asp Appl Biol 72:89–98

He Z, Gentry TJ, Schadt CW, Wu L, Liebich J, Chong SC, Huang Z, Wu W, Gu B, Jardine P, Criddle C, Zhou J (2007) GeoChip: a comprehensive microarray for investigating biogeochemical, ecological and environmental processes. ISME J 1:67–77

Henco K, Heibey M (1990) Quantitative PCR – the determination of template copy numbers by temperature-gradient gel-electrophoresis (TGGE). Nucleic Acid Res 18:6733–6734

Heuer H, Krsek M, Baker P, Smalla K, Wellington EMH (1997) Analysis of actinomycete communities by specific amplification of genes encoding 16S rRNA and gel-electrophoretic separation in denaturing gradients. Appl Environ Microbiol 63:3233–3241

Ito T, Nielsen JL, Okabe S, Watanabe Y, Nielsen PH (2002) Phylogenetic identification and substrate uptake patterns of sulfate-reducing bacteria inhabiting an oxic-anoxic sewer biofilm determined by combining microautoradiography and fluorescent in situ hybridization. Appl Environ Microbiol 68:356–364

Jones DL, Oburger E (2011) Solubilization of phosphorus by soil microorganisms. In: Bünemann EK, Oberson A, Frossard E (eds) Phosphorus in action: biological processes in soil phosphorus cycling, vol 26, Soil biology. Springer, Heidelberg, Berlin. doi:10.1007/978-3-642-15271-9_7

Kong YH, Nielsen JL, Nielsen PH (2005) Identity and ecophysiology of uncultured actinobacterial polyphosphate-accumulating organisms in full-scale enhanced biological phosphorus removal plants. Appl Environ Microbiol 71:4076–4085

Kragelund L, Hosbond C, Nybroe O (1997) Distribution of metabolic activity and phosphate starvation response of *lux*-tagged *Pseudomonas fluorescens* reporter bacteria in the barley rhizosphere. Appl Environ Microbiol 63:4920–4928

Mardis ER (2008) The impact of next-generation sequencing technology on genetics. Trends Genet 24:133–141

Marschner P, Crowley D, Yang CH (2004) Development of specific rhizosphere bacterial communities in relation to plant species, nutrition and soil type. Plant Soil 261:199–208

Massol-Deya AA, Odelson DA, Hickey RF, Tiedje JM (1995) Bacterial community fingerprinting of amplified 16S and 16-23S ribosomal DNA gene sequences and restriction endonuclease analysis (ARDRA). In: Akkermans ADL, van Elsas JD, Bruijin FJ (eds) Molecular microbial ecology. Kluwer, Dordrecht, pp 1–8

Misson J, Thibaud MC, Bechtold N, Raghothama K, Nussaume L (2004) Transcriptional regulation and functional properties of Arabidopsis Pht1;4, a high affinity transporter contributing greatly to phosphate uptake in phosphate deprived plants. Plant Mol Biol 55:727–741

Misson J, Raghothama KG, Jain A, Jouhet J, Block MA, Bligny R, Ortet P, Creff A, Somerville S, Rolland N, Doumas P, Nacry P, Herrerra-Estrella L, Nussaume L, Thibaud M-C (2005) A genome-wide transcriptional analysis using *Arabidopsis thaliana* Affymetrix gene chips determined plant responses to phosphate deprivation. Proc Natl Acad Sci USA 102:11934–11939

Mitsukawa N, Okumura S, Shirano Y, Sato S, Kato T, Harashima S, Shibata D (1997) Overexpression of an *Arabidopsis thaliana* high-affinity phosphate transporter gene in tobacco cultured cells enhances cell growth under phosphate-limited conditions. Proc Natl Acad Sci USA 94:7098–7102

Monteiro JM, Vollú RE, Coelho MRR, Alciano CS, Blank AF, Seldin L (2009) Comparison of the bacterial community and characterization of plant growth-promoting rhizobacteria from different genotypes of *Chrysopogon zizanioides* (L.) Roberty (Vetiver) rhizospheres. J Microbiol 47:363–370

Mudge SR, Rae AL, Diatloff E, Smith FW (2002) Expression analysis suggests novel roles for members of the Pht1 family of phosphate transporters in Arabidopsis. Plant J 31:341–353

Nagy R, Karandashov V, Chague W, Kalinkevich K, Tamasloukht M, Xu GH, Jakobsen I, Levy AA, Amrhein N, Bucher M (2005) The characterization of novel mycorrhiza-specific phosphate transporters from *Lycopersicon esculentum* and *Solanum tuberosum* uncovers functional redundancy in symbiotic phosphate transport in solanaceous species. Plant J 42:236–250

Nannipieri P, Giagnoni L, Landi L, Renella G (2011) Role of phosphatase enzymes in soil. In: Bünemann EK, Oberson A, Frossard E (eds) Phosphorus in action: biological processes in soil phosphorus cycling, vol 26, Soil biology. Springer, Heidelberg, Berlin. doi:10.1007/978-3-642-15271-9_9

Neumann G, Martinoia E (2002) Cluster roots – an underground adaptation for survival in extreme environments. Trends Plant Sci 7:162–167

Neumann G, Römheld V (1999) Root excretion of carboxylic acids and protons in phosphorus-deficient plants. Plant Soil 211:121–130

Ozawa K, Osaki M, Matsui H, Honma M, Tadano T (1995) Purification and properties of acid phosphatase secreted from lupin roots under phosphorus-deficiency conditions. Soil Sci Plant Nutr 41:461–469

Pernthaler A, Pernthaler J, Amann R (2002) Fluorescence in situ hybridization and catalyzed reporter deposition for the identification of marine bacteria. Appl Environ Microbiol 68:3094–3101

Prosser JI, Rangel-Castro JI, Killham K (2006) Studying plant-microbe interactions using stable isotope technologies. Curr Opin Biotechnol 17:98–102

Rangel-Castro JI, Killham K, Ostle N, Nicol GW, Anderson IC, Scrimgeour CM, Ineson P, Meharg A, Prosser JI (2005) Stable isotope probing analysis of the influence of liming on root exudate utilization by soil microorganisms. Environ Microbiol 7:828–838

Rogers SO, Bendich AJ (1985) Extraction of DNA from milligram amounts of fresh herbarium and minifield plant tissues. Plant Mol Biol 5:69–76

Sakurai M, Wasaki J, Tomizawa Y, Shinano T, Osaki M (2008) Analysis of bacterial communities on alkaline phosphatase gene in organic matter applied soil. Soil Sci Plant Nutr 54:62–71

Schriml LM, Padilla-Nash HM, Coleman A, Moen P, Nash WG, Menninger J, Jones G, Ried T, Dean M (1999) Tyramide signal amplification (TSA)-FISH applied to mapping PCR-labeled probes less than 1 kb in size. Biotechniques 27:608

Sheffield VC, Cox DR, Lerman LS, Myers RM (1989) Attachment of a 40-base-pair G+C-rich sequence (GC-clamp) to genomic DNA fragments by the polymerase chain reaction results in improved detection of single-base changes. Proc Natl Acad Sci USA 86:232–236

Suzuki K, Okazaki K, Tawaraya K, Osaki M, Shinano T (2009) Gas chromatography-mass spectrometry associated grobal analysis of rice root exudates under aseptical conditions. Soil Sci Plant Nutr 55:505–513

Torsvik V, Øvreås L (2002) Microbial diversity and function in soil: from genes to ecosystems. Curr Opin Microbiol 5:240–245

Tscherko D, Hammesfahr U, Marx MC, Kandeler E (2004) Shifts in rhizosphere microbial communities and enzyme activity of *Poa alpina* across an alpine chronosequence. Soil Biol Biochem 36:1685–1698

Uhde-Stone C, Zinn KE, Ramirez-Yáñez M, Li A, Vance CP, Allan DL (2003) Nylon filter arrays reveal differential gene expression in proteoid roots of white lupin in response to phosphorus deficiency. Plant Physiol 131:1064–1079

Unno Y, Okubo K, Wasaki J, Shinano T, Osaki M (2005) Plant growth promotion abilities and micro-scale bacterial dynamics in the rhizosphere of lupin analyzed by phytate utilization ability. Environ Microbiol 7:396–404

Wang Y-H, Gravin DF, Kochian LV (2002) Rapid induction of regulatory and transporter genes in response to phosphorus, potassium, and iron deficiencies in tomato roots. Evidence for cross talk and root/rhizosphere-mediated signals. Plant Physiol 130:1361–1370

Wasaki J, Ando M, Ozawa K, Omura M, Osaki M, Ito H, Matsui H, Tadano T (1997) Properties of secretory acid phosphatase from lupin roots under phosphorus-deficient conditions. Soil Sci Plant Nutr 43:981–986

Wasaki J, Omura M, Ando M, Shinano T, Osaki M, Tadano T (1999) Secreting portion of acid phosphatase in roots of lupin (*Lupinus albus* L.) and a key signal for the secretion from the roots. Soil Sci Plant Nutr 45:937–945

Wasaki J, Omura M, Ando M, Dateki H, Shinano T, Osaki M, Ito H, Matsui H, Tadano T (2000) Molecular cloning and root specific expression of secretory acid phosphatase from phosphate deficient lupin (*Lupinus albus* L.). Soil Sci Plant Nutr 46:427–437

Wasaki J, Yonetani R, Kuroda S, Shinano T, Yazaki J, Fujii F, Shimbo K, Yamamoto K, Sakata K, Sasaki T, Kishimoto N, Kikuchi S, Yamagishi M, Osaki M (2003) Transcriptomic analysis of metabolic changes by phosphorus stress in rice plant roots. Plant Cell Environ 26:1515–1523

Wasaki J, Rothe A, Kania A, Neumann G, Römheld V, Shinano T, Osaki M, Kandeler E (2005) Root exudation, P acquisition and microbial diversity in the rhizosphere of *Lupinus albus* as affected by P supply and atmospheric CO_2 concentration. J Environ Qual 34:2157–2166

Wasaki J, Shinano T, Onishi K, Yonetani R, Yazaki J, Fujii F, Shimbo K, Ishikawa M, Shimatani Z, Nagata Y, Hashimoto A, Ohta T, Sato Y, Miyamoto C, Honda S, Kojima K, Sasaki T, Kishimoto N, Kikuchi S, Osaki M (2006) Transcriptomic analysis indicates putative metabolic changes caused by manipulation of phosphorus availability in rice leaves. J Exp Bot 57:2049–2059

Wasaki J, Kojima S, Maruyama H, Haase S, Osaki M, Kandeler E (2008) Localization of acid phosphatase activities in roots of white lupin plants grown under phosphorus-deficient conditions. Soil Sci Plant Nutr 54:95–102

Wasaki J, Maruyama H, Tanaka M, Yamamura T, Dateki H, Shinano T, Ito S, Osaki M (2009) Overexpression of *LASAP2* gene for secretory acid phosphatase of white lupin improves P uptake and growth of tobacco plants. Soil Sci Plant Nutr 55:107–113

Weisskopf L, Fromin N, Tomasi N, Arango M, Martinoia E (2005) Secretion activity of white lupin's cluster roots influences bacterial abundance, function and community structure. Plant Soil 268:181–194

Weisskopf L, Abou-Mansour E, Fromin N, Tomasi N, Santelia D, Edelkott I, Neumann G, Aragno M, Tabacchi R, Martinoia E (2006) White lupin has developed a complex strategy to limit microbial degradation of secreted citrate required for phosphate acquisition. Plant Cell Environ 29:919–927

Wu P, Ma L, Hou X, Wang M, Wu Y, Liu F, Deng XW (2003) Phosphate starvation triggers distinct alterations of genome expression in Arabidopsis roots and leaves. Plant Physiol 132:1260–1271

Chapter 5
Modelling Phosphorus Dynamics in the Soil–Plant System

Andrea Schnepf, Daniel Leitner, Sabine Klepsch, Sylvain Pellerin, and Alain Mollier

5.1 Introduction

The importance of phosphorus (P) as both sparingly mobile essential nutrient and pollutant is reflected by the large number of P models at different scales and with different purposes. P dynamics have been studied at a wide range of spatial scales. Examples include the global scale (Harrison et al. 2005), watershed scale (Radcliffe et al. 2009), ecosystem scale (Schlecht and Hiernaux 2005), farming systems scale (Schils et al. 2007), field scale (Schoumans and Groenendijk 2000; Torbert et al. 2008), whole plant scale (Mollier et al. 2008), soil profile scale (Roose and Fowler 2004), and single root scale (Kirk 1999; Roose et al. 2001). The positions of the various P models in a space–time diagram (Fig. 5.1) illustrate the main temporal and spatial scales of application. Most of the models are mechanistic and deterministic; the degree of empiricalness generally increases with spatial scale.

The applications for which P models have been used include carbon uptake by the terrestrial biosphere (Wang et al. 2007), effect of tectonic uplift and erosion on P availability in soil (Porder et al. 2007), water quality due to agricultural

A. Schnepf (✉) and D. Leitner
Department of Forest and Soil Sciences, Institute of Soil Science, BOKU – University of Natural Resources and Life Sciences Vienna, Peter Jordan Strasse 82, 1190 Vienna, Austria
e-mail: andrea.schnepf@boku.ac.at; daniel.leitner@boku.ac.at

S. Klepsch
Department of Forest and Soil Sciences, Institute of Soil Science, BOKU – University of Natural Resources and Life Sciences Vienna, Peter Jordan Strasse 82, 1190 Vienna, Austria
and
Health & Environment Department, Environmental Resources & Technologies, AIT Austrian Institute of Technology, 2444 Seibersdorf, Austria
e-mail: sabine.klepsch@boku.ac.at

S. Pellerin and A. Mollier
Department for the Environment and Agronomy, Joint Research Unit for Soil–Plant Transfer and the Cycle of Nutrients and Trace Elements (TCEM), National Institute for Agricultural Research (INRA), 71, Avenue Edouard Bourlaux, BP 81, 33883 Villenave d'Ornon Cedex, France
e-mail: pellerin@bordeaux.inra.fr; mollier@bordeaux.inra.fr

E.K. Bünemann et al. (eds.), *Phosphorus in Action*, Soil Biology 26,
DOI 10.1007/978-3-642-15271-9_5,

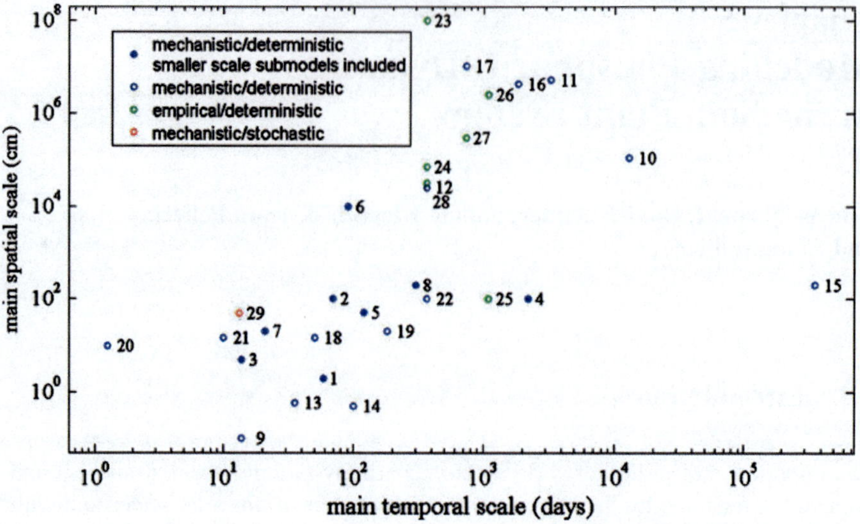

Fig. 5.1 Space–time diagram of commonly used P models. *Closed symbols* indicate that this specific model includes smaller-scale submodels. The *colour* illustrates whether the model is mechanistic or empirical, deterministic or stochastic. References for model numbers: *1* Ptashnyk et al. (2010), *2* Shi and Erickson (2001), *3* Schnepf and Roose (2006), *4* Grant et al. (2004), *5* Roose and Fowler (2004), *6* Mollier et al. (2008), *7* Bhadoria et al. (2002), *8* Dunbabin et al. (2009), *9* Hofflmann et al. (1994), *10* Torbert et al. (2008), *11* Radcliffe et al. (2009), *12* Schoumans and Groenendijk (2000), *13* Kirk (1999), *14* Roose et al. (2001), *15* Porder et al. (2007), *16* Arheimer et al (2004), *17* Migliaccio et al. (2007), *18* Huguenin-Elie et al. (2009), *19* Reginato et al. (2000), *20* Overman and Scholtz (1999), *21* Grant and Robertson (1997), *22* Landis and Fraser (2008), *23* Harrison et al. (2005), *24* Schils et al. (2007), *25* Chung et al. (2003), *26* Dalzell et al. (2004), *27* Djodjic et al. (2002), *28* Giasson et al. (2002), *29* Ge et al. (2000)

management practices (Arheimer et al. 2004; Chung et al. 2003; Dalzell et al. 2004; Djodjic et al. 2002; Grant et al. 2004; Migliaccio et al. 2007), P uptake from soil by root systems (Leitner et al. 2010a; Roose and Fowler 2004), root P uptake and P efficiency (Bhadoria et al. 2002; Huguenin-Elie et al. 2009; Kirk 1999; Reginato et al. 2000), crop response to soil P levels (Mollier et al. 2008), competing root systems (Smethurst and Comerford 1993), manure management and fertiliser optimisation (Dunbabin et al. 2009; Giasson et al. 2002; Teklić et al. 2002), P as amendment in in-situ stabilisation of lead (Pb)-contaminated soils (Shi and Erickson 2001), P sorption in soil (Overman and Scholtz 1999; Van Der Zee and Van Riemsdijk 1991), P uptake by root hairs and mycorrhizal fungi (Grant and Robertson 1997; Landis and Fraser 2008; Leitner et al. 2010b; Schnepf and Roose 2006; Schnepf et al. 2008b), biodiversity (Fitter et al. 2005) and forest growth response (Gillespie and Pope 1990; Kirschbaum et al. 1998). Reviews describing models for plant solute uptake are given by Tinker and Nye (2000), Darrah et al. (2006) and Luster et al. (2009).

Many of these models are based on generic software, and it is often not apparent what the underlying equations and assumptions are. This could lead to inappropriate applications and erroneous model outputs. Moreover, the applicability of such models is limited to the specific purpose for which each was designed, making it problematic to extend a particular model for a slightly different aim. This calls for documenting the underlying theory and mathematical basis of any model in a form that can be understood by non-experts. Furthermore, nowadays a set of mathematical equations can be solved relatively easily by flexible software packages (e.g. Comsol Multiphysics, FlexPDE).

5.1.1 Building a Mathematical Model

A mathematical model is a simplified description of reality in terms of mathematical equations. For many scientific problems, models can help to quantify expected results, compare the effects of alternative theories, describe the effect of complex factors, explain how underlying processes contribute to the observed results, extrapolate results to other situations, or predict future events (Smith and Smith 2007). In order to make the model most meaningful and least prone to errors, it should be as simple as possible, but not any simpler. In the words of Einstein (1934): "It can scarcely be denied that the supreme goal of all theory is to make the irreducible basic elements as simple and as few as possible without having to surrender the adequate representation of a single datum of experience."

Each model is built for a specific purpose and starts from prior knowledge and hypotheses about the system. This will influence the type of model chosen and the corresponding mathematical equations used to describe the system. Models are generally classified as deterministic or stochastic and as mechanistic or empirical. Each model needs data for parameterisation; if some of the data are incompletely known; model calibration is required to determine these values accurately from available measurements (Janssen and Heuberger 1995).

Finally, accuracy, sensitivity and uncertainty analyses should be made for quantitative model evaluation (Saltelli et al. 2000; Smith and Smith 2007). Model development can be an iterative process if the model is found to be inappropriate after evaluation. In this chapter, we focus on building mathematical models based on the objectives and prior knowledge.

5.1.2 Aims of This Chapter

We illustrate model building for understanding P dynamics in the plant–soil system using three case studies. The studies focus on the traits that enhance plant P uptake from soil: mycorrhizal associations and root architecture, as well as on crop responses to soil P levels.

5.2 Modelling Case Studies

5.2.1 *P Uptake by Mycorrhizal Roots*

5.2.1.1 Aim of the Model

The soil volume that a plant root can exploit for sparingly soluble nutrients such as P increases enormously due to mycorrhizal roots. Non-mycorrhizal parts of a root typically have a depletion zone that is less than 5 mm wide, whereas the depletion zone of mycorrhizal root parts can reach several centimetres into the soil. Thus, mycorrhizal fungi can increase soil P availability to plants. It may be possible to exploit symbioses between various crop plants and mycorrhizal fungi to reduce the use of mineral fertilisers in agricultural management (Frossard et al. 2000). The model presented here helps to estimate how much P fertiliser could be substituted in this way, a topic especially important because of the anticipated insufficient future mineral phosphate supply (Lambers et al. 2006). This examplary simulation study is designed to quantify the effect of arbuscular mycorrhizal fungi on plant P nutrition.

In this case study, we demonstrate how to quantify soil phosphate (P_i) depletion and P_i influx into a plant root colonised by arbuscular mycorrhizal fungi using the model of Schnepf and Roose (2006) and Schnepf et al. (2008a, b). Three arbuscular mycorrhizal fungi with different growth strategies are considered. The first growth strategy describes a fungus with tip splitting proportional to the hyphal tip density (linear branching); the second assumes that branching ceases at a given maximal hyphal tip density (nonlinear branching); and the third describes a fungus that develops a highly interconnected mycelium (anastomosis).

We consider a single root with different fungal mycelia that correspond to the three growth strategies. As described in Schnepf and Roose (2006) and Schnepf et al. (2008b), P_i transport in soil is described by the diffusion equation. Furthermore, we assume that P_i is taken up by root and hyphae according to Michaelis–Menten kinetics. Previous simulations were performed in one-dimensional Cartesian coordinates in order to be consistent with validation experiments performed in compartment systems (e.g. as used by Li et al. 1991). The present study focuses on a single mycorrhizal root and therefore uses cylindrical coordinates.

5.2.1.2 Model Description

The chosen modelling approach is based on coupling a fungal growth and P_i uptake model with a classical single root model. The classical single root model (Barber 1995; Tinker and Nye 2000) is extended by a sink term for P_i uptake from soil due to arbuscular mycorrhizal hyphae [see (5.1)]. The model is given by the following equations:

$$(\theta + b)\frac{\partial c}{\partial t} = \frac{1}{r}\frac{\partial}{\partial r}\left(D\theta f r\frac{\partial c}{\partial r}\right) - 2r_{\mathrm{h}}\pi\rho(r,t)\frac{F_m c}{K_m + c}, \quad (5.1)$$

$$c = c_0 \quad \text{at} \quad t = 0\,, \tag{5.2}$$

$$D\theta f \frac{\partial c}{\partial r} = \frac{F_m c}{K_m + c}, \quad \text{at} \quad r = r_0\,, \tag{5.3}$$

$$D\theta f \frac{\partial c}{\partial r} = 0, \quad \text{at} \quad r = r_1, \tag{5.4}$$

where c is concentration of phosphate in soil solution, t time, r radial distance from root axis, θ volumetric water content, b buffer power, f impedance factor, D diffusion coefficient in water, c_0 initial P_i concentration in soil solution, r_h hyphal radius, r_0 root radius, r_1 mean half distance between roots, F_m maximal P_i influx into root, K_m Michaelis–Menten constant and ρ hyphal length density.

The hyphal length density, ρ, is determined using a continuous and spatially explicit hyphal population growth model (Schnepf et al. 2008a). It calculates hyphal tip and length densities based on elongation of the region just behind the hyphal tip, branching due to tip splitting, anastomosis, and tip and hyphal death. Calibration of this model to experimental data (Jakobsen et al. 1992) indicated that all three presented growth strategies occurred. The parameters found in that calibration study (Schnepf et al. 2008a) are used in the following simulations. Model equations for the hyphal tip density n and the hyphal length density ρ are given by:

$$\frac{\partial n}{\partial t} = -\frac{1}{r}\frac{\partial}{\partial r}(rnv) + F, \tag{5.5}$$

$$\frac{\partial \rho}{\partial t} = nv - d\rho\,, \tag{5.6}$$

$$F = \underbrace{b_n n \left(1 - \frac{n}{n_{\max}}\right)}_{branching} - \underbrace{a_1 n^2 - a_2 n\rho}_{anastomosis} - \underbrace{d_n n}_{tipdeath}\,, \tag{5.7}$$

$$n = 0, \rho = 0 \quad \text{at} \quad t = 0, \tag{5.8}$$

$$n = f_b(t) \quad \text{at} \quad r = r_0\,, \tag{5.9}$$

where n is hyphal tip density, ρ hyphal length density, v tip elongation rate, d hyphal death rate, b_n branching rate, $n_{\max}$ maximal tip density, a_1 and a_2 tip-tip and tip-side anastomosis rates, d_n tip death rate, f_b tip density at the root interface and F is the rate of creation or destruction of hyphal tips.

Equations (5.5)–(5.9) were solved with a Lax–Wendroff-scheme (Morton and Mayers 1994). Equations (5.1)–(5.4) were solved with the finite element method using Comsol multiphysics.

5.2.1.3 Results

Figure 5.2 shows a two-dimensional visualisation of the colony shapes resulting from the different fungal growth strategies. Simulations are based on a discrete L-system model, which approximates the continuous model given by (5.5)–(5.9). The hyphal length densities and P_i depletion at different distances from the root after 21 days are shown in Fig. 5.3. The different fungal growth strategies clearly yield different patterns of extraradical hyphal length densities and thus different P_i depletion radii of the mycorrhizal roots. The radius of the zone where more than half of the initial P_i concentration has been depleted is 1.5 cm for the linear branching strategy, 3 cm for nonlinear branching and 6 cm for anastomosis.

Figure 5.4 shows the resulting influx of P_i into root and hyphae. In all cases, influx into fungus exceeds that into root by an order of magnitude. Influx into root decreases quickly with time, whereas influx into fungus increases as the colony grows. For the two nonlinear growth strategies, this can decrease with time because the fungal colony covers a large area that is already depleted and, thus, additional P_i is taken up only at the front of the colony. This is shown more explicitly in Fig. 5.4, where P_i influx into hyphae per unit hyphal length, i.e. colony efficiency, is plotted against time.

5.2.1.4 Discussion and Outlook

The results show that, as soon as the fungal colony is sufficiently large, mycorrhizal P_i influx is up to an order of magnitude larger than root P_i influx. This is particularly pronounced in the linear branching case, where the fungal hyphae compete for P_i inside the root depletion zone. The anastomosis case, in contrast, makes more use of the soil volume near the front of the fungal colony, outside the root P_i depletion zone, so that the root contribution is higher. However, fungal P_i uptake dominates overall P_i uptake in all cases, supporting the idea that roots can completely rely on the fungus for their P_i nutrition (Smith et al. 2003). The hyphal length density in the

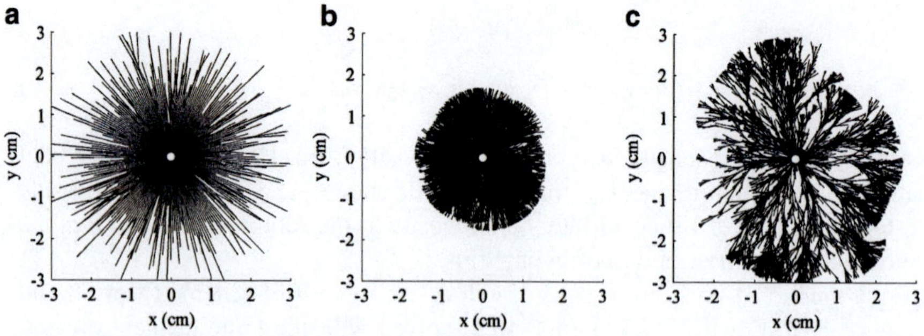

Fig. 5.2 Visualisation of (**a**) linear branching, (**b**) nonlinear branching and (**c**) anastomosis growth strategies of arbuscular mycorrhizal fungi after 14 days

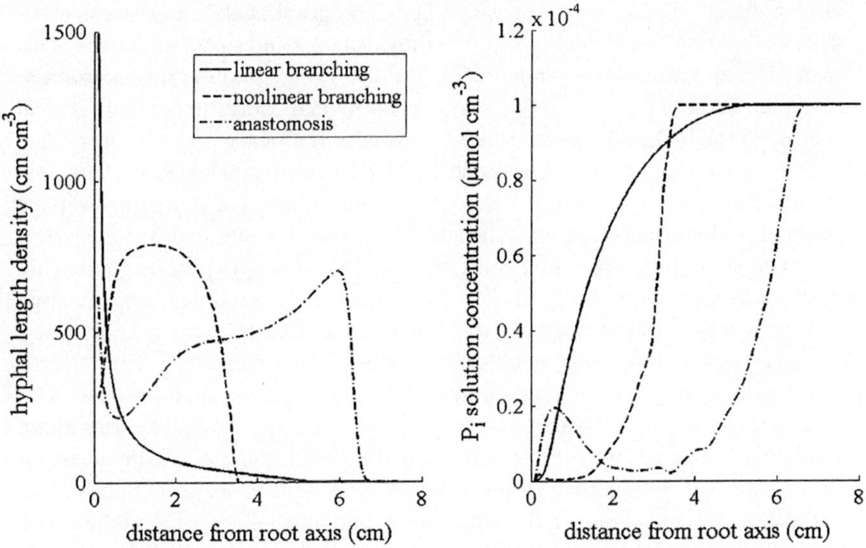

Fig. 5.3 Hyphal length densities (*left*) and P_i solution concentration around the root (*right*) corresponding to the linear branching, nonlinear branching and anastomosis growth strategies of arbuscular mycorrhizal fungi, assuming that root and fungi have the same P_i uptake parameters

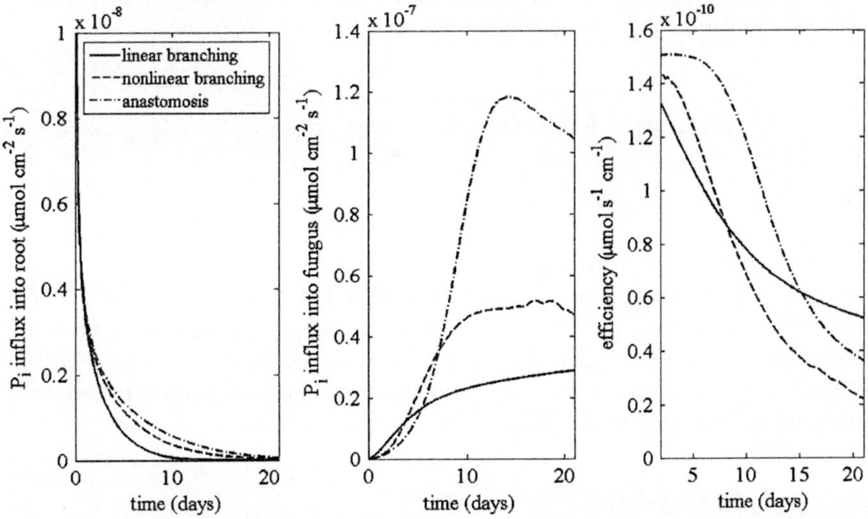

Fig. 5.4 P_i influx per unit root surface area due to root (*left*) and hyphae (*centre*) corresponding to the linear branching, nonlinear branching and anastomosis growth strategies of arbuscular mycorrhizal fungi. P_i influx per unit hyphal length (*right*)

linear branching case is high near the root surface but low further away. Accordingly, it is initially less efficient than the two other growth strategies. The nonlinear branching and anastomosis strategies result in a higher P_i influx into hyphae, but they are less efficient when relating this influx to the colony size produced.

The predicted hyphal length densities and P_i concentrations in Schnepf et al. (2008b) were based on simulations in one-dimensional Cartesian coordinates. This geometry is appropriate for a validation experiment using a compartment system in which a membrane separates a root from a purely hyphal compartment (e.g. as used by Li et al. 1991). In the present case study, we used cylindrical geometry appropriate for an individual root. This changes the shape of predicted hyphal length density and P_i depletion because, in cylindrical geometry, the same number of hyphae grow into a soil volume (which increases with distance from the root surface). Compared to Cartesian geometry, the linear branching strategy has a lower hyphal length density further away from the root surface. For the nonlinear branching strategy, the maximal tip density is reached faster near the root surface than further away and, similarly, anastomosis occurs more likely near the root surface than further away. This reduces hyphal length density near the root surface compared to Cartesian geometry.

Schnepf et al. (2008b) simulated differences with regard to P_i uptake sites along the individual hyphae. Compared with published values of P influx into mycorrhizal plants and soil P depletion, their results suggest that P_i uptake occurs not only at the tip but also at parts of the mycelium that are metabolically active. Nonetheless, a spatially explicit model for the spread of mycorrhizal mycelium and active parts is still missing. Assuming Michaelis–Menten kinetics for uptake might also oversimplify the actual uptake process. The molecular and biochemical characterisation of the corresponding P_i transport systems is currently being extensively studied (Bucher 2007; Raghothama and Karthikeyan 2005). Thus, more experimental data for parameterisation and validation are required.

5.2.2 *P Uptake by a Root System*

5.2.2.1 Aim of the Model

This case study is designed to estimate the impact of all individual roots to the overall root system dynamics. We use the model of Leitner et al. (2010a) to study (1) the effect of root system architecture and branching structure on overall root system P_i uptake and soil P_i depletion, and (2) root system development as affected by P_i concentrations in soil. The latter includes feedback between soil P_i concentration and root system development, including chemotropism.

5.2.2.2 Model Description

The focus here is on P_i uptake, neglecting water movement. We consider a maize root system growing in a conical pot (height 10 cm, top diameter 10 cm, bottom diameter 6 cm). The simulation starts with five germinating seeds at the top of the pot. The P_i concentration of the left half of the pot differs from that on the right half, and our aim is to quantify the effect of this difference on root system development and P_i uptake.

We simulate the development of root system architecture with the three-dimensional L-system model of Leitner et al. (2010a). The Matlab code is freely available online (Schnepf and Klepsch 2010). This model is similar to other root architecture models such as RootMap (Claassen et al. 2006; Diggle 1988) or RootTyp (Pages et al. 2004) in that root growth parameters are predetermined for each topological order. All parameter values are given by mean value and standard deviation [see Leitner et al. (2010a) for a full table of model parameters]. The model also provides a simple way to include different kinds of tropisms such as gravitropism, chemotropism and thigmotropism. The latter can be used to define a geometry, e.g. a pot, within which root growth is confined. All tropisms are implemented by a random optimisation algorithm, which is independent of the spatial discretisation along the root axis. The outcome of the root growth model is a root system composed of individual root segments. The information stored for each segment includes length, radius, position and age.

We describe P_i diffusion in soil and nutrient uptake by the root system with (5.10)–(5.12) (Barber 1995; Tinker and Nye 2000):

$$(\theta + b)\frac{\partial c}{\partial t} = \nabla \cdot (D\theta f \nabla c) - F\,, \tag{5.10}$$

$$c = c_0 \quad \text{at} \quad t = 0\,, \tag{5.11}$$

$$(D\theta f \nabla c) \cdot \mathbf{n} = 0 \quad \text{at} \quad \mathbf{x} \in \partial\Omega, \tag{5.12}$$

where c is phosphate concentration in soil solution, t time, $\mathbf{x}$ space vector, θ volumetric water content, b buffer power, f impedance factor, D diffusion coefficient in water, c_0 initial P_i concentration in soil solution, F volumetric sink term for root P_i uptake, $\partial\Omega$ domain boundary, and $\mathbf{n}$ outer normal vector. The domain Ω is a conical pot with a height of 10 cm, a top diameter of 10 cm and a bottom diameter of 6 cm (see Fig. 5.5). Boundary condition (5.12) ensures that P_i cannot diffuse out of the domain.

The diffusion equation, (5.10), is coupled with the root system growth model via the sink term F, which represents the average root uptake in a representative elementary volume (REV). We determine F by summing over the P_i uptake of every root segment within each REV,

$$F = \frac{1}{\text{REV}} \sum_{s=1}^{N} 2 a_s \pi l_s F_s, \tag{5.13}$$

where a_s is radius of root segment, F_s influx into a root segment, l_s root segment length, N number of root segments within REV. Water and solute transport models with sink terms for root uptake generally neglect depletion zones around individual roots (Claassen et al. 2006). In order to calculate the flux into a root segment, we account for the dynamic development of a P_i depletion zone according to the approximate analytical solution of Roose et al. (2001) given by (5.14):

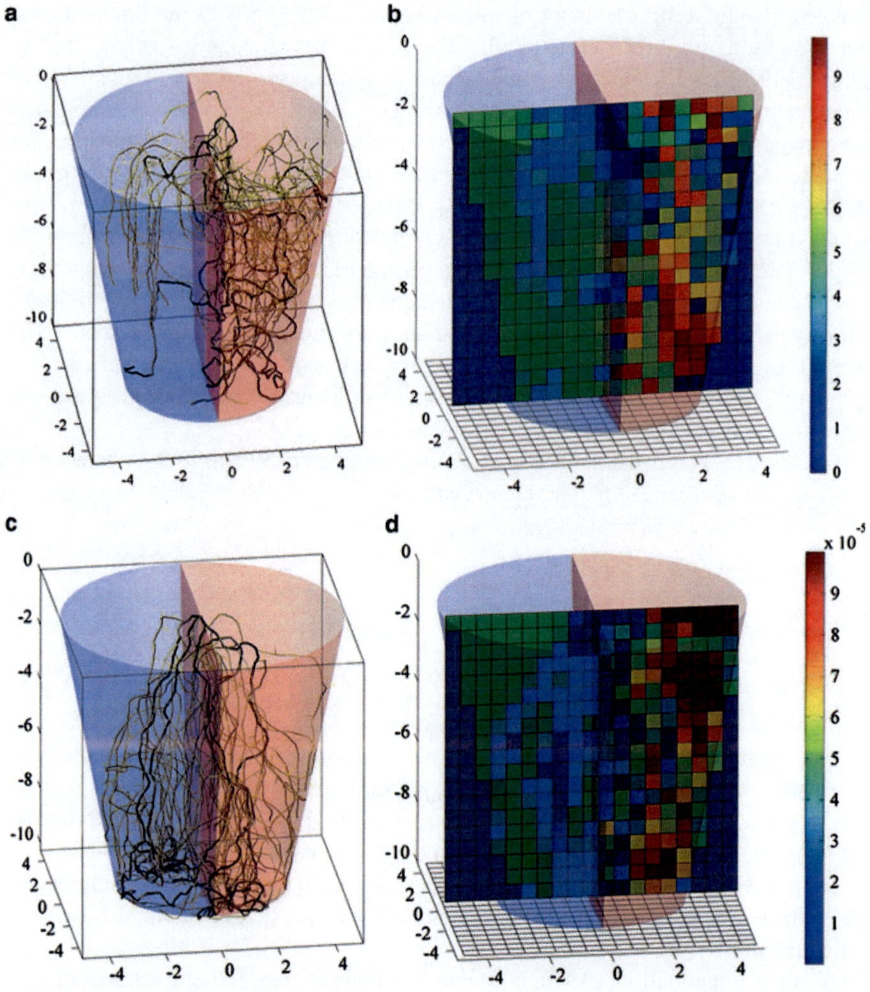

Fig. 5.5 Root system development and soil P_i depletion after 20 days. Initial P_i concentration in pot 1×10^{-4} μmol cm^{-3} (*right half*) and 0.5×10^{-4} μmol cm^{-3} (*left half*). (**a**) Root system growth according to chemotropism. (**b**) Soil depletion due to root system shown in (**a**). (**c**) Root system growth according to gravitropism only. (**d**) Soil depletion due to root system shown in (**c**)

$$F_s(t_s, a_s) = \frac{2F_m c_\infty}{1 + c_\infty + L(t_s) + \sqrt{4c_\infty + [1 - c_\infty + L(t_s)]^2}},$$
$$c_\infty = c_0/K_m, \tag{5.14}$$
$$L(t_s, a_s) = \frac{F_m a_s}{2\theta D f K_m} \ln\left(4e^{-\gamma} \frac{\theta D f}{(\theta + b) a_s^2} t_s + 1\right),$$

where $\gamma = 0.5772$ is Euler's constant, t_s age of root segment, F_m maximal P_i influx into root and K_m Michaelis–Menten constant. Equations (5.13) and (5.14) yield a volumetric sink term F that depends on (1) the age of each root segment, (2) the number of root segments in the REV and (3) the size of the REV.

The cubic REVs act as spatial discretisation used for the numerical solution of (5.10)–(5.12). For every REV a sink term is created according to (5.13). The REV must be large enough to regard the roots in a single REV as root densities, but small enough to represent root density variations of the root system. Here, we choose an REV size of 0.5^3 cm^3. For the numerical solution, we use the Crank–Nicholson finite difference scheme. The nonlinear sink term is solved by using fixed-point iteration. The time step Δt is chosen such that the Courant–Friedrichs–Lewy (CFL) condition for the diffusive case, $\Delta t \leq \Delta x^2/2D$, is fulfilled. Note that the CFL condition is a necessary condition for the convergence of numerical solutions of partial differential equations (Morton and Mayers 1994).

In this case study, we examine the effect of inhomogeneous initial P_i distribution on root system development, soil P_i depletion and P_i uptake. Root growth parameters for a maize root system are taken from Leitner et al. (2010a), and the soil parameters and P_i uptake parameters for maize from Roose et al. (2001). We consider a pot in which the initial concentration in the left half is half that in the right. This is illustrated by the blue and red colour in Fig. 5.5. At the top of the pot, five seeds are initially present and grow according to the model of Leitner et al. (2010a). All simulations assume that gravitropism, i.e. the tendency of the root to grow downwards, occurs. We also study the effect when the root system does or does not additionally follow chemotropism, i.e. the tendency of the roots to grow towards higher P_i concentrations.

5.2.2.3 Results

Figure 5.5a, b shows the root system and soil P_i depletion after 20 days, when roots tend to grow towards higher P_i concentrations (chemotropism). Figure 5.5c, d shows the same when the root system only follows gravitropism but not chemotropism. In this example, overall root length is the same in both cases and only the positions of the roots differ. In the case of chemotropism, the root system is denser in the right part of the pot, where the initial P_i concentration was higher. Therefore, depletion is less in the left part of the pot. In the case of no chemotropism, root length densities are similar in both halves, equally depleting both parts of the pot.

Chemotropism enhanced P_i depletion in the region with higher P_i concentration and thus increased overall P_i uptake by that root system as compared to the root system without chemotropism (see Fig. 5.6).

5.2.2.4 Discussion and Outlook

This case study uses a model that considers two spatial scales: the root system scale and the single root scale. Uptake and depletion caused by individual roots is described on the single root scale, whereas overall P_i transport and uptake is described on root system scale. The sink term is created by averaging over a REV.

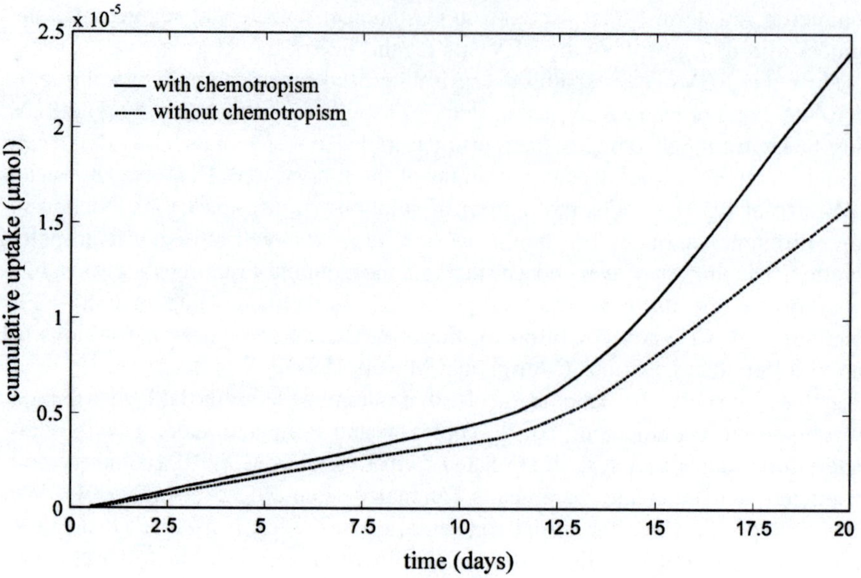

Fig. 5.6 Cumulative P_i uptake by root systems, with and without chemotropism

We consider dynamic development of the depletion zone around each root, but neglect inter-root competition. This is valid for sparingly soluble nutrients such as P_i.

This example quantifies the effect of chemotropism on P_i uptake. The overall root length of the root systems was the same in the two cases considered (chemotropism and no chemotropism). However, root surface area was 4.6 times denser in the high concentration region in the case of chemotropism, whereas it was approximately equal in both pot halves in the case of gravitropism only. Thus, chemotropism yielded a 1.5-fold increase of P_i uptake.

In addition to the plastic responses of the root system to nutrient-rich patches, growth and uptake rates can be affected (Claassen et al. 2006). Further challenges for future model development include coupling the model to a model of soil water movement, upscaling of additional single root traits (e.g. root exudation and mycorrhizas) to the root system scale, and the explicit modelling of organic P dynamics.

5.2.3 *P Uptake and Crop Response to Soil P Levels*

5.2.3.1 Aim of the Model

Most nutrient models combine the equation describing the radial movement of ions from soil to root surface by diffusion and mass flow with an equation relating root uptake to the ion concentration at the root surface. In the former, the interactions of the ion with the soil solid phase are also considered. An integration procedure allows the calculation of nutrient uptake by the whole root system (Barber 1995; Tinker and Nye 2000). Such models were successfully evaluated on short periods and were useful tools for investigation of the mechanisms at the rhizosphere scale. Nevertheless, these classical models often failed to predict nutrient uptake over long periods because the feedback effects between nutrient uptake and plant functions were poorly accounted for.

The aim of this case study is to illustrate how to include crop growth response in nutrient uptake models and the ability of such models to provide a basis for assessing target values for soil nutrient concentration. In order to do so, we developed a mechanistic model that combined an ecophysiological model and a root nutrient uptake model (Mollier et al. 2008). A mechanistic model for the simultaneous simulation of P_i supply, its uptake by the root system and the crop growth response is presented. The dynamic link between these processes was explicitly taken into account. This model was used to simulate P_i uptake and maize crop response in field conditions under three levels of soil P_i availability (high P, intermediate P and low P). All parameters and input variables are given in Mollier et al. (2008). Secondly, we simulated P_i uptake and crop growth for a wide range of soil P_i concentration to determine target values for soil P_i availability.

5.2.3.2 Model Description

The proposed model consists of three modules that are closely connected (Fig. 5.7). The first module deals with crop growth based on crop phenology and biomass accumulation depending on climatic conditions, and on crop P demand derived from potential crop growth depending on the environmental conditions. The second module describes P_i supply from the soil, considering the ion P_i concentration in soil solution and the soil buffer capacity. The third module deals with crop P_i uptake depending on crop P demand and P_i uptake capacity determined by the soil P_i supply and root length density distribution in the soil profile. The three modules are integrated to simulate the feedback loop of effective P_i uptake on crop growth. The actual shoot and root growth are adjusted according P_i uptake. Thus, the model tightly couples crop growth with soil processes.

Module 1: Modelling Crop Growth and Crop P Demand

The crop growth module simulates crop phenology and dry matter accumulation as a function of daily temperature and the photosynthetically active radiation absorbed by the canopy (PAR_a) and its conversion into dry biomass (Monteith 1977). The daily biomass produced is partitioned between shoot and root, assuming that shoot demand for carbohydrates is satisfied first. The remaining carbohydrates are allocated to the root system. The daily crop P demand is derived from the potential leaf expansion rate allowed by temperature and the carbohydrate availability using a

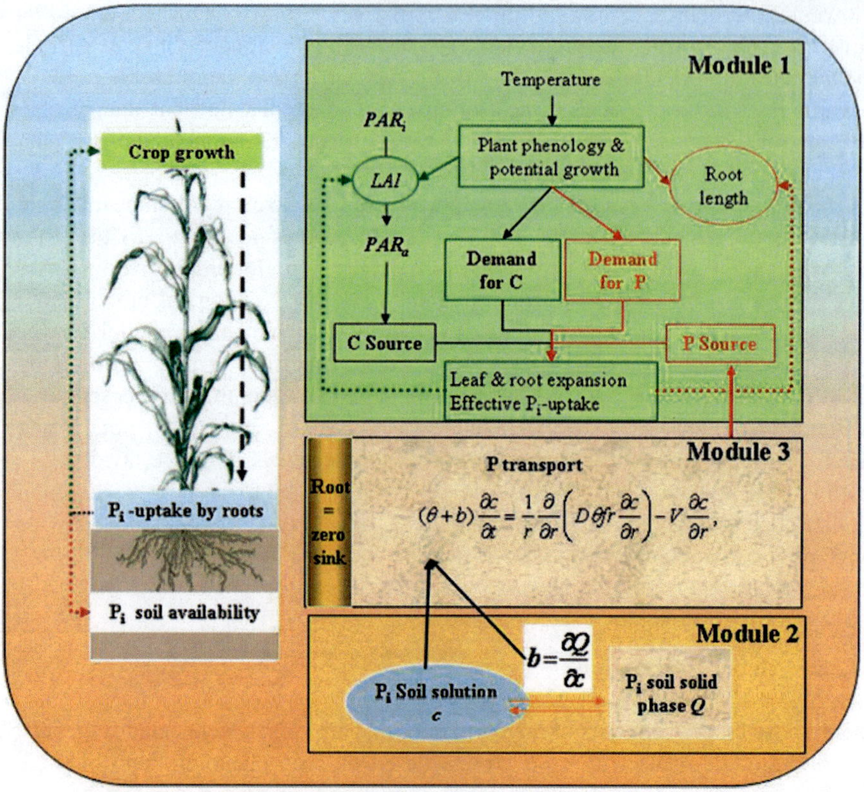

Fig. 5.7 Schematic representation of P_i uptake model including crop P response (adapted from Mollier et al. 2008)

close relationship between leaf area index (LAI) and total crop P_i uptake obtained under non-limiting conditions.

Module 2: Modelling Soil P Supply to Crops

Phosphate soil availability refers to the concentration of P_i in soil solution (c mg mL^{-1}) plus the amount of P_i (Q mg cm^{-3}) associated to the solid phase, which is in equilibrium with soil solution P_i. The relationship between c and Q is described by a Freundlich equation (McGechan and Lewis 2002):

$$Q = k_f c^n. \tag{5.15}$$

The soil buffer capacity is the ratio of changes in solid P_i to those in solution:

$$b = \frac{\partial Q}{\partial c}, \tag{5.16}$$

where k_f and n are Freundlich coefficients determined from a sorption/desorption experiment and b is the soil buffer capacity.

Module 3: Modelling P Uptake by the Root System According to Crop P Demand and Soil P Supply

The two-dimensional soil domain is subdivided into square control volumes (5 × 5 cm). Each of these is characterised by soil properties and root length density. The equations for P_i transport by mass flow and diffusion within the solution of the soil cylinder around the root and P_i uptake are used for each control volume:

$$(\theta + b)\frac{\partial c}{\partial t} = \frac{1}{r}\frac{\partial}{\partial r}\left(D\theta f r\frac{\partial c}{\partial r}\right) - V\frac{\partial c}{\partial r}, \tag{5.17}$$

$$c = c_0 \quad \text{at} \quad t = 0\,. \tag{5.18}$$

The boundary condition at the root surface follows from crop P demand (S_{sr} mg cm^{-2} day^{-1}):

$$-D\theta f\frac{\partial c}{\partial r} + Vc = -\frac{S_{sr}}{2\pi r_0 \Delta z L_{rv}} \quad \text{at} \quad r = r_0, \tag{5.19}$$

$$-D\theta f\frac{\partial c}{\partial r} + Vc = 0 \quad \text{at} \quad r = r_1, \tag{5.20}$$

where c is phosphate concentration in soil solution, t time, r radial distance from root axis, θ volumetric water content, b buffer power, f impedance factor, D diffusion coefficient in water, V flux of water toward root, c_0 initial P_i concentration in soil solution, r_0 root radius, $r_1 = 1/\sqrt{\pi L_{rv}}$ mean half distance between roots calculated for each control volume, L_{rv} root length density and Δz thickness of control volume.

Required uptake cannot occur when the diffusion and mass flow processes in soil cannot replenish enough P_i to the root. We assume that the maximum uptake rate equals the maximum possible rate of transport (by diffusion and mass flow) to the root, i.e. the root behaves as a zero sink. The maximum nutrient uptake rate per unit surface area S_{sm} (mg cm^{-2} day^{-1}) is derived from the steady-rate approximate solution for the concentration profile around the root for the zero sink condition (De Willigen and Van Noordwijk 1994):

$$S_{\mathrm{sm}} = 2\pi \Delta z L_{\mathrm{rv}} D \frac{(\rho^2 - 1)}{2G(\rho, \nu)}\overline{P}, \tag{5.21}$$

where $\overline{P}$ is the average P_i in soil surrounding the root, ρ the dimensionless radius and $G(\rho,\nu)$ is a geometry function (see Mollier et al. 2008).

Actual P_i uptake may or may not satisfy crop P demand. We assume that the actual P_i uptake rate S_s equals the required P_i uptake rate S_{sr} as long as S_{sr} is less than the maximum P_i uptake rate S_{sm}, otherwise S_{sr} equals the maximum P_i uptake rate S_{sm}. The total P_i uptake by the entire root system is the sum of P_i uptake from all control volumes.

Integration and Feedback Loop

Shoot and root growth are calculated from carbohydrate assimilation and the actual P_i uptake. If crop P demand is satisfied, effective crop growth is only limited by carbohydrate assimilation and climatic conditions. If actual P_i uptake is less than required, leaf area expansion is reduced, so that more carbohydrates are allocated to the root system. The new root distribution in the soil is derived from the biomass allocated to the root and is distributed in the soil based on the diffusion-type root growth model proposed by (De Willigen et al. 2002). The nutrient transport equation is explicitly solved with a time step that is restricted according to the ratio of the Courant and Peclet numbers (Daus et al. 1985).

5.2.3.3 Results

Figure 5.8a shows predicted cumulated P_i uptake as function of soil P_i levels. In this example, P_i uptake was reduced as soil P_i level decreased. This reduction was more pronounced for low-P treatment. The simulated crop growth response to P_i uptake is shown in Fig. 5.8b, c. For high-P treatment, leaf area expansion and root growth were only governed by climatic conditions and biomass partitioning within the plant. For intermediate-P treatment, leaf area expansion was slightly reduced, whereas root growth was almost not affected. For low-P treatment, both shoot and root growths were reduced in response to low P_i uptake. Moreover, limitation

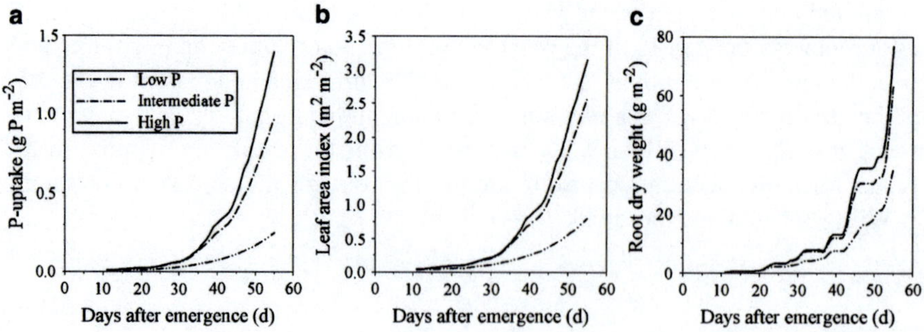

Fig. 5.8 Simulation of (**a**) cumulative P_i uptake by crop (g P_i m^{-2}), (**b**) leaf area index (m^2 m^{-2}) and (**c**) root biomass (g m^{-2}) versus time elapsed since emergence, for three P_i soil levels (low, intermediate and high P)

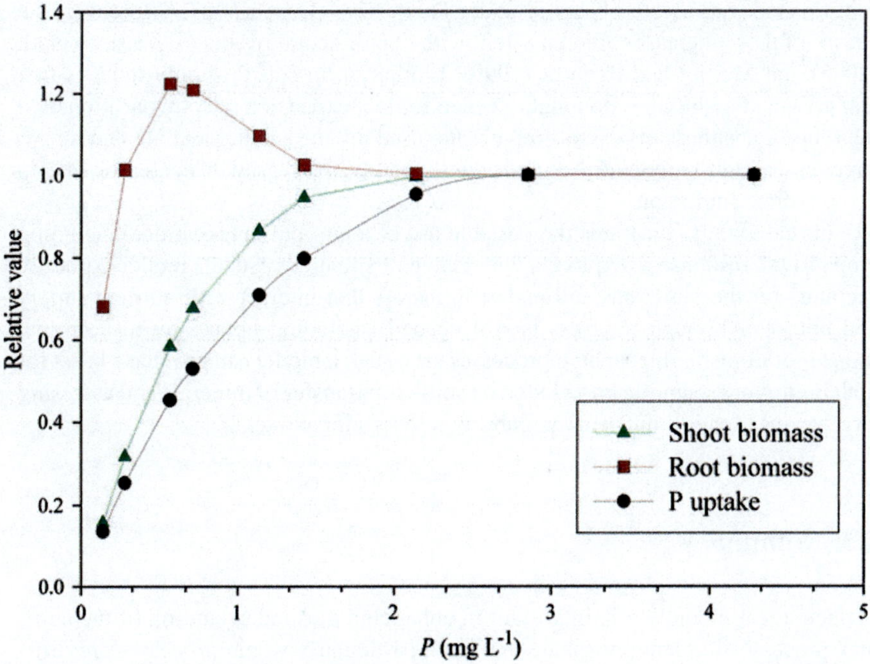

Fig. 5.9 Relative values of P_i uptake, shoot and root biomass (simulated value until 60 days after emergence divided by simulated value under high P_i conditions) as a function of P_i soil solution concentration (in mg P L^{-1}) for a sandy soil

in soil P_i supply reduced first leaf area, and assimilates no longer needed for leaf expansion rate were allocated to the root. Consequently, under low-P treatment the root growth was less affected than shoot growth and root-to-shoot ratio was increased.

In Fig. 5.9, the relative P_i uptake, shoot dry biomass and root biomass predicted by the proposed model were plotted versus the concentration of P_i in the soil solution. As P_i decreased under a threshold value, both P_i uptake and shoot growth were reduced, whereas relative root growth was maintained or increased. Such crop response allowed the root system to increase the soil volume explored and consequently the access to soil P_i. However, for very low P_i supply, such crop response could not fully counter the P_i shortage. The threshold value for P_i is dependent on crops, soil P_i supply properties (mainly b, θ and f) and climatic conditions.

5.2.3.4 Discussion and Outlook

The presented model is a mechanistic model that explicitly includes a feedback loop between nutrient uptake and crop growth. The simulated crop responses are consistent with those commonly reported when P_i is limiting. Increases in root-to-shoot ratio were observed under P_i shortage (Lynch 2007). Moreover, the causes of these increases are consistent with experimental results of Wissuwa et al. (2005) and Mollier and Pellerin (1999). Limitation in soil P_i supply reduces first leaf area, and assimilates no longer needed for leaf expansion rate are partitioned to the roots. Once these excess assimilates are used up, the smaller leaf area no longer supplies enough carbohydrates and, consequently, root growth decreases due to carbohydrate limitation.

This case study illustrates the possible use of a nutrient uptake model including crop growth response to explore a wide range of situations. Such a model explicitly accounts for the numerous soil and crop factors that interact with nutrient supply and uptake. Although progress is still needed (including uptake by mycorrhizal roots, root tropism, rhizosphere processes etc.) such a model can provide a basis for both improving scientific knowledge on soil–crop transfer of minerals and assessing target values for soil nutrient availability in a specific context.

5.3 Summary

Mathematical modelling is important in enhancing our understanding of the complex processes in plant phosphate nutrition, particularly where processes are difficult to assess experimentally. This chapter illustrates the model building process for plant phosphate uptake models, focusing on different plant traits that enhance phosphate uptake.

All models presented in this chapter are mechanistically based as well as deterministic, and are based on partial differential equations. Mechanistic mathematical models help to enhance our understanding of phenomena occurring across different spatial and temporal scales (Roose and Schnepf 2008). Close collaboration between experimentalists and modellers is necessary for model validation and parameterisation and will further enhance scientific progress. In this respect, modelling could assist in crop management and breeding of crops with traits that are beneficial, for example in low nutrient environments.

References

Arheimer B, Andersson L, Larsson M, Lindström G, Olsson J, Pers BC (2004) Modelling diffuse nutrient flow in eutrophication control scenarios. Water Sci Technol 49:37–45

Barber SA (1995) Soil nutrient bioavailability: a mechanistic approach. Wiley, New York

Bhadoria PS, Steingrobe B, Claassen N, Liebersbach H (2002) Phosphorus efficiency of wheat and sugar beet seedlings grown in soils with mainly calcium, or iron and aluminium phosphate. Plant Soil 246:41–52

Bucher M (2007) Functional biology of plant phosphate uptake at root and mycorrhiza interfaces. New Phytol 173:11–26

Chung SO, Kim HS, Kim JS (2003) Model development for nutrient loading from paddy rice fields. Agric Water Manage 62:1–17

Claassen N, De Willigen P, Diederik J, Doussan C, Dunbabin V, Heinen M, Hopmans JW, Kirk G, Kuzyakov Y, Mayer U, Mollier A, Nietfeld H, Nowack B, Oswald S, Roose T, Schnepf A, Schulin R, Seuntjens P, Silk WK, Steingrobe B, van Beinum W, Van Bodegom P (2006) Rhizosphere models. In: Luster J et al (eds) Handbook of methods used in rhizosphere research, part II. Swiss Federal Research Institute WSL, Birmensdorf, pp 487–517

Dalzell BJ, Gowda PH, Mulla DJ (2004) Modeling sediment and phosphorus losses in an agricultural watershed to meet TMDLs. J Am Water Resour Assoc 40:533–543

Darrah PR, Jones DL, Kirk GJD, Roose T (2006) Modelling the rhizosphere: a review of methods for 'upscaling' to the whole-plant scale. Eur J Soil Sci 57:13–25

Daus AD, Frind EO, Sudicky EA (1985) Comparative error analysis in finite-element formulations of the advection-dispersion equation. Adv Water Resour 8:86–95

De Willigen P, Van Noordwijk M (1994) Mass flow and diffusion of nutrients to a root with constant or zero-sink uptake. II. Zero-sink uptake. Soil Sci 157:171–175

De Willigen P, Heinen M, Mollier A, Van Noordwijk M (2002) Two-dimensional growth of a root system modelled as a diffusion process. I. Analytical solutions. Plant Soil 240:225–234

Diggle AJ (1988) Rootmap – a model in three-dimensional coordinates of the growth and structure of fibrous root systems. Plant Soil 105:169–178

Djodjic F, Montas H, Shirmohammadi A, Bergström L, Ulén B (2002) A decision support system for phosphorus management at a watershed scale. J Environ Qual 31:937–945

Dunbabin VM, Armstrong RD, Officer SJ, Norton RM (2009) Identifying fertiliser management strategies to maximise nitrogen and phosphorus acquisition by wheat in two contrasting soils from Victoria, Australia. Aust J Soil Res 47:74–90

Einstein A (1934) On the method of theoretical physics. Philos Sci 1:163–169

Fitter AH, Gilligan CA, Hollingworth K, Kleczkowski A, Twyman RM, Pitchford JW (2005) Biodiversity and ecosystem function in soil. Funct Ecol 19:369–377

Frossard E, Condron LM, Oberson A, Sinaj S, Fardeau JC (2000) Processes governing phosphorus availability in temperate soils. J Environ Qual 29:15–23

Ge ZY, Rubio G, Lynch JP (2000) The importance of root gravitropism for inter-root competition and phosphorus acquisition efficiency: results from a geometric simulation model. Plant Soil 218:159–171

Giasson E, Bryant RB, Bills NL (2002) Environmental and economic optimization of dairy manure management: a mathematical programming approach. Agron J 94:757–766

Gillespie AR, Pope PE (1990) Rhizosphere acidification increases phosphorus recovery of black locust. II. Model predictions and measured recovery. Soil Sci Soc Am J 54:538–541

Grant RF, Robertson JA (1997) Phosphorus uptake by root systems: mathematical modelling in ecosys. Plant Soil 188:279–297

Grant RF, Amrani M, Heaney DJ, Wright R, Zhang M (2004) Mathematical modeling of phosphorus losses from land application of hog and cattle manure. J Environ Qual 33:210–231

Harrison JA, Seitzinger SP, Bouwman AF, Caraco NF, Beusen AHW, Vörösmarty CJ (2005) Dissolved inorganic phosphorus export to the coastal zone: results from a spatially explicit, global model. Global Biogeochem Cycles 19:1–15

Hoffmann C, Ladewig E, Claassen N, Jungk A (1994) Phosphorus uptake of maize as affected by ammonium and nitrate nitrogen – Measurements and model calculations. Journal of Plant Nutrition and Soil Science 157:225–232

Huguenin-Elie O, Kirk GJD, Frossard E (2009) The effects of water regime on phosphorus responses of rainfed lowland rice cultivars. Ann Bot 103:211–220

Jakobsen I, Abbott LK, Robson AD (1992) External hyphae of vesicular-arbuscular mycorrhizal fungi associated with *Trifolium subterraneum L.* 1. Spread of hyphae and phosphorus inflow into roots. New Phytol 120:371–380

Janssen PHM, Heuberger PSC (1995) Calibration of process-oriented models. Ecol Modell 83:55–66

Kirk GJD (1999) A model of phosphate solubilization by organic anion excretion from plant roots. Eur J Soil Sci 50:369–378

Kirschbaum MUF, Medlyn BE, King DA, Pongracic S, Murty D, Keith H, Khanna PK, Snowdon P, Raison RJ (1998) Modelling forest-growth response to increasing CO_2 concentration in relation to various factors affecting nutrient supply. Glob Change Biol 4:23–41

Lambers H, Shane MW, Cramer MD, Pearse SJ, Veneklaas EJ (2006) Root structure and functioning for efficient acquisition of phosphorus: matching morphological and physiological traits. Ann Bot 98:693–713

Landis FC, Fraser LH (2008) A new model of carbon and phosphorus transfers in arbuscular mycorrhizas. New Phytol 177:466–479

Leitner D, Klepsch S, Bodner G, Schnepf A (2010a) A dynamic root system growth model based on L-systems. Tropisms and coupling to nutrient uptake from soil. Plant Soil 332:177–192

Leitner D, Klepsch S, Ptashnyk M, Marchant A, Kirk GJD, Schnepf A, Roose T (2010b) A dynamic model of nutrient uptake by root hairs. New Phytol 185:792–802

Li X-L, George E, Marschner H (1991) Extension of the phosphorus depletion zone in VA-mycorrhizal white clover in a calcareous soil. Plant Soil 136:41–48

Luster J, Göttlein A, Nowack B, Sarret G (2009) Sampling, defining, characterising and modeling the rhizosphere – the soil science tool box. Plant Soil 321:457–482

Lynch JP (2007) Roots of the second green revolution. Aust J Bot 55:493–512

McGechan MB, Lewis DR (2002) Sorption of phosphorus by soil, part 1: principles, equations and models. Biosystems Eng 82:1–24

Migliaccio KW, Chaubey I, Haggard BE (2007) Evaluation of landscape and instream modeling to predict watershed nutrient yields. Environ Modell Softw 22:987–999

Mollier A, Pellerin S (1999) Maize root system growth and development as influenced by phosphorus deficiency. J Exp Bot 50:487–497

Mollier A, De Willigen P, Heinen M, Morel C, Schneider A, Pellerin S (2008) A two-dimensional simulation model of phosphorus uptake including crop growth and P-response. Ecol Modell 210:453–464

Monteith JL (1977) Climate and the efficiency of crop production in Britain. Philos Trans R Soc London B 281:277–294

Morton KW, Mayers DF (1994) Numerical solution of partial differential equations. Cambridge University Press, Cambridge

Overman AR, Scholtz RV (1999) Langmuir-Hinshelwood model of soil phosphorus kinetics. Commun Soil Sci Plant Anal 30:109–119

Pages L, Vercambre G, Drouet JL, Lecompte F, Collet C, Le Bot J (2004) RootTyp: a generic model to depict and analyse the root system architecture. Plant Soil 258:103–119

Porder S, Vitousek PM, Chadwick OA, Chamberlain CP, Hilley GE (2007) Uplift, erosion, and phosphorus limitation in terrestrial ecosystems. Ecosystems 10:158–170

Ptashnyk M, Roose T, Kirk GJD (2010) Diffusion of strongly sorbed solutes in soil: a dual-porosity model allowing for slow access to sorption sites and time-dependent sorption reactions. Eur J Soil Sci 61:108–119

Radcliffe DE, Lin Z, Risse LM, Romeis JJ, Jackson CR (2009) Modeling phosphorus in the Lake Allatoona watershed using SWAT. I. Developing phosphorus parameter values. J Environ Qual 38:111–120

Raghothama KG, Karthikeyan AS (2005) Phosphate acquisition. Plant Soil 274:37–49

Reginato JC, Palumbo MC, Moreno IS, Bernardo IC, Tarzia DA (2000) Modeling nutrient uptake using a moving boundary approach: comparison with the Barber-Cushman model. Soil Sci Soc Am J 64:1363–1367

Roose T, Fowler AC (2004) A mathematical model for water and nutrient uptake by plant root systems. J Theor Biol 228:173–184

Roose T, Schnepf A (2008) Mathematical models of plant-soil interaction. Philos Trans R Soc London A 366:4597–4611

Roose T, Fowler AC, Darrah PR (2001) A mathematical model of plant nutrient uptake. J Math Biol 42:347–360

Saltelli A, Chan K, Scott EM (eds) (2000) Sensitivity analysis. Wiley series on probability and statistics. Wiley, Chichester

Schils RLM, De Haan MHA, Hemmer JGA, Van Den Pol-van DA, De Boer JA, Evers AG, Holshof G, Van Middelkoop JC, Zom RLG (2007) DairyWise, a whole-farm dairy model. J Dairy Sci 90:5334–5346

Schlecht E, Hiernaux P (2005) Beyond adding up inputs and outputs: process assessment and upscaling in modelling nutrient flows. Nutr Cycl Agroecosyst 70:303–319

Schnepf A, Klepsch S (2010) Mathematics and rhizotechnology. Mathematical methods for upscaling of rhizosphere control mechanisms. BOKU – University of Natural Resources and Applied Life Sciences, Vienna. http://www.boku.ac.at/marhizo/. Accessed 23 July 2010

Schnepf A, Roose T (2006) Modelling the contribution of arbuscular mycorrhizal fungi to plant phosphate uptake. New Phytol 171:669–682

Schnepf A, Roose T, Schweiger P (2008a) Growth model for arbuscular mycorrhizal fungi. J R Soc Interface 5:773–784

Schnepf A, Roose T, Schweiger P (2008b) Impact of growth and uptake patterns of arbuscular mycorrhizal fungi on plant phosphorus uptake – a modelling study. Plant Soil 312:85–99

Schoumans OF, Groenendijk P (2000) Modeling soil phosphorus levels and phosphorus leaching from agricultural land in the Netherlands. J Environ Qual 29:111–116

Shi Z, Erickson LE (2001) Mathematical model development and simulation of in situ stabilization in lead-contaminated soils. J Hazard Mater 87:99–116

Smethurst PJ, Comerford NB (1993) Potassium and phosphorus uptake by competing pine and grass: observations and model verification. Soil Sci Soc Am J 57:1602–1610

Smith J, Smith P (2007) Introduction to environmental modelling. Oxford University Press, New York

Smith SE, Smith FA, Jakobsen I (2003) Mycorrhizal fungi can dominate phosphate supply to plants irrespective of growth responses. Plant Physiol 130:16–20

Teklić T, Vukadinović V, Lončarić Z, Rengel Z, Dropulić D (2002) Model for optimizing fertilization of sugar beet, wheat, and maize grown on pseudogley soils. J Plant Nutr 25:1863–1879

Tinker PB, Nye PH (2000) Solute movement in the rhizosphere. Oxford University Press, New York

Torbert HA, Gerik TJ, Harman WL, Williams JR, Magre M (2008) EPIC evaluation of the impact of poultry litter application timing on nutrient losses. Commun Soil Sci Plant Anal 39:3002–3031

Van Der Zee SEATM, Van Riemsdijk WH (1991) Model for the reaction kinetics of phosphate with oxides and soil. In: Bolt GH et al (eds) Interactions at the soil colloid–soil solution interface. Kluwer Academic, Dordrecht, pp 205–239

Wang YP, Houlton BZ, Field CB (2007) A model of biogeochemical cycles of carbon, nitrogen, and phosphorus including symbiotic nitrogen fixation and phosphatase production. Global Biogeochem Cycles 21:1–15

Wissuwa M, Gamat G, Ismail AM (2005) Is root growth under phosphorus deficiency affected by source or sink limitations? J Exp Bot 56:1943–1950

Part II
Processes

Chapter 6
Role of Mycorrhizal Symbioses in Phosphorus Cycling

Jan Jansa, Roger Finlay, Håkan Wallander, F. Andrew Smith, and Sally E. Smith

6.1 Introduction

6.1.1 Mycorrhizal Symbiosis: Definition, Partners, Diversity

Mycorrhizal symbioses are associations of plant roots or rhizoids with fungi that, at least under some conditions, are beneficial to both partners. Arbuscular mycorrhizal (AM) symbiosis was established at the dawn of terrestrial plant evolution, some 400–500 million years ago, between ancestral vascular plants (*Cooksonia*, *Rhynia*, *Aglaophyton*) and fungi belonging to the phylum Glomeromycota (Pirozynski and Dalpé 1989; Redecker et al. 2000; Schüßler et al. 2001). AM symbiosis has been identified in thousands of plant species among all major plant lineages including bryophytes, ferns, gymno- and angiosperms (Brundrett 2009; Wang and Qiu 2006). It is the most widespread type of mycorrhizal symbiosis with respect to the number of plant species it involves (Trappe 1987; Wang and Qiu 2006) and can be found in virtually all ecosystems on Earth. Despite its broad host range, the fungal diversity is limited to a few hundred species (Redecker and Raab 2006), inferring that the

J. Jansa (✉)
Institute of Plant, Animal and Agroecosystem Sciences, ETH Zurich, Eschikon 33, 8315 Lindau, Switzerland
e-mail: jjansa@ethz.ch

R. Finlay
Department of Forest Mycology and Pathology, Uppsala BioCenter, Swedish University of Agricultural Sciences, 75007 Uppsala, Sweden
e-mail: roger.finlay@mykopat.slu.se

H. Wallander
Microbial Ecology, Department of Ecology, Lund University, 22362 Lund, Sweden
e-mail: hakan.wallander@mbioekol.lu.se

F.A. Smith, and S.E. Smith
School of Agriculture, Food and Wine, University of Adelaide, Waite Campus DX 650 636, Adelaide, SA 5005, Australia
e-mail: andrew.smith@adelaide.edu.au; sally.smith@adelaide.edu.au

E.K. Bünemann et al. (eds.), *Phosphorus in Action*, Soil Biology 26,
DOI 10.1007/978-3-642-15271-9_6, © Springer-Verlag Berlin Heidelberg 2011

association is particularly species-unspecific. Quantitative preferences for root colonization have, however, been repeatedly found between some plant and the fungal species (Jansa et al. 2002; Öpik et al. 2009; Sýkorová et al. 2007; Vandenkoornhuyse et al. 2003) and it is possible that a broader specificity between some fungal and plant genotypes does exist. This remains difficult to test experimentally, given the current limitations in the understanding of AM fungal genomic organization and molecular diversity (Croll and Sanders 2009; Martin et al. 2008).

Evolutionarily more recent types of mycorrhizal associations include ectomycorrhizas (ECM), ericoid mycorrhizas (ERM), and orchid mycorrhizas (Cairney 2000). These types are restricted to narrower groups of plant taxa, and involve fungi from the phyla Basidiomycota, Ascomycota, and Zygomycota. Some of these fungi form quite species-specific associations (e.g., *Cortinarius* and *Suillus* associate with rather a narrow range of plant species, such as *Pseudotsuga*, *Betula*, *Larix*, and *Pinus*), whereas others (e.g., *Cenococcum* spp.) may colonize roots of a broad range of plant species (Bruns et al. 2002; Smith and Read 2008; Tedersoo et al. 2008). A few plant species (such as eucalyptuses, willows, and alders) have the capacity to interact with both AM and ECM fungi. This can give rise to root systems simultaneously colonized by mycorrhizal fungi belonging to different types (Adams et al. 2006; van der Heijden and Vosátka 1999). The different types of mycorrhizal symbioses have also shown predominance for different plant biomes (Read and Perez-Moreno 2003; Smith and Read 2008 and references therein). Plants in polar regions and at high altitudes are usually not mycorrhizal, but often have roots extensively colonized by dark septate fungal endophytes with unclear function (Mandyam and Jumpponen 2005; Newsham et al. 2009). By definition, endophytes live solely within the plant tissues, and often fulfill their entire lifecycle including reproduction inside the plant body (Faeth and Fagan 2002). Mycorrhizal fungi, in contrast, inhabit and interconnect two kinds of environment, namely the inner volume of the plant roots and the surrounding soil (Jansa and Gryndler 2010). The status of many root-inhabiting fungi is unclear because it is inherently difficult to demonstrate absence of hyphal growth outside the roots. Heathlands (here we refer to heathlands at both high latitudes and altitudes as well as nutrient-limited and wildfire-prone ecosystems in Mediterranean climates, known as fynbos, chaparral, maquis or matorral in different parts of the world) are usually dominated by ERM, and Taiga (coniferous boreal forest) is dominated by the ECM (Read et al. 2004; Thormann et al. 1999). Most trees in deciduous forests establish ECM symbiosis, although the understorey plants are primarily colonized by AM fungi (Helgason et al. 2002). Most plants in grasslands and in tropical forests also establish AM symbiosis (Castillo et al. 2006; Treseder and Cross 2006). Plants inhabiting highly weathered and severely phosphorus (P)-impoverished soils (as in Western Australia) are often devoid of mycorrhizas; these plants have other adaptations, such as cluster roots, that fulfill their P requirements (Lambers et al. 2008). Under field conditions, the root system of a single plant is usually colonized by different mycorrhizal fungal species simultaneously (Burke et al. 2005; Jansa et al. 2003b; Merryweather and Fitter 1998; Miller et al. 1991). This diversity could have important consequences for acquisition of nutrients as well as for maximizing symbiotic benefits for the plants (Baxter and Dighton 2001; Jansa et al. 2008; Koide 2000).

6.1.2 Mycorrhizal Functioning

The unequivocal importance of mycorrhizal symbioses to plants (and soils) is inherently difficult to demonstrate because the mycorrhizal condition is normal, and absence is rare under natural settings (Merryweather and Fitter 1995). Establishing non-mycorrhizal control treatments in pot experiments or in the field is a great challenge and may potentially introduce experimental artefacts (Jones and Smith 2004; Kahiluoto et al. 2000a). Soil sterilization by steaming or autoclaving, for example, changes chemical soil properties (Serrasolses et al. 2008). It also eliminates other soil organisms, introducing further confounding effects. While these artefacts should always be considered in the interpretation of experimental results, there have been numerous independent studies comparing the performance of mycorrhizal and non-mycorrhizal plants under a range of environmental conditions. These experiments have collectively demonstrated that plants benefit from their mycorrhizal associations through improved nutrient acquisition, mainly of elements with low mobility in the soil (e.g., P, zinc, and copper), and through greater resistance to drought and biotic stresses (Clark and Zeto 2000; Jansa et al. 2003a; Marschner 1995; Redon et al. 2009; Sikes et al. 2009).

The benefits of mycorrhizal colonization to the plants result from expansion and/or complementation of the root function. Mycorrhizal fungi colonize two environments: the inner root volume (sometimes extending to the hyphal sheath on the root surface) and the surrounding soil, thus directly connecting the root system with a greater soil volume. By increasing soil contact, the plants are able to acquire resources from zones lying far beyond the direct reach of the roots and the root hairs. This effect is not trivial because the hyphae of some AM fungi can extend many centimeters away from the root surface, unlike root hairs, which only extend a few millimeters (Jakobsen et al. 1992; Jansa et al. 2003a, 2005). ECM fungal mycelium can bridge even greater distances, particularly if the fungi form thick rhizomorphs (Allen et al. 2003). P transport through the ECM hyphae over distances up to 40 cm has been documented (Finlay and Read 1986; Timonen et al. 1996). In contrast, some mycorrhizal fungi establish very dense mycelial networks in the vicinity of the roots, growing only a few millimeters from their host (Smith et al. 2004). These mycelial networks significantly expand the capacity of the plant to acquire mineral nutrients from the soil. In addition, in the case of ECM fungi, these mycelial networks also play an important role in acquisition of water by the plants from the soil (Allen 2007; George and Marschner 1996). Root colonization by mycorrhizal fungi and their mycelial networks may also be important for interactions between plants and soil-borne pathogens. This is due to (1) changes in plant nutritional status upon mycorrhiza development, (2) direct competition for plant carbohydrates between the mycorrhizal fungi and the pathogens, and/or (3) changes in the activity and composition of the microbial communities in the rhizosphere (Fitter 2005; Graham 2001; Newsham et al. 1995; Toljander et al. 2007).

Many ERM and ECM fungi are also thought to be involved in mineralization of organic nutrients (Bueé et al. 2007; Finlay 2008; Tibbett and Sanders 2002) and bioweathering of recalcitrant inorganic nutrients from carbonates, micas, and apatites (Blum et al. 2002; Wallander 2000). It appears, however, that even in some of the well-documented case studies, other soil microorganisms (particularly the prokaryotes) might have played important roles in the release of nutrients from minerals or organic compounds. Therefore, the contribution of mycorrhizal fungi to these processes remains, in most cases, poorly quantified (Finlay 2008; Koele et al. 2009). Evidence for direct involvement of AM fungi in mineralization of significant amounts of organic P is still inconsistent. In particular, the role of other soil microorganisms associated with the mycorrhizal mycelium, which might be very important under nonsterile soil conditions, is not properly understood (Finlay 2008; Joner and Jakobsen 1995; Joner and Johansen 2000; Koide and Kabir 2000; Tarafdar and Marschner 1994; Toljander et al. 2006). The presence of mycorrhizal fungi also significantly modifies soil conditions in the rooting zone (aggregation, wettability, biological activity), modulates intra- and interspecific competition within the plant community, and affects soil microbial communities (Barea 2000; Facelli et al. 1999; Johansson et al. 2004; Rillig and Mummey 2006).

Mycorrhizal fungi are invariably heterotrophic organisms, and they mostly derive the organic carbon needed for their growth, respiration, and biological maintenance directly from their host plants in the form of recently fixed photosynthates. Some of the fungi (especially those forming ECM and ERM associations) are currently considered capable of limited saprophytic growth (Azcón-Aguilar and Barea 1995; Gibson and Mitchell 2004; Koide et al. 2008; Zeller et al. 2008), though under natural conditions its widespread realization has recently been questioned (Baldrian 2009; Taylor and Alexander 2005). The dependence of the fungi on the host plant for organic carbon can, depending on environmental conditions (e.g., availability of soil P, availability of light and CO_2 to the plants, plant density, and possibly other factors), cause the association to vary from highly beneficial to apparently parasitic (Johnson et al. 1997; Schroeder and Janos 2004; Smith and Smith 1996; West et al. 1993; Whitbeck 2001). For example, under very low soil P availability (<0.2 mg P kg^{-1}, extractable with 0.5 M $NaHCO_3$) caused either by low total P levels or high P sorption in highly weathered soils, mycorrhizal symbiosis may be less beneficial in terms of net P acquisition than if small amounts of P were added to the soil (Bolan 1991; Bolan et al. 1983). Therefore, under very low soil P availability, plants specialized in modes of P mobilization and uptake other than mycorrhizal symbiosis (e.g., those forming cluster roots) can dominate the plant communities (Lambers et al. 2008). In contrast, under conditions of very high soil P availability (>50 mg kg^{-1}, extractable with 0.5 M $NaHCO_3$), such as following application of high rates of water-soluble P-fertilizers, the plant can gain access to enough P with its root system, resulting in little or no net demand for mycorrhizal P uptake (Kahiluoto et al. 2000b; Sorensen et al. 2003). In such situations, the extent of mycorrhizal development in roots is often reduced (Bolan et al. 1984; Jansa et al. 2009; Youpensuk et al. 2008). In addition, depending on the plant species and soil and climatic conditions, the association with AM fungi could have negative effects on plant growth (Jifon et al. 2002; Morgan et al. 2005; Ryan and Graham 2002; Kahiluoto et al. 2000b). Under conditions where P availability limits plant growth ($NaHCO_3$-extractable P levels typically between 5 and 20 mg kg^{-1}), mineral nutrient acquisition benefits conferred to the plants by the AM fungi will usually far outweigh the costs of carbon supply to the fungus and will result in a net growth benefit to the plant (Jakobsen 1995; Li et al. 2005; Morgan et al. 2005; Ortas et al. 2002). These studies indicate that the benefits to the plants of association with AM fungi across a range of environmental conditions (Fig. 6.1) are best described by a bell-shaped (or unimodal) response curve (Bolan et al. 1984; Gange and Ayres 1999; Picone 2002). Even so, plants may benefit from mycorrhizal symbiosis even though the growth and/or nutritional benefits (such as net P uptake) may not be apparent, and special techniques such as isotopic labeling are necessary to demonstrate mycorrhizal function in these cases (Fitter 2005; Grace et al. 2009; Smith et al. 2004, 2009).

Although negative effects of some ECM fungi on host plant growth have been occasionally reported (Burgess et al. 1993; Corrêa et al. 2008), the great majority of

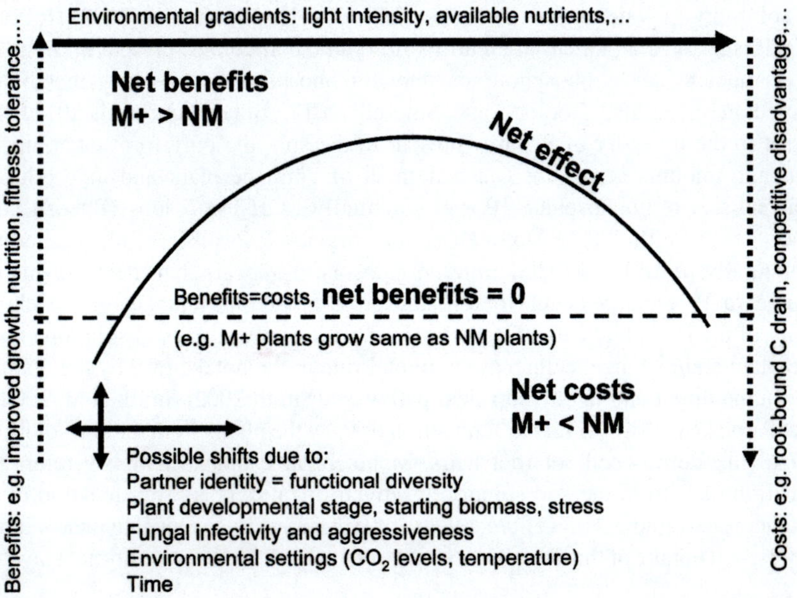

Fig. 6.1 Conceptual model of whole-plant effects of the interactions between a plant, mycorrhizal fungi, and the environment. This scheme delineates interdependencies between mycorrhizal costs and benefits, resulting in a continuum of outcomes, ranging from highly beneficial to potentially detrimental effects. This scheme is based mainly on the evidence gathered for arbuscular mycorrhizas, but also appears to be generally applicable to ectomycorrhizas. Processes in other mycorrhizal types (especially orchid mycorrhizas and mycorrhizas of achlorophylous plants) may follow different trajectories. *M+* mycorrhizal plant, *NM* non-mycorrhizal plant

studies on both ECM and ERM associations report growth benefits to the host plants upon colonization of the roots by the symbiotic fungi (Choi et al. 2005; Diedhiou et al. 2005; Finlay et al. 1992; Jansa and Vosátka 2000). The extent of root colonization by the different fungi and the magnitude of benefits, however, depend on a broader environmental context and not only on the P availability (Hoeksema et al. 2010). Additional factors that determine the extent of root colonization and mycorrhizal benefits are: soil nitrogen (N) levels, seed size and nutrient reserves contained in the seed, plant age and growth rate, and the identity of both plant and fungal species (Corrêa et al. 2006; Duponnois et al. 2008; Egerton-Warburton and Allen 2001).

6.2 Different Forms of P in the Soil and Their Accessibility to Mycorrhizas

6.2.1 *Forms of P*

Phosphorus is present in different forms in the soil. Inorganic forms (crystalline apatites; amorphous phosphates of calcium, potassium, iron and aluminum, and other phosphates; inorganic polyphosphate; and orthophosphate) differ greatly in their solubility in water and in their chemical reactivity (Dou et al. 2009; Holford 1997). P is also a component of an array of organic compounds present in the soil, such as nucleic acids, phospholipids, inositol phosphates, and many metabolic intermediates (see also Doolette and Smernik 2011; Bünemann et al. 2011). In contrast to the diversity of P forms present in the soil, the only form taken up in significant amounts across the plasmalemma of both the plant and mycorrhizal fungal cells is orthophosphate (P_i), preferentially as $H_2PO_4^-$ ions (Rausch and Bucher 2002; Smith 2002). Although various organic P forms have been reported as potentially utilizable by some microorganisms, it appears that their enzymatic cleavage to P_i actually occurs before the cross-membrane uptake, in the close vicinity of the microbial cells (Heath 2005). Therefore, P_i in the soil solution close to the plant and/or fungal cells plays a pivotal role in the uptake of P by the plants, both via the direct and the mycorrhizal pathways (Smith 2002; Smith et al. 2004). Direct P uptake pathway refers to acquisition of P_i by the plants from the soil solution through rhizodermis cells or root hairs. Mycorrhizal P uptake pathway refers to acquisition of P from the soil solution by mycorrhizal hyphae, translocation of P throught the extraradical mycelium, release of P from the mycorrhizal hyphae within the roots, and uptake of this released P by the root cells (usually in the cortical layer).

6.2.2 *Kinetics of P Acquisition by Hyphae*

Uptake of P from the soil solution to the cells (either of plant roots or mycorrhizal fungal hyphae) is mediated by P_i transporters (Rausch and Bucher 2002; Smith 2002; Tatry et al. 2009). These transporters are large proteins with several transmembrane domains and are responsible for the proton or sodium symport of phosphate molecules across the membrane and against a steep electrochemical gradient (Karandashov and Bucher 2005; Smith 2002). These proteins have been most studied in AM fungi, but recently, two P_i transporters from ECM fungus *Hebeloma cylindrosporum* have also been characterized (Tatry et al. 2009). The first mycorrhizal P_i transporter was identified from cDNA libraries of *Medicago truncatula* roots colonized by *Glomus versiforme*, using hybridization with a probe derived from a yeast P_i transporter (Harrison and van Buuren 1995). Further experiments, including expression of this P_i transporter in a yeast P_i transporter (*pho84*)-mutant, indicated Michaelis–Menten kinetics with an apparent K_m value of 18 μM. This value is one to two orders of magnitude higher than that predicted for high-affinity P transporter systems of AM fungi (Schweiger and Jakobsen 1999; Smith et al. 2001; Thomson et al. 1990). The discrepancy, however, could easily be due to problems with heterologous gene expression, as suggested earlier (Smith et al. 2001). Recently reported K_m values for the two P_i transporters of the ECM fungus *Hebeloma cylindrosporum* (4 and 55 μM) were closer to the values predicted from earlier hydroponic experiments (Tatry et al. 2009; van Tichelen and Colpaert 2000). All other reported details on P_i transporters from different AM fungi only refer to the so-called high-affinity transporter family, important for uptake of P_i from the soil solution, where the P_i concentration does normally not exceed 10 μM (Harrison 1999; Marschner 1995). The other (low-affinity) P_i transporter system, apparently operating in germ tubes of *Gigaspora margarita* (Thomson et al. 1990), has not yet been characterized at the gene level.

The described P_i transporters of both AM and ECM fungi show a high degree of structural conservation and other similarities to the P_i transporters of other organisms (Karandashov and Bucher 2005; Schachtman et al. 1998). The energetic expenditure of this high-affinity P_i acquisition is not fully resolved, but the estimates are between two and four protons per molecule of phosphate (Jennings 1996; Leggewie et al. 1997; Rausch and Bucher 2002 and references therein), with the proton gradient being generated by H^+-ATPases (Requena et al. 2003; Smith and Read 2008 and references therein). The other aspect of P_i uptake kinetics is the P_i inflow per unit of hyphal biomass or hyphal surface. Previous studies indicate some interspecies variation in both the AM and ECM fungal groups (Smith and Read 2008; van Tichelen and Colpaert 2000). This might relate to the kinetic parameters of the different P_i transporters and their expression patterns in the fungal mycelium in the soil. The uptake of P_i into the hyphae may further be modulated by possible differences between the different fungal species and genotypes in the intensity of hyphal proliferation and the dynamics of the hyphal networks.

Improved acquisition of P by mycorrhizal plants appears to be derived from several characteristics. The mycorrhizal hyphae are capable of penetrating smaller soil pores (5–30 μm) than the roots (>50–100 μm), thus expanding access to the soil. In addition, due to their size, the formation of a prominent P depletion zone around the individual hyphae is minimal. Mycorrhizal hyphae are also more efficient at spatial exploration of the soil volume (up to 50 m hyphae g^{-1}) as compared to roots (up to 0.1 m roots g^{-1}) and have a lower carbon cost per unit of hyphal surface as compared to the root surface (Gregory 2006; Jansa et al. 2003a, 2005; Li et al. 1991; Schnepf et al. 2008; Tinker 1975). In contrast, a poorly studied aspect is the dynamics of the hyphal networks (de Vries et al. 2009; Fitter et al. 2004). It appears that short-lived hyphal structures, like the branched-absorbing structures described for AM fungi from the genus *Glomus*, can provide a highly flexible pathway for acquisition of P and other nutrients to the plants from the soil. These structures have rather high turnover rates (7–35 days from initiation to death under axenic conditions) as compared to backbone hyphal strands, which retain cytoplasm over 3 months under axenic culture conditions (Bago et al. 1998, 2004; de Vries et al. 2009). Hyphal turnover of AM fungi from genus *Glomus*, as estimated by isotopic signature of carbon supplied to the plants and recovered in the AM hyphae, was only about 5–6 days (Staddon et al. 2003). These results are congruent with the time-span of the fine and short-lived hyphal structures, as presented above, as well as with previous estimates (5–7 days) based on microscopy of soil hyphae (Friese and Allen 1991). The rates of hyphal turnover in ECM and ERM networks are not well known, but are probably lower than in the AM fungi. Estimates from a carbon-flux study comparable to the study of Staddon et al. (2003) indicated an average lifespan for ECM mycelium of about 9 days, whereas rhizomorphs of some ECM fungi were observed to live for a number of months (Godbold et al. 2006 and references therein). Great levels of variability with respect to the hyphal growth and/or turnover rates have been recognized within each mycorrhizal type and between different fungal taxa (Downes et al. 1992; Godbold et al. 2006; Wallander 2006). Possibly, part of this variability could be explained by the different P acquisition strategies of the different mycorrhizal fungi. For example, soil P mining (defined here as accessing recalcitrant P sources through solubilizing or hydrolyzing exudates according to Lambers et al. 2008) would assume longer-lived mycelium in the same soil patch, whereas P scavenging (defined here as collecting the easily available P beyond the reach of roots) could more efficiently be carried out by fungi with fast hyphal turnover, expanding rapidly into uncolonized soil patches. This hypothesis remains to be tested experimentally within each of the mycorrhizal types as well as between the different types.

6.2.3 Access to Recalcitrant P Forms, Weathering, and Mineralization

Varying amounts of information are known about different mycorrhizal types and their effects on release of P from recalcitrant forms present in soil. Many ECM fungi have been shown to be able to release P_i from poorly soluble P sources such as apatite. This function has been demonstrated in numerous pure culture, pot, and microcosm studies, and was recently reviewed by Rosling (2009). The relevance of pure culture studies has, however, been questioned because large amounts of added organic carbon can induce production of acids at higher rates than under natural conditions. Pot and microcosm studies are also difficult to interpret because other microorganisms (some of them having the capacity to solubilize sorbed phosphate and/or the capacity to produce exocellular phosphatases) are usually present in the system (Jones and Smith 2004; Vessey 2003). These factors imply that the ECM could potentially take up P_i primarily released by the other microbes and transfer it to plants. In extreme cases, this could lead to measurable elevation of plant P uptake from recalcitrant sources, even when the involved mycorrhizal fungus was incapable of solubilization of the recalcitrant P on its own. However, Smits et al. (2008) recently demonstrated fungal-induced weathering of apatite in sterile microcosms with *Pinus sylvestris* seedlings colonized by the ECM fungus *Paxillus involutus*. Fungal colonization of apatite grains (Fig. 6.2) increased weathering rates threefold, and ^{14}C simultaneously supplied to the plant was preferentially allocated to apatite patches colonized by the fungus. The proposed mechanism for apatite dissolution is enhanced acidification and chelation of calcium cations from the apatite through fungal exudation of oxalic acid. Complexation of calcium with oxalic acid would then lead to formation of calcium oxalate crystals on the surface of the fungal hyphae in contact with the apatite. These crystals have previously been observed (Allen et al. 1996; Landeweert et al. 2001), as have the elevated levels of oxalic acid in the ectomycorrhizal mats of forest soils (Griffiths et al. 1994).

This relationship is supported by the observation that apatite grains introduced into forest soil usually become heavily colonized by ECM hyphae (Hagerberg et al. 2003; Turpault et al. 2009). Furthermore, Wallander and Thelin (2008) demonstrated that this fungal colonization became more intense when P levels of Norway spruce needles dropped below 1.5 mg P g^{-1} dry weight. This value is close to P-limiting conditions (1.3 mg P g^{-1} dry weight) according to Linder (1995), suggesting that carbon allocation to fungal-colonized nutrient patches is regulated by the nutrient status of the tree. In spite of these studies, direct evidence that ECM fungi enhance the rates of apatite weathering is still limited. For example, Turpault et al. (2009) incubated apatite grains in mesh bags for 4 years in a beech forest in

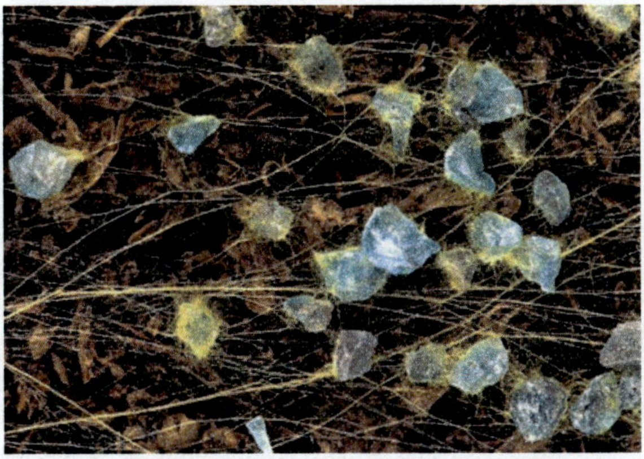

Fig. 6.2 Apatite grains (1 mm diameter) colonized by ectomycorrhizal fungus *Paxillus involutus* in sterile microcosms. Reproduced from Smits et al. (2008), with permission

western France. Half of the bags were placed in trenched plots to which roots and mycorrhizal hyphae had no access and the other half were placed at different depths in soil accessible to roots and mycorrhizal hyphae. Apatite grains in untrenched plots were heavily colonized by fungal hyphae (presumably ECM) and, in contrast to the grains without roots, showed many weathering marks (as assessed by electron microscopy). However, with exception of the treatment at a soil depth of 25 cm (0.2% apatite weight loss over 4 years without mycorrhizal roots versus 0.5% weight loss with the roots), the apatite dissolution measured as a loss of mass was seemingly unaffected by the presence of mycorrhizal roots when compared to the trenched plots established in the same soil. Similarly, in a study by Wallander and Thelin (2008), apatite grains from P-limited forests did not dissolve significantly faster than apatite grains from P-sufficient forests, based on the amount of rare earth elements (La, Nd, Sm, Eu, Tb and Yb) from the apatite that accumulated in mycorrhizal roots surrounding the mesh bags. Both the above studies, however, were carried out under acidic soil conditions (pH $\leq$4.3), which can preclude strong mycorrhizal effects on apatite solubilization through mycorrhizosphere acidification.

Although it seems indisputable that ECM fungi have the potential to release P from phosphorus-containing minerals under laboratory conditions, the extent to which this has an influence on field weathering rates of apatite is still the subject of much debate (Hutchens 2009; Rosling et al. 2009; Sverdrup et al. 2002; van Scholl et al. 2008). Uptake of elements from apatite by ECM fungi under field conditions has been previously demonstrated (Blum et al. 2002; Hagerberg et al. 2003), but the quantitative role of this process, operating on very long time scales (for contemporary science), is difficult to estimate.

Different approaches have been used to examine the roles played by different mycorrhizal fungi in mobilizing P (and N) from organic substrates and these have been reviewed by Read and Perez-Moreno (2003), among others. These approaches range from axenic systems in which the fungal hyphae are exposed to identified model compounds, to microcosm and field studies using more natural biological substrates. ERM fungi produce a range of organic polymer-degrading enzymes that can attack molecules such as chitin, lignins, polyphenols, and tannins, which either contain N or protect access to organically bound or spatially inaccessible N and P sources (organic or inorganic). Detected enzymes include lignases, polyphenol oxidases, laccase, and catechol oxidase, and the ability to degrade hydrolyzable polyphenols appears to be more extensively developed in ERM than in ECM fungi (Bending and Read 1996a, b). In addition to degrading structural components of plant litter, the ERM fungi also produce enzymes that hydrolyze P-containing molecules. Experiments performed under axenic conditions using DNA as a sole P source (Leake and Miles 1996; Myers and Leake 1996) have shown that phosphodiesters can be used by the ERM fungi as sole P sources without the intervention of other saprotrophs. These findings were corroborated by studies of the ERM fungi isolated from *Woollsia pungens* roots (Chen et al. 1999). All four of the studied isolates were able to utilize various organic compounds as a sole carbon and N source, and two of the isolates were able to grow on DNA or inositol sodium hexaphosphate as sole sources of P, with higher biomass production than

Hymenoscyphus (now *Rhizoscyphus*) *ericae*. The relevance of the above results for the rates of acquisition of P by the ERM fungi and associated host plants from complex organic substrates under nonsterile soil conditions needs, in many cases, to be refined by coupling enzymatic assays and isotope labeling, as in the N studies of Wurzburger and Hendrick (2006, 2009).

Similar studies to those on ERM listed above have also been conducted on ECM fungi involving "natural" substrates. In these studies, pollen grains or dead nematodes were added to microcosms containing mycorrhizal (*Paxillus involutus*) or non-mycorrhizal *Betula pendula* seedlings (Perez-Moreno and Read 2001a, b). More than 96% of the P in pollen (measured as P concentration in the pollen-enriched soil patches) added to mycorrhizal microcosms was removed and, on average, 25% of this was transferred to the mycorrhizal seedlings (calculated from the system P budget, comparing mycorrhizal and non-mycorrhizal plants and assuming no significant contribution of seed P to the plant P content). In contrast, only 25% of the P in pollen added to non-mycorrhizal microcosms was removed and only 7% of this ended up in the non-mycorrhizal plants, suggesting that ECM fungal hyphae play an important role in resource capture from organic substrates. In the study with nematodes, 65% of the P originally present in the nematodes was removed from the site of addition and 73% of this was transferred to the mycorrhizal *B. pendula* plants. In non-mycorrhizal systems, the plants gained half as much P as the mycorrhizal systems, representing only 22% of the total originally present in the nematodes. In earlier studies, Bending and Read (1995a, b) examined the structure and function of the ECM fungal mycelium in relation to nutrient mobilization from forest litter. In microcosms containing *Pinus sylvestris* seedlings, colonization of organic material from the fermentation horizon by *Suillus bovinus* reduced concentrations of P by 22%, but colonization by *Thelephora terrestris* had no effect. Activities of nutrient-mobilizing enzymes in birch litter colonized by *Paxillus involutus* were studied and phosphomonoesterase activity increased 28–50 days but decreased again between 50 and 98 days after the initial colonization of the organic patches by the fungal mycelia. The final levels of activity were below those of uncolonized litter, but in these unsterile substrates it was not possible to distinguish between the activities of mycorrhizal fungi and those of saprotrophs. Another study addressing the utilization of inositol hexaphosphate by ECM fungi (Colpaert et al. 1997) demonstrated substantial extracellular acid phosphatase activity associated with mycelia of *Thelephora terrestris* and *Suillus luteus* that was correlated with mycelial biomass and increasing P nutrition of the mycorrhizal plants. Phytase activity of the mycelium could not be detected, but activity at the surface of mycorrhizal roots was higher than that at the surface of non-mycorrhizal roots, though the relative contributions of plant roots and fungi to hydrolysis of soluble inositol hexaphosphate were unclear. Tibbett and Sanders (2002) have shown that colonization of willow roots by *Hebeloma syrjense* resulted in substantial improvement of P capture by the plants from plant litter (8% of the added P transferred to the shoots within 35 days in mycorrhizal plants as compared with only 1% in the non-mycorrhizal plants). This may be due to either the short-circuiting of the organic P re-cycling between soil and plants via mycorrhizal hyphae, or secondarily through the effects of the ECM on other components of the system involved in P cycling (such as bacteria, saprophytic fungi, mites, collembolans etc.). In this context, it is interesting to mention an earlier study that documented efficient transfer of P from the mycelium of the saprophytic fungus *Hypholoma fasciculare* to the ECM fungi *Suillus variegatus* or *Paxillus involutus* (Lindahl et al. 1999). In this study, up to 25% of the ^{32}P contained in the hyphae of *Hypholoma* appeared in the mycorrhizal plants, whereas transfer from the mycorrhizal fungi to the saprophyte was at least one order of magnitude lower.

Studies of the extraradical hyphae of *Glomus intraradices* (Koide and Kabir 2000) have demonstrated that this AM fungus can hydrolyze organic P and transfer it through the mycelium. The magnitude of these processes under unsterile soil conditions, however, remains poorly quantified, and the importance of organic P mineralization by the AM fungi themselves has been called into question by other experiments (Joner and Jakobsen 1995; Joner and Johansen 2000). In another study, wheat was grown in chambers composed of several compartments. These compartments permitted both root and mycorrhizal hyphae growth, or, blocked root access and allowed for only hyphal growth. The soils in different compartments were then supplemented with large amounts of P (200 mg kg^{-1} soil) in inorganic or organic forms. Control chambers without mycorrhizal fungi were also established. Elevated phosphatase activity was observed in the root-free soil colonized by *Glomus mosseae* when compared to AM-free soil, particularly upon organic P addition (Tarafdar and Marschner 1994). Similarly designed experiments using red clover and *Glomus versiforme* suggest that AM fungal colonization of a root-free soil amended with organic P makes a significant contribution to plant uptake of P from sources such as lecithin, RNA, and sodium phytate (Feng et al. 2003). On the other hand, the study by Antibus et al. (1997) demonstrated that field-collected AM roots of red maple had consistently lower levels of phosphatase activity than ECM roots of the same plant species.

Tarafdar et al. (2001) showed that fungal (*Aspergillus* spp.) acid phosphatases were more efficient than plant enzymes at mobilizing P from lecithin and phytate. This could be interpreted, on one hand, as proof of the capacity of fungi in general to efficiently hydrolyze organic P in the soil or, on the other hand, as a demonstration of the capacity of certain specific fungal groups (particularly the soil saprophytes) to hydrolyze such substrates. It is now clear from a range of studies that ERM and ECM fungi have the saprotrophic capacity to intervene in microbial mobilization–immobilization cycles and to sequester both N and P from the organic complexes formed during the decomposition of microbial, faunal, and plant remains. In heathland and forest ecosystems, these are the dominant sources of both N and P, and the enzymatic capacity to sequester these nutrients from complex organic substrates is probably most highly developed in the ERM fungi (Smith and Read 2008). The evidence for AM fungi is less clear and is complicated by the need to distinguish between the physiological activity of the AM hyphae and that of other fungi or bacteria that might be associated with them. Experiments by Hodge et al. (2001) demonstrated accelerated decomposition and N uptake from organic material associated with AM hyphae, but the potential contribution of other soil saprotrophs was unclear. Further experiments by Leigh et al. (2009) have shown uptake of P and N associated with patches of organic material, but again, additional uptake of P from bone-meal and Terragreen substrate present within the microcosms cannot be ruled out. The ambiguity here indicates the need for further experiments investigating the potential role of mycorrhiza-associated bacteria. The role of mycorrhizal fungal hyphae as primers of soil microbial activity has been discussed by a number of authors (e.g., Jones et al. 2004; Talbot et al. 2008; Toljander et al. 2007), but so far our understanding of how this regulates nutrient acquisition and transfer within the mycorrhizosphere is limited.

6.2.4 Mycorrhizas as Compound-Specific Filters

In an overwhelming number of studies, the role of mycorrhizal symbiosis in plant acquisition of P has been documented under a wide range of environmental conditions, in different ecosystems and for different host plants (Arihara and Karasawa 1998; Cardoso and Kuyper 2006; Jansa et al. 2009 and references therein; Smith and Read 2008 and references therein). Nevertheless, to regard mycorrhizas only as P pumps would be utterly incorrect, particularly if considering ECM and ERM fungi in comparison to the most abundant form of mycorrhizal symbiosis formed by the AM fungi. In addition to their role in P nutrition, ECM, ERM and, to a lesser extent, AM fungi are involved in acquisition of N by plants, both from inorganic and organic sources (Finlay et al. 1992; Johansen et al. 1992; Mäder et al. 2000; Read et al. 2004). Involvement of ECM in plant water uptake has been shown (Allen 2007; Plamboeck et al. 2007) and the mycorrhizas are also known to alleviate deficiencies in micronutrients such as zinc and copper. In addition, at least some genotypes of mycorrhizal fungi have the capacity to protect their host plants from acquisition of soil pollutants such as radiocaesium (de Boulois et al. 2008; Joner et al. 2004; Ladeyn et al. 2008) and heavy metals (Joner et al. 2000; Martino et al. 2000; Sharples et al. 1999; Sudová et al. 2008), while maintaining the P and/or zinc supply to the plants (Joner et al. 2004; Soares and Siqueira 2008).

6.3 Translocation of P Within the Hyphae and Its Release to the Plants

6.3.1 Transport Within the Hyphae

The P taken up by mycorrhizal fungi from the soil solution is used to meet the physiological demand of the fungus, with the remainder transported to the plants or stored in the hyphae. Transport of P to the plant implies a long-distance transfer through the hyphal network, which, in both ECM and AM fungi, is assumed to involve polyphosphates (Bucking and Heyser 2003; Ezawa et al. 2002). The short-chain polyphosphates, resulting from depolymerization of longer polyphosphate chains transferred over long distances, appear to be the immediate source of P for the plants (Ohtomo and Saito 2005; Solaiman et al. 1999; Takanishi et al. 2009). Rapid transfer rates of P via mycorrhizal fungal hyphae have been measured either using radioisotope labeling or microscopy (Bago et al. 2002; Cooper and Tinker 1978; Cox et al. 1980; Nielsen et al. 2002; Rhodes and Gerdemann 1978; Timonen et al. 1996). Together with the capacity of some ECM and ERM fungi to release P from recalcitrant sources, this rapid transfer represents what has been referred to as a mycorrhizal short-circuit in soil–plant P cycling, bypassing release from minerals or organic sources by free-living soil microorganisms (Johnson et al. 2005; Pankow et al. 1991).

6.3.2 Release of P to the Plant

The mechanics of how the P is released from the fungi to the plants has not yet been described for any mycorrhizal type. It is also not yet known whether this process is through passive leakage or an active transport system. It is likely that some fungi can, to different extents, retain P in their mycelium during translocation, resulting in partial immobilization and perhaps storage of P on the way from the soil to the plants (Boddington and Dodd 1999; Chilvers and Harley 1980; Harley and McCready 1981; Solaiman et al. 1999) – either as a result of slow transfer within the hyphae or due to limited release to the plant. In either case, P cycling between soil and plant is slowed down and the P could potentially also be released back to soil from the fungal hyphae or transferred to other soil organisms upon hyphal death due to soil disturbance, parasitism, or grazing. On the other hand, transitional storage of P in the fungal mycelium may function as a buffer ensuring continuous supply of P into long-lived plants such as trees or perennial herbs in a changing environment (Genet et al. 2000; Lussenhop and Fogel 1999; Read 1984). Thus, in the long run, the rapid provision of P to the host plants by fungi may not necessarily be the most beneficial system.

The plant side of the transfer, namely the mycorrhiza-inducible P_i transporters expressed in the close vicinity of fungal structures such as hyphal coils or arbuscules, has already been characterized for several plant species establishing AM symbiosis (e.g., Glassop et al. 2005; Javot et al. 2007; Nagy et al. 2005; Paszkowski et al. 2002; Rausch et al. 2001). Similar transporters are also likely to exist in ECM and ERM plants, but have not yet been characterized at the molecular level.

6.3.3 Consequences of Mycorrhizal P Acquisition for the Plants and for Maintenance of Mutualism

The efficiency of P acquisition by the mycorrhizal fungi, the temporary P immobilization in the fungal biomass, and the controlled P release to the plants all have important consequences for plant P nutritional status and growth. Although large improvements of plant P uptake have been reported upon mycorrhizal colonization of the roots by certain AM fungal species, association with other fungi may yield negative growth responses, and thus qualify the relationship as parasitic (Johnson et al. 1997; Smith et al. 2004). The potential for fungal control of P release to the plants points to one possible mechanism for maintaining the mycobiont diversity and mutualistic nature of the symbiosis through diversification of the benefits of carbon trading between the plant and the mycorrhizal fungi (Cowden and Peterson 2009; Helgason and Fitter 2009; Kiers and van der Heijden 2006).

In cases where the roots are colonized simultaneously by several species and/or genotypes of the mycorrhizal fungi, there is some evidence for preferential carbon distribution to more beneficial mycorrhizal symbionts (Bever et al. 2009; Fitter 2006). This preference is probably dependent upon the rates of P transfer within the specific root cells or fragments colonized by the different fungi. This is corroborated by physiological studies that showed that the release of P from intraradical hyphae of *Gigaspora* increased upon glucose addition (Solaiman and Saito 2001). However, there is also experimental evidence that some AM fungi can gain carbon from plants in spite of the net P gain of the host plant being very limited, such as under P- and N-sufficient conditions (Hoeksema et al. 2010; Pearson and Jakobsen 1993; Smith et al. 2003, 2004, 2009). Additionally, the existence of mycoheterotrophic plants, where the plants apparently receive both mineral nutrients and carbon from the fungus (Bidartondo et al. 2002; Imhof 2009; Taylor et al. 2004), is difficult to reconcile with the hypothesis of preferential carbon allocation to the most beneficial fungal symbiont. Possibly, some of the processes are regulated at the ecosystem level and not at a single plant level. This may mean that the answers are hidden in plant community ecology, source–sink relationships and so called "common mycorrhizal networks" interconnecting different plants (Bever 1999; Bever et al. 2009), although little evidence is so far available about mycorrhiza-mediated transfer of P and carbon between plants (Newman and Ritz 1986; Philip and Simard 2008; Robinson and Fitter 1999; Selosse et al. 2006; Yao et al. 2003). Furthermore, experimental data have recently been gathered showing a more inconspicuous contribution of mycorrhizas to the P acquisition by the plants without detectable plant growth or net P uptake improvements (Smith et al. 2003, 2004). These effects are, however, only seen in carefully designed radioisotope experiments, using non-mycorrhizal mutant genotypes, or expression analyses of fungal and plant P transporters (Burleigh and Bechmann 2002; Grace et al. 2009; Li et al. 2008; Poulsen et al. 2005; Smith et al. 2009). Results of the above studies showed that the symbiosis may be fully functional, even if traditional measurements of mycorrhizal "benefits" such as growth or net P uptake improvements were not able to detect its contribution (Facelli et al. 2009). They also indicate that the definition of the symbiosis, currently often implicating measurable benefits to both partners, may need to be broadened to cover these cases of association with no obvious "benefits" (Cavagnaro et al. 2004; Jones and Smith 2004).

6.4 Functional Diversity of Mycorrhizas with Respect to P Uptake

In addition to major functional differences between the different mycorrhizal types, variation in P uptake patterns and efficiency have been recognized between species and genotypes of both AM and the ECM fungi (Boddington and Dodd 1999; Cairney 1999; Cavagnaro et al. 2005; Jansa et al. 2005; Munkvold et al. 2004; Smith et al. 2003). In these studies, radioisotope labeling has proven to be a particularly important approach (see also Frossard et al. 2011). The recorded functional diversity among the different mycorrhizal fungi appears to be important for understanding mycorrhizal functioning in the field. This is because roots of most plants are colonized by a mixture of mycorrhizal fungal species, usually, but not always, belonging to the same type (van der Heijden and Vosátka 1999; van der Heijden et al. 1998).

It has been postulated that fungi differing in P acquisition strategies could complement each other when sharing the same root system, resulting in greater symbiotic benefits than those conferred by each of the fungi in isolation (Koide 2000). Although this may well be the case, direct experimental evidence for functional complementarity in P acquisition within mycorrhizal communities is still limited (Jansa et al. 2008; Maherali and Klironomos 2007). Better understanding of the components of the mycorrhizal uptake pathway in different mycorrhizal fungi (e.g., numbers of alternative P transporters, their regulation and expressional dynamics, dynamics of the polyphosphate pool in the fungal hyphae) and how the functional diversity is structured within fungal communities, will be necessary before more general conclusions can be drawn. A combination of approaches spanning molecular biology (Burleigh et al. 2002; Grace et al. 2009; Jansa et al. 2008; Tatry et al. 2009), isotopic labeling (Jakobsen et al. 1992; Jansa et al. 2005), carefully designed pot experiments (Maherali and Klironomos 2007; Smith et al. 2004), and modeling (Antoninka et al. 2009; Deressa and Schenk 2008; Schnepf et al. 2008, 2011) is essential for further progress in this area. Likewise, interactions between mycorrhizal P uptake, carbon costs, and cycling of other nutrients such as N at the whole plant or plant community levels must be considered for an ecologically relevant picture (Grelet et al. 2009; Johnson et al. 2010; Smith et al. 2009).

6.5 Human Impact on the Mycorrhizal Pathway of P Acquisition by Plants

Agricultural activities, pollution, climate change, and other anthropogenic environmental influences affect mycorrhizal symbioses and their role in P acquisition by plants. Substantial information has been accumulated on these influences over the years; however, only a fraction is discussed here. For a more detailed overview, the reader is referred to other sources (Allen et al. 2003; Allison and Treseder 2008; Drigo et al. 2008; Jansa et al. 2006).

In agricultural systems, the application of water-soluble mineral P and N fertilizers usually reduces the dependency of plants on nutrient uptake via the mycorrhizal pathway. This forced selection may eventually promote fungi that are less beneficial to plants or promote an abundance of plants that are less mycorrhiza-dependent (Covacevich et al. 2007; Jansa et al. 2006; Johnson 1993). Crop breeding efforts for high yields have been shown to inadvertently select for lower mycorrhizal dependencies, probably through selection for greater dependency on mineral fertilizer inputs (Hetrick et al. 1993; Tawaraya 2003; Zhu et al. 2001). These effects are not limited to agricultural plants and soils as nutrients, and pollutants often leach into natural ecosystems from agricultural fields, inappropriate waste management, and/or industrial activities. These effects are especially pronounced in forests and heathlands. The activity of fungi sensitive to elevated nutrient availability and pollutants decreases, and plant species or genotypes are favored that can better tolerate the pollution and/or depend less on nutrient acquisition via mycorrhizal fungi (Dighton 1995; Egerton-Warburton and Allen 2000; Kieliszewska-Rokicka 1999; Rejšek 1991; Robertson et al. 2007).

Elevation of CO_2 levels in the atmosphere may transiently increase the biomass and metabolic activities of mycorrhizal fungi, such as P transfer from soil to plants, due to a reduction of carbon limitation and through creating a greater requirement for P and other nutrients by the plants (Alberton and Kuyper 2009; Millard et al. 2007). In a longer perspective, however, this may result in a more rapid exploitation of soil nutrient reserves and an increased sensitivity of ecosystems to disturbance events (Pritchard et al. 2008). Global warming and redistribution of precipitation patterns may also change the activity of soil microorganisms in general and mycorrhizal fungi in particular (Fitter et al. 2004), but the future prospects remain rather blurred – usually the availability of water and not soil nutrients appears to be the primary driver of the expected ecosystem changes (Aerts et al. 2009; Allison and Treseder 2008; Heinemeyer et al. 2007). If water availability is not limiting the microbial activity, decay of roots and plant debris may be accelerated by global warming, which could in turn speed up mycorrhizal P capture from these sources (e.g., Carleton and Read 1991) thus speeding up the P cycling. Nevertheless, genotypic differences in temperature tolerance and acclimation between different mycorrhizal fungi might also contribute to the variability in the observed responses in mycorrhizal community composition and functioning to climatic changes (Malcolm et al. 2008).

A final aspect of human activity worth mentioning here is the phenomenon of plant invasion. It has been proposed that non-mycorrhizal alien plants, such as *Hakea* spp. that form cluster roots, could gain a P-acquisition advantage in ecosystems normally dominated by mycorrhizal plants (Allsopp and Holmes 2001; Sousa et al. 2007). For alien plants like *Centaurea maculosa* in Northern American grasslands, a different mode of action has been proposed: invasive plants could tap into the existing mycorrhizal networks and divert the flux of resources, such as P, to individuals with this ability (Batten et al. 2008; Callaway et al. 2004; Zabinski et al. 2002). Another scenario is that the native mycorrhizal fungi could be suppressed by (non-mycorrhizal) alien plants (e.g., *Alliaria petiolata* in North America), which would in turn result in suppression of the native plants that rely on their (native) mycorrhizal symbionts (Callaway et al. 2008; Mitchell et al. 2006). Although experimental evidence for these processes and their importance in the phenomenon of ecological invasions is still equivocal, it is becoming more and more apparent that underground processes, nutrient balances, and associated microflora including mycorrhizal fungi might all be important players in plant invasions, and thus relevant topics for further study.

6.6 Conclusions

Mycorrhizal symbioses are highly diverse in their relationships with plants and ecosystems and are involved in different P cycling processes. Although evidence is accumulating to suggest that some ECM fungi have the capacity to induce or accelerate weathering of P-bearing minerals, thus affecting P inputs into the biological P cycling within the ecosystems, these processes are still largely unquantified. It is therefore unclear how important these processes are compared to the other input pathways such as aerial deposition, sedimentation, abiotic weathering, and anthropogenic inputs via fertilization and pollution, and how they vary amongst different ecosystems and on different geographic and time scales (Johnson et al. 2010; Newman 1995; Smits et al. 2008). To solve these issues, precise measurement of P fluxes should be continued, using radio- and stable isotope methods (Frossard et al. 2011), as well as complementary modeling efforts and long-term observations (Rosling 2009; Rosling et al. 2009; Schnepf et al. 2011). Soil and environmental conditions must always be taken into account when interpreting experimental results (e.g., little biotic effect on solubilization of apatite through acidification should be expected for strongly acidic soil conditions). Additionally, chemical reactivity and other relevant information (chemical composition, crystallinity, grain size, provenance) of compounds such as apatite used for experimental studies should always be assessed and reported to allow strict reproducibility of results.

The importance of mycorrhizal symbiosis in plant P uptake has been established in hundreds of studies published over many decades (Smith and Read 2008). This extends from traditional comparisons of small (non-mycorrhizal) and big (mycorrhizal) plants, where the differences in net P uptake are easily demonstrable; through more inconspicuous cases, where improvements of net P uptake are not always translated into improved plant growth (Jansa et al. 2005); to the extreme cases, where net P uptake of the plant is not different between mycorrhizal and non-mycorrhizal plants, although the mycorrhizal P uptake pathway is fully operational (Grace et al. 2009; Smith et al. 2004, 2009).

In spite of some well-documented model cases, the mechanisms governing the transfer of P within the common mycorrhizal networks interconnecting different plants, as well as the involvement of other soil microorganisms in soil–mycorrhiza-plant P transfer are still only fragmentarily understood (Finlay 2008; Lindahl et al. 1999; Toljander et al. 2006). Extensive use of mineral P fertilizers and breeding for high-yielding crop varieties under fertilized soil conditions has seemingly led to selection of genotypes that are less responsive to mycorrhizal symbiosis than some of the older cultivars (Hetrick et al. 1993). This might be of concern regarding efficient use of natural resources in the future (Sawers et al. 2008).

Major differences in the accessibility of different soil P pools to different mycorrhizal types explain the distribution of the different mycorrhizal types in ecosystems: AM fungi appear to access mainly orthophosphate in the soil solution, ECM can access both inorganic and organic P in soil, and ERM appear to be able to access mainly the P contained in organic substrates (Read and Perez-Moreno 2003). Direct acquisition of P from decaying plant biomass or other organic substrates via mycorrhizal pathway effectively short-circuits the soil–plant P cycling, in which the largest portion of P mineralization would otherwise have to be accomplished by free-living soil decomposers before being accessed by roots or the mycorrhizal hyphae (Carleton and Read 1991). Through efficient P recycling from organic forms, capture of P_i from soil solution, and by improving soil mechanical stability, mycorrhizal symbiosis has the potential to contribute substantially to reduction of P loss through leaching and erosion (Asghari et al. 2005; Cardoso and Kuyper 2006). On the other hand, in agricultural and other ecosystems where management includes intentional removal of some plant products (green biomass, grains, fibers, wood), mycorrhizal symbiosis alone cannot guarantee long-term sustainability of production without adequate P inputs in the form of mineral or organic fertilizers, plant residues, or products resulting from wastewater treatment or municipal waste processing (Jansa et al. 2006; Oberson et al. 2011).

Acknowledgements Funding by Swiss National Science Foundation (project 31003A-125491 granted to Jan Jansa) is gratefully acknowledged. We thank Mark Smits for providing the high-resolution photograph of apatite grains colonized by *Paxillus involutus*, and to Mineralogical Magazine for granting permission to reproduce this picture. Editorial and language corrections by Angela Erb and useful suggestions made by the reviewers are thankfully acknowledged.

References

Adams F, Reddell P, Webb MJ, Shipton WA (2006) Arbuscular mycorrhizas and ectomycorrhizas on *Eucalyptus grandis* (Myrtaceae) trees and seedlings in native forests of tropical north-eastern Australia. Aust J Bot 54:271–281

Aerts R, Callaghan TV, Dorrepaal E, van Logtestijn RSP, Cornelissen JHC (2009) Seasonal climate manipulations result in species-specific changes in leaf nutrient levels and isotopic composition in a sub-arctic bog. Funct Ecol 23:680–688

Alberton O, Kuyper TW (2009) Ectomycorrhizal fungi associated with *Pinus sylvestris* seedlings respond differently to increased carbon and nitrogen availability: implications for ecosystem responses to global change. Glob Change Biol 15:166–175

Allen MF (2007) Mycorrhizal fungi: highways for water and nutrients in arid soils. Vadose Zone J 6:291–297

Allen MF, Figueroa C, Weinbaum BS, Barlow SB, Allen EB (1996) Differential production of oxalate by mycorrhizal fungi in arid ecosystems. Biol Fertil Soils 22:287–292

Allen MF, Swenson W, Querejeta JI, Egerton-Warburton LM, Treseder KK (2003) Ecology of mycorrhizae: a conceptual framework for complex interactions among plants and fungi. Annu Rev Phytopathol 41:271–303

Allison SD, Treseder KK (2008) Warming and drying suppress microbial activity and carbon cycling in boreal forest soils. Glob Change Biol 14:2898–2909

Allsopp N, Holmes PM (2001) The impact of alien plant invasion on mycorrhizas in mountain fynbos vegetation. S Afr J Bot 67:150–156

Antibus RK, Bower D, Dighton J (1997) Root surface phosphatase activities and uptake of ^{32}P-labelled inositol phosphate in field-collected gray birch and red maple roots. Mycorrhiza 7:39–46

Antoninka A, Wolf JE, Bowker M, Classen AT, Johnson NC (2009) Linking above- and below-ground responses to global change at community and ecosystem scales. Glob Change Biol 15:914–929

Arihara J, Karasawa T (1998) Phosphorus nutrition in cropping systems through arbuscular mycorrhizal management. In: Ae N, Arihara J, Okada K, Srinivasan A (eds) International workshop on plant nutrient acquisition. Tsukuba, Japan. Conference proceedings. Springer, Berlin, pp 319–337

Asghari HR, Chittleborough DJ, Smith FA, Smith SE (2005) Influence of arbuscular mycorrhizal (AM) symbiosis on phosphorus leaching through soil cores. Plant Soil 275:181–193

Azcón-Aguilar C, Barea JM (1995) Saprophytic growth of arbuscular mycorrhizal fungi. In: Varma A, Hock B (eds) Mycorrhiza: structure, function, molecular biology and biotechnology. Springer, Berlin, pp 391–407

Bago B, Azcón-Aguilar C, Goulet A, Piché Y (1998) Branched adsorbing structures (BAS): a feature of the extraradical mycelium of symbiotic arbuscular mycorrhizal fungi. New Phytol 139:375–388

Bago B, Pfeffer PE, Zipfel W, Lammers P, Shachar-Hill Y (2002) Tracking metabolism and imaging transport in arbuscular mycorrhizal fungi. Metabolism and transport in AM fungi. Plant Soil 244:189–197

Bago B, Cano C, Azcón-Aguilar C, Samson J, Coughlan AP, Piché Y (2004) Differential morphogenesis of the extraradical mycelium of an arbuscular mycorrhizal fungus grown monoxenically on spatially heterogeneous culture media. Mycologia 96:452–462

Baldrian P (2009) Ectomycorrhizal fungi and their enzymes in soils: is there enough evidence for their role as facultative soil saprotrophs? Oecologia 161:657–660

Barea JM (2000) Rhizosphere and mycorrhiza of field crops. In: Balazs E, Galante E, Lynch JM, Schepers JS, Toutant JP, Werner D, Werry PAT (eds) Conference on biological resource management – connecting science and policy, Paris. France. Conference proceedings. Springer, Berlin, pp 81–92

Batten KM, Scow KM, Espeland EK (2008) Soil microbial community associated with an invasive grass differentially impacts native plant performance. Microb Ecol 55:220–228

Baxter JW, Dighton J (2001) Ectomycorrhizal diversity alters growth and nutrient acquisition of grey birch (*Betula populifolia*) seedlings in host-symbiont culture conditions. New Phytol 152:139–149

Bending GD, Read DJ (1995a) The structure and function of the vegetative mycelium of ectomycorrhizal plants. V. Foraging behaviour and translocation of nutrients from exploited litter. New Phytol 130:401–409

Bending GD, Read DJ (1995b) The structure and function of the vegetative mycelium of ectomycorrhizal plants. VI. Activities of nutrient mobilizing enzymes in birch litter colonized by *Paxillus involutus* (Fr.) Fr. New Phytol 130:411–417

Bending GD, Read DJ (1996a) Effects of the soluble polyphenol tannic acid on the activities of ericoid and ectomycorrhizal fungi. Soil Biol Biochem 28:1595–1602

Bending GD, Read DJ (1996b) Nitrogen mobilization from protein-polyphenol complex by ericoid and ectomycorrhizal fungi. Soil Biol Biochem 28:1603–1612

Bever JD (1999) Dynamics within mutualism and the maintenance of diversity: inference from a model of interguild frequency dependence. Ecol Lett 2:52–62

Bever JD, Richardson SC, Lawrence BM, Holmes J, Watson M (2009) Preferential allocation to beneficial symbiont with spatial structure maintains mycorrhizal mutualism. Ecol Lett 12:13–21

Bidartondo MI, Redecker D, Hijri I, Wiemken A, Bruns TD, Dominguez L, Sersic A, Leake JR, Read DJ (2002) Epiparasitic plants specialized on arbuscular mycorrhizal fungi. Nature 419:389–392

Blum JD, Klaue A, Nezat CA, Driscoll CT, Johnson CE, Siccama TG, Eagar C, Fahey TJ, Likens GE (2002) Mycorrhizal weathering of apatite as an important calcium source in base-poor forest ecosystems. Nature 417:729–731

Boddington CL, Dodd JC (1999) Evidence that differences in phosphate metabolism in mycorrhizas formed by species of *Glomus* and *Gigaspora* might be related to their life-cycle strategies. New Phytol 142:531–538

Bolan NS (1991) A critical review on the role of mycorrhizal fungi in the uptake of phosphorus by plants. Plant Soil 134:189–207

Bolan NS, Robson AD, Barrow NJ (1983) Plant and soil factors including mycorrhizal infection causing sigmoidal response of plants to applied phosphorus. Plant Soil 73:187–201

Bolan NS, Robson AD, Barrow NJ (1984) Increasing phosphorus supply can increase the infection of plant roots by vesicular arbuscular mycorrhizal fungi. Soil Biol Biochem 16:419–420

Brundrett MC (2009) Mycorrhizal associations and other means of nutrition of vascular plants: understanding the global diversity of host plants by resolving conflicting information and developing reliable means of diagnosis. Plant Soil 320:37–77

Bruns TD, Bidartondo MI, Taylor DL (2002) Host specificity in ectomycorrhizal communities: what do the exceptions tell us? Integr Comp Biol 42:352–359

Bucking H, Heyser W (2003) Uptake and transfer of nutrients in ectomycorrhizal associations: interactions between photosynthesis and phosphate nutrition. Mycorrhiza 13:59–68

Bueé M, Courty PE, Mignot D, Garbaye J (2007) Soil niche effect on species diversity and catabolic activities in an ectomycorrhizal fungal community. Soil Biol Biochem 39:1947–1955

Bünemann EK, Prusisz B, Ehlers K (2011) Characterization of phosphorus forms in soil microorganisms. In: Bünemann EK, Oberson A, Frossard E (eds) Phosphorus in action: biological processes in soil phosphorus cycling. Soil biology, vol 26. Springer, Heidelberg. doi:10.1007/978-3-642-15271-9_2

Burgess TI, Malajczuk N, Grove TS (1993) The ability of 16 ectomycorrhizal fungi to increase growth and phosphorus uptake of *Eucalyptus globulus* Labill and *E. diversicolor* F. Muell. Plant Soil 153:155–164

Burke DJ, Martin KJ, Rygiewicz PT, Topa MA (2005) Ectomycorrhizal fungi identification in single and pooled root samples: terminal restriction fragment length polymorphism (T-RFLP) and morphotyping compared. Soil Biol Biochem 37:1683–1694

Burleigh SH, Bechmann IE (2002) Plant nutrient transporter regulation in arbuscular mycorrhizas. Plant Soil 244:247–251

Burleigh SH, Cavagnaro T, Jakobsen I (2002) Functional diversity of arbuscular mycorrhizas extends to the expression of plant genes involved in P nutrition. J Exp Bot 53:1593–1601

Cairney JWG (1999) Intraspecific physiological variation: implications for understanding functional diversity in ectomycorrhizal fungi. Mycorrhiza 9:125–135

Cairney JWG (2000) Evolution of mycorrhiza systems. Naturwissenschaften 87:467–475

Callaway RM, Thelen GC, Barth S, Ramsey PW, Gannon JE (2004) Soil fungi alter interactions between the invader *Centaurea maculosa* and North American natives. Ecology 85:1062–1071

Callaway RM, Cipollini D, Barto K, Thelen GC, Hallett SG, Prati D, Stinson K, Klironomos J (2008) Novel weapons: invasive plant suppresses fungal mutualists in America but not in its native Europe. Ecology 89:1043–1055

Cardoso IM, Kuyper TW (2006) Mycorrhizas and tropical soil fertility. Agric Ecosyst Environ 116:72–84

Carleton TJ, Read DJ (1991) Ectomycorrhizas and nutrient transfer in conifer-feather moss ecosystems. Can J Bot 69:778–785

Castillo CG, Borie F, Godoy R, Rubio R, Sieverding E (2006) Diversity of mycorrhizal plant species and arbuscular mycorrhizal fungi in evergreen forest, deciduous forest and grassland ecosystems of Southern Chile. J Appl Bot Food Qual 80:40–47

Cavagnaro TR, Smith FA, Hay G, Carne-Cavagnaro VL, Smith SE (2004) Inoculum type does not affect overall resistance of an arbuscular mycorrhiza-defective tomato mutant to colonisation but inoculation does change competitive interactions with wild-type tomato. New Phytol 161:485–494

Cavagnaro TR, Smith FA, Smith SE, Jakobsen I (2005) Functional diversity in arbuscular mycorrhizas: exploitation of soil patches with different phosphate enrichment differs among fungal species. Plant Cell Environ 28:642–650

Chen A, Chambers SM, Cairney JWG (1999) Utilisation of organic nitrogen and phosphorus sources by mycorrhizal endophytes of *Woollsia pungens* (Cav.) F. Muell. (Epacridaceae). Mycorrhiza 8:181–187

Chilvers GA, Harley JL (1980) Visualization of phosphate accumulation in beech mycorrhizas. New Phytol 84:319–326

Choi DS, Quoreshi AM, Maruyama Y, Jin HO, Koike T (2005) Effect of ectomycorrhizal infection on growth and photosynthetic characteristics of *Pinus densiflora* seedlings grown under elevated CO_2 concentrations. Photosynthetica 43:223–229

Clark RB, Zeto SK (2000) Mineral acquisition by arbuscular mycorrhizal plants. J Plant Nutr 23:867–902

Colpaert JV, van Laere A, van Tichelen KK, van Assche JA (1997) The use of inositol hexaphosphate as a phosphorus source by mycorrhizal and non-mycorrhizal Scots pine (*Pinus sylvestris*). Funct Ecol 11:407–415

Cooper KM, Tinker PB (1978) Translocation and transfer of nutrients in vesicular-arbuscular mycorrhizas. II. Uptake and translocation of phosphorus, zinc and sulphur. New Phytol 81:43–52

Corrêa A, Strasser RJ, Martins-Loução MA (2006) Are mycorrhiza always beneficial? Plant Soil 279:65–73

Corrêa A, Strasser RJ, Martins-Loução MA (2008) Response of plants to ectomycorrhizae in N-limited conditions: which factors determine its variation? Mycorrhiza 18:413–427

Covacevich F, Echeverria HE, Aguirrezabal LAN (2007) Soil available phosphorus status determines indigenous mycorrhizal colonization of field and glasshouse-grown spring wheat from Argentina. Appl Soil Ecol 35:1–9

Cowden CC, Peterson CJ (2009) A multi-mutualist simulation: applying biological market models to diverse mycorrhizal communities. Ecol Modell 220:1522–1533

Cox G, Moran KJ, Sanders F, Nockolds C, Tinker PB (1980) Translocation and transfer of nutrients in vesicular-arbuscular mycorrhizas. 3. Polyphosphate granules and phosphorus translocation. New Phytol 84:649–659

Croll D, Sanders IR (2009) Recombination in *Glomus intraradices*, a supposed ancient asexual arbuscular mycorrhizal fungus. BMC Evol Biol 9:13

de Boulois HD, Joner EJ, Leyval C, Jakobsen I, Chen BD, Roos R, Thiry Y, Rufyikiri G, Delvaux B, Declerck S (2008) Role and influence of mycorrhizal fungi on radiocesium accumulation by plants. J Environ Radioactiv 99:785–800

de Vries FT, Bååth E, Kuyper TW, Bloem J (2009) High turnover of fungal hyphae in incubation experiments. FEMS Microbiol Ecol 67:389–396

Deressa TG, Schenk MK (2008) Contribution of roots and hyphae to phosphorus uptake of mycorrhizal onion (*Allium cepa* L.) – a mechanistic modeling approach. J Plant Nutr Soil Sci 171:810–820

Diedhiou AG, Gueye O, Diabate M, Prin Y, Duponnois R, Dreyfus B, Ba AM (2005) Contrasting responses to ectomycorrhizal inoculation in seedlings of six tropical African tree species. Mycorrhiza 16:11–17

Dighton J (1995) Nutrient cycling in different terrestrial ecosystems in relation to fungi. Can J Bot 73:S1349–S1360

Doolette AL, Smernik RJ (2011) Soil organic phosphorus speciation using spectroscopic techniques. In: Bünemann EK, Oberson A, Frossard E (eds) Phosphorus in action: biological processes in soil phosphorus cycling. Soil biology, vol 26. Springer, Heidelberg. doi:10.1007/978-3-642-15271-9_1

Dou ZX, Ramberg CF, Toth JD, Wang Y, Sharpley AN, Boyd SE, Chen CR, Williams D, Xu ZH (2009) Phosphorus speciation and sorption-desorption characteristics in heavily manured soils. Soil Sci Soc Am J 73:93–101

Downes GM, Alexander IJ, Cairney JWG (1992) A study of aging of spruce [*Picea sitchensis* (Bong) Carr] ectomycorrhizas. 1. Morphological and cellular changes in mycorrhizas formed by *Tylospora fibrillosa* (Burt) Donk and *Paxillus involutus* (Batsch ex Fr) Fr. New Phytol 122:141–152

Drigo B, Kowalchuk GA, van Veen JA (2008) Climate change goes underground: effects of elevated atmospheric CO_2 on microbial community structure and activities in the rhizosphere. Biol Fertil Soils 44:667–679

Duponnois R, Kisa M, Prin Y, Ducousso M, Plenchette C, Lepage M, Galiana A (2008) Soil factors influencing the growth response of *Acacia holosericea* A. Cunn. ex G. Don to ectomycorrhizal inoculation. New Forest 35:105–117

Egerton-Warburton LM, Allen EB (2000) Shifts in arbuscular mycorrhizal communities along an anthropogenic nitrogen deposition gradient. Ecol Appl 10:484–496

Egerton-Warburton L, Allen MF (2001) Endo- and ectomycorrhizas in *Quercus agrifolia* Nee. (Fagaceae): patterns of root colonization and effects on seedling growth. Mycorrhiza 11:283–290

Ezawa T, Smith SE, Smith FA (2002) P metabolism and transport in AM fungi. Plant Soil 244:221–230

Facelli E, Facelli J, McLaughlin MJ, Smith SE (1999) An experimental study of the interactive effects of mycorrhizal symbiosis, intraspecific competition and resource availability using *Trifolium subterraneum* L. cv Mt Barker. New Phytol 141:535–547

Facelli E, Smith SE, Smith FA (2009) Mycorrhizal symbiosis – overview and new insights into roles of arbuscular mycorrhizas in agro- and natural ecosystems. Australas Plant Pathol 38:338–344

Faeth SH, Fagan WF (2002) Fungal endophytes: common host plant symbionts but uncommon mutualists. Integr Comp Biol 42:360–368

Feng G, Song YC, Li XL, Christie P (2003) Contribution of arbuscular mycorrhizal fungi to utilization of organic sources of phosphorus by red clover in a calcareous soil. Appl Soil Ecol 22:139–148

Finlay RD (2008) Ecological aspects of mycorrhizal symbiosis: with special emphasis on the functional diversity of interactions involving the extraradical mycelium. J Exp Bot 59:1115–1126

Finlay RD, Read DJ (1986) The structure and function of the vegetative mycelium of ectomycorrhizal plants. II. The uptake and distribution of phosphorus by mycelial strands interconnecting host plants. New Phytol 103:157–165

Finlay RD, Frostegård Å, Sonnerfeldt AM (1992) Utilization of organic and inorganic nitrogen sources by ectomycorrhizal fungi in pure culture and in symbiosis with *Pinus contorta* Dougl. Ex Louds. New Phytol 120:105–115

Fitter AH (2005) Darkness visible: reflections on underground ecology. J Ecol 93:231–243

Fitter AH (2006) What is the link between carbon and phosphorus fluxes in arbuscular mycorrhizas? A null hypothesis for symbiotic function. New Phytol 172:3–6

Fitter AH, Heinemeyer A, Husband R, Olsen E, Ridgway KP, Staddon PL (2004) Global environmental change and the biology of arbuscular mycorrhizas: gaps and challenges. Can J Bot 82:1133–1139

Friese CF, Allen MF (1991) The spread of VA mycorrhizal fungal hyphae in the soil: inoculum types and external hyphal architecture. Mycologia 83:409–418

Frossard E, Achat DL, Bernasconi SM, Bünemann EK, Fardeau J-C, Jansa J, Morel C, Rabeharisoa L, Randriamanantsoa L, Sinaj S, Tamburini F, Oberson A (2011) The use of tracers to investigate phosphate cycling in soil–plant systems. In: Bünemann EK, Oberson A, Frossard E (eds) Phosphorus in action: biological processes in soil phosphorus cycling. Soil biology, vol 26. Springer, Heidelberg. doi:10.1007/978-3-642-15271-9_3

Gange AC, Ayres RL (1999) On the relation between arbuscular mycorrhizal colonization and plant 'benefit'. Oikos 87:615–621

Genet P, Prevost A, Pargney JC (2000) Seasonal variations of symbiotic ultrastructure and relationships of two natural ectomycorrhizae of beech (*Fagus sylvatica*/*Lactarius blennius* var. *viridis* and *Fagus sylvatica*/*Lactarius subdulcis*). Trees Struct Funct 14:465–474

George E, Marschner H (1996) Nutrient and water uptake by roots of forest trees. J Plant Nutr Soil Sci 159:11–21

Gibson BR, Mitchell DT (2004) Nutritional influences on the solubilization of metal phosphate by ericoid mycorrhizal fungi. Mycol Res 108:947–954

Glassop D, Smith SE, Smith FW (2005) Cereal phosphate transporters associated with the mycorrhizal pathway of phosphate uptake into roots. Planta 222:688–698

Godbold DL, Hoosbeek MR, Lukac M, Cotrufo MF, Janssens IA, Ceulemans R, Polle A, Velthorst EJ, Scarascia-Mugnozza G, De Angelis P, Miglietta F, Peressotti A (2006) Mycorrhizal hyphal turnover as a dominant process for carbon input into soil organic matter. Plant Soil 281:15–24

Grace EJ, Cotsaftis O, Tester M, Smith FA, Smith SE (2009) Arbuscular mycorrhizal inhibition of growth in barley cannot be attributed to extent of colonization, fungal phosphorus uptake or effects on expression of plant phosphate transporter genes. New Phytol 181:938–949

Graham JH (2001) What do root pathogens see in mycorrhizas? New Phytol 149:357–359

Gregory PJ (2006) Plant roots: growth, activity and interaction with soils. Blackwell, Oxford

Grelet GA, Johnson D, Paterson E, Anderson IC, Alexander IJ (2009) Reciprocal carbon and nitrogen transfer between an ericaceous dwarf shrub and fungi isolated from *Piceirhiza bicolorata* ectomycorrhizas. New Phytol 182:359–366

Griffiths RP, Baham JE, Caldwell BA (1994) Soil solution chemistry of ectomycorrhizal mats in forest soil. Soil Biol Biochem 26:331–337

Hagerberg D, Thelin G, Wallander H (2003) The production of ectomycorrhizal mycelium in forests: relation between forest nutrient status and local mineral sources. Plant Soil 252:279–290

Harley JL, McCready CC (1981) Phosphate accumulation in *Fagus* mycorrhizas. New Phytol 89:75–80

Harrison MJ (1999) Biotrophic interfaces and nutrient transport in plant fungal symbioses. J Exp Bot 50:1013–1022

Harrison MJ, van Buuren ML (1995) A phosphate transporter from the mycorrhizal fungus *Glomus versiforme*. Nature 378:626–629

Heath RT (2005) Microbial turnover of organic phosphorus in aquatic systems. In: Turner BL, Frossard E, Baldwin DS (eds) Organic phosphorus in the environment. CABI, Wallingford, pp 185–204

Heinemeyer A, Hartley IP, Evans SP, de la Fuente JAC, Ineson P (2007) Forest soil CO_2 flux: uncovering the contribution and environmental responses of ectomycorrhizas. Glob Change Biol 13:1786–1797

Helgason T, Fitter AH (2009) Natural selection and the evolutionary ecology of the arbuscular mycorrhizal fungi (Phylum Glomeromycota). J Exp Bot 60:2465–2480

Helgason T, Merryweather JW, Denison J, Wilson P, Young JPW, Fitter AH (2002) Selectivity and functional diversity in arbuscular mycorrhizas of co-occurring fungi and plants from a temperate deciduous woodland. J Ecol 90:371–384

Hetrick BAD, Wilson GWT, Cox TS (1993) Mycorrhizal dependence of modern wheat cultivars and ancestors – a synthesis. Can J Bot 71:512–518

Hodge A, Campbell CD, Fitter AH (2001) An arbuscular mycorrhizal fungus accelerates decomposition and acquires nitrogen directly from organic material. Nature 413:297–299

Hoeksema JD, Chaudhary VB, Gehring CA, Johnson NC, Karst J, Koide RT, Pringle A, Zabinski C, Bever JD, Moore JC, Wilson GWT, Klironomos JN, Umbanhowar J (2010) A meta-analysis of context-dependency in plant response to inoculation with mycorrhizal fungi. Ecol Lett 13:394–407

Holford ICR (1997) Soil phosphorus: its measurement, and its uptake by plants. Aust J Soil Res 35:227–239

Hutchens E (2009) Microbial selectivity on mineral surfaces: possible implications for weathering processes. Fungal Biol Rev 23:115–121. doi:10.1016/j.fbr.2009.10.002

Imhof S (2009) Arbuscular, ecto-related, orchid mycorrhizas-three independent structural lineages towards mycoheterotrophy: implications for classification? Mycorrhiza 19:357–363

Jakobsen I (1995) Transport of phosphorus and carbon in VA mycorrhizas. In: Varma A, Hock B (eds) Mycorrhiza: structure, function, molecular biology and biotechnology. Springer, Berlin, pp 297–324

Jakobsen I, Abbott LK, Robson AD (1992) External hyphae of vesicular-arbuscular mycorrhizal fungi associated with *Trifolium subterraneum* L. 1. Spread of hyphae and phosphorus inflow into roots. New Phytol 120:371–380

Jansa J, Gryndler M (2010) Biotic environment of the arbuscular mycorrhizal fungi in soil. In: Koltai H, Kapulnik Y (eds) Arbuscular Mycorrhizas: Physiology and Function, 2nd edn. Springer, Heidelberg, 209–236. doi: 10.1007/978-90-481-9489-6_10

Jansa J, Vosátka M (2000) *In vitro* and *post vitro* inoculation of micropropagated Rhododendrons with ericoid mycorrhizal fungi. Appl Soil Ecol 15:125–136

Jansa J, Mozafar A, Anken T, Ruh R, Sanders IR, Frossard E (2002) Diversity and structure of AMF communities as affected by tillage in a temperate soil. Mycorrhiza 12:225–234

Jansa J, Mozafar A, Frossard E (2003a) Long-distance transport of P and Zn through the hyphae of an arbuscular mycorrhizal fungus in symbiosis with maize. Agronomie 23:481–488

Jansa J, Mozafar A, Kuhn G, Anken T, Ruh R, Sanders IR, Frossard E (2003b) Soil tillage affects the community structure of mycorrhizal fungi in maize roots. Ecol Appl 13:1164–1176

Jansa J, Mozafar A, Frossard E (2005) Phosphorus acquisition strategies within arbuscular mycorrhizal fungal community of a single field site. Plant Soil 276:163–176

Jansa J, Wiemken A, Frossard E (2006) The effects of agricultural practices on arbuscular mycorrhizal fungi. In: Frossard E, Blum WEH, Warkentin BP (eds) Function of soils for human societies and the environment. Geological Society, London, pp 89–115

Jansa J, Smith FA, Smith SE (2008) Are there benefits of simultaneous root colonization by different arbuscular mycorrhizal fungi? New Phytol 177:779–789

Jansa J, Oberholzer H-R, Egli S (2009) Environmental determinants of the arbuscular mycorrhizal fungal infectivity of Swiss agricultural soils. Eur J Soil Biol 45:400–408

Javot H, Penmetsa RV, Terzaghi N, Cook DR, Harrison MJ (2007) A *Medicago truncatula* phosphate transporter indispensable for the arbuscular mycorrhizal symbiosis. Proc Natl Acad Sci U S A 104:1720–1725

Jennings DH (1996) The physiology of fungal nutrition. Cambridge University Press, Cambridge

Jifon JL, Graham JH, Drouillard DL, Syvertsen JP (2002) Growth depression of mycorrhizal *Citrus* seedlings grown at high phosphorus supply is mitigated by elevated CO_2. New Phytol 153:133–142

Johansen A, Jakobsen I, Jensen ES (1992) Hyphal transport of ^{15}N-labelled nitrogen by a vesicular-arbuscular mycorrhizal fungus and its effect on depletion of inorganic soil N. New Phytol 122:281–288

Johansson JF, Paul LR, Finlay RD (2004) Microbial interactions in the mycorrhizosphere and their significance for sustainable agriculture. FEMS Microbiol Ecol 48:1–13

Johnson NC (1993) Can fertilization of soil select less mutualistic mycorrhizae? Ecol Appl 3:749–757

Johnson NC, Graham JH, Smith FA (1997) Functioning of mycorrhizal associations along the mutualism-parasitism continuum. New Phytol 135:575–585

Johnson D, Ijdo M, Genney DR, Anderson IC, Alexander IJ (2005) How do plants regulate the function, community structure, and diversity of mycorrhizal fungi? J Exp Bot 56:1751–1760

Johnson NC, Wilson GWT, Bowker MA, Wilson JA, Miller RM (2010) Resource limitation is a driver of local adaptation in mycorrhizal symbioses. Proc Natl Acad Sci U S A 107:2093–2098

Joner EJ, Jakobsen I (1995) Uptake of ^{32}P from labelled organic matter by mycorrhizal and non-mycorrhizal subterranean clover (*Trifolium subterraneum* L). Plant Soil 172:221–227

Joner E, Johansen A (2000) Phosphatase activity of external hyphae of two arbuscular mycorrhizal fungi. Mycol Res 104:81–86

Joner EJ, Briones R, Leyval C (2000) Metal-binding capacity of arbuscular mycorrhizal mycelium. Plant Soil 226:227–234

Joner EJ, Roos P, Jansa J, Frossard E, Leyval C, Jakobsen I (2004) No significant contribution of arbuscular mycorrhizal fungi to transfer of radiocesium from soil to plants. Appl Environ Microbiol 70:6512–6517

Jones MD, Smith SE (2004) Exploring functional definitions of mycorrhizas: are mycorrhizas always mutualisms? Can J Bot 82:1089–1109

Jones DL, Hodge A, Kuzyakov Y (2004) Plant and mycorrhizal regulation of rhizodeposition. New Phytol 163:459–480

Kahiluoto H, Ketoja E, Vestberg M (2000a) Creation of a non-mycorrhizal control for a bioassay of AM effectiveness 1. Comparison of methods. Mycorrhiza 9:241–258

Kahiluoto H, Ketoja E, Vestberg M (2000b) Promotion of utilization of arbuscular mycorrhiza through reduced P fertilization 1. Bioassays in a growth chamber. Plant Soil 227:191–206

Karandashov V, Bucher M (2005) Symbiotic phosphate transport in arbuscular mycorrhizas. Trends Plant Sci 10:22–29

Kieliszewska-Rokicka B (1999) Phosphate status and acid phosphatase activity in soil and ectomycorrhizas in two mature stands of Scots pine (*Pinus sylvestris* L.) exposed to different levels of anthropogenic pollution. Acta Soc Bot Pol 68:311–317

Kiers ET, van der Heijden MGA (2006) Mutualistic stability in the arbuscular mycorrhizal symbiosis: exploring hypotheses of evolutionary cooperation. Ecology 87:1627–1636

Koele N, Turpault MP, Hildebrand EE, Uroz S, Frey-Klett P (2009) Interactions between mycorrhizal fungi and mycorrhizosphere bacteria during mineral weathering: budget analysis and bacterial quantification. Soil Biol Biochem 41:1935–1942

Koide RT (2000) Functional complementarity in the arbuscular mycorrhizal symbiosis. New Phytol 147:233–235

Koide RT, Kabir Z (2000) Extraradical hyphae of the mycorrhizal fungus *Glomus intraradices* can hydrolyse organic phosphate. New Phytol 148:511–517

Koide RT, Sharda JN, Herr JR, Malcolm GM (2008) Ectomycorrhizal fungi and the biotrophy–saprotrophy continuum. New Phytol 178:230–233

Ladeyn I, Plassard C, Staunton S (2008) Mycorrhizal association of maritime pine, *Pinus pinaster*, with *Rhizopogon roseolus* has contrasting effects on the uptake from soil and root-to-shoot transfer of ^{137}Cs, ^{85}Sr and ^{95m}Tc. J Environ Radioactiv 99:853–863

Lambers H, Raven JA, Shaver GR, Smith SE (2008) Plant nutrient-acquisition strategies change with soil age. Trends Ecol Evol 23:95–103

Landeweert R, Hoffland E, Finlay RD, Kuyper TW, van Breemen N (2001) Linking plants to rocks: ectomycorrhizal fungi mobilize nutrients from minerals. Trends Ecol Evol 16:248–254

Leake JR, Miles W (1996) Phosphodiesters as mycorrhizal P sources. 1. Phosphodiesterase production and the utilization of DNA as a phosphorus source by the ericoid mycorrhizal fungus *Hymenoscyphus ericae*. New Phytol 132:435–443

Leggewie G, Willmitzer L, Riesmeier J (1997) Two cDNA from potato are able to complement a phosphate uptake-deficient yeast mutant: identification of phosphate transporters from higher plants. Plant Cell 9:381–392

Leigh J, Hodge A, Fitter AH (2009) Arbuscular mycorrhizal fungi can transfer substantial amounts of nitrogen to their host plant from organic material. New Phytol 181:199–207

Li XL, George E, Marschner H (1991) Extension of the phosphorus depletion zone in VA-mycorrhizal white clover in a calcareous soil. Plant Soil 136:41–48

Li HY, Zhu YG, Marschner P, Smith FA, Smith SE (2005) Wheat responses to arbuscular mycorrhizal fungi in a highly calcareous soil differ from those of clover, and change with plant development and P supply. Plant Soil 277:221–232

Li HY, Smith SE, Ophel-Keller K, Holloway RE, Smith FA (2008) Naturally occurring arbuscular mycorrhizal fungi can replace direct P uptake by wheat when roots cannot access added P fertiliser. Funct Plant Biol 35:124–130

Lindahl B, Stenlid J, Olsson S, Finlay R (1999) Translocation of ^{32}P between interacting mycelia of a wood-decomposing fungus and ectomycorrhizal fungi in microcosm systems. New Phytol 144:183–193

Linder S (1995) Foliar analysis for detecting and correcting nutrient imbalance in Norway spruce. Ecol Bull 44:178–196

Lussenhop J, Fogel R (1999) Seasonal change in phosphorus content of *Pinus strobus–Cenococcum geophilum* ectomycorrhizae. Mycologia 91:742–746

Mäder P, Vierheilig H, Streitwolf-Engel R, Boller T, Frey B, Christie P, Wiemken A (2000) Transport of ^{15}N from a soil compartment separated by a polytetrafluoroethylene membrane to plant roots via the hyphae of arbuscular mycorrhizal fungi. New Phytol 146:155–161

Maherali H, Klironomos JN (2007) Influence of phylogeny on fungal community assembly and ecosystem functioning. Science 316:1746–1748

Malcolm GM, Lopez-Gutierrez JC, Koide RT, Eissenstat DM (2008) Acclimation to temperature and temperature sensitivity of metabolism by ectomycorrhizal fungi. Glob Change Biol 14:1169–1180

Mandyam K, Jumpponen A (2005) Seeking the elusive function of the root-colonising dark septate endophytic fungi. Stud Mycol 53:173–189

Marschner H (1995) Mineral nutrition of higher plants. Academic, London

Martin F, Gianinazzi-Pearson V, Hijri M, Lammers P, Requena N, Sanders IR, Shachar-Hill Y, Shapiro H, Tuskan GA, Young JPW (2008) The long hard road to a completed *Glomus intraradices* genome. New Phytol 180:747–750

Martino E, Turnau K, Girlanda M, Bonfante P, Perotto S (2000) Ericoid mycorrhizal fungi from heavy metal polluted soils: their identification and growth in the presence of zinc ions. Mycol Res 104:338–344

Merryweather J, Fitter A (1995) Phosphorus and carbon budgets – mycorrhizal contribution in *Hyacinthoides non-scripta* (L) Chouard Ex Rothm under natural conditions. New Phytol 129:619–627

Merryweather J, Fitter A (1998) The arbuscular mycorrhizal fungi of *Hyacinthoides non-scripta* – II. Seasonal and spatial patterns of fungal populations. New Phytol 138:131–142

Millard P, Sommerkorn M, Grelet GA (2007) Environmental change and carbon limitation in trees: a biochemical, ecophysiological and ecosystem appraisal. New Phytol 175:11–28

Miller SL, Koo CD, Molina R (1991) Characterization of red alder ectomycorrhizae – a preface to monitoring belowground ecological responses. Can J Bot 69:516–531

Mitchell CE, Agrawal AA, Bever JD, Gilbert GS, Hufbauer RA, Klironomos JN, Maron JL, Morris WF, Parker IM, Power AG, Seabloom EW, Torchin ME, Vazquez DP (2006) Biotic interactions and plant invasions. Ecol Lett 9:726–740

Morgan JAW, Bending GD, White PJ (2005) Biological costs and benefits to plant–microbe interactions in the rhizosphere. J Exp Bot 56:1729–1739

Munkvold L, Kjøller R, Vestberg M, Rosendahl S, Jakobsen I (2004) High functional diversity within species of arbuscular mycorrhizal fungi. New Phytol 164:357–364

Myers MD, Leake JR (1996) Phosphodiesters as mycorrhizal P sources. II. Ericoid mycorrhiza and the utilization of nuclei as a phosphorus and nitrogen source by Vaccinium macrocarpon. New Phytol 132:445–451

Nagy R, Karandashov V, Chague W, Kalinkevich K, Tamasloukht M, Xu GH, Jakobsen I, Levy AA, Amrhein N, Bucher M (2005) The characterization of novel mycorrhiza-specific phosphate transporters from *Lycopersicon esculentum* and *Solanum tuberosum* uncovers functional redundancy in symbiotic phosphate transport in solanaceous species. Plant J 42:236–250

Newman EI (1995) Phosphorus inputs to terrestrial ecosystems. J Ecol 83:713–726

Newman EI, Ritz K (1986) Evidence on the pathways of phosphorus transfer between vesicular-arbuscular mycorrhizal plants. New Phytol 104:77–87

Newsham KK, Fitter AH, Watkinson AR (1995) Multi-functionality and biodiversity in arbuscular mycorrhizas. Trends Ecol Evol 10:407–411

Newsham KK, Upson R, Read DJ (2009) Mycorrhizas and dark septate root endophytes in polar regions. Fungal Ecol 2:10–20

Nielsen JS, Joner EJ, Declerck S, Olsson S, Jakobsen I (2002) Phospho-imaging as a tool for visualization and noninvasive measurement of P transport dynamics in arbuscular mycorrhizas. New Phytol 154:809–819

Oberson A, Pypers P, Bünemann EK, Frossard E (2011) Management impacts on biological phosphorus cycling in cropped soils. In: Bünemann EK, Oberson A, Frossard E (eds) Phosphorus in action: biological processes in soil phosphorus cycling. Soil biology, vol 26. Springer, Heidelberg. doi:10.1007/978-3-642-15271-9_17

Ohtomo R, Saito M (2005) Polyphosphate dynamics in mycorrhizal roots during colonization of an arbuscular mycorrhizal fungus. New Phytol 167:571–578

Öpik M, Metsis M, Daniell TJ, Zobel M, Moora M (2009) Large-scale parallel 454 sequencing reveals host ecological group specificity of arbuscular mycorrhizal fungi in a boreonemoral forest. New Phytol 184:424–437

Ortas I, Ortakçi D, Kaya Z, Çinar A, Önelge N (2002) Mycorrhizal dependency of sour orange in relation to phosphorus and zinc nutrition. J Plant Nutr 25:1263–1279

Pankow W, Boller T, Wiemken A (1991) The significance of mycorrhizas for protective ecosystems. Experientia 47:391–394

Paszkowski U, Kroken S, Roux C, Briggs SP (2002) Rice phosphate transporters include an evolutionarily divergent gene specifically activated in arbuscular mycorrhizal symbiosis. Proc Natl Acad Sci U S A 99:13324–13329

Pearson JN, Jakobsen I (1993) Symbiotic exchange of carbon and phosphorus between cucumber and three arbuscular mycorrhizal fungi. New Phytol 124:481–488

Perez-Moreno J, Read DJ (2001a) Nutrient transfer from soil nematodes to plants: a direct pathway provided by the mycorrhizal mycelial network. Plant Cell Environ 24:1219–1226

Perez-Moreno J, Read DJ (2001b) Exploitation of pollen by mycorrhizal mycelial systems with special reference to nutrient recycling in boreal forests. Proc Biol Sci 268:1329–1335

Philip LJ, Simard SW (2008) Minimum pulses of stable and radioactive carbon isotopes to detect belowground carbon transfer between plants. Plant Soil 308:23–35

Picone C (2002) Managing mycorrhizae for sustainable agriculture in the tropics. In: Vandermeer JH (ed) Tropical agroecosystems. CRC, Boca Raton, pp 95–132

Pirozynski KA, Dalpé Y (1989) Geological history of the Glomaceae with particular reference to mycorrhizal symbiosis. Symbiosis 7:1–35

Plamboeck AH, Dawson TE, Egerton-Warburton LM, North M, Bruns TD, Querejeta JI (2007) Water transfer via ectomycorrhizal fungal hyphae to conifer seedlings. Mycorrhiza 17:439–447

Poulsen KH, Nagy R, Gao LL, Smith SE, Bucher M, Smith FA, Jakobsen I (2005) Physiological and molecular evidence for Pi uptake via the symbiotic pathway in a reduced mycorrhizal colonization mutant in tomato associated with a compatible fungus. New Phytol 168:445–453

Pritchard SG, Strand AE, McCormack ML, Davis MA, Oren R (2008) Mycorrhizal and rhizomorph dynamics in a loblolly pine forest during 5 years of free-air-CO_2-enrichment. Glob Change Biol 14:1252–1264

Rausch C, Bucher M (2002) Molecular mechanisms of phosphate transport in plants. Planta 216:23–37

Rausch C, Daram P, Brunner S, Jansa J, Laloi M, Leggewie G, Amrhein N, Bucher M (2001) A phosphate transporter expressed in arbuscule-containing cells in potato. Nature 414:462–466

Read DJ (1984) The structure and function of the vegetative mycelium of mycorrhizal roots. In: Jennings DH, Rayner ADM (eds) The ecology and physiology of the fungal mycelium. Cambridge University Press, Cambridge, pp 215–240

Read DJ, Perez-Moreno J (2003) Mycorrhizas and nutrient cycling in ecosystems – a journey towards relevance? New Phytol 157:475–492

Read DJ, Leake JR, Perez-Moreno J (2004) Mycorrhizal fungi as drivers of ecosystem processes in heathland and boreal forest biomes. Can J Bot 82:1243–1263

Redecker D, Raab P (2006) Phylogeny of the Glomeromycota (arbuscular mycorrhizal fungi): recent developments and new gene markers. Mycologia 98:885–895

Redecker D, Kodner R, Graham LE (2000) Glomalean fungi from the Ordovician. Science 289:1920–1921

Redon PO, Beguiristain T, Leyval C (2009) Differential effects of AM fungal isolates on *Medicago truncatula* growth and metal uptake in a multimetallic (Cd, Zn, Pb) contaminated agricultural soil. Mycorrhiza 19:187–195

Rejšek K (1991) Acid phosphomonoesterase activity of ectomycorrhizal roots in Norway spruce pure stands exposed to pollution. Soil Biol Biochem 23:667–671

Requena N, Breuninger M, Franken P, Ocon A (2003) Symbiotic status, phosphate, and sucrose regulate the expression of two plasma membrane H+-ATPase genes from the mycorrhizal fungus *Glomus mosseae*. Plant Physiol 132:1540–1549

Rhodes LH, Gerdemann JW (1978) Translocation of calcium and phosphate by external hyphae of vesicular-arbuscular mycorrhizae. Soil Sci 126:125–126

Rillig MC, Mummey DL (2006) Mycorrhizas and soil structure. New Phytol 171:41–53

Robertson SJ, McGill WB, Massicotte HB, Rutherford PM (2007) Petroleum hydrocarbon contamination in boreal forest soils: a mycorrhizal ecosystems perspective. Biol Rev 82:213–240

Robinson D, Fitter A (1999) The magnitude and control of carbon transfer between plants linked by a common mycorrhizal network. J Exp Bot 50:9–13

Rosling A (2009) Trees, mycorrhiza and minerals – field relevance of *in vitro* experiments. Geomicrobiol J 26:389–401

Rosling A, Roose T, Hermann AM, Davidson FA, Finlay RD, Gadd GM (2009) Approaches to modelling mineral weathering by fungi. Fungal Biol Rev 23:138–144. doi:10.1016/j.fbr.2009.09.003

Ryan MH, Graham JH (2002) Is there a role for arbuscular mycorrhizal fungi in production agriculture? Plant Soil 244:263–271

Sawers RJH, Gutjahr C, Paszkowski U (2008) Cereal mycorrhiza: an ancient symbiosis in modern agriculture. Trends Plant Sci 13:93–97

Schachtman DP, Reid RJ, Ayling SM (1998) Phosphorus uptake by plants: from soil to cell. Plant Physiol 116:447–453

Schnepf A, Roose T, Schweiger P (2008) Impact of growth and uptake patterns of arbuscular mycorrhizal fungi on plant phosphorus uptake – a modelling study. Plant Soil 312:85–99

Schnepf A, Leitner D, Klepsch S, Pellerin S, Mollier A (2011) Modelling phosphorus dynamics in the soil–plant system. In: Bünemann EK, Oberson A, Frossard E (eds) Phosphorus in action: biological processes in soil phosphorus cycling. Soil biology, vol 26. Springer, Heidelberg. doi:10.1007/978-3-642-15271-9_5

Schroeder MS, Janos DP (2004) Phosphorus and intraspecific density alter plant responses to arbuscular mycorrhizas. Plant Soil 264:335–348

Schüßler A, Schwarzott D, Walker C (2001) A new fungal phylum, the Glomeromycota: phylogeny and evolution. Mycol Res 105:1413–1421
Schweiger P, Jakobsen I (1999) The role of mycorrhizas in plant P nutrition: fungal uptake kinetics and genotype variation. In: Gissel-Nielsen G, Jensen A (eds) Plant nutrition – molecular biology and genetics. Kluwer, Dordrecht, pp 277–289
Selosse MA, Richard F, He XH, Simard SW (2006) Mycorrhizal networks: des liaisons dangereuses? Trends Ecol Evol 21:621–628
Serrasolses I, Romanya J, Khanna PK (2008) Effects of heating and autoclaving on sorption and desorption of phosphorus in some forest soils. Biol Fertil Soils 44:1063–1072
Sharples JM, Meharg AA, Chambers SM, Cairney JWG (1999) Arsenate sensitivity in ericoid and ectomycorrhizal fungi. Environ Toxicol Chem 18:1848–1855
Sikes BA, Cottenie K, Klironomos JN (2009) Plant and fungal identity determines pathogen protection of plant roots by arbuscular mycorrhizas. J Ecol 97:1274–1280
Smith FW (2002) The phosphate uptake mechanism. Plant Soil 245:105–114
Smith SE, Read DJ (2008) Mycorrhizal symbiosis, 3rd edn. Academic, New York
Smith FA, Smith SE (1996) Mutualism and parasitism: diversity in function and structure in the "arbuscular" (VA) mycorrhizal symbiosis. Adv Bot Res 22:1–43
Smith SE, Dickson S, Smith FA (2001) Nutrient transfer in arbuscular mycorrhizas: how are fungal and plant processes integrated? Aust J Plant Physiol 28:683–694
Smith SE, Smith FA, Jakobsen I (2003) Mycorrhizal fungi can dominate phosphate supply to plants irrespective of growth responses. Plant Physiol 133:16–20
Smith SE, Smith FA, Jakobsen I (2004) Functional diversity in arbuscular mycorrhizal (AM) symbioses: the contribution of the mycorrhizal P uptake pathway is not correlated with mycorrhizal responses in growth or total P uptake. New Phytol 162:511–524
Smith FA, Grace EJ, Smith SE (2009) More than a carbon economy: nutrient trade and ecological sustainability in facultative arbuscular mycorrhizal symbioses. New Phytol 182:347–358
Smits MM, Bonneville S, Haward S, Leake JR (2008) Ectomycorrhizal weathering, a matter of scale. Mineral Mag 72:135–138
Soares CRFS, Siqueira JO (2008) Mycorrhiza and phosphate protection of tropical grass species against heavy metal toxicity in multi-contaminated soil. Biol Fertil Soils 44:833–841
Solaiman MZ, Saito A (2001) Phosphate efflux from intraradical hyphae of *Gigaspora margarita in vitro* and its implication for phosphorus translocation. New Phytol 151:525–533
Solaiman MZ, Ezawa T, Kojima T, Saito M (1999) Polyphosphates in intraradical and extraradical hyphae of an arbuscular mycorrhizal fungus, *Gigaspora margarita*. Appl Environ Microbiol 65:5604–5606
Sorensen JN, Larsen J, Jakobsen I (2003) Management strategies for capturing the benefits of mycorrhizas in the production of field-grown vegetables. Acta Hortic 627:65–71
Sousa MF, Facanha AR, Tavares RM, Lino-Neto T, Geros H (2007) Phosphate transport by proteoid roots of *Hakea sericea*. Plant Sci 173:550–558
Staddon PL, Ramsey CB, Ostle N, Ineson P, Fitter AH (2003) Rapid turnover of hyphae of mycorrhizal fungi determined by AMS microanalysis of ^{14}C. Science 300:1138–1140
Sudová R, Doubková P, Vosátka M (2008) Mycorrhizal association of *Agrostis capillaris* and *Glomus intraradices* under heavy metal stress: combination of plant clones and fungal isolates from contaminated and uncontaminated substrates. Appl Soil Ecol 40:19–29
Sverdrup H, Hagen-Thorn A, Holmqvist J, Walman P, Warfinge P, Walse C, Alveteg M (2002) Biochemical processes and mechanisms. In: Sverdrup H, Stjernquist I (eds) Developing principles and models for sustainable forestry in Sweden. Managing forest ecosystems, vol 5. Kluwer, Dordrecht, pp 91–196
Sýkorová Z, Wiemken A, Redecker D (2007) Co-occurring *Gentiana verna* and *Gentiana acaulis* and their neighboring plants in two Swiss upper montane meadows harbor distinct arbuscular mycorrhizal fungal communities. Appl Environ Microbiol 73:5426–5434

Takanishi I, Ohtomo R, Hayatsu M, Saito M (2009) Short-chain polyphosphate in arbuscular mycorrhizal roots colonized by *Glomus* spp.: a possible phosphate pool for host plants. Soil Biol Biochem 41:1571–1573

Talbot JM, Allison SD, Treseder KK (2008) Decomposers in disguise: mycorrhizal fungi as regulators of soil C dynamics in ecosystems under global change. Funct Ecol 22:955–963

Tarafdar JC, Marschner H (1994) Phosphatase activity in the rhizosphere and hyphosphere of VA mycorrhizal wheat supplied with inorganic and organic phosphorus. Soil Biol Biochem 26:387–395

Tarafdar JC, Yadav RS, Meena SC (2001) Comparative efficiency of acid phosphatase originated from plant and fungal sources. J Plant Nutr Soil Sci 164:279–282

Tatry MV, Kassis EE, Lambilliotte R, Corratge C, van Aarle I, Amenc LK, Alary R, Zimmermann S, Sentenac H, Plassard C (2009) Two differentially regulated phosphate transporters from the symbiotic fungus *Hebeloma cylindrosporum* and phosphorus acquisition by ectomycorrhizal *Pinus pinaster*. Plant J 57:1092–1102

Tawaraya K (2003) Arbuscular mycorrhizal dependency of different plant species and cultivars. Soil Sci Plant Nutr 49:655–668

Taylor AFS, Alexander I (2005) The ectomycorrhizal symbiosis: life in the real world. Mycologist 19:102–112

Taylor DL, Bruns TD, Hodges SA (2004) Evidence for mycorrhizal races in a cheating orchid. Proc Biol Sci 271:35–43

Tedersoo L, Jairus T, Horton BM, Abarenkov K, Suvi T, Saar I, Koljalg U (2008) Strong host preference of ectomycorrhizal fungi in a Tasmanian wet sclerophyll forest as revealed by DNA barcoding and taxon-specific primers. New Phytol 180:479–490

Thomson BD, Clarkson DT, Brain P (1990) Kinetics of phosphorus uptake by the germ tubes of the vesicular arbuscular mycorrhizal fungus, *Gigaspora margarita*. New Phytol 116:647–653

Thormann MN, Currah RS, Bayley SE (1999) The mycorrhizal status of the dominant vegetation along a peatland gradient in southern boreal Alberta, Canada. Wetlands 19:438–450

Tibbett M, Sanders FE (2002) Ectomycorrhizal symbiosis can enhance plant nutrition through improved access to discrete organic nutrient patches of high resource quality. Ann Bot 89:783–789

Timonen S, Finlay RD, Olsson S, Söderström B (1996) Dynamics of phosphorus translocation in intact ectomycorrhizal systems: non-destructive monitoring using a beta-scanner. FEMS Microbiol Ecol 19:171–180

Tinker PB (1975) Soil chemistry of phosphorus and mycorrhizal effects on plant growth. In: Sanders FE, Mosse B, Tinker PB (eds) Endomycorrhizas. Academic, London, pp 353–371

Toljander JF, Artursson V, Paul LR, Jansson JK, Finlay RD (2006) Attachment of different soil bacteria to arbuscular mycorrhizal fungal extraradical hyphae is determined by hyphal vitality and fungal species. FEMS Microbiol Lett 254:34–40

Toljander JF, Lindahl BD, Paul LR, Elfstrand M, Finlay RD (2007) Influence of arbuscular mycorrhizal mycelial exudates on soil bacterial growth and community structure. FEMS Microbiol Ecol 61:295–304

Trappe JM (1987) Phylogenetic and ecologic aspects of mycotrophy in the angiosperms from an evolutionary standpoint. In: Safir GR (ed) Ecophysiology of VA mycorrhizal plants. CRC, Boca Raton, pp 5–25

Treseder KK, Cross A (2006) Global distributions of arbuscular mycorrhizal fungi. Ecosystems 9:305–316

Turpault M-P, Nys C, Calvaruso C (2009) Rhizosphere impact on the dissolution of test minerals in a forest ecosystem. Geoderma 153:147–154

van der Heijden EW, Vosátka M (1999) Mycorrhizal associations of *Salix repens* L. communities in succession of dune ecosystems. II. Mycorrhizal dynamics and interactions of ectomycorrhizal and arbuscular mycorrhizal fungi. Can J Bot 77:1833–1841 van der Heijden MGA, Klironomos JN, Ursic M, Moutoglis P, Streitwolf-Engel R, Boller T, Wiemken A, Sanders IR (1998) Mycorrhizal fungal diversity determines plant biodiversity, ecosystem variability and productivity. Nature 396:69–72

van Scholl L, Kuyper TW, Smits MM, Landeweert R, Hoffland E, van Breemen N (2008) Rock-eating mycorrhizas: their role in plant nutrition and biogeochemical cycles. Plant Soil 303:35–47

van Tichelen KK, Colpaert JV (2000) Kinetics of phosphate absorption by mycorrhizal and non-mycorrhizal Scots pine seedlings. Physiol Plant 110:96–103

Vandenkoornhuyse P, Ridgway KP, Watson IJ, Fitter AH, Young JPW (2003) Co-existing grass species have distinctive arbuscular mycorrhizal communities. Mol Ecol 12:3085–3095

Vessey JK (2003) Plant growth promoting rhizobacteria as biofertilizers. Plant Soil 255:571–586

Wallander H (2000) Uptake of P from apatite by *Pinus sylvestris* seedlings colonised by different ectomycorrhizal fungi. Plant Soil 218:249–256

Wallander H (2006) External mycorrhizal mycelia – the importance of quantification in natural ecosystems. New Phytol 171:240–242

Wallander H, Thelin G (2008) The stimulating effect of apatite on ectomycorrhizal growth diminishes after PK fertilization. Soil Biol Biochem 40:2517–2522

Wang B, Qiu YL (2006) Phylogenetic distribution and evolution of mycorrhizas in land plants. Mycorrhiza 16:299–363

West HM, Fitter AH, Watkinson AR (1993) Response of *Vulpia ciliata* ssp. *ambigua* to removal of mycorrhizal infection and to phosphate application under natural conditions. J Ecol 81:351–358

Whitbeck JL (2001) Effects of light environment on vesicular-arbuscular mycorrhiza development in *Inga leiocalycina*, a tropical wet forest tree. Biotropica 33:303–311

Wurzburger N, Hendrick RL (2006) Rhododendron thickets alter N cycling and soil extracellular enzyme activities in southern Appalachian hardwood forests. Pedobiologia 50:563–576

Wurzburger N, Hendrick RL (2009) Plant litter chemistry and mycorrhizal roots promote a nitrogen feedback in a temperate forest. J Ecol 97:528–536

Yao Q, Li XL, Ai WD, Christie P (2003) Bi-directional transfer of phosphorus between red clover and perennial ryegrass via arbuscular mycorrhizal hyphal links. Eur J Soil Biol 39:47–54

Youpensuk S, Lordkaew S, Rerkasem B (2008) Arbuscular mycorrhizal fungi associated with tangerine (*Citrus reticulata*) in Chiang Mai province, northern Thailand, and their effects on the host plant. Sci Asia 34:259–264

Zabinski CA, Quinn L, Callaway RM (2002) Phosphorus uptake, not carbon transfer, explains arbuscular mycorrhizal enhancement of *Centaurea maculosa* in the presence of native grassland species. Funct Ecol 16:758–765

Zeller B, Brechet C, Maurice JP, Le Tacon F (2008) Saprotrophic versus symbiotic strategy during truffle ascocarp development under holm oak. A response based on ^{13}C and ^{15}N natural abundance. Ann Forest Sci 65:607

Zhu YG, Smith SE, Barritt AR, Smith FA (2001) Phosphorus (P) efficiencies and mycorrhizal responsiveness of old and modern wheat cultivars. Plant Soil 237:249–255

Chapter 7
Solubilization of Phosphorus by Soil Microorganisms

David L. Jones and Eva Oburger

7.1 Introduction

Agricultural production remains highly reliant on the application of phosphorus (P) fertilizers derived from phosphate rock. Due to increasing demand and dwindling stocks, it is predicted that current global reserves of phosphate rock may be depleted within 50–100 years (Cordell et al. 2009). Furthermore, continued agricultural expansion has led to co-saturation of many ecosystems with both N and P, resulting in the degradation of terrestrial, freshwater and marine resources (Tilman et al. 2001). This concern has highlighted the imperative need to better understand the plant–soil–microbial P cycle, with an aim of reducing our reliance on mineral fertilizers. This has led to increased interest in the harnessing of microorganisms to support P cycling in agroecosystems. It is well known that some microbes in soil have the potential to greatly enhance the rate of organic P (P_o) or inorganic (P_i) cycling (i.e. by solubilizing insoluble organic- and mineral-bound P). This chapter aims to identify which P solubilizing organisms (PSM) exist in soil, the types of P they can utilize, the mechanisms by which this occurs and the potential for managing them in an agricultural context. Our aim is to encompass all types of soil microorganisms; however, we will not cover mycorrhizas as these are comprehensively covered by Jansa et al. (2011).

D.L. Jones (✉)
Environment Centre Wales, Bangor University, Gwynedd LL57 2UW, UK
e-mail: d.jones@bangor.ac.uk

E. Oburger
Department of Forest and Soil Sciences, BOKU – University of Natural Resources and Applied Life Sciences, Peter Jordan Strasse 82, 1190 Vienna, Austria
e-mail: eva.oburger@boku.ac.at

E.K. Bünemann et al. (eds.), *Phosphorus in Action*, Soil Biology 26,
DOI 10.1007/978-3-642-15271-9_7,

7.2 P in the Soil Environment

Compared to other essential macronutrients (with the exception of N), P is one of the less-abundant elements in the lithosphere (0.1% of total). However, sufficient P nutrition is of crucial importance for all microorganisms due to its central role in energy transfer (e.g. ATP), cell structure (phospholipids), metabolism and signaling (see Bünemann et al. 2011; Frossard et al. 2011). In soils, concentrations of available P in soil solution are typically low (<0.01 to 1 mg L^{-1} in highly fertile soils) due to the comparatively low content of P in the parent material, but also due to the high reactivity of P_i that results in strong retention by the soil's mineral matrix. This has led to microorganisms developing a wide range of strategies to enhance P availability in soil. Although plants can only take up P_i (i.e. HPO_4^{2-}, $H_2PO_4^{-}$), fungi and bacteria can also potentially take up low molecular weight (LMW) organic P (P_o) (e.g. sugar-P; Schwöppe et al. 2003). In contrast, protozoa can take up and assimilate high molecular weight (HMW) P_o, implying that there are only a few microbially unavailable organic P pools in soil and suggesting that different P sources can potentially provide ecological niches for different species (Foster and Dormaar 1991). To date, most research has focussed on the biological manipulation of P_i availability in soil rather than P_o.

In a recent review of P chemistry in soils, Sims and Pierzynski (2005) identified the major processes of the soil P cycle that affect soil solution P concentrations as (1) dissolution–precipitation (mineral equilibria), (2) sorption–desorption (interactions between P in solution and soil solid surfaces), and (3) mineralization–immobilization (biologically mediated conversions of P between inorganic and inorganic forms). With a significant proportion of total soil P being organically bound, the role of microorganisms in P turnover should not be underestimated. Furthermore, microbes (like plants) actively or passively release protons, CO_2 and secondary organic metabolites (e.g. sugars, organic acid anions, amino acids, siderophores, enzymes, phenols) that may all contribute to the solubilization of P from soil minerals (via processes 1 and 2). Overall, between 1–50% of soil bacteria and 0.5–0.1% of soil fungi can be classified as P-solubilizing microorganisms (PSM) (Gyaneshwar et al. 2002; Kucey et al. 1989). Although the number of bacteria in soil classed as PSM generally outnumber those of fungi, the fungal isolates generally exhibit a greater P-solubilizing capacity in both liquid and solid media (Banik and Dey 1982; Gyaneshwar et al. 2002).

7.2.1 Sources of Soil P Capable of Microbial Solubilization

The chemical and physical form of P in soil is clearly an important regulator of the efficiency (i.e. gross release and/or solubilization) of PSM to mobilize P bound in the soil's solid phase. For example, phosphate rock contains different types of P minerals with variable solubility. Typically, PSM are selected for their ability to dissolve phosphate rock in vitro; however, phosphate rock may not reflect the form of P found in many soils. This is probably one of the main reasons why PSM show differential responses in situ. Here, we briefly describe the dominant forms of P in soil.

7.2.1.1 Inorganic P

Total P content in top soils (0–15 cm) typically ranges from 50 to 3,000 mg kg^{-1} depending on parent material, soil type, vegetation cover and soil management (Sims and Pierzynski 2005), with P_i comprising 35–70% of total soil P (Sample et al. 1980). The chemical forms of P in soil differ not only with parent material, soil pH and vegetation cover, but also with time and the extent of pedogenesis (Walker and Syers 1976; Foth and Ellis 1997; Fig. 7.1). The organic P pool increases with soil development but tends to decline again in highly weathered, older soils. Consequently, soil development and therefore the distribution of P between the P_o and P_i pools, as well as the composition of P forms, have a major impact on P accessibility for the microbial community, which ultimately determines the success of PSM in the field.

Calcium-phosphates (mainly different forms of apatites, such as fluoroapatite [$Ca_5(PO_4)_3F$], hydroxyapatite [$Ca_5(PO_4)_3OH$], and francolites [carbonate-fluorapatites of variable chemical composition $(Ca, Mg, Sr, Na)_{10}(PO_4, SO_4, CO_3)_6F_{2-3}$]; Benmore et al. 1983) represent the primary mineral source of P_i in unweathered or moderately weathered soils with neutral to alkaline pH, whereas Fe and Al

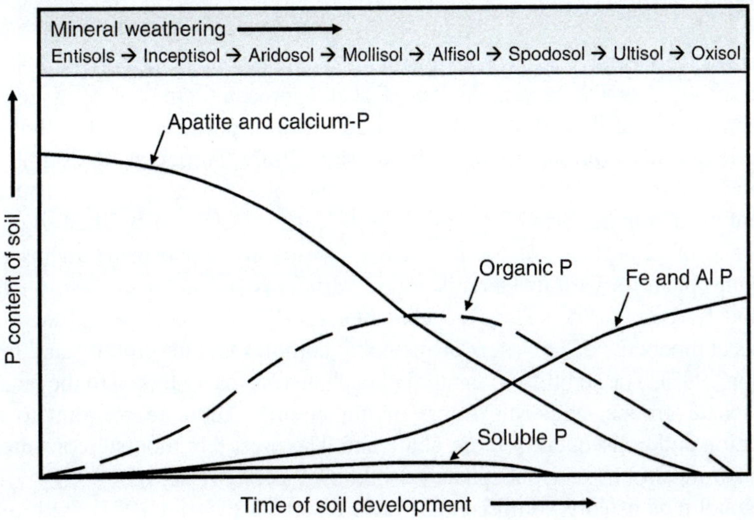

Fig. 7.1 Relative distribution of the major forms of soil P as related to soil development and the major US Soil Taxonomy Orders over a timeframe of millions of years. Adapted from Foth and Ellis (1997)

phosphates and P_i bound and/or occluded by Fe and Al oxy(hydr)oxides predominate in acidic and more progressively weathered soils (Sims and Pierzynski 2005).

Due to the increasing stability of Ca-P minerals at acidic pH, localized acidification by PSM can result in the solubilization of Ca phosphates and the release of P_i (see also Sect. 7.3.1). Consequently, various acidifying PSM have been applied to accelerate dissolution of phosphate rock (i.e. collective term for igneous, metamorphous or marine sedimentary rock containing Ca-P-bearing minerals) prior to addition to soil (e.g. through the addition of microbial consortia in composts or via addition of individual PSM; Arcand and Schneider 2006; Odongo et al. 2007; Aria et al. 2010).

In neutral and particularly in acidic soils, Al and Fe oxides and hydroxides exert a great impact on P availability, because various identified Fe and Al phosphates, such as wavellite [$Al_3(OH)_3(PO_4)_2{\cdot}5H_2O$], variscite ($AlPO_4{\cdot}2H_2O$), strengite ($FePO_4{\cdot}2H_2O$), etc. (for an extensive review see Harris 2002) are generally rare in occurrence. Due to the increased positive surface charge of the Fe and Al oxides with decreasing pH, strong covalent bonds (chemisorption) are formed with the negatively charged P, rendering it rather recalcitrant to exchange reactions. Nevertheless, LMW organic anions (e.g. gluconate, oxalate, etc.) released by PSM are capable of competing with P_i for sorption sites. Also, changes in pH might directly or indirectly affect the oxides' surface potential and consequently P_i solubility (see also Sects. 7.3.1 and 7.3.2).

7.2.1.2 Organic P

Soil Organic Matter

On average, between 30% and 65% of total P is present as P_o in mineral soils. In organic soils (>20–30% organic matter), P_o can approach up to 90% of the total P. The main identified P_o compounds in soil are inositol phosphates, phospholipids and nucleic acids (Quiquampoix and Mousain 2005; Turner et al 2002). The abundance of inositol phosphates is highly variable; however, they frequently represent the dominant form of P_o in soil ($\geq$80% of P_o; Dalal 1977). They comprise a sequence of phosphate monoesters, from inositol monophosphate to insositol hexakisphosphate in various stereoisomeric variations (*myo, scyllo, neo,* D-*chiro*) (Celi and Barberis 2005). They are characterized by high acidity and are often found as components of polymers or insoluble complexes with proteins and lipids (Harrison 1987). The stability of inositol phosphates is closely linked to the number of phosphate groups, rendering esters of higher order more recalcitrant to biodegradation and consequently more abundant. The most common stereoisomer in soil is *myo*-inositol hexakisphosphate (also known as phytic acid).

Phospholipids usually comprise 0.5–7.0% of total P_o (Dalal 1977), with phosphoglycerides being the most dominant form. Nucleic acids and their derivatives account for less than 3% of total P_o; they are rapidly mineralized, re-synthesized, combined with other soil constituents or incorporated into microbial biomass

(Anderson and Malcolm 1974). Other, less abundant forms of P_o include sugar-P (Anderson 1961), monophosphorylated carboxylic acids (Anderson and Malcolm 1974) and teichoic acids (a major cell wall component of many Gram-positive bacteria; Zhang et al. 1998).

A major characteristic of P biogeochemistry is that only 1% of the total soil P (ca. 400–4,000 kg P ha^{-1} in the top 30 cm) is incorporated into living plant biomass during each growing season (ca. 10–30 kg P ha^{-1}), reflecting its low availability for plant uptake (Blake et al. 2000; Quiquampoix and Mousain 2005). Concentrations of microbial P in soils can range between 0.75 (sandy Spodosol, fertilized pine plantation) and 106 mg kg^{-1} (calcaric Regosol, permanent grassland) in mineral top soils and between 50 (haplic Podzol, Norway spruce forest) and 169 mg kg^{-1} (typic Udivitrand, indigenous New Zealand forest) in organic litter layers, and have been found to comprise between 0.5% and 26% of total soil P (Oberson and Joner 2005). Microbial P generally decreases with increasing soil depth and decreasing soil organic matter (SOM) content. The P-containing compounds in microorganisms are reviewed extensively elsewhere (Bünemann et al. 2011). Briefly, P-containing compounds in bacteria and fungi have been found to include nucleic acids (30–65% of total microbial P), phospholipids, acid-soluble P_i and P_o compounds (i.e. phosphate esters, phosphorylated coenzymes, 15–20%), polyphosphates, as well as teichoic acid (only Gram-positive bacteria). Both polyphosphates and teichoic acid have been reported to serve as P storage compounds (Alexander 1977; Gächter and Meyer 1993). Several authors have reported that immobilization of P by microbes is regulated more by C limitation than by P limitation (Bünemann et al. 2004; Oehl et al. 2001a). Therefore, P concentrations in the microbial biomass seem to be closely linked to C dynamics in soil (Achat et al. 2009). Fertilization with P often results in a decline in microbial P (Clarholm 1993; Grierson et al. 1998), whereas other authors have reported the opposite trend (Joergensen and Scheu 1999) or observed rather small or no effects (Bünemann et al. 2004). Furthermore, apart from substrate-driven changes, it appears that seasonal variations in microbial P are mostly related to changes in gravimetric water content in the soil (Chen et al. 2003), causing reduced microbial biomass P during dry periods.

Organic Soil Additives

In an effort to enhance soil quality and divert waste from landfill, the application of organic waste (e.g. treated municipal sewage sludge – so-called biosolids, compost, animal manures) to land has significantly increased in recent years. Organic soil amendments may increase P nutrition of plants and microbes; however, P availability in these heterogeneous materials is highly dependent on the chemical forms of P present, as well as on the complex interaction of the added material with the soil. There is increasing interest in the potential beneficial interaction of biofertilizers (i.e. PSM) with these organic wastes to provide optimal nutrient delivery to crops. Synchrotron-based analysis of selected biosolids and manure originated from various animal stocks revealed that P in biosolids was mainly present as P_i in the form of variscite (Al-P, 86% of total P) and the less-soluble hydroxylapatite (Ca-P, 14%). Manure contained dicalcium phosphate dehydrate (12–65%), struvite (ammonium magnesium phosphate, 12–68%) and variscite (0–18%) (Ajiboye et al. 2007). Organic P was mainly present as Ca-phytate (20–70%) (Ajiboye et al. 2007). The P_i fraction in compost is bound to Ca as apatite or octacalcium-phosphates (Frossard et al. 2002). Furthermore, the distribution of P_i between Al, Fe and Ca fractions in organic amendments also depends on further additives such as lime and metal salts [e.g. $FeCl_3$, $Al_2(SO_4)_3$]. Despite their reducing effect on potential heavy metal toxicity, these additives have been found to effectively reduce P solubility (Maguire et al. 2006). Metal salts induce P sorption to precipitated Al or Fe hydroxides rather than immobilization as Al or Fe phosphate. The addition of lime induces the formation of recalcitrant Ca phosphates (e.g. hydroxylapatite, tricalcium phosphate) and, consequently, reduces P sorption to Fe hydroxide surfaces (Shober and Sims 2009). The solubility of P in organic soil amendments when applied to soils will be mainly governed by the soil solution equilibrium and the soil and substrate pH, rendering Fe and Al phosphates more, and Ca phosphates less, recalcitrant with decreasing pH. The chemical form of P in the biosolids can also be manipulated to improve their use in the field (e.g. through the addition of stabilizing agents such as CaO, $FeSO_4$ etc.; Huang et al. 2008). Nevertheless, despite being an additional potential P source, organic amendments provide a significant amount of easily available carbon and nitrogen, resulting in an almost direct response in increasing microbial activity (biomass build up, turnover, respiration and substrate mineralization) (Giller et al. 1998; Kao et al. 2006; Saha et al. 2008), accelerating dynamics not only in C- and N-, but also in P cycling. The increasing microbial activity may potentially solubilize organic as well as inorganic P contained in the additives, but also P forms present in the indigenous soil.

7.3 P-Solubilizing Mechanisms

Availability of P_i in soil is mainly governed by the dissolution properties of the P-bearing minerals (which are determined largely by pH) as well as by solution equilibrium reactions (sorption and desorption). In contrast, the availability of P derived from P_o is mainly governed by microbial activity (mineralization and enzymatic hydrolysis). Like higher plants, microorganisms have the potential to modify their immediate chemical environment via the uptake and release of organic and inorganic ions and molecules. The main P solubilization mechanisms employed by soil microorganisms include: (1) release of complexing or mineral dissolving compounds (e.g. organic acid anions, siderophores, protons, hydroxyl ions, CO_2), (2) liberation of extracellular enzymes (biochemical P_o mineralization) and (3) the release of P_o during substrate degradation (biological P_o mineralization) (McGill and Cole 1981). P incorporated in the microbial biomass may be temporarily immobilized but remains in a bioavailable form that can be released via microbial turnover (re-mineralization). Therefore, microorganisms play an important role in all three major components of the soil P cycle (i.e. dissolution–precipitation, sorption–desorption, and mineralization–immobilization).

Microbial P mobilization strategies, uptake and subsequent release, as well as redistribution of P in soil may all affect the success of PSM in improving plant growth. In particular, PSM may also compete with plants for any P released. Soil solution P only increases when P solubilization (from soil minerals) and gross mineralization (mineralization of SOM, re-mineralization of P_o and P_i held in the microbial biomass) exceed P immobilization (uptake and incorporation in the biomass) and sorption to soil minerals. All these processes are driven by a wide range of chemical and physical soil properties (e.g. mineral composition, SOM, texture, structure, temperature, water content) and vegetation properties, making temporal and spatial predictions of P availability in soil from added PSM difficult. The mechanisms and processes involved in P mobilization by PSM in soil are discussed in Sects. 7.3.1–7.3.6.

7.3.1 P Release Mediated by Changes in pH

The release of protons or hydroxide ions by microorganisms can significantly alter the soil solution pH in the close vicinity of the exuding organisms, inducing changes in mineral nutrient availability. Whereas only a few reports of microbial P solubilization by alkalinization exist, microbial P solubilization via acidification is well documented for several fungi and bacterial species (Gyaneshwar et al. 1999; Illmer and Schinner 1992; Ben Farhat et al. 2009) and is often found to be particularly successful when P is associated with Ca. Typically, the release of protons is also linked to the extrusion of organic acid anions into the external media (Arvieu et al. 2003; Casarin et al. 2003). The amount of protons released into the external medium is often significantly influenced by N supply. In general, a greater reduction in pH together with more solubilized P can be observed with NH_4^+ as the sole N source compared to NO_3^-, due to the extrusions of protons to compensate for NH_4^+ uptake (Roos and Luckner 1984; Illmer et al. 1995; Sharan et al. 2008). In contrast, Reyes et al. (1999) found a decrease in P solubilization by *Penicillium rugulosum* from various P-bearing minerals (hydroxyapatite, $FePO_4$, $AlPO_4$) when higher concentrations of NH_4^+ were supplied. The authors attributed these findings to the repressive effect of easily metabolized N sources on secondary metabolic biosynthesis in fungi. The same study also showed that the assimilation of amino acids (in this case arginine) as the sole N source can also lead to a decrease in pH and enhanced mobilization of P.

For some microorganisms, NH_4^+-driven proton release seems to be the sole mechanism to promote P solubilization. Asea et al. (1988) tested two fungi, *Penicillium bilaii* and *Penicillium fuscum*, for their ability to solubilize phosphate rock in the presence of NH_4^+ or without N addition, and showed that only *P. bilaii* maintained the ability to decrease the pH and mobilize P when no N was supplied.

In a study of *Pseudomonas fluorescens*, the form of C supply (e.g. glucose versus fructose) rather than N supply (e.g. NH_4^+ versus NO_3^-) had the greatest effect on proton release (Park et al. 2009). This indicates that for different species, different mechanisms are responsible for proton release, only partly depending on the presence of NH_4^+. Furthermore, Asea et al. (1988) reported a direct relationship between mobilized P and pH for *P. fuscum*, whereas *P. bilaii* solubilized more P than could be accounted for by pH change, indicating the presence of an additional solubilization mechanism.

Despite the demonstration of pH changes as a potential P solubilization mechanism, we must keep in mind that most published studies are carried out in vitro and that conditions in the field may not be as conducive to significant acidification occurring (e.g. due to a lack of labile N and C, which limits their activity in the bulk soil). Furthermore, particularly calcareous soils possess a high pH buffer capacity, which could limit the P-solubilizing effect. Gyaneshwar et al. (1999) showed that the numbers of culturable microorganisms inducing a pH-reduction zone in the growth medium were drastically reduced (from 10^4–10^6 to 10^2) when a buffered medium with phosphate rock was used as the sole P source. Furthermore, the buffered culture medium resulted in a reduced recovery of PSM, with only one out of ten soil samples containing PSM. This indicates that early estimates of effective PSM in soil (e.g. 10^3–10^6 g^{-1}; Kucey et al. 1989) might significantly overestimate reality. On the other hand, the majority of soil microbes cannot be cultured but might potentially be capable of solubilizing different forms of P_i, making real numbers of effective PSM in soil hard to predict. Nevertheless, since microbial abundance and activity is particularly high in the rhizosphere, the combined effort of proton (or hydroxide) extrusion by plants and microbes might be sufficient to increase P availability for both groups of organisms. Additionally, higher concentrations of CO_2 in the rhizosphere originated from plant root and microbial respiration might also contribute to a local drop in pH.

7.3.2 P Release Mediated by Organic Acid Anions

Acidification alone often does not fully explain the solubilization of mineral P (Asea et al. 1988; Whitelaw et al. 1999). LMW organic acid anions (carboxylates) released by microbes have been frequently found in P_i solubilization studies (Illmer et al. 1995; Reyes et al. 1999; Patel et al. 2008). Reported organic acid anions secreted by PSM include gluconic, 2-ketogluconic, citric, malic, malonic, oxalic, succinic, lactic, tartaric and glycolic acids (Kucey et al. 1989; Gyaneshwar et al. 2002). Though they are commonly referred to as organic "acids" in the literature, it has been pointed out by many authors that organic anions would be the more appropriate terminology (Hinsinger 2001; Jones et al. 2003; Parker et al. 2005). Due to the low acid dissociation constants (pK_a) values of many organic acid anions (Table 7.1), these LMW compounds are present as dissociated anions in the cytosol of cells (pH ~7), as well as in soil across a wide pH range. Nevertheless, organic

Table 7.1 Acid dissociation constants (pK_a) of some organic acids implicated in P solubilization

Acid	Number of carboxylic groups	pK_a 1	pK_a 2	pK_a 3
Citric	3	3.15	4.77	6.40
Malic	2	3.40	5.13	–
Oxalic	2	1.27	4.28	–
Gluconate	1	3.86	–	–

Table 7.2 Stability constants of some organic acid–metal complexes determined at a 1:1 metal–ligand ratio at 25°C and zero ionic strength

Organic anion	Number of carboxylic groups	Fe^{3+}	Al^{3+}	Ca^{2+}
Citrate	3	11.5	7.9	4.9
Malate	2	7.1	6.0	2.7
Oxalate	2	7.5	6.1	4.9
Gluconate	1	37.2	2.0	1.2

Adapted from Martell and Smith (1977) and Jones (1998)

anion release is often found to be accompanied by medium acidification, but it is important to note that it is not the organic anions that cause acidification, because they are already in their dissociated forms when released into the soil. It is rather the secretion of protons compensating the loss of net negative charges that can cause a drop in pH.

The P-solubilizing effect of organic anions is either caused by their negative charge or by their metal complexation properties (Table 7.2). Being anions, they can mobilize P_i from metal oxide surfaces via ligand exchange or by ligand enhanced dissolution of Fe or Al oxides and Ca phosphates, where the weakening of mineral bonds due to prior organic anion adsorption and/or chelation liberates the occluded P. Additionally, organic anion adsorption on metal oxide surfaces decreases the positive surface potential (Filius et al. 1997), facilitating the release of adsorbed P.

Gluconic and 2-ketogluconic acid are frequently reported to be released by bacteria (Rodriguez and Fraga 1999), and *Gluconacetobacter diazotrophicus* mutants lacking their production capacity have been shown to partially lose their P mobilization ability (Intorne et al. 2009). In contrast, gluconic, citric and oxalic acid are often found to be released by fungi (Reyes et al. 1999; Whitelaw et al. 1999). In general, tri-carboxylic anions such as citrate show a higher potential in solubilizing P_i than do di-carboxylic acids (gluconate, oxalate, etc.). There is increasing evidence that the mobilization of P_i by particularly citrate (other organic acid anions to a minor extent) is accompanied by a significant increase of Fe and Al in the solution, indicating that mineral dissolution is the main mobilizing mechanism (Gerke et al. 2000; Oburger and coworkers, unpublished data). Furthermore, oxalate has been shown to be particularly efficient in mobilizing P in calcareous soils (Ström et al. 2005) due to its high affinity to form Ca precipitates. To our knowledge, the behaviour of gluconic acid and 2-ketogluconic acid (unlike citric and oxalic acid) in soils has not yet been thoroughly investigated.

The P mobilization efficiency of organic anions released by microbes and plants is determined largely by soil properties (e.g. sorption sites, pH), as well as by the quantity and characteristics of the compounds released. Reported concentrations of organic acid anions released by cultured PSM range from a few micromolar (Illmer et al. 1995) to 100 mM (Reyes et al. 1999; Gyaneshwar et al. 1999; Patel et al. 2008). Detected concentrations in bacterial and fungal cultures differ greatly with incubation conditions (Illmer et al. 1995) and with different C sources (Reyes et al. 1999; Patel et al. 2008). It should be noted that organic acid anion exudation patterns might be completely different in soil (e.g. due to the lack of available C to synthesize the organic acid anions). In soil, sorption to metal oxide surfaces will decrease the free organic anion concentration in solution, making predictions of actual soil solution concentrations difficult. It has also been shown that the P-solubilizing effect of organic acid anions is significantly reduced in soils rich in carbonate or in Fe and Al (hydr)oxides (Ström et al. 2005; Oburger et al. 2009). Additionally, LMW carboxylates produced by microbes (and plant roots) can also serve as a labile C substrate for the microbial community, thus removing them from solution and reducing their P mobilization potential. The half-life of organic acid anions in soil typically ranges from 0.5 to 12 h, suggesting that organic acid anions need to be continually produced by PSM to maintain P dissolution over the lifetime of a crop (Jones et al. 2003). However, microbial breakdown of organic acid anions have been found to be drastically reduced in high-sorbing soils (Oburger et al. 2009; Oburger and Jones 2009; van Hees et al. 2003), indicating the importance of sorption processes to organic acid bioavailability and functional efficiency. Alongside the direct stimulation of microbial growth and P solubilization from inorganic sources, organic acid anions such as citrate, malate and oxalate can improve the solubility of P_o (e.g. phytate) making it more susceptible to enzymatic hydrolysis (Otani and Ae 1999; Tang et al. 2006).

7.3.3 Exopolysaccharide-Mediated Release of P

To our knowledge, the role of HMW (non-enzymatic) microbial exudates (i.e. mucilage, exopolysaccharides) in P solubilization from soil constituents has not yet been directly investigated in situ. Gaume et al. (2000) showed that maize root mucilage adsorbed onto synthetic ferrihydrite significantly decreased consecutive P adsorption, but the investigated mucilage components were not able to mobilize significant amounts of already adsorbed P. Nevertheless, microbial mucilages can have an indirect effect on P availability through their important role in soil aggregation and in increasing pore connectivity in soil (Aspiras et al. 1971), thereby facilitating soil water retention and movement (Ionescu and Belkin 2009). Exopolysaccharides (EPS) and biosurfactants are produced by microorganisms largely in response to biofilm formation and stress. Studies on microbially produced EPS have shown their ability to complex metals in soil ($Al^{3+} > Cu^{2+} > Zn^{2+} > Fe^{3+} > Mg^{2+} > K^{+}$; Ochoa-Loza et al. 2001), from which it can be deduced that they must in some way influence P solubility in soil. In pure culture, microbial EPS has been shown to stimulate the dissolution of tricalcium phosphate synergistically with organic acid anions (Yi et al. 2008). Furthermore, the rate of dissolution appears dependent on the microbial source and concentration of EPS. Although there is some evidence to suggest that EPS production is stimulated under P deficiency, this does not appear to be a universal phenomenon in bacteria (Dephilippis et al. 1991, 1993). Furthermore, EPS production seems to be more dependent on the rate of N supply than on available P (Danhorn and Fuqua 2003; Wielbo and Skorupska 2008).

7.3.4 Siderophore-Mediated Release of P

Siderophores are complexing agents that have a high affinity for iron and are produced by almost all microorganisms in response to iron deficiency. There are approximately 500 known siderophores, with the majority of them being used by a wide range of microorganisms and plants and some of them being exclusively used by the microbial species and strains that produce them (Crowley 2007). Many studies have reported the release of siderophores from PSM (Vassilev et al. 2006; Caballero-Mellado et al. 2007; Hamdali et al. 2008c); however, siderophore production has not been widely implicated as a P-solubilization mechanism. Considering the dominance of mineral dissolution over ligand exchange by organic acid anions as a P-solubilizing mechanism (Parker et al. 2005), the potential role of siderophores in enhancing P availability should be obvious. However, there is an impressive body of literature concerning Fe mobilization by microbial siderophores, but to the best of our knowledge only one study exists that has investigated the effect of microbial siderophores on P availability. More than two decades ago, Reid et al. (1985) investigated the ability to increase Fe and P diffusion of two siderophores (desferrioxamine-B, desferriferrichrome) and the iron-chelating agent EDDHA as compared to water using a root simulation technique. They found that desferriferrichrome increased P diffusion 13-fold compared to water whereas different concentrations of desferrioxamine-B exhibited only a small effect (Fig. 7.2). Considering the occurrence of Fe phosphates in soil and, probably even more important, the large P sorption capacity of Fe (hydr)oxides and considering the needs of microorganisms for Fe, the lack of knowledge about siderophore-enhanced P solubilization is quite surprising.

7.3.5 Enzyme-Mediated Release of P

The biochemical mineralization of P_o is mediated by either cell-wall-bound or free phosphatase enzymes, whose release is mainly driven by P demand. The role of these enzymes in P cycling is reviewed extensively elsewhere (Nannipieri et al. 2011); however, here we aim to present evidence relevant to the behaviour of PSM in soil. Typically, extracellular phosphatases rather than intracellular or membrane-

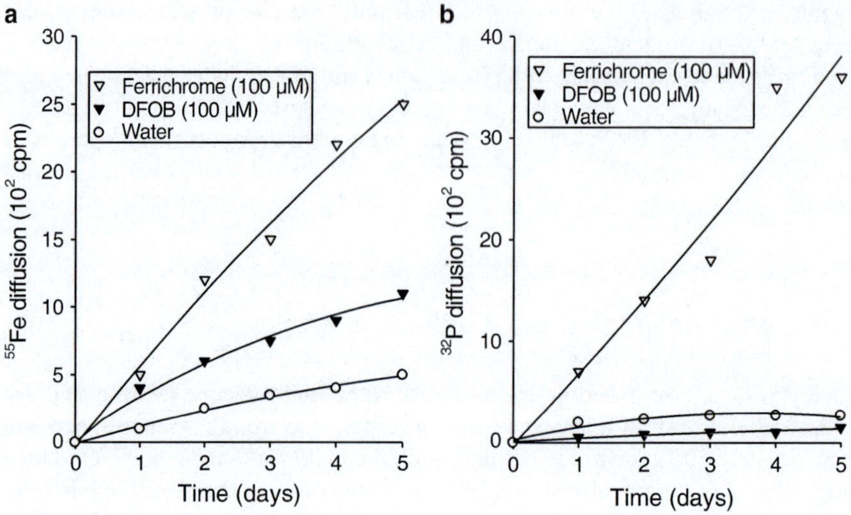

Fig. 7.2 Effects of the microbial siderophores desferrioxamine-B (*DFOB*) and ferrichrome on the diffusion of (**a**) ^{55}Fe or (**b**) ^{32}P towards a root in a low pH soil. Water is shown as a control. Fe and P were added to the soil as $FeCl_3$ and KH_2PO_4, respectively. Overall the results indicate that some microbial siderophores (e.g. ferrichrome) can stimulate both Fe and P solubilization in soil. Adapted from Reid et al. (1985)

bound phosphatases are thought to be responsible for inducing large changes in soil solution P concentration. However, experimental differentiation between exo- and endo-enzyme activity still remains problematic.

Phosphatases or phosphohydrolases describe a broad group of enzymes that catalyze the hydrolysis of both esters and anhydrides of H_3PO_4 (Tabatabai 1994). Its activities have been shown to be inhibited by increasing concentrations of orthophosphate (end-product) as well as other polyvalent anions (e.g. MoO_4^{2-}, AsO_4^{3-}) and high concentrations of several metals [Zn, Hg, Cu, Mn (II), Fe (II)]. Lower concentrations of divalent cations (e.g. Ca, Mg, Zn, Co) have been found to act as enzyme activators (Quiquampoix and Mousain 2005). Furthermore, adsorption to soil mineral or organomineral surfaces can also significantly alter enzyme conformation and activity. Although sorption to the solid phase reduces enzymatic activity, it can also help protect the enzymes from microbial attack or thermal inactivation (Huang et al. 2005). Typically, phosphatases are held most strongly to clay-sized particles. These results clearly show that the activity of enzymes released from PSM are not simply related to their release rate but are also strongly influenced by soil properties such as mineral composition, SOM and pH.

Among the variety of phosphatase enzyme classes released by PSM, phosphomonoesterases (often just called phosphatases) are the most abundant and best studied. Depending on their pH optima, these enzymes are divided into acid and alkaline phosphomonoesterases and both can be produced by PSM depending upon the external conditions (Kim et al. 1998; Jorquera et al. 2008). Typically, acid phosphatases predominate in acid soils, whereas alkaline phosphatases are more abundant in neutral and alkaline soils (Eivazi and Tabatabai 1977; Juma and Tabatabai 1977, 1988; Renella et al. 2006). Although plant roots can produce acid phosphatases they rarely produce large quantities of alkaline phosphatases, suggesting that this is a potential niche for PSM (Juma and Tabatabai 1988; Criquet et al. 2004). Laboratory studies have shown a gross mineralization potential of 1–4 mg P kg^{-1} soil day^{-1} (Lopez-Hernandez et al. 1998; Oehl et al. 2001b); however, so far it has proved impossible to distinguish between enzymatic (biochemical) and biological (microbial turnover) mineralization. It is also difficult to differentiate between root- and PSM-produced phosphatases (Richardson et al. 2009a, b) but some evidence suggests that phosphatases of microbial origin possess a greater affinity for P_o compounds than those derived from plant roots (Tarafdar et al. 2001). The relationship between PSM introduced into soil, phosphatase activity and the subsequent mineralization of P_o still remains poorly understood (Chen et al. 2003). Controversial results have been reported about the correlation between increased phosphatase activity and P_i concentrations in soil solution, with several authors reporting no relationship between the two (Criquet et al. 2002, 2004; Olander and Vitousek 2000). Other groups found a positive correlation (Tate and Salcedo 1988; Rojo et al 1990; George et al. 2002) and some observed a negative relation between P_i concentrations and phosphatase activity (Ali et al. 2009). Considering the interactive complexity of biological, chemical and biochemical processes of P mobilization in soils, these controversial findings are not surprising but highlight the uncertainty about predicting the benefits of PSM introduced into soil.

7.3.6 Release of P Held in P-Solubilizing Microorganisms

Although some of the P released by PSM will be captured by plants and other soil organisms, it is inevitable that a large proportion will be immobilized in the PSM. Release of P immobilized by PSM primarily occurs when cells die due to changes in environmental conditions, starvation or predation. Environmental changes, such as drying–rewetting or freezing–thawing, can result in so-called flush-events, a sudden increase in available P in the solution due to an unusually high proportion of microbial cell lysis (Turner et al. 2003; Butterly et al. 2009). Grierson et al. (1998) found that about 30–45% of microbial P (0.8–1 mg kg^{-1}) was released in a sandy spodosol in an initial flush after drying–rewetting cycles within the first 24 h. However, the availability of P after these flush events is also likely to be highly dependent on the P sorption properties of the soil because a large proportion could become subsequently immobilized on the solid phase.

P is also released when microorganisms are grazed by microbivores (e.g. nematodes, protozoa). Cole et al. (1978) showed that significant net P mineralization occurred within 1 week in the presence of bacterial grazers, whereas the absence of predators resulted in a constantly high P immobilization and no net P release after more than 3 weeks. SOM availability and its C:P ratio has also been shown to have a significant impact on microbial P immobilization/re-mobilization dynamics (Chauhan et al. 1979, 1981). Fresh organic matter inputs, particularly easily available C sources, tend to increase microbial P followed by a subsequent decline and increase in soil solution P if the substrate is depleted. However, the time elapsed between P immobilization and re-mineralization is determined by substrate quality and soil properties, with dynamics being less pronounced with more recalcitrant organic matter (Oehl et al. 2001a, b). Overall, however, the release of P held in PSM is poorly understood and certainly warrants further research.

7.4 P-Solubilizing Organisms

7.4.1 Bacteria

It has been known for a long time that significant variation in the ability to solubilize P in soil exists within the bacterial community. Those that are known to enhance P availability includes species of the common soil bacteria *Pseudomonas*, *Azotobacter*, *Burkholderia*, *Bacillus* and *Rhizobium*. The recent isolation of a supersolubilizer (*Serratia marcescens*) has also suggested that selected bacteria could be used to develop environmentally friendly processes for fertilizer production (Ben Farhat et al. 2009). Recent work on *P. fluorescens* strains isolated from a range of agricultural fields has suggested that significant variation also exists within a single bacterial species (Browne et al. 2009). Interestingly, in the study by Browne et al. (2009) P fertilizer regime and crop type appeared to have little effect on the abundance of pseudomonads in the rhizosphere (Fig. 7.3). However, it was clear that significant variation exists within a single species linked to a single phylogenetic lineage. Identification of the underlying genetic mechanisms, combined with work to enhance their rhizosphere competence, may therefore provide more efficient biofertilizer agents. Evidence from *G. diazotrophicus* also suggests that bacteria may operate more than one P solubilization mechanism simultaneously (Intorne et al. 2009).

A study by Hariprasad et al. (2009) has also indicated that the selection for one bacterial trait may not always be the best use of inoculation technology. For example, they showed that rhizobacteria producing P-solubilizing indole acetic acid (PSIRB) may prove better than either P-solubilizing rhizobacteria (PSRB) or indole-acetic-acid-producing rhizobacteria (IRB) in isolation.

Many bacteria can also colonize the surface of mycorrhizal hyphae in soil (hyphosphere) and, therefore, may contribute indirectly to P uptake by mycorrhizas (and ultimately the plant) if they express P-solubilizing activity (Gonzalez-Chavez et al. 2008). These bacteria can be found embedded in hyphal mucilage, on the hyphoplane, between hyphal wall layers and even inside hyphae and are known to include several common PSM species (Mansfeld-Giese et al. 2002). Further work is required to elucidate their quantitative role in supplying P in comparison to the mycorrhizas.

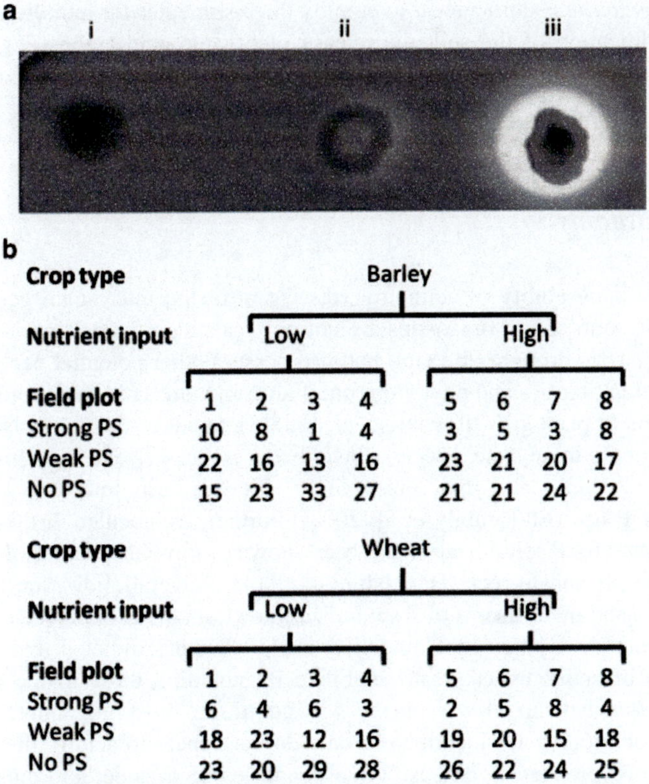

Fig. 7.3 P solubilization by different isolates of *Pseudomonas fluorescens* obtained from agricultural fields receiving either high or low inputs of fertilizer and under either a barley or wheat crop. (**a**) Plate assay demonstrating the three $Ca_3(PO_4)_2$ solubilization phenotypes used to classify isolates: (1) No $Ca_3(PO_4)_2$ solubilization, (2) weak $Ca_3(PO_4)_2$ solubilization and (3) strong $Ca_3(PO_4)_2$ solubilization. All 752 isolated from the field were then spot-inoculated (six repeats) on modified NBRIP agar, incubated for 6 days at 30°C. (**b**) Sampling strategy employed in the study and the numbers of isolates with each P solubilization (*PS*) phenotype (strong, weak, or no P solubilization activity) for each field plot. A total of 47 isolates were evaluated per plot. Adapted from Browne et al. (2009)

7.4.2 *Non-mycorrhizal Fungi*

A range of non-mycorrhizal soil fungi have been screened and selected for their P-solubilizing capacity. Of those identified, many are commonly found in agricultural soils such as *Penicillium* spp., *Mucor* spp. and *Aspergillus* spp., which has been shown to increase plant growth by 5–20% after inoculation (Dwivedi et al. 2004; Babana and Antoun 2006; Wakelin et al. 2007; Gunes et al. 2009). In addition, a range of *Trichoderma* spp. have also been identified and found to stimulate plant growth both in the laboratory and field (Rudresh et al. 2005). As with many ectomycorrhizal fungi, P-solubilizing non-mycorrhizal fungi (e.g. *Emericella rugulosa*, *Penicillium* spp.) appear to employ three strategies for mobilizing soil P, namely acidification of the soil, the release of organic acid anions (e.g. citrate, oxalate, gluconate) and the release of acid and alkaline phosphatases and phytase (Yadav and Tarafdar 2007; Xiao et al. 2009).

7.4.3 Actinomycetes

The P-solubilizing ability of actinomycetes has attracted interest in recent years because this group of soil organisms are not only capable of surviving in extreme environments (e.g. drought, fire etc.) but also possess other potential benefits (e.g. production of antibiotics and phytohormone-like compounds etc.) that could simultaneously benefit plant growth (Fabre et al. 1988; Hamdali et al. 2008a). Numerous P-solubilizing actinomycete species have been isolated from the rhizosphere (Barreto et al. 2008) and their presence in soil has been linked to enhanced efficiency of P use (El-Tarabily et al. 2008). Further, re-inoculation of soil with isolates selected for P solubilization has been shown to stimulate plant growth when supplied with phosphate rock (Hamdali et al. 2008b). Overall, however, the taxonomic groups and mechanisms of P solubilization within the actinomycetes remain poorly elucidated. A study by Hamdali et al. (2008c) has indicted that approximately 20% of actinomycetes can solubilize P, including those in the common genera *Streptomyces* and *Micromonospora*. In contrast to most fungi, most of the P-solubilizing actinomycetes identified to date do not appear to acidify the external medium. However, they do release large quantities of organic acid anions (e.g. citrate, formiate, lactate, malate, succinate), which are implicated in the P dissolution process (Hoberg et al. 2005), and possibly other P dissolution-promoting organic substances (Hamdali et al. 2010). After uptake, the P is stored in polyphosphate within the mycelium (Hamdali et al. 2010). One exception was reported by Abdulla (2009), who showed that P solubilization occurred concomitantly with acidification. A marine study has also suggested that actinomycetes may enhance P availability through the release of phosphatases; however, the significance of this in soil remains unknown (Sahu et al. 2007). Field trials inoculating P-poor soils have shown significant yield benefits, although whether this was due to P or other beneficial effects of the actinomyctes remains unknown. One potential application of actinomycetes is to harness their thermo-tolerant properties to enhance P availability during the composting of municipal and animal wastes (Chang and Yang 2009).

7.4.4 Protozoa

As protozoa represent a major mechanism for regulating bacterial and fungal numbers in soil, they both directly and indirectly influence soil P cycling (Alphei et al. 1996). If P-solubilizing microorganisms are introduced into soil, then protozoal grazing can be expected to dramatically reduce their effectiveness (Rosenberg et al. 2009; Pedersen et al. 2009). Although protozoa have the capacity to take up and assimilate SOM, ultimately making P more bioavailable, the likelihood of managing protozoal numbers in soil to harness this potential remains remote (due to difficulties and cost of mass production of a protozoal biofertilizer and uncertainty surrounding their potentially deleterious impact on microbial food webs in soil).

7.4.5 *Mesofaunal Interactions*

Mesofauna are known to enhance P availability and cycling in a range of soils; however, due to difficulties in their practical handling they are rarely used as a management tool to directly manipulate nutrient availability in agricultural soils (Lopez-Hernandez et al. 1993). Of significance, however, are the positive interactions that may occur between mesofauna and PSM in soil. For example, Sreenivas and Narayanasamy (2009) showed that the earthworm, *Eisenia fetida*, enhanced the P-solubilizing ability of the fungus *Aspergillus awamori*, resulting in increases in both soluble P_i and soluble P_o. The mechanistic basis for this response currently remains unknown. Similarly, Wan and Wong (2004) showed that earthworms promoted growth and phosphatase production in the P-solubilizing bacteria *Bacillus megaterium* and that this subsequently enhanced P_i availability in soil. Mba (1994, 1997) identified earthworm casts as being a particular site of enhanced PSM activity. In contrast, nematodes can be expected to dramatically reduce the amount of PSM inoculated into soil (Pedersen et al. 2009). Although this may reduce the effectiveness of P solubilization, it may also stimulate the release of P immobilized in the PSM. For more information on P mesofaunal interactions, the reader should consult Chapuis-Lardy et al. (2011).

7.5 Significance of PSM in the Field and Potential for Management

It is notable that the conclusions of many publications on the subject state that PSM hold great potential for development as a biofertilizer that can enhance soil fertility and promote plant growth. In addition, others state that PSM could constitute a novel and non-polluting biofertilizer product useful for the development of sustainable agriculture. Potentially this could be true, however, are we promising too much, too soon? For any PSM product to be a commercial success and to be accepted by farmers, a major perception change will be required within the agrochemical and agricultural industry. Typically, the industry is sceptical of products that cannot demonstrate a clear and positive benefit and that may be technically difficult to administer to fields. For PSM to be accepted requires that the technology is robust enough to be rolled out across wide geographical zones encompassing different soil types, crops and abiotic stresses. Furthermore, there must be a tangible economic benefit for farmer adoption because there are likely to be few legislative drivers to encourage farmer adoption of PSM technology. The costs associated with the environmental licensing of PSM products, particularly if they are genetically engineered, may also be prohibitive. In our view, PSM technology is still in its infancy and requires further optimization and refinement before commercial release, at least into developed world markets. Critical evidence in support of our view is presented below.

Typically, plant growth response trials with PSM have been carried out under controlled conditions that are rarely representative of those in the field (e.g. in small pots in the absence of mesofauna, under optimal conditions for plant growth and with a high inoculation dose). It is known from bitter experience with plant-growth-promoting rhizobacteria (PGPR) and N_2 fixation inoculants, however, that the positive growth responses obtained in the greenhouse often fail to reflect those subsequently obtained in the field, the latter of which can show zero or even negative yield responses (Streeter 1994). This is highlighted by Okon and Labandera-Gonzalez (1994), who concluded that of the published N_2-fixing *Azospirillum* field trials, 30–40% showed no positive response and, when a response was reported, the yield gain was very low (5–30% in comparison to uninoculated controls). This highlights the uncertain world of microbial inoculants. Success is largely determined by the ability of the inoculum to remain alive long enough in soil to have an appreciable benefit and to be able to compete with the indigenous microbial community (Denton et al. 2003). Despite this, some PSM isolates have been successfully translated from the laboratory to the field whilst others have failed (Fernández et al. 2007). For example, field studies have shown that PSM application can enhance foliar P concentration and increase efficiency of P use (Sud and Jatav 2007; Malboobi et al. 2009). This effect is often dramatic when undertaken with poor soils and low grade phosphate rock fertilizers (Sharma and Prasad 2003). Although not noted by Malboobi et al. (2009), their results revealed a high degree of dependence on geographical location and fertilizer dose. This context-specific view is also taken by Sahin et al. (2004), who showed that the beneficial effects of PSM on plant growth varied significantly depending on environmental conditions, bacterial strains, and plant and soil conditions. The success of PSM has also been shown to be influenced by the use of agrochemicals and the addition of organic fertilizers (Sutaliya and Singh 2005; Das and Debnath 2006). This suggests that blanket recommendation for farmers will be difficult to formulate and that depressions in yield may also be possible, an outcome clearly not welcomed by farmers and that will undermine adoption of the technology. It is also clear that PSM cannot provide a complete replacement for conventional fertilizers but simply a way of reducing our reliance on them (i.e. the use of PSM as part of an integrated nutrient management regime; Jilani et al. 2007; Sharma et al. 2009). Overall, across a range of field trials, PSM application typically results in marginal increases in crop yield

Table 7.3 Effect of the presence (+) and absence (−) of the P-solubilizing microorganism *Penicillium radicum* inoculation on grain yield and protein levels in wheat grown under field conditions

	Grain yield (tons ha^{-1})		Seed protein (kg ha^{-1})	
	(−)	(+)	(−)	(+)
P applied (kg/ha)				
0	2.1	2.3	213	226
5	2.6	3.1	254	299
10	3.2	3.3	328	315
15	3.3	4.1	307	407
20	3.5	4.1	379	375
Statistical analysis				
Least significant difference ($P = 0.05$)				
Phosphate	0.6		64	
P. radicum	0.4		41	
Significance of effects				
Phosphate	$P < 0.001$		$P < 0.001$	
P. radicum	$P < 0.05$		n.s.	
P × *P. radicum*	n.s.		n.s.	

P fertilizer was added as single superphosphate. Values represent mean ($n = 3$). *n.s.* indicates not significant ($P > 0.05$). Adapted from Whitelaw et al. (1997)

(0–20%) (Table 7.3; Sahin et al. 2004; Chen et al. 2008). Although there are only a few reports of the potential savings, some studies have speculated that inoculation with PSM may be equivalent to a saving of between 100 and 150 kg P ha^{-1} in some high-intensity horticultural production systems (Gunes et al. 2009). The potential P savings in more conventional cropping systems, however, are expected to be much less. Cost saving may also be made if the PSM repress fungal diseases, thereby reducing the application of fungicides (Khan and Khan 2001). In a study in India, the application of PSM induced yield increases of 0.1–0.2 tons ha^{-1} in rice and 0.1–0.5 tons ha^{-1} in wheat (Dwivedi et al. 2004).

In the case of legumes in particular there is great potential to co-inoculate with N_2-fixing *Rhizobium* sp. and PSM. Where this has been attempted, synergistic effects have been reported whereby the co-inoculants enhance crop nodulation, growth and nutrient uptake and improve the use and availability of added chemical fertilizers (Sahin et al. 2004; Dadhich et al. 2006; Elkoca et al. 2008; Rugheim and Abdelgani 2009; Sammauria et al. 2009). This co-inoculation improved yield by as much as 30% in soybean and 10% in sugar beet and barley (Sahin et al. 2004; Govindan and Thirumurugan 2005).

Similarly, field trials have indicated that synergistic effects may exist between arbuscular mycorrhizal fungi (AMF) and PSM with co-application consistently increasing crop growth, foliar N and P concentrations, grain quality and yield (Khan and Zaidi 2007). In one of the most successful trials undertaken in Mali, co-inoculating seeds with PSM and AMF showed that it was possible to obtain wheat grain yields comparable to those produced from conventional inorganic N and P fertilizers (Babana and Antoun 2006). However, the uncertainly of the approach is highlighted by Wu et al. (2005), who showed that AMF inhibited any PSM effect.

One of the major problems with interpreting field trial data is that the trials are rarely matched with a mechanistic understanding of soil P cycling (i.e. measurements of P availability and rates of cycling). Consequently, it is often difficult to decide whether any observed increase in yield is due to a direct P-solubilizing effect or some other effect (e.g. repression of pathogens, microbial production of indole acetic acid or siderophores). This is highlighted in a classic study by deFreitas et al. (1997), who showed that P solubilization was not the main mechanism responsible for the positive growth response upon inoculation with PSM. Specifically, PSM might increase root growth and/or mycorrhizal colonization, which subsequently enhances soil exploration and subsequent P capture (Richardson et al. 2009a, b). A more thorough review of the benefits of PGPR on root growth and function is provided in Vessey (2003) and Lugtenberg and Kamilova (2009).

As discussed by Oberson et al. (2011), there is also potential to manage the agronomic regime to influence P dynamics in soil. There is clear evidence that some of these strategies (e.g. organic residue addition) can directly affect the size, activity and structure of the microbial community and its associated P-solubilizing potential (Bolton et al. 1985; Hu et al. 2009). For example, organic manures and composts typically cause an initial stimulation in microbial activity and in P_o and P_i solubilization activity in soil (Takeda et al. 2009; Hu et al. 2009). However, longer-term trials have shown that organic residues may in some cases reduce biological P solubilization (Martens et al. 1992; Garcia-Gil et al. 2000). Similarly, crop types could be used that are known to promote PSM activity in soil. For example, Oliveira et al. (2009) provided evidence to suggest that maize cultivar varieties may differentially stimulate PSM in soil. Similarly, Souchie and Abboud (2007) have shown a soil type × genotype interaction regulating the abundance and type of PSM in pigeonpea. This opens the potential for influencing PSM activity in soil; however, translating this information into a reliable means of enhancing crop P acquisition remains a long-term goal.

7.6 Conclusions and Future Research Directions

This chapter has highlighted the wide range of non-mycorrhizal microorganisms that possess an innate capacity to enhance P cycling in soil. Presumably, this functional group of microorganisms use their capacity to mobilize P to gain a competitive advantage in an environment where resources can be growth limiting. There appears to be two main strategies used by PSM for enhancing P availability in soil, namely (1) the enhanced dissolution of P-containing minerals through a combination of soil acidification and the release of metal complexing agents (predominantly organic acid anions and siderophores), and (2) the enzymatic breakdown and subsequent release of P from organic P. In terms of P cycling in natural environments, it is likely that strategy (2) is most important in terms of the annual flux of P through the plant–soil system. However, in highly P-limiting environments it is likely that strategy (1) becomes more important for mobilizing highly insoluble mineral-bound P. Furthermore, P temporarily immobilized in the microbial biomass may leave a significant proportion of P_o in a potentially bioavailable form. Due to the in vitro procedures used to select PSM from soil (e.g. Petri-dish culture), most research has focussed on the organisms that can accelerate the dissolution of phosphate rock. Rarely are PSM isolated that have the capacity to mobilize both organic and inorganic P (including Al- and Fe-P) in soil. To a large extent, this limits their ability to work across a wide range of soil types with vastly different properties. In terms of sustainable agriculture there is an urgent need to find new ways to make soil P more available to crop plants. Consequently, PSM isolated in the laboratory have been multiplied ex situ and then inoculated back into soil, mostly in greenhouse studies and often at very high dose rates. Typically, inoculation trials in pots show higher foliar P concentrations and a positive growth response, especially when conditions are optimized to show an effect. However, the results from field application of PSM are much more erratic. This is because many PSM are not selected for their rhizosphere competence or for their ability to survive extreme conditions, which are typical in many P-limiting soils. One major problem when interpreting the results of PSM field trials is the lack of consideration or quantification of P dynamics in the soil. Therefore, it remains difficult to differentiate between a direct P effect and an indirect effect induced by the addition of large amounts of PSM into soil (e.g. suppression of pathogens, stimulation of SOM cycling upon the death of the PSM inoculum). In addition, the indirect effects of PSM on plant growth (e.g. a hormone-induced stimulation of root growth or suppression of root pathogens that are not linked to any P-solubilization mechanism) also deserve more attention (Richardson et al. 2009b). If PSM are to be adopted by industry and farmers, a greater mechanistic understanding of PSM behaviour in soil is required. Specifically, a thorough understanding is required of their rhizosphere ecology, genetic stability and the mechanisms associated with enhancing P availability in soils and with promoting plant growth .

References

Abdulla H (2009) Bioweathering and biotransformation of granitic rock minerals by actinomycetes. Microb Ecol 58:753–761

Achat DL, Bakker MR, Morel C (2009) Process-based assessment of phosphorus availability in a low phosphorus sorbing forest soil using isotopic dilution methods. Soil Sci Soc Am J 73:2131–2142

Ajiboye B, Hu Y, Flaten DN (2007) Phosphorus speciation of sequential extracts of organic amendments using nuclear magnetic resonance and X-ray absorption near-edge structure spectroscopies. J Environ Qual 36:1563–1576

Alexander M (1977) Introduction to soil microbiology, 2nd edn. Wiley, New York, p 467

Ali MA, Louche J, Legname E, Duchemin M, Plassard C (2009) *Pinus pinaster* seedlings and their fungal symbionts show high plasticity in phosphorus acquisition in acidic soils. Tree Physiol 29:1587–1597

Alphei J, Bonkowski M, Scheu S (1996) Protozoa, Nematoda and Lumbricidae in the rhizosphere of *Hordelymus europaeus* (Poaceae): faunal interactions, response of microorganisms and effects on plant growth. Oecologia 106:111–126

Anderson G (1961) Estimation of purines and pymiridines in soil humic acid. Soil Sci 91:156–161

Anderson G, Malcolm RE (1974) The nature alkali-soluble soil organic phosphates. J Soil Sci 13:216–224

Arcand MM, Schneider KM (2006) Plant- and microbial-based mechanisms to improve the agronomic effectiveness of phosphate rock: a review. Ann Braz Acad Sci 78:791–807

Aria MM, Lakzian A, Haghnia GH, Berenji AR, Besharati H, Fotovat A (2010) Effect of *Thiobacillus*, sulfur, and vermicompost on the water-soluble phosphorus of hard rock phosphate. Bioresour Technol 101:551–554

Arvieu JC, Leprince F, Plassard C (2003) Release of oxalate and protons by ectomycorrhizal fungi in response to P-deficiency and calcium carbonate in nutrient solution. Ann For Sci 60: 815–821

Asea PEA, Kucey RMN, Stewart JWB (1988) Inorganic phosphate solubilization by two *Penicillium* species in solution culture and soil. Soil Biol Biochem 20:459–464

Aspiras RB, Allen ON, Harris RF, Chesters G (1971) The role of microorganisms in the stabilization of soil aggregates. Soil Biol Biochem 3:347–353

Babana AH, Antoun H (2006) Effect of Tilemsi phosphate rock-solubilizing microorganisms on phosphorus uptake and yield of field-grown wheat (*Triticum aestivum* L.) in Mali. Plant Soil 287:51–58

Banik S, Dey BK (1982) Available phosphate content of an alluvial soil as influenced by inoculation of some isolated phosphate solubilizing bacteria. Plant Soil 69:353–364

Barreto TR, da Silva ACM, Soares ACF, de Souza JT (2008) Population densities and genetic diversity of Actinomycetes associated to the rhizosphere of *Theobroma cacao*. Braz J Microbiol 39:464–470

Ben Farhat M, Farhat A, Bejar W, Kammoun R, Bouchaala K, Fourati A, Antoun H, Bejar S, Chouayekh H (2009) Characterization of the mineral phosphate solubilizing activity of *Serratia marcescens* CTM 50650 isolated from the phosphate mine of Gafsa. Arch Microbiol 191:815–824

Benmore RA, Coleman ML, McArthur JM (1983) Origin of sedimentary francolite from its sulphur and carbon composition. Nature 302:516–518

Blake L, Mercik S, Koerschens M, Moskal S, Poulton PR, Goulding KWT, Weigel A, Powlson DS (2000) Phosphorus content in soil, uptake by plants and balance in three European long-term field experiments. Nutr Cycl Agroecosys 56:263–275

Bolton H, Elliott LF, Papendick RI, Bezdicek DF (1985) Soil microbial biomass and selected soil enzyme-activities – effect of fertilization and cropping practices. Soil Biol Biochem 17: 297–302

Browne P, Rice O, Miller SH, Burke J, Dowling DN, Morrissey JP, O'Gara F (2009) Superior inorganic phosphate solubilization is linked to phylogeny within the *Pseudomonas fluorescens* complex. Appl Soil Ecol 43:131–138

Bünemann E, Bossio DA, Smithson PC, Frossard E, Oberson A (2004) Microbial community composition and substrate use in a highly weathered soil as affected by crop rotation and P fertilization. Soil Biol Biochem 36:889–901

Bünemann EK, Prusisz B, Ehlers K (2011) Characterization of phosphorus forms in soil microorganisms. In: Bünemann E, Oberson A, Frossard E (eds) Phosphorus in action: biological processes in soil phosphorus cycling. Soil biology, vol 26. Springer, Heidelberg. doi:10.1007/978-3-642-15271-9_2

Butterly CR, Bünemann EK, McNeill AM, Baldock JA, Marschner P (2009) Carbon pulses but not phosphorus pulses are related to decrease in microbial biomass during repeated drying and rewetting of soils. Soil Biol Biochem 41:1406–1416

Caballero-Mellado J, Onofre-Lemus J, Estrada-de los Santos P, Martinez-Aguilar L (2007) The tomato rhizosphere, an environment rich in nitrogen-fixing *Burkholderia* species with capabilities of interest for agriculture and bioremediation. Appl Environ Microbiol 73:5308–5319

Casarin V, Plassard C, Souche G, Arvieu JC (2003) Quantification of oxalate ions and protons released by ectomycorrhizal fungi in rhizosphere soil. Agronomie 23:461–469

Celi L, Barberis E (2005) Abiotic stabilization of organic phosphorus in the environment. In: Turner BL, Frossard E, Baldwin DS (eds) Organic phosphorus in the environment. CABI, Wallingford, pp 113–132

Chang CH, Yang SS (2009) Thermo-tolerant phosphate-solubilizing microbes for multi-functional biofertilizer preparation. Bioresour Technol 100:1648–1658

Chapuis-Lardy L, Le Bayon R-C, Brossard M, López-Hernández D, Blanchart E (2011) Role of soil macrofauna in phosphorus cycling. In: Bünemann E, Oberson A, Frossard E (eds) Phosphorus in action: biological processes in soil phosphorus cycling. Soil biology, vol 26. Springer, Heidelberg. doi:10.1007/978-3-642-15271-9_8

Chauhan BS, Stewart JWB, Paul EA (1979) Effect of carbon additions on soil labile inorganic, organic and microbially held phosphate. Can J Soil Sci 59:387–396

Chauhan BS, Stewart JWB, Paul EA (1981) Effect of labile inorganic-phosphate status and organic-carbon additions on the microbial uptake of phosphorus in soils. Can J Soil Sci 61:373–385

Chen CR, Condron LM, Davis MR, Sherlock RR (2003) Seasonal changes in soil phosphorus and associated microbial properties under adjacent grassland and forest in New Zealand. For Ecol Manage 117:539–557

Chen ZX, Ma S, Liu L (2008) Studies on phosphorus solubilizing activity of a strain of phosphobacteria isolated from chestnut type soil in China. Bioresour Technol 99:6702–6707

Clarholm M (1993) Microbial biomass P, labile P and acid-phosphatase-activity in the humus layer of a spruce forest, after repeated additions of fertilizers. Biol Fertil Soils 16:287–292

Cole CV, Elliot ET, Hunt HW, Coleman DC (1978) Trophic interactions in soil as they affect energy and nutrient dynamics. V. Phosphorus transformations. Microb Ecol 4:381–387

Cordell D, Drangert JO, White S (2009) The story of phosphorus: global food security and food for thought. Glob Environ Change 19:292–305

Criquet S, Tagger S, Vogt G, Le Petit J (2002) Endoglucanase and β-glycosidase activities in an evergreen oak litter: annual variation and regulating factors. Soil Biol Biochem 34:1111–1120

Criquet S, Ferre E, Farner EM, Le Petit J (2004) Annual dynamics of phosphatase activities in an evergreen oak litter – influence of biotic and abiotic factors. Soil Biol Biochem 36:1111–1118

Crowley DE (2007) Microbial siderophores in the plant rhizosphere. In: Barton LL, Abadia J (eds) Iron nutrition in plants and rhizospheric microorganisms. Springer, Dordrecht, pp 169–198

Dadhich SK, Somani LL, Verma A (2006) Improved soybean yield, nutrient uptake and P enrichment in soil due to co-inoculation of phosphate solubilizing bacteria and VAM fungi in a clay loam soil. Indian J Microbiol 46:405–407

Dalal RC (1977) Soil organic phosphorus. Adv Agron 29:83–117

Danhorn T, Fuqua C (2003) The effect of phosphorus concentration on the adherence of *Agrobacterium tumefaciens* to surfaces. In: Abstracts of the general meeting of the American Society for Microbiology 103. pp I–100

Das AC, Debnath A (2006) Effect of systemic herbicides on N_2-fixing and phosphate solubilizing microorganisms in relation to availability of nitrogen and phosphorus in paddy soils of West Bengal. Chemosphere 65:1082–1086

deFreitas JR, Banerjee MR, Germida JJ (1997) Phosphate-solubilizing rhizobacteria enhance the growth and yield but not phosphorus uptake of canola (*Brassica napus* L). Biol Fertil Soils 24:358–364

Denton MD, Reeve WG, Howieson JG, Coventry DR (2003) Competitive abilities of common field isolates and a commercial strain of *Rhizobium leguminosarum* bv. *trifolii* for clover nodule occupancy. Soil Biol Biochem 35:1039–1048

Dephilippis R, Sili C, Tassinato G, Vincenzini M, Materassi R (1991) Effects of growth-conditions on exopolysaccharide production by *Cyanospira capsulata*. Bioresour Technol 38:101–104

Dephilippis R, Margheri MC, Pelosi E, Ventura S (1993) Exopolysaccharide production by a unicellular cyanobacterium isolated from a hypersaline habitat. J Appl Phycol 5:387–394

Dwivedi BS, Singh VK, Dwivedi V (2004) Application of phosphate rock, with or without *Aspergillus awamori* inoculation, to meet phosphorus demands of rice-wheat systems in the Indo-Gangetic plains of India. Aust J Exp Agric 44:1041–1050

Eivazi F, Tabatabai MA (1977) Phosphatases in soils. Soil Biol Biochem 9:167–172

Elkoca E, Kantar F, Sahin F (2008) Influence of nitrogen fixing and phosphorus solubilizing bacteria on the nodulation, plant growth, and yield of chickpea. J Plant Nutr 31:157–171

El-Tarabily KA, Nassar AH, Sivasithamparam K (2008) Promotion of growth of bean (*Phaseolus vulgaris* L.) in a calcareous soil by a phosphate-solubilizing, rhizosphere-competent isolate of *Micromonospora endolithica*. Appl Soil Ecol 39:161–171

Fabre B, Armau E, Etienne G, Legendre F, Tiraby G (1988) A simple screening method for insecticidal substances from actinomycetes. J Antibiot 41:212–219

Fernández LA, Zalba P, Gomez MA, Sagardoy MA (2007) Phosphate-solubilization activity of bacterial strains in soil and their effect on soybean growth under greenhouse conditions. Biol Fertil Soils 43:805–809

Filius JD, Hiemstra T, Van Riemsdijk WH (1997) Adsorption of small weak organic acids on goethite: modeling of mechanisms. J Colloid Interface Sci 195:368–380

Foster RC, Dormaar JF (1991) Bacteria-grazing amebas in situ in the rhizosphere. Biol Fertil Soils 11:83–87

Foth HD, Ellis BG (1997) Soil fertility. CRC, Boca Raton

Frossard E, Skrabal P, Sinaj S, Bangerter F, Traore O (2002) Forms and exchangeability of inorganic phosphate in composted solid organic wastes. Nutr Cycl Agroecosys 62:103–113

Frossard E, Achat DL, Bernasconi SM, Bünemann EK, Fardeau J-C, Jansa J, Morel C, Rabeharisoa L, Randriamanantsoa L, Sinaj S, Tamburini F, Oberson A (2011) The use of tracers to investigate phosphate cycling in soil–plant systems. In: Bünemann E, Oberson A, Frossard E (eds) Phosphorus in action: biological processes in soil phosphorus cycling. Soil biology, vol 26. Springer, Heidelberg. doi:10.1007/978-3-642-15271-9_3

Gächter R, Meyer JS (1993) The role of microorganisms in mobilization and fixation in sediments. Hydrobiologia 253:103–121

Garcia-Gil JC, Plaza C, Soler-Rovira P, Polo A (2000) Long-term effects of municipal solid waste compost application on soil enzyme activities and microbial biomass. Soil Biol Biochem 32:1907–1913

Gaume A, Weidler PG, Frossard E (2000) Effect of maize root mucilage on phosphate adsorption and exchangeability on a synthetic ferrihydrite. Biol Fertil Soils 31:525–532

George TS, Gregory PJ, Robinson JS, Buresh RJ (2002) Changes in phosphorus concentrations and pH in the rhizosphere of some agroforestry and crop species. Plant Soil 246:65–73

Gerke J, Beißner L, Römer W (2000) The quantitative effect of chemical phosphate mobilization by carboxylate anions on P uptake by a single root. I. The basic concept and determination of soil parameters. J Plant Nutr Soil Sci 163:207–212

Giller KE, Witter E, McGrath SP (1998) Toxicity of heavy metals to microorganisms and microbial processes in agricultural soils: a review. Soil Biol Biochem 30:1389–1414

Gonzalez-Chavez MDA, Newsam R, Linderman R, Dodd J, Valdez-Carrasco JM (2008) Bacteria associated with the extraradical mycelium of an arbuscular mycorrhizal fungus in an As/Cu polluted soil. Agrociencia 42:1–10

Govindan K, Thirumurugan V (2005) Synergistic association of *Rhizobium* with phosphate-solubilizing bacteria under different sources of nutrient supply on productivity and soil fertility in soybean (*Glycine max*). Indian J Agron 50:214–217

Grierson PF, Comerford NB, Jokela EJ (1998) Phosphorus mineralization kinetics and response of microbial phosphorus to drying and rewetting in a Florida Spododol. Soil Biol Biochem 30:1323–1331

Gunes A, Ataoglu N, Turan M, Esitken A, Ketterings QM (2009) Effects of phosphate-solubilizing microorganisms on strawberry yield and nutrient concentrations. J Plant Nutr Soil Sci 172:385–392

Gyaneshwar P, Parekh LJ, Archana G, Poole PS, Collins MD, Huston RA, Naresh Kumar G (1999) Involvement of a phosphate starvation inducible glucose dehydrogenase in soil phosphate solubilization by *Enterobacter asburiae*. FEMS Microbiol Lett 171:223–229

Gyaneshwar P, Naresh Kumar G, Parekh LJ, Poole PS (2002) Role of soil microorganisms in improving P nutrition of plants. Plant Soil 245:83–93

Hamdali H, Hafidi M, Virolle MJ, Ouhdouch Y (2008a) Growth promotion and protection against damping-off of wheat by two rock phosphate solubilizing actinomycetes in a P-deficient soil under greenhouse conditions. Appl Soil Ecol 40:510–517

Hamdali H, Hafidi M, Virolle MJ, Ouhdouch Y (2008b) Rock phosphate-solubilizing Actinomy-cetes: screening for plant growth-promoting activities. World J Microbiol Biotechnol 24: 2565–2575

Hamdali H, Bouizgarne B, Hafidi M, Lebrihi A, Virolle MJ, Ouhdouch Y (2008c) Screening for rock phosphate solubilizing Actinomycetes from Moroccan phosphate mines. Appl Soil Ecol 38:12–19

Hamdali H, Smirnov A, Esnault C, Ouhdouch Y, Virolle MJ (2010) Physiological studies and comparative analysis of rock phosphate solubilization abilities of Actinomycetales originating from Moroccan phosphate mines and of *Streptomyces lividans*. Appl Soil Ecol 44:24–31

Hariprasad P, Navya HM, Nayaka SC, Niranjana SR (2009) Advantage of using PSIRB over PSRB and IRB to improve plant health of tomato. Biol Control 50:307–316

Harris WG (2002) Phosphate minerals. In: Dixon JB, Schulze DG (eds) Soil mineralogy with environmental applications, SSSA book series 7. SSSA, Madison, pp 637–665

Harrison AF (1987) Soil organic phosphorus: a review of world literature. CABI, Wallingford

Hinsinger P (2001) Bioavailability of soil inorganic P in the rhizosphere as affected by root-induced chemical changes: a review. Plant Soil 237:173–195

Hoberg E, Marschner P, Lieberei R (2005) Organic acid exudation and pH changes by *Gordonia* sp and *Pseudomonas fluorescens* grown with P adsorbed to goethite. Microbiol Res 160: 177–187

Hu J, Lin X, Wang J, Chu H, Yin R, Zhang J (2009) Population size and specific potential of P-mineralizing and -solubilizing bacteria under long-term P-deficiency fertilization in a sandy loam soil. Pedobiologia 53:49–58

Huang Q, Liang W, Cai P (2005) Adsorption, desorption and activities of acid phosphatase on various colloidal particles from an Ultisol. Colloid Surface B 45:209–241

Huang XL, Chen Y, Shenker M (2008) Chemical fractionation of phosphorus in stabilized biosolids. J Environ Qual 37:1949–1958

Illmer P, Schinner F (1992) Solubilization of inorganic phosphates by microorganisms isolated from forest soils. Soil Biol Biochem 24:389–395

Illmer P, Barbato A, Schinner F (1995) Solubilization of hardly-soluble $AlPO_4$ with P-solubilizing microorganisms. Soil Biol Biochem 27:265–270

Intorne AC, de Oliveira MVV, Lima ML, da Silva JF, Olivares FL, de Souza GA (2009) **Identification and characterization of** *Gluconacetobacter diazotrophicus* mutants defective in the solubilization of phosphorus and zinc. Arch Microbiol 191:477–483

Ionescu M, Belkin S (2009) Overproduction of exopolysaccharides by an *Escherichia coli* K-12 rpoS mutant in response to osmotic stress. Appl Environ Microbiol 75:483–492

Jansa J, Finlay R, Wallander H, Smith FA, Smith SE (2011) Role of mycorrhizal symbioses in phosphorus cycling. In: Bünemann E, Oberson A, Frossard E (eds) Phosphorus in action: biological processes in soil phosphorus cycling. Soil biology, vol 26. Springer, Heidelberg. doi:10.1007/978-3-642-15271-9_6

Jilani G, Akram A, Ali RM, Hafeez FY, Shamsi IH, Chaudhry AN, Chaudhry AG (2007) **Enhancing crop growth, nutrients availability, economics and beneficial rhizosphere micro-flora through organic and biofertilizers. Ann Microbiol 57:177–184**

Joergensen RG, Scheu S (1999) Response of soil microorganisms to the addition of carbon, nitrogen and phosphorus in a forest Rendzina. Soil Biol Biochem 31:859–866

Jones DL (1998) Organic acids in the rhizosphere – a critical review. Plant Soil 205:25–44

Jones DL, Dennis PG, Owen AG, van Hees PAW (2003) Organic acid behaviour in soils misconceptions and knowledge gaps. Plant Soil 248:31–41

Jorquera MA, Hernandez MT, Rengel Z, Marschner P, Mora MD (2008) Isolation of culturable phosphobacteria with both phytate-mineralization and phosphate-solubilization activity from the rhizosphere of plants grown in a volcanic soil. Biol Fertil Soils 44:1025–1034

Juma NG, Tabatabai MA (1977) Effects of trace-elements on phosphatase-activity in soils. Soil Sci Soc Am J 41:343–346

Juma NG, Tabatabai MA (1988) Hydrolysis of organic phosphates by corn and soybean roots. Plant Soil 107:31–38

Kao PH, Huang CC, Hseu ZY (2006) Response of microbial activities to heavy metals in a neutral loamy soil treated with biosolid. Chemosphere 64:63–70

Khan MR, Khan SM (2001) Biomanagement of Fusarium wilt of tomato by the soil application of certain phosphate-solubilizing microorganisms. Int J Pest Manage 47:227–231

Khan MS, Zaidi A (2007) Synergistic effects of the inoculation with plant growth-promoting rhizobacteria and an arbuscular mycorrhizal fungus on the performance of wheat. Turk J Agric For 31:355–362

Kim KY, Jordan D, McDonald GA (1998) *Enterobacter agglomerans*, phosphate solubilizing bacteria, and microbial activity in soil: effect of carbon sources. Soil Biol Biochem 30: 995–1003

Kucey RMN, Jenzen HH, Leggett ME (1989) Microbially mediated increases in plant available phosphorus. Adv Agron 42:199–228

Lopez-Hernandez D, Lavelle P, Fardeau JC, Nino M (1993) Phosphorus transformations in two P-sorption contrasting tropical soils during transit through *Pontoscolex corethrurus* (Glossoscolecidae, Oligochaeta). Soil Biol Biochem 25:789–792

Lopez-Hernandez D, Bossard M, Frossard E (1998) P-isotopic exchange values in relation to Po mineralization in soils with very low P-sorbing capacities. Soil Biol Biochem 30:1663–1670

Lugtenberg B, Kamilova F (2009) Plant-growth-promoting rhizobacteria. Annu Rev Microbiol 63:541–556

Maguire RO, Hesterberg D, Gernat A, Anderson K, Wineland M, Grimes J (2006) Liming poultry manures to decrease soluble phosphorus and suppress the bacteria population. J Environ Qual 35:849–857

Malboobi MA, Behbahani M, Madani H, Owlia P, Deljou A, Yakhchali B, Moradi M, Hassanabadi H (2009) Performance evaluation of potent phosphate solubilizing bacteria in potato rhizosphere. World J Microbiol Biotechnol 25:1479–1484

Mansfeld-Giese K, Larsen J, Bodker L (2002) Bacterial populations associated with mycelium of the arbuscular mycorrhizal fungus *Glomus intraradices*. FEMS Microbiol Ecol 41:133–140

Martell AE, Smith RM (1977) Critical stability constants. Other organic ligands, vol 3. Plenum, New York

Martens DA, Johanson JB, Frankenberger WT (1992) Production and persistence of soil enzymes with repeated addition of organic residues. Soil Sci 153:53–61

Mba CC (1994) Field studies on two rock phosphate solubilizing Actinomycete isolates as biofertilizer sources. Environ Manage 18:263–269

Mba CC (1997) Rock phosphate solubilizing *Streptosporangium* isolates from casts of tropical earthworms. Soil Biol Biochem 29:381–385

McGill WB, Cole CV (1981) Comparative aspects of cycling of organic C, N, S and P through soil organic matter. Geoderma 26:267–268

Nannipieri P, Giagnoni L, Landi L, Renella G (2011) Role of phosphatase enzymes in soil. In: Bünemann E, Oberson A, Frossard E (eds) Phosphorus in action: biological processes in soil phosphorus cycling. Soil biology, vol 26. Springer, Heidelberg. doi:10.1007/978-3-642-15271-9_9

Oberson A, Joner EJ (2005) Microbial turnover of phosphorus in soil. In: Turner BL, Frossard E, Baldwin DS (eds) Organic phosphorus in the environment. CABI, Wallingford, pp 133–164

Oberson A, Pypers P, Bünemann EK, Frossard E (2011) Management impacts on biological phosphorus cycling in cropped soils. In: Bünemann E, Oberson A, Frossard E (eds) Phosphorus in action: biological processes in soil phosphorus cycling. Soil biology, vol 26. Springer, Heidelberg. doi:10.1007/978-3-642-15271-9_17

Oburger E, Jones DL (2009) Substrate mineralization studies in the laboratory show different microbial C partitioning dynamics than in the field. Soil Biol Biochem 41:1951–1956

Oburger E, Kirk GJD, Wenzel WW, Puschenreiter M, Jones DL (2009) Interactive effects of organic acids in the rhizosphere. Soil Biol Biochem 41:449–457

Ochoa-Loza FJ, Artiola JF, Maier RM (2001) Stability constants for the complexation of various metals with a rhamnolipid biosurfactant. J Environ Qual 30:479–485

Odongo NE, Hyoung-Ho K, Choi HC, van Straaten P, McBride BW, Romney DL (2007) Improving rock phosphate availability through feeding, mixing and processing with composting manure. Bioresour Technol 98:2911–2918

Oehl F, Oberson A, Probst M, Fliessbach A, Roth HR, Frossard E (2001a) Kinetics of microbial phosphorus uptake in cultivated soils. Biol Fertil Soils 34:31–41

Oehl F, Oberson A, Sinaj S, Frossard E (2001b) Organic phosphorus mineralization studies using isotopic dilution techniques. Soil Sci Soc Am J 65:780–787

Okon Y, Labandera-Gonzalez CA (1994) (1994) Agronomic applications of *Azospirillum* – an evaluation of 20 years worldwide field inoculation. Soil Biol Biochem 26:1591–1601

Olander LP, Vitousek PM (2000) Regulation of soil phosphatase and chitinase activity by N and P availability. Biogeochemistry 49:175–190

Oliveira CA, Alves VMC, Marriel IE, Gomes EA, Scotti MR, Carneiro NP, Guimaraes CT, Schaffert RE, Sa NMH (2009) Phosphate solubilizing microorganisms isolated from rhizosphere of maize cultivated in an oxisol of the Brazilian Cerrado Biome. Soil Biol Biochem 41:1782–1787

Otani T, Ae N (1999) Extraction of organic phosphorus in Andosols by various methods. Soil Sci Plant Nutr 45:151–161

Park KH, Lee CY, Son HJ (2009) Mechanism of insoluble phosphate solubilization by *Pseudomonas fluorescens* RAF15 isolated from ginseng rhizosphere and its plant growth-promoting activities. Lett Appl Microbiol 49:222–228

Parker DR, Reichmann SM, Crowley DE (2005) Metal chelation in the rhizosphere. In: Zobel RW (ed) Roots and soil management: interactions between roots and the soil. Agronomy monograph no. 48. American Society of Agronomy, Madison, pp 57–93

Patel DK, Archana G, Naresh Kumar G (2008) Variation in the nature of organic acid secretion and mineral phosphate solubilization by *Citrobacter* sp. DHRSS in the presence of different sugars. Curr Microbiol 65:168–174

Pedersen AL, Nybroe O, Winding A, Ekelund F, Bjornlund L (2009) Bacterial feeders, the nematode *Caenorhabditis elegans* and the flagellate *Cercomonas longicauda*, have different effects on outcome of competition among the *Pseudomonas* biocontrol strains CHA0 and DSS73. Microb Ecol 57:501–509

Quiquampoix H, Mousain D (2005) Enzymatic hydrolysis of organic phosphorus. In: Turner BL, Frossard E, Baldwin DS (eds) Organic phosphorus in the environment. CABI, Wallingford, pp 89–112

Reid RK, Reid CPP, Szaniszlo PJ (1985) Effects of synthetic and microbially produced chelates on the diffusion of iron and phosphorus to a simulated root in soil. Biol Fertil Soils 1:45–52

Renella G, Egamberdiyeva D, Landi L, Mench M, Nannipieri P (2006) Microbial activity and hydrolase activities during decomposition of root exudates released by an artificial root surface in Cd-contaminated soils. Soil Biol Biochem 38:702–708

Reyes I, Bernier L, Simard RR, Antoun H (1999) Effect of nitrogen source on the solubilization of different inorganic phosphates by an isolate of *Penicillium rugulosum* and two UV-induced mutants. FEMS Microbiol Ecol 28:281–290

Richardson AE, Hocking PJ, Simpson RJ, George TS (2009a) Plant mechanisms to optimise access to soil phosphorus. Crop Pasture Sci 60:124–143

Richardson AE, Barea JM, McNeill AM, Prigent-Combaret C (2009b) Acquisition of phosphorus and nitrogen in the rhizosphere and plant growth promotion by microorganisms. Plant Soil 321:305–339

Rodriguez H, Fraga R (1999) Phosphate solubilizing bacteria and their role in plant growth promotion. Biotechnol Adv 17:319–339

Rojo MJ, Carcedo SG, Mateos MP (1990) Distribution and characterization of phosphatase and organic phosphorus in soil fractions. Soil Biol Biochem 22:169–174

Roos W, Luckner M (1984) Relationships between proton extrusion and fluxes of ammonium ions and organic acids in *Penicillium cyclopium*. J Gen Microbiol 130:1007–1014

Rosenberg K, Bertaux J, Krome K, Hartmann A, Scheu S, Bonkowski M (2009) Soil amoebae rapidly change bacterial community composition in the rhizosphere of *Arabidopsis thaliana*. ISME J 3:675–684

Rudresh DL, Shivaprakash MK, Prasad RD (2005) Tricalcium phosphate solubilizing abilities of *Trichoderma* spp. in relation to P uptake and growth and yield parameters of chickpea (*Cicer arietinum* L.). Can J Microbiol 51:217–222

Rugheim AME, Abdelgani ME (2009) Substituting chemical fertilizers with microbial fertilizers for increasing productivity of Faba bean (*Vicia faba* L.) in arid lands. International Symposium on Environmental Science and Technology, June 2005, Shanghai, China

Saha S, Prakash V, Kundu S, Kumar N, Mina BL (2008) Soil enzymatic activity as affected by long term application of farm yard manure and mineral fertilizer under a rainfed soybean-wheat system in N-W Himalaya. Eur J Soil Biol 44:309–315

Sahin F, Cakmakci R, Kantar F (2004) Sugar beet and barley yields in relation to inoculation with N_2-fixing and phosphate solubilizing bacteria. Plant Soil 265:123–129

Sahu MK, Sivakumar K, Thangaradjou T, Kannan L (2007) Phosphate solubilizing Actinomycetes in the estuarine environment: an inventory. J Environ Biol 28:795–798

Sammauria R, Yadav RS, Nagar KC (2009) Performance of cluster bean (*Cyamopsis tetragonoloba*) as influenced by nitrogen and phosphorus fertilization and biofertilizers in Western Rajasthan. Indian J Agron 54:319–323

Sample EC, Soper RJ, Racz GJ (1980) Reaction of phosphate fertilizers in soils. In: Khasawneh FE, Sample AC, Kamprath EJ (eds) The role of phosphorus in agriculture. ASA–CSSA–SSSA, Madison, pp 263–310

Schwöppe C, Winkler HH, Neuhaus HE (2003) Connection of transport and sensing by UhpC, the sensor for external glucose-6-phosphate in *Escherichia coli*. Eur J Biochem 270:1450–1457

Sharan A, Shikha, Darmwal NS (2008) Efficient phosphorus solubilization by mutant strain of *Xanthomonas campestris* using different carbon, nitrogen and phosphorus sources. World J Microbiol Biotechnol 24:3087–3090

Sharma SN, Prasad R (2003) Yield and P uptake by rice and wheat grown in a sequence as influenced by phosphate fertilization with diammonium phosphate and Mussoorie rock phosphate with or without crop residues and phosphate solubilizing bacteria. J Agric Sci 141: 359–369

Sharma SN, Prasad R, Shivay YS, Dwivedi MK, Kumar S, Kumar D (2009) Effect of rates and sources of phosphorus on productivity and economics of rice (*Oryza sativa*) as influenced by crop-residue incorporation. Indian J Agron 54:42–46

Shober AL, Sims JT (2009) Evaluating phosphorus release from biosolids and manure-amended soils under anoxic conditions. J Environ Qual 38:309–318

Sims JT, Pierzynski GM (2005) Chemistry of phosphorus in soil. In: Tabatabai AM, Sparks DL (eds) Chemical processes in soil, SSSA book series 8. SSSA, Madison, pp 151–192

Souchie EL, Abboud ACD (2007) Phosphate solubilization by microorganisms from the rhizosphere of Pigeonpea genotypes grown in different soil classes. Semina Cienc Agrar 28:11–18

Sreenivas C, Narayanasamy G (2009) Role of earthworm (*Eisenia fetida*) and phosphate solubilising microorganism (*Aspergillus awamori*) in vermi-phosphocomposting. Res Crops 10:293–300

Streeter JG (1994) Failure of inoculant rhizobia to overcome the dominance of indigenous strains for nodule formation. Can J Microbiol 40:513–522

Ström L, Owen AG, Godbold DL, Jones DL (2005) Organic acid behaviour in a calcareous soil implications for rhizosphere nutrient cycling. Soil Biol Biochem 37:2046–2054

Sud KC, Jatav MK (2007) Response of potato to phosphorus and phosphorus solubilizing bacteria in brown hill soils of Shimla. Potato J 34:109–110

Sutaliya R, Singh RN (2005) Effect of planting time, fertility level and phosphate-solubilizing bacteria on growth, yield and yield attributes of winter maize (*Zea mays*) under rice (*Oryza sativa*)-maize cropping system. Indian J Agron 50:173–175

Tabatabai MA (1994) Soil enzymes. In: Weaver RW (ed) Methods of soil analysis. Part 2. Microbial and biochemical properties. SSSA book series 5. SSSA, Madison

Takeda M, Nakamoto T, Miyazawa K, Murayama T, Okada H (2009) Phosphorus availability and soil biological activity in an Andosol under compost application and winter cover cropping. Appl Soil Ecol 42:86–95

Tang J, Leung A, Leung C, Lim BL (2006) Hydrolysis of precipitated phytate by three distinct families of phytases. Soil Biol Biochem 38:1316–1324

Tarafdar JC, Yadav RS, Meena SC (2001) Comparative efficiency of acid phosphatase originated from plant and fungal sources. J Plant Nutr Soil Sci 164:279–282

Tate KR, Salcedo I (1988) Phosphorus control of soil organic matter accumulation and cycling. Biogeochemistry 5:99–107

Tilman D, Fargione J, Wolff B, D'Antonio C, Dobson A, Howarth R, Schindler D, Schlesinger WH, Simberloff D, Swackhamer D (2001) Forecasting agriculturally driven global environmental change. Science 292:281–284

Turner BL, McKelvie ID, Haygarth PM (2002) Characterisation of water-extractable soil organic phosphorus by phosphatase hydrolysis. Soil Biol Biochem 34:27–35

Turner B, Driessen J, Haygarth P, McKelvie I (2003) Potential contribution of lysed bacterial cells to phosphorous solubilisation in two rewetted Australian pasture soils. Soil Biol Biochem 35:187–189

Van Hees PAW, Vinogradoff SI, Edwards AC, Godbold DL, Jones DL (2003) Low molecular weight organic acid adsorption in forest soils: effects on solution concentration and biodegradation rates. Soil Biol Biochem 35:1025–1026

Vassilev N, Vassileva M, Niklaeva I (2006) Simultaneous P-solubilizing and biocontrol activity of microorganisms: potential and future trends. Appl Microbiol Biotechnol 71:137–144

Vessey JK (2003) Plant growth promoting rhizobacteria as biofertilizers. Plant Soil 255:571–586

Wakelin SA, Gupta VVSR, Harvey PR, Ryder MH (2007) The effect of Penicillium fungi on plant growth and phosphorus mobilization in neutral to alkaline soils from southern Australia. Can J Microbiol 53:106–115

Walker TW, Syers JK (1976) Fate of phosphorus during pedogenesis. Geoderma 15:1–19

Wan JHC, Wong MH (2004) Effects of earthworm activity and P-solubilizing bacteria on P availability in soil. J Plant Nutr Soil Sci 167:209–213

Whitelaw MA, Harden TJ, Bender GL (1997) Plant growth promotion of wheat inoculated with *Penicillium radicum* sp. Nov. Aust J Soil Res 35:291–300

Whitelaw MA, Harden TJ, Helyar KR (1999) Phosphate solubilization in solution culture by the soil fungus *Penicillium radicum*. Soil Biol Biochem 31:655–665

Wielbo J, Skorupska A (2008) Influence of phosphate and ammonia on the growth, exopolysaccharide production and symbiosis of *Rhizobium leguminosarum* bv. *trifolii* TA1 with clover (*Trifolium pratense*). Acta Biol Hung 59:115–127

Wu SC, Cao ZH, Li ZG, Cheung KC, Wong MH (2005) Effects of biofertilizer containing N-fixer, P and K solubilizers and AM fungi on maize growth: a greenhouse trial. Geoderma 125: 155–166

Xiao C, Chi R, He H, Qiu G, Wang D, Zhang W (2009) Isolation of phosphate-solubilizing fungi from phosphate mines and their effect on wheat seedling growth. Appl Biochem Biotechnol 159:330–342

Yadav BK, Tarafdar JC (2007) Ability of *Emericella rugulosa* to mobilize unavailable P compounds during Pearl millet [*Pennisetum glaucum* (L.) R. Br.] crop under arid condition. Indian J Microbiol 47:57–63

Yi YM, Huang WY, Ge Y (2008) Exopolysaccharide: a novel important factor in the microbial dissolution of tricalcium phosphate. World J Microbiol Biotechnol 24:1059–1065

Zhang X, Amelung WYY, Zech W (1998) Amino sugar signature of particle-size fractions in soils of the native prairie as affected by climate. Soil Sci 163:220–229

Chapter 8
Role of Soil Macrofauna in Phosphorus Cycling

Lydie Chapuis-Lardy, Renée-Claire Le Bayon, Michel Brossard, Danilo López-Hernández and Eric Blanchart

8.1 Introduction

Soils play a major role in the providing of goods and services by ecosystems to humans. They are involved in the main biogeochemical cycles (H_2O, C, N, P). They support agriculture, forestry, and pasture systems, and participate in climate regulation and detoxification. The importance of soils for ecosystem goods and services is based on their biological functioning, i.e., the whole biological functions carried out by soil biota in interactions with physical and chemical components of soil. These functions ensure, e.g., soil organic matter dynamics, nutrient recycling, soil structure and water retention. They are carried out by organisms of different sizes (from bacteria to macrofauna) and functional roles (decomposers, microbial regulators, soil engineers, and predators) (Lavelle et al. 2006).

Invertebrates of the macrofauna are key species for soil functioning. They participate in litter decomposition, mix organic and mineral matter, create and maintain soil structure by digging burrows and modifying aggregation, regulate microbial diversity and activity, and protect plants against pests and diseases (Lavelle et al. 2006). Following Jones et al. (1994), ecosystem engineers are

L. Chapuis-Lardy (✉)
Institut de Recherche pour le Développement (IRD), UMR 210 Eco&Sols, BP 1386, CP 18524, Dakar, Sénégal
e-mail: lydie.lardy@ird.fr

R.-C. Le Bayon
Soil and Vegetation Laboratory, Institute of Biology, University of Neuchâtel, Rue Emile-Argand 11, CP 158, 2009 Neuchâtel, Switzerland

M. Brossard and E. Blanchart
Institut de Recherche pour le Développement (IRD), UMR 210 Eco&Sols, Campus SupAgro, 2 place Viala, 34 060 Montpellier CEDEX 2, France

D. López-Hernández
Laboratorio de Estudios Ambientales, Instituto de Zoología y Ecología Tropical, Facultad de Ciencias, Universidad Central de Venezuela, Caracas, Venezuela

E.K. Bünemann et al. (eds.), *Phosphorus in Action*, Soil Biology 26,
DOI 10.1007/978-3-642-15271-9_8,

organisms able to modify the physical environment and availability of nutrients for other organisms. Lavelle (1997) defined the organisms fulfilling these functions in soil as soil engineers. They develop very complex interactions with other soil biota. For instance, the gut and biogenic structures (burrows, casts, nests) of earthworms and termites are specific habitats in which soil microbial activities are either stimulated or attenuated. Freshly egested earthworm casts are generally characterized by an intense mineralization of organic matter and the release of nutrients for plants. Conversely, the mineralization in old casts is reduced, which allows for long-term carbon and nutrient storage in soil (Martin 1991).

It is now recognized that soil engineers can improve soil fertility due to their role in nutrient dynamics at different spatial and temporal scales (review in Brown et al. 1999). They have been shown to affect the availability of nitrogen (N) and phosphorus (P), the main growth-limiting nutrients (e.g., Lee and Wood 1971; Bignell and Eggleton 2000; Brown et al. 2000; Brossard et al. 2007; Le Bayon and Milleret 2009). Numerous studies showed an increase in plant growth in the presence of earthworms, although the mechanisms are not fully understood; one of them being an increase in nutrient availability (Brown et al. 1999).

The objective of this chapter is to highlight, in a non-exhaustive view of published data, the role of soil macrofauna on P cycling. We will focus on the main soil engineers, i.e., earthworms and termites, whose roles in soil functioning have been frequently studied.

8.2 Earthworms

8.2.1 Phosphorus Contents and Forms in Earthworm Biogenic Structures

The first scientific studies on earthworm ecology, in the 1940s, related earthworm activity and soil fertility. Earthworms may affect soil structure and processes in different ways, according to their feeding and burrowing behaviors (synthesis in Brown et al. 2000). Three broad functional groups of earthworms have been described by Bouché (1977): epigeic, endogeic, and anecic. Epigeic earthworms live in the litter layer, consume plant litter and rarely ingest soil. Endogeics are mostly soil organic matter feeders and burrow extensively both horizontally and vertically within the soil. Anecics feed on particulate organic matter mixed with soil particles and often form deep, primarily vertical, burrows in which they bury surface litter. Numerous studies report results on chemical or physicochemical properties of earthworm casts, i.e., the by-products of gut passage, in comparison with surrounding soil. Most of these studies show that casts are characterized by higher P content, especially water-soluble or available (Bray- or Truog-P) P content (e.g., Nijhawan and Kanwar 1952; Bates 1960; Gupta and Sakal 1967; Graff 1970; Sharpley and Syers 1976; review in Edwards 1981; de Vleeschauwer and Lal 1981, review in Lal 1987; Mulongoy and Bedoret 1989; López-Hernández et al. 1993). More recently, Kuczak et al. (2006) studied different P fractions (available P, moderately available P, and resistant P) in earthworm casts and soil in Amazonia. They observed that casts were characterized by higher total P and percentage of labile P pools (available and moderately available P). Guggenberger et al. (1996) found higher levels of alkali-extractable organic P in earthworm casts than in the surrounding oxisol. At this same site, Jiménez et al. (2003) worked in the field and analyzed inorganic and organic P fractions, including microbial P, in casts and surrounding soil. Under field conditions where earthworms ingest soil plus organic residues, total P concentrations were higher in casts than in surrounding soil. This increase was distributed over all inorganic and organic P fractions, including increased microbial P.

Because castings and earthworm-processed soils have the potential for higher P availabilities than uningested soils, the role of earthworm activities over the long term in soil P dynamics and the transfer of P into growing plants still needs consideration, especially in low-P status soils (e.g., Brown et al. 1999; Patrón et al. 1999; Chapuis-Lardy et al. 2009).

8.2.2 Phosphorus Dynamics and Availability in Earthworm Casts

Earthworm activity creates a series of geochemical and biological effects that are especially important for the cycling of P in soils (Brossard et al. 1996; Chapuis-Lardy et al. 1998, 2009; Nziguheba and Bünemann 2005; Le Bayon and Milleret 2009).

The availability of P is strongly influenced by the physical adsorption or fixation of P in soil (Frossard et al. 1995 and references therein). The phosphate ion (hereafter abbreviated as PO_4) concentration in the soil solution (C_P) may also decrease as a direct consequence of P removal by root uptake. This depletion gives rise to a replenishment of P from the solid phase, which is time- and C_P-dependent. This PO_4 exchange reflects the buffer power of the soil for PO_4 and varies with the composition and the physicochemical conditions of the soil. Isotopic exchange kinetics (IEK) utilizing ^{32}P as a tracer has been extensively used to specifically assess inorganic P availability in a variety of soils (e.g., Morel et al. 1994; Fardeau 1996; Sinaj et al. 2001). Such approaches (see details in Frossard et al. 2011) have been used to evaluate P availability in earthworm casts or in soil in the presence or absence of earthworms.

We briefly focus here on the case of *Pontoscolex corethrurus* (Glossoscolecidae, Oligocheta), a widespread tropical endogeic earthworm. López-Hernández et al. (1993) reported higher P availability in fresh casts derived from two tropical soils with contrasting P sorption capacity. Chapuis and Brossard (1995) found higher C_P and $E_{(1\ min)}$ (the amount of PO_4 that is isotopically exchanged within 1 min) values in casts egested by *P. corethrurus* fed with a Peruvian Ultisol. These changes in P availability were ascribed to (1) the higher pH of the gut content

(Barois and Lavelle 1986), (2) changes in sorption complexes induced by competition for sorbing sites between PO_4 and carboxyl groups of a mucus glycoprotein produced by the earthworm in its gut (López-Hernández et al. 1993), and (3) an increase in microbial activity during digestion (López-Hernández et al. 1993). More recently, IEK experiments were carried out to measure, describe, and compare the PO_4 exchange between liquid and solid phases of soil suspensions as a function of C_P and time, for two Malagasy soils and the associated casts egested by *P. corethrurus* (Chapuis-Lardy et al. 2009). The question arises whether the earthworm-induced modifications can change the ability of the soil solid phase to buffer C_P. The endogeic worm *P. corethrurus* increased the ability of PO_4 bound to the soil solid phase to exchange with PO_4 in soil solution (Fig. 8.1), because soil ingestion induced transformation of inorganic P to a more rapidly exchangeable form.

Tuffen et al. (2002) studied the transfer of ^{32}P in soil and between plants inoculated with arbuscular mycorrhizal fungi (AMF) in the presence of a temperate endogeic earthworm (*Aporrectodea caliginosa*). These authors reported enhanced ^{32}P transfer in soil and also between AMF-inoculated donor and receiver plants in the presence of earthworms.

Casts of earthworms usually contained larger amounts of organic P (e.g., for *P. corethrurus*, Chapuis-Lardy et al. 1998; Patrón et al. 1999), probably derived from a selective ingestion of soil particles (Chapuis-Lardy et al. 1998). Several

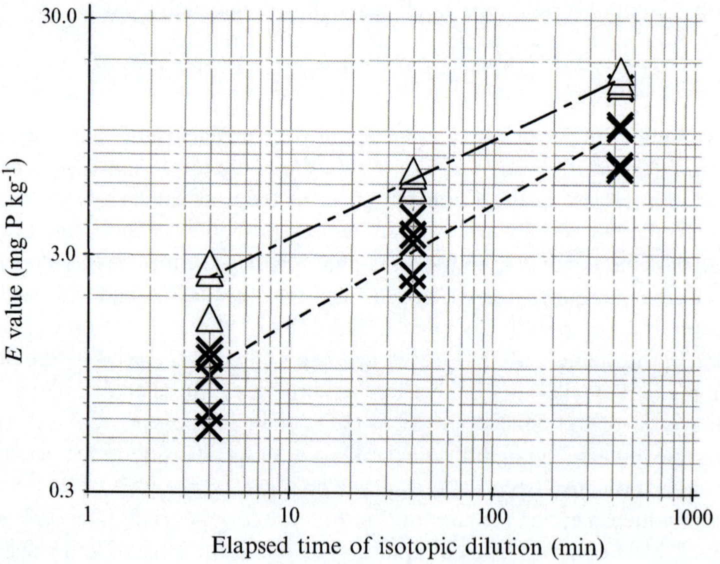

Fig. 8.1 Kinetics of isotopically exchangeable P ions (E, mg kg^{-1} soil) transferred between solid and liquid phases and the associated regression lines (E vs. time t) in control soil (from Andranomanelatra site) (*crosses, dashed line*) and casts of *Pontoscolex corethrurus* (*triangles, semi-dashed line*). The x- and y-axes are represented in log–log scales. Modified from Chapuis-Lardy et al. (2009)

studies reported increased enzymatic activity in earthworm casts and the stimulation of microbial activity (Sharpley and Syers 1976; James 1991; López-Hernández et al. 1993; Chapuis and Brossard 1995; Brossard et al. 1996; Le Bayon and Binet 2006). However, Zhang et al. (2000) observed an increase of inorganic P in soils in the presence of earthworms, despite a decrease of acid and alkaline phosphatase activities in earthworm casts, suggesting that P-containing organic matter is digested in the gut rather than in the casts. This confirms the results of Devliegher and Verstraete (1996), who showed that the increase of inorganic P in soils in the presence of earthworms was due to the production of alkaline phosphatases by earthworms (Satchell and Martin 1984; Park et al. 1990) and a stimulation of microbial (acid) phosphatase production by earthworm activity (Satchell and Martin 1984). Devliegher and Verstraete (1996) concluded for the anecic earthworm *Lumbricus terrestris* that the increase of organic P mineralization and inorganic P availability in the presence of earthworms is more of a gut-associated process than a cast-associated process. Overall, the higher P availability observed in soils in the presence of earthworms may be linked to a rapid turnover of an increased organic P content. However, depending of the dominant species, earthworm activity could also result in the protection of organic P in stable macroaggregates and the increased fixation of inorganic P in Fe and Al hydroxides (Scheu and Parkinson 1994a, b; Suarez et al. 2004). It is concluded that earthworms markedly change the biogeochemical status of P (availability, organic P pool, enzymatic activities) in their guts and in the soil where they are active (drilosphere, sensu Lavelle et al. 1997), including casts and burrow linings. However, their impact on P dynamics and availability in the soil depends on the particular properties of soil, the organic P source, and the specific burrowing behavior and food preferences of worms.

The assumptions on earthworm-mediated transformations of organic P and related effects on P cycling in soils could be efficiently verified using soil labeling with radioactive P, especially if performed before chemical extraction (see methodological aspects in Frossard et al. 2011).

8.2.3 Surface-Cast Erosion and Phosphorus Transfer

The earthworm ecological category, and especially the feeding behavior, determines the type of surface-casts egested (Lee 1985; Curry and Schmidt 2007) as well as the architecture of the earthworm-induced macropores (Jégou et al. 2000; Bastardie et al. 2003). Such biogenic structures are involved in the regulation of soil physicochemical processes. However, how earthworms affect soil erosion is poorly understood (Blanchart et al. 2004). Surface-casts may be impacted by raindrops, and the fine soil particles they contain easily detached and transported during rainfall events (Le Bayon et al. 2002; Mariani et al. 2007). Structural stability of the biogenic structures is, together with rainfall intensity and field slope, an important determinant of soil erosion. Hence, earthworms can be proposed as important moderators of soil erosion. Where stability is low and rainfall intense, material from the biogenic structures could be transported downstream by water runoff, thus leading to a significant loss of soil particles and associated P. Several studies have been performed to better understand the processes involved in surface-cast erosion and P transfer. Nevertheless, comparison between these investigations are difficult because of very variable experimental conditions (in situ annual monitoring or simulated rainfall events), different chemical analyses of P (colorimetric methods, anion exchange resins, sequential extractions, loss by ignition), and different expression of the results (actual measurements or estimations).

Sharpley and Syers (1976, 1977) and Sharpley et al. (1979) were the first to report a contribution of earthworm casts to P enrichment in runoff waters. They estimated under a temperate permanent pasture in New Zealand that surface-casts of the endogeic *A. caliginosa* may account for 45% of the annual transfer of P (0.25 kg of particulate P ha^{-1}, and 75% of the annual soil losses, i.e., 800 kg of soil ha^{-1}). In a tropical forest in Ivory Coast, the contribution of surface-casts to soil losses reached 1,200 kg ha^{-1} $year^{-1}$, which corresponded to a maximum release of 0.23 kg of total P ha^{-1} $year^{-1}$ (Nooren et al. 1995). Focusing on the impact of raindrops that cause the disintegration of fresh and dry old casts of *Martiodrilus carimaguensis* in Colombia, Mariani et al. (2007) used simulated rainfall events and demonstrated a significant effect of soil drying on cast dispersion. They estimated that the mass of fresh casts dispersed could range between 4 and 32 tons ha^{-1} $year^{-1}$, which might be equivalent to 0.025 and 0.354 kg of available P ha^{-1} $year^{-1}$ in savannah and pasture, respectively. However, opposite results were reported in a temperate agro-ecosystem in France by Le Bayon and Binet (2001). Rainstorm events were simulated on a maize plot with a gentle slope of 4.5% populated by two dominant species of earthworms, the endogeic *A. caliginosa* and the anecic *L. terrestris* (Binet and Le Bayon 1999). Despite a high cast abundance on the soil surface (25% of the area), amounts of P recovered in runoff waters were twice as high without surface-casts (0.35 and 0.86 kg of particulate P ha^{-1} $rainfall^{-1}$ with and without casts, respectively). Earthworm casts were thus proposed to act as a physical brake on soil erosion by creating a surface roughness, and this was recently confirmed by Jouquet et al. (2008) in steep-slope ecosystems of North Vietnam. Nevertheless, Le Bayon and Binet (2001) observed that once the breaking-down point of the physical resistance of casts is reached, all surface-casts may then be quickly disintegrated and washed away. Hence, transfer of P can occur over a short distance through successive suspension/deposition of soil particles in the water runoff.

Simultaneous investigations were conducted on annual variations of earthworm surface-casting and P transfer (Le Bayon et al. 2002), including a 2-month period in spring when earthworm activities are enhanced (Le Bayon and Binet 1999). Working at such different temporal scales highlighted the possibility that the contribution of earthworm surface-casts to soil and P transfers could be over-estimated. For instance, a total of 0.05–0.20 kg of particulate P ha^{-1} $year^{-1}$ were actually measured in runoff waters during one year, whereas 0.25–0.49 kg of particulate P ha^{-1} $year^{-1}$ were estimated to be lost during the 2-month study of intensive earthworm activity in spring. Moreover, belowground egestions, which may represent large amounts of casting activities (Decaëns et al. 1999; Bohlen et al. 2004), and earthworm burrows need to be taken into account for a better understanding of P fluxes (Jensen et al. 2000).

In conclusion, despite the high variability of the data presented in different studies, a contribution of earthworm casts and burrows to the transfer of P has been demonstrated. This is intimately linked both to biotic factors (earthworm species, ecological group, abundance, biomass, vegetation cover, etc.) and also abiotic parameters (slope, climate, intensity and frequency of the rainfall events, aggregate stability, etc.).

8.3 Termites

8.3.1 Phosphorus Contents in Termite Mounds

Termites play an important role in the transformation processes of organic compounds in savannahs and tropical forest ecosystems (e.g., Lee and Wood 1971; Bignell and Eggleton 2000; Holt and Lepage 2000; López-Hernández 2001). Data on the P content of termite mounds in relation to adjacent soils are sometimes contradictory. Extensive studies of mounds and other structures of various Australian and African termites showed little difference in P concentration between termite structures and the soil from which they were built (Lee and Wood 1971; Leprun and Roy-Noël 1977). However, Okello-Oloya et al. (1985) for Australian *Amitermes* mounds, and Nutting et al. (1987) for subterranean North American desert termites, found that extractable P was significantly greater in mounds than in associated soils. López-Hernández et al. (2006) reported, for neo-tropical termites, significantly higher levels of both total and available P in termite mounds of *Nasutitermes ephratae*, a common plant-debris-feeding termite, compared with the adjacent soil. Moreover, in a comprehensive P-fractionation study of *Nasutitermes*, López-Hernández (2001) showed that all P levels were significantly higher in the mounds than in the adjacent soils. Similar results were found when comparing $E_{(t)}$ (the amount of PO_4 that is isotopically exchanged within t minutes) values of isotopically exchangeable P (López-Hernández et al. 1989a). Rückamp et al. (2010) also reported P enrichment in various termite mounds collected from seven dominant Brazilian ecosystems.

The relative P enrichment of the mound may have two main sources: (1) a biological origin from salivary, fecal, and plant debris used in the mound building, and (2) a mineral origin from an accumulation of clay within the mound. As a result of different feeding habits, termites collect different materials to build their nest structures (López-Hernández et al. 2006). Wood et al. (1983) differentiate plant-debris-feeding termites, which employ the selection of certain particle sizes (e.g., clays with high P-fixing capacity from the subsoil) to cement soil particles with saliva alone or with saliva and feces (Lee and Wood 1971; Wood et al. 1983), from soil-feeding termites that mainly employ ingestion of organic-rich fractions mixed with soil particles (mostly from the topsoil). Then, they reported that available P appears to be proportionally more abundant in mounds of soil-feeding termites than in the mounds of plant-debris-feeding termites when compared to adjacent soils.

8.3.2 *Phosphorus Dynamics and Availability in Termite Mounds*

Fardeau and Frossard (1992) showed that P isotopically exchangeable in 1 min [$E_{(1\ min)}$] increased from 2.2 mg kg^{-1} in the bulk soil to 87.0 mg kg^{-1} in the centre of a *Trinervitermes geminatus* nest. The P sorption processes are strongly reduced in organic-matter-enriched mounds compared with adjacent soils for both grass-feeders *T. geminatus* and *N. ephratae* (Table 8.1). On the other hand, *Macrotermes* species forage mostly from the organic carbon-deficient subsoil and preferably selected finer soil particles for the nest construction (Leprun and Roy-Noël 1976; Pomeroy 1983; López-Hernández and Febres 1984; López-Hernández et al. 2006). Their mounds built with clay-enriched materials have very high P-sorbing capacities and consequently a lower P availability than the adjacent topsoil (Table 8.1). For the subterranean species (*Ancistrotermes cavithorax* and *Microtermes toumodiensis*), there was more P sorption in nest structures than in associated soils (Table 8.1). The low concentration of water-soluble P (C_P) found in the subsoil reflects strong P sorption in this soil layer, probably due to the presence of the materials enriched with high P-sorbing clay.

The higher levels of water-soluble inorganic P (C_P) in the termite nests presented in Table 8.1 may also result from the transformation of organic P through enzymatic activity in the fresh biostructures, as observed for earthworm casts (Sharpley and Syers 1976; Le Bayon and Binet 2006). Comparisons of phosphatase activities between termite nests and adjacent soils have already been made for several termite species (López-Hernández et al. 1989b; Roose-Amsaleg et al. 2005). Phosphatase activities in the soil and mounds for the South American *N. ephratae* ranged from 0.56 to 2.32 μmol *p*-nitrophenol g^{-1} h^{-1}. Comparison of phosphatase activities for both mound and associated soils showed no statistical difference (López-Hernández

Table 8.1 Isotopically exchangeable P [$E_{(1\ min)}$, mg kg^{-1} soil] and water-soluble P (C_P, mg L^{-1}) in termite nests and adjacent soils

Termite species	Nest		Soil	
	$E_{(1\ min)}$	C_P	$E_{(1\ min)}$	C_P
Macrotermes bellicosus	0.29	0.002	0.34	0.021
Trinervitermes geminatus	3.11	0.221	0.19	0.010
Cubitermes severus	0.33	0.016	0.13	0.006
Nasutitermes ephratae	1.88	0.124	0.46	0.006
Ancistrotermes cavithorax	0.33	0.006	0.20	0.008
Microtermes toumodiensis	0.45	0.009	0.24	0.009

Adapted from López-Hernández et al. (2006)

et al. 1989a, b). Increases in enzyme activities and microbial biomass are expected to occur in mounds as a result of the increase in their organic matter content. Therefore, enzyme activity in the *N. ephratae* mounds might be inhibited as a consequence of the relatively high available P found in those structures. In contrast to the results of López-Hernández et al. (1989a), Roose-Amsaleg et al. (2005) found that soil-feeding termites affect soil phosphatase activity, with lower activities in mature nests than in surrounding soils. Comminution and mixing by termite activity allows for a greater surface area of substrate to come into contact with appropriate microorganisms and enzymes (Garnier-Sillam et al. 1987). Roose-Amsaleg et al. (2005) found that phosphatase activities tend to decrease with ageing from fresh to old and mature nests, and concluded that the released inorganic P could have a retroactive effect, inhibiting phosphatase activities as previously reported by López-Hernández et al. (1989a).

So far, it seems that termite feeding and nest-building habits have a strong influence on P availability parameters in mound, nest, and gallery structures. Humivorous and plant-feeding species cause a substantial increase of PO_4 bioavailability in nests, associated with a reduction of P sorption, whereas *Macrotermes* sp. (which in many cases build very large mounds) reduce PO_4 availability, with an associated increase of P sorption, by transporting finer soil materials from deeper soil layers.

8.3.3 *Termite Mounds and Phosphorus Transfer*

Phosphorus transport from the mound into the soil can happen in several ways. Termites might directly relocate nest materials into the soil. Furthermore, relocation can happen when termite mounds with dying colonies decay and erode. Although intact, inhabited termite mounds often have a dense surface that is particularly impermeable to water (e.g., Lal 1988; Contour-Ansel et al. 2000; Jouquet et al. 2004), the partial erosion and leaching from inhabited termite mounds may add P to the adjacent soil. However, erosion rates can be variable for a given species: Roose (1981) reported for *T. geminatus* that erosion of mounds can vary from 810 to 2,670 kg ha^{-1} $year^{-1}$, representing 10–33% of the mound's mass (Brossard et al. 2007).

Coventry et al. (1988) measured the nutrient content of a range of termite mounds and used a lower estimated rate of erosion of the mounds to calculate the amounts of nutrients returned to soil each year. Given annual erosion losses from the mounds of 3.5% (Bonell et al. 1986), they estimated that mound-building termites are responsible for the return of at least 15 g ha^{-1} of weak acid-extractable P per year. These estimates did not take into account the return of P as a result of erosion of structures other than mounds, such as feeding galleries. The amounts of soil contained in such structures may be of a similar magnitude to the amounts of soil incorporated in mounds (Lee and Wood 1971) and, if taken into account, might double the above estimates.

Finally, considering water infiltration as a major process, in particular in semi-arid and savannah areas, the influence of large macropores derived from soil macrofauna activity has been studied for about two decades (e.g., Elkins et al. 1986; Léonard et al. 2004). However, the results vary depending on measurement methods, and the processes involved deserve further consideration. Janeau and Valentin (1987) showed for *Trinervitermes* spp. that termite activity reduced the rate of water infiltration as a consequence of declines in soil surface porosity. However, Roose (1981), in a study on the sealing crusts in *T. geminatus*-affected soils, reported an increase in soil infiltration induced by termite activity. Léonard and Rajot (2001) showed that the influence of the large macropores made by termites was better described as a runoff interception process than as ponded infiltration. The P transfers through these processes (infiltration, runoff) are all related to P associated with soil particles. For soils with low-P status, the PO_4 pool in soil solution involved in these transfers is not well defined and very difficult to measure.

8.4 Conclusion

Macrofauna affects soil P cycling in a way that differs according to functional groups (i.e., feeding and construction behavior). However, P contents are usually larger in biogenic structures than in the surrounding soil, ultimately leading to higher P availability through a rapid turnover of organic P and/or enhanced PO_4 exchange. Further research efforts should be directed at investigating macrofauna-mediated processes in agricultural systems because these are potentially important for enhancing soil P availability to cultivated plants. Isotopic labeling techniques have the potential to elucidate soil P dynamics and should be extensively used to identify such processes. Controlled laboratory studies must be conducted to better understand the impact of gut passage, the microbial populations associated with guts, and the egested materials on P forms and availability in the biogenic structures. Experiments with a wide gamut of species and ecological strategies comprising a community will permit upscaling from the individual to community and population levels. Companion field studies should aim to better understand the volume of soil impacted and the dynamics of the biogenic structures produced, on a landscape scale. Coupled modeling approaches can also be useful to fully understand the complex interplay of biogeochemical interactions and to upscale from cast levels to soil profile and the entire ecosystem (Standing et al. 2007; Blanchart et al. 2009).

References

Barois I, Lavelle P (1986) Changes in respiration rate and some physicochemical properties of a tropical soil during transit through *Pontoscolex corethrurus* (Glossoscolecidae, Oligochaeta). Soil Biol Biochem 18:539–541

Bastardie F, Capowiez Y, de Dreuzy JR, Cluzeau D (2003) X-ray tomographic and hydraulic characterization of burrowing by three earthworm species in repacked soil cores. Appl Soil Ecol 24:3–16

Bates JAR (1960) Studies on Nigerian forest soils. I. The distribution of organic matter in the profile and in various soil fractions. J Soil Sci 11:246–256

Bignell DE, Eggleton P (2000) Termites in ecosystems. In: Abe T, Bignell DE, Higashi M (eds) Termites: evolution, sociality, symbioses, ecology. Kluwer, Dordrecht, pp 363–387

Binet F, Le Bayon RC (1999) Space-time dynamics in situ of earthworm casts under temperate cultivated soils. Soil Biol Biochem 31:85–93

Blanchart E, Albrecht A, Brown G, Decaëns T, Duboisset A, Lavelle P, Mariani L, Roose E (2004) Effects of tropical endogeic earthworms on soil erosion. Agric Ecosyst Environ 104:303–315

Blanchart E, Marilleau N, Chotte JL, Drogoul A, Perrier E, Cambier C (2009) SWORM: an agent-based model to simulate the effect of earthworms on soil structure. Eur J Soil Sci 60:13–21

Bohlen J, Parmelee RW, Blair JM (2004) Integrating the effects of earthworms on nutrient cycling across spatial and temporal scales. In: Edwards CA (ed) Earthworm ecology, 2nd edn. CRC, Boca Raton, pp 183–200

Bonell M, Coventry RJ, Holt JA (1986) Erosion of termite mounds under natural rainfall in semiarid tropical northeastern Australia. Catena 13:11–28

Bouché MB (1977) Stratégies lombriciennes. Ecol Bull 25:122–132

Brossard M, Lavelle P, Laurent JY (1996) Digestion of a vertisol by the endogeic earthworm *Polypheretima elongate*, Megascolecidae, increases soil phosphate extractability. Eur J Soil Biol 32:107–111

Brossard M, López-Hernández D, Lepage M, Leprun JC (2007) Nutrient storage in soils and nests of mound-building *Trinervitermes* termites in Central Burkina Faso: consequences for soil fertility. Biol Fertil Soils 43:437–447

Brown GG, Pashanasi B, Villenave C, Patron JC, Senapati BK, Giri S, Barois I, Lavelle P, Blanchart E, Blakemore RJ, Spain AV, Boyer J (1999) Effects of earthworms on plant production in the tropics. In: Lavelle P, Brussaard L, Hendrix P (eds) Earthworm management in tropical agroecosystems. CABI, Wallingford, pp 87–148

Brown GG, Barois I, Lavelle P (2000) Regulation of soil organic matter dynamics and microbial activity in the drilosphere and the role of interactions with other edaphic functional domains. Eur J Soil Biol 26:177–198

Chapuis L, Brossard M (1995) Modifications et stabilité du phosphore échangeable d'un ferralsol ingéré par un ver géophage. CR Acad Sci II A 320:587–592

Chapuis-Lardy L, Brossard M, Lavelle P, Schouller E (1998) Phosphorus transformations in a Ferralsol through ingestion by *Pontoscolex corethrurus*, a geophagous earthworm. Eur J Soil Biol 34:61–67

Chapuis-Lardy L, Ramiandrisoa RS, Randriamanantsoa L, Morel C, Rabeharisoa L, Blanchart E (2009) Modification of P availability by endogeic earthworms (Glossoscolecidae) in Ferralsols of the Malagasy Highlands. Biol Fertil Soils 45:415–422

Contour-Ansel D, Garnier-Sillam E, Lachaux M, Croci V (2000) High performance liquid chromatography studies on the polysaccharides in the walls of the mounds of two species of termite in Senegal, *Cubitermes oculatus* and *Macrotermes subhyalinus*: their origin and contribution to structural stability. Biol Fertil Soils 31:508–516

Coventry RJ, Holt JA, Sinclair DF (1988) Nutrient cycling by mound-building termites in low fertility soils of semi-arid tropical Australia. Aust J Soil Res 26:375–390

Curry JP, Schmidt O (2007) The feeding ecology of earthworms – a review. Pedobiologia 50:463–477

de Vleeschauwer D, Lal R (1981) Properties of worm casts in some tropical soils. Soil Sci 129:259–271

Decaëns T, Rangel AF, Asakawa N, Thomas RJ (1999) Carbon and nitrogen dynamics in ageing earthworm casts in grasslands of the eastern plains of Colombia. Biol Fertil Soils 28:20–28

Devliegher W, Verstraete W (1996) *Lumbricus terrestris* in a soil core experiment: effects of nutrient-enrichment processes (NEP) and gut-associated processes (GAP) on the availability of plant nutrients and heavy metals. Soil Biol Biochem 28:489–496

Edwards CA (1981) Earthworms, soil fertility and plant growth. In: Apelhof M (ed) Proceedings of workshop on the role of earthworms in the stabilization of organic residues. Western Michigan University, Kalamazoo, pp 61–85

Elkins N, Sabol G, Ward T, Whitford W (1986) The influence of subterranean termites on the hydrological characteristics of a Chihuahuan desert ecosystem. Oecologia 68:521–528

Fardeau JC (1996) Dynamics of phosphate in soils – an isotopic outlook. Fertil Res 45:91–100

Fardeau JC, Frossard E (1992) Processus de transformations du phosphore dans les sols d'Afrique de l'Ouest semi aride: application au phosphore assimilable. In: Tiessen H, Frossard E (eds) Phosphorus cycles in terrestrial and aquatic systems of semi arid Africa. SCOPE/UNEP proceedings. University of Saskatchewan, Saskatoon, pp 108–128

Frossard E, Brossard M, Hedley MJ, Metherell A (1995) Reactions controlling the cycling of P in soils. In: Tiessen H (ed) Phosphorus in the global environment. Transfers, cycles and management – SCOPE 54. Wiley, New York, pp 107–137

Frossard E, Achat DL, Bernasconi SM, Bünemann EK, Fardeau J-C, Jansa J, Morel C, Rabeharisoa L, Randriamanantsoa L, Sinaj S, Tamburini F, Oberson A (2011) The use of tracers to investigate phosphate cycling in soil–plant systems. In: Bünemann EK, Oberson A, Frossard E (eds) Phosphorus in action: biological processes in soil phosphorus cycling. Soil biology, vol 26. Springer, Heidelberg. doi:10.1007/978-3-642-15271-9_3

Garnier-Sillam E, Villemein G, Toutain F, Renoux J (1987) Contribution à l'étude du role des termites dans l'humification des sols forestiers tropicaux. In: Fedoroff N, Bresson JM, Courty MA (eds) Soil micromorphology. AFES, Plaisir, pp 331–335

Graff O (1970) Phosphorus content in earthworm casts. Landbauforsch Volk 20:33–36

Guggenberger G, Haumaier L, Thomas RJ, Zech W (1996) Assessing the organic phosphorous status of an Oxisol under tropical pastures following native savanna using ^{31}P NMR spectroscopy. Biol Fertil Soils 23:332–339

Gupta ML, Sakal R (1967) The role of earthworms on the availability of nutrients in garden and cultivated soils. J Indian Soc Soil Sci 15:149–151

Holt JA, Lepage M (2000) Termite and soil properties. In: Abe T, Bignell DE, Higashi M (eds) Termites: evolution, sociality, symbiosis, ecology. Kluwer, Dordrecht, pp 389–407

James SW (1991) Soil, nitrogen, phosphorus, and organic matter processing by earthworms in tallgrass prairie. Ecology 72:2101–2109

Janeau JL, Valentin C (1987) Relations entre les termitières *Trinervitermes* sp. et la surface du sol: réorganisations, ruissellement et érosion. Rev Écol Biol Sol 24:637–647

Jégou D, Cluzeau D, Hallaire V, Balesdent J, Tréhen P (2000) Burrowing activity of the earthworms *Lumbricus terrestris* and *Aporrectodea giardi* and consequences on C transfers in soil. Eur J Soil Biol 36:27–34

Jensen MB, Olsen TB, Hansen HCB, Magid J (2000) Dissolved and particulate phosphorus in leachate from structured soil amended with fresh cattle faeces. Nutr Cycl Agroecosyst 56:253–261

Jiménez JJ, Cepeda A, Decaëns T, Oberson A, Friesen DK (2003) Phosphorus fractions and dynamics in surface earthworm casts under native and improved grasslands in a Colombian savanna oxisol. Soil Biol Biochem 35:715–727

Jones CG, Lawton JH, Shachak M (1994) Organisms as ecosystem engineers. Oikos 69: 373–386

Jouquet P, Tessier D, Lepage M (2004) The soil structural stability of termite nests: role of clays in *Macrotermes bellicosus* (Isoptera, Macrotermitinae) mound soils. Eur J Soil Biol 40:23–29

Jouquet P, Podwojewski P, Bottinelli N, Mathieu J, Ricoy M, Orange D, Tran TT, Valentin C (2008) Above-ground earthworm casts affect water runoff and soil erosion in Northern Vietnam. Catena 74:13–21

Kuczak CN, Fernandes ECM, Lehmann J, Rondon MA, Luizao FJ (2006) Inorganic and organic phosphorus pools in earthworm casts (Glossoscolecidae) and a Brazilian rainforest Oxisol. Soil Biol Biochem 38:553–560

Lal R (1987) Tropical ecology and physical edaphology. Wiley, Chichester

Lal R (1988) Effects of macrofauna on soil properties in tropical ecosystems. Agric Ecosyst Environ 24:101–116

Lavelle P (1997) Faunal activities and soil processes: adaptive strategies that determine ecosystem function. Adv Ecol Res 27:93–132

Lavelle P, Bignell DE, Lepage M, Volters V, Roger P, Ineson P, Heal W, Dillion S (1997) Soil function in a changing world: the role of invertebrate ecosystem engineers. Eur J Soil Biol 33:159–193

Lavelle P, Decaëns T, Aubert M, Barot S, Blouin M, Bureau F, Margerie P, Mora P, Rossi JP (2006) Soil invertebrates and ecosystem services. Eur J Soil Biol 42:S3–S15

Le Bayon RC, Binet F (1999) Rainfall effects on erosion of earthworm casts and phosphorus transfer by water runoff. Biol Fertil Soils 30:7–13

Le Bayon RC, Binet F (2001) Earthworm surface casts affect soil erosion by runoff water and phosphorus transfer in a temperate maize crop. Pedobiologia 45:430–442

Le Bayon RC, Binet F (2006) Earthworms change the distribution and availability of phosphorus in organic substrates. Soil Biol Biochem 38:235–246

Le Bayon RC, Milleret R (2009) Effects of earthworms on phosphorus dynamics – a review. Dyn Soil Dyn Plant 3(2):21–27

Le Bayon RC, Moreau S, Gascuel-Odoux C, Binet F (2002) Annual variations in earthworm surface-casting activity and soil transport by water runoff under a temperate maize agroecosytem. Geoderma 106:121–135

Lee KE (1985) Earthworms, their ecology and relationships with soils and land use. Academic, Bowen Hills

Lee KE, Wood TG (1971) Termites and soils. Academic, London

Léonard J, Rajot JL (2001) Influence of termites on runoff and infiltration: quantification and analysis. Geoderma 104:17–40

Léonard J, Perrier E, Rajot JL (2004) Biological macropores effect on runoff and infiltration: a combined experimental and modelling approach. Agric Ecosyst Environ 104:277–285

Leprun J-C, Roy-Noël J (1976) Minéralogie des argiles et répartition des nids épigés de deux espèces du genre *Macrotermes* au Sénégal occidental (Presqu'île du Cap-Vert). Insectes Soc 23:535–547

Leprun J-C, Roy-Noël J (1977) Les caractères analytiques distinctifs des matériaux des nids du genre *Macrotermes* au Sénégal occidental. Leurs rapports avec les sols. Pedobiologia 17: 361–368

López-Hernández D (2001) Nutrient dynamics (C, N and P) in termite mounds of *Nasutitermes ephratae* from savannas of the Orinoco llanos (Venezuela). Soil Biol Biochem 33:747–753

López-Hernández D, Febres A (1984) Changements chimiques et granulométriques produits dans des sols de Côte d'Ivoire par la présence de trois espèces de termites. Rev Écol Biol Sol 21:477–489

López-Hernández D, Niño M, Fardeau JC, Nannipieri P, Chacón P (1989a) Phosphorus accumulation in savanna termite mound in Venezuela. J Soil Sci 40:635–640

López-Hernández D, Niño M, Nannipieri P, Fardeau JC (1989b) Phosphatase activity in *Nasutitermes ephratae* termite nest. Biol Fertil Soils 7:134–137

López-Hernández D, Lavelle P, Fardeau JC, Niño M (1993) Phosphorous transformations in two P-sorption contrasting tropical soils during transit through *Pontoscolex corethrurus* (Glossoscolecidae, Oligochaeta). Soil Biol Biochem 25:789–792

López-Hernández D, Brossard M, Fardeau JC, Lepage M (2006) Effect of different termite feeding groups on P sorption and P availability in African and South American savannas. Biol Fertil Soils 42:207–214

Mariani L, Jiménez JJ, Torres EA, Amezquita E, Decaëns T (2007) Rainfall impact effects on ageing casts of a tropical anecic earthworm. Eur J Soil Sci 58:1525–1534

Martin A (1991) Short- and long-term effects of the endogeic earthworm *Millsonia anomala* (Omodeo) (Megascolecidae, Oligochaeta) of tropical savannas on soil organic matter. Biol Fertil Soils 11:234–238

Morel C, Tiessen H, Moir J, Stewart JWB (1994) Phosphorus transformations and availability due to crop rotations and mineral fertilization assessed by an isotopic exchange method. Soil Sci Soc Am J 58:1439–1445

Mulongoy K, Bedoret A (1989) Properties of worm casts and surface soils under various plant covers in the humid tropics. Soil Biol Biochem 21:197–203

Nijhawan SD, Kanwar JS (1952) Physicochemical properties of earthworm castings and their effect on the productivity of soil. Indian J Agric Sci 22:357–373

Nooren CAM, van Breemen N, Stoorvogel JJ, Jongmans AG (1995) The role of earthworms in the formation of sandy surface soils in a tropical forest in Ivory Coast. Geoderma 65:135–148

Nutting WL, Haverty MI, Lafage JP (1987) Physical and chemical alteration by two subterranean termite species in Sonoran Desert grassland. J Arid Environ 12:233–239

Nziguheba G, Bünemann EK (2005) Organic phosphorus dynamics in tropical agroecosystems. In: Turner BL, Frossard E, Baldwin DS (eds) Organic phosphorus in the environment. CABI, Wallingford, pp 243–268

Okello-Oloya T, Spain AV, John RD (1985) Selected chemical characteristics of the mounds of two species of *Amitermes* (Isoptera, Termitinae) and their adjacent surface soil from Northeastern Australia. Rev Écol Biol Sol 22:291–311

Park SC, Smith TJ, Bisesi MS (1990) Hydrolysis of bis(4-nitro phenyl) phosphate by the earthworm *Lumbricus terrestris*. Soil Biol Biochem 22:729–730

Patrón JC, Sánchez P, Brown GG, Brossard M, Barois I, Gutiérrez C (1999) Phosphorus in soil and *Brachiaria decumbens* plants as affected by the geophagous earthworm *Pontoscolex corethrurus* and P fertilization. Pedobiologia 43:547–556

Pomeroy DE (1983) Some effects of mound building termites on soils of Uganda. J Soil Sci 27:377–394

Roose E (1981) Dynamique actuelle de sols ferrallitiques et ferrugineux tropicaux d'Afrique occidentale. Étude expérimentale des transferts hydrologiques et biologiques de matières sous végétations naturelles ou cultivées. Trav Doc ORSTOM 130. ORSTOM, Paris

Roose-Amsaleg C, Mora P, Harry M (2005) Physical, chemical and phosphatase activities characteristics in soil-feeding termite nests and tropical rainforest soils. Soil Biol Biochem 37:1910–1917

Rückamp D, Amelung W, Theisz N, Bandeira AG, Martius C (2010) Phosphorus forms in Brazilian termite nests and soils: relevance of feeding guild and ecosystems. Geoderma 155:269–279

Satchell JE, Martin K (1984) Phosphatase activity in earthworm faeces. Soil Biol Biochem 16:191–194

Scheu S, Parkinson D (1994a) Effects of earthworms on nutrient dynamics, carbon turnover and microorganisms in soil from cool temperate forests of the Canadian Rocky Mountains: laboratory studies. Appl Soil Ecol 1:113–125

Scheu S, Parkinson D (1994b) Effects of an invasion of an Aspen woodland (Alberta, Canada) by *Dendrobaena octaedra* on plant growth. Ecology 75:2348–2361

Sharpley AN, Syers JK (1976) Potential role of earthworms casts for the phosphorus enrichment of run-off water. Soil Biol Biochem 8:341–346

Sharpley AN, Syers JK (1977) Seasonal variation in casting activity and in the amounts and release to solution of phosphorous forms in earthworm casts. Soil Biol Biochem 9:227–231

Sharpley AN, Syers JK, Springett JA (1979) Effect of surface-casting earthworms on the transport of phosphorus and nitrogen in surface runoff from pasture. Soil Biol Biochem 11:459–462

Sinaj S, Buerkert A, El-Hajj G, Bationo A, Traore H, Frossard E (2001) Effects of fertility management strategies on phosphorus bioavailability in four West African soils. Plant Soil 233:71–83

Standing D, Baggs EM, Wattenbach M, Smith P, Killham K (2007) Meeting the challenge of scaling up processes in the plant–soil–microbe system. Biol Fertil Soils 44:245–257

Suarez ER, Fahey TJ, Groffman PM, Bohlen PJ, Fisk MC (2004) Effects of exotic earthworms on soil phosphorus cycling in two broadleaf temperate forests. Ecosystems 7:28–44

Tuffen F, Eason WR, Scullion J (2002) The effect of earthworms and arbuscular mycorrhizal fungi on growth of and ^{32}P transfer between *Allium porrum* plants. Soil Biol Biochem 34:1027–1036

Wood TG, Johnson RA, Anderson JM (1983) Modification of soils in Nigerian savanna by soil-feeding *Cubitermes*, Termitidae. Soil Biol Biochem 15:575–579

Zhang BG, Li GT, Shen TS, Wang JK, Sun Z (2000) Changes in microbial biomass C, N and P and enzyme activities in soil incubated with the earthworms *Metaphire guillelmi* or *Eisenia fetida*. Soil Biol Biochem 32:2055–2062

Chapter 9
Role of Phosphatase Enzymes in Soil

P. Nannipieri, L. Giagnoni, L. Landi, and G. Renella

9.1 Introduction

Phosphatases have been extensively studied in soil, as shown by some reviews (Ramirez-Martinez 1968; Speir and Ross 1978; Malcom 1983; Tabatabai 1994), because they catalyse the hydrolysis of ester–phosphate bonds, leading to the release of phosphate (P), which can be taken up by plants or microorganisms (Cosgrove 1967; Halstead and McKercher 1975; Quiquampoix and Mousain 2005). It has been shown that the activities of phosphatases (like those of many hydrolases) depend on several factors such as soil properties, soil organism interactions, plant cover, leachate inputs and the presence of inhibitors and activators (Speir and Ross 1978).

Phosphatases are enzymes catalysing the hydrolysis of both esters and anhydrides of phosphoric acid (Schmidt and Laskowski 1961) and, according to the Nomenclature Committee of the International Union of Biochemistry and Molecular Biology, they can be classified as phosphoric monoester hydrolases or phosphomonoesterases (EC 3.1.3), phosphoric diester hydrolases or phosphodiesterases (EC 3.1.4), triphosphoric monoester hydrolases (EC 3.1.5) and enzymes acting on phosphoryl-containing anhydrides (EC 3.6.1) and on P–N bonds (EC 3.9). Phosphatases can also be subdivided according to their regulation (e.g. calmodulin), the requirements of metal cations for their activity (e.g. Mg^{2+} and Ca^{2+}) and their sensitivity to various phosphatase inhibitors. Phosphomonoesterases include acid and alkaline phosphomonoesterase (which hydrolyse monoester bonds including mononucleotides and sugar phosphates), phosphoprotein phosphatases (which hydrolyse phosphoester bonds of phosphoserines, phosphothreonines or phosphotyrosines), phytases (EC 3.1.3.26 for 4-phytase and EC 3.1.3.8 for 3-phytase, which hydrolyse all six phosphate groups from inositol hexaphosphate) and nucleotidases. Acid and alkaline phosphomonoesterases do not hydrolyse phosphates of phytic

P. Nannipieri (✉), L. Giagnoni, L. Landi, and G. Renella
Department of Plant, Soil and Environmental Sciences, University of Firenze, Piazzale delle Cascine 18, 50144 Firenze, Italy
e-mail: paolo.nannipieri@unifi.it

E.K. Bünemann et al. (eds.), *Phosphorus in Action*, Soil Biology 26,
DOI 10.1007/978-3-642-15271-9_9, © Springer-Verlag Berlin Heidelberg 2011

acid (*myo*-inositol hexaphosphates) but they can hydrolyse lower-order inositol phosphates (Cosgrove 1980). Phosphodiesterases hydrolyse one or two ester bonds in phosphodiester compounds and include nucleases, which catalyse the hydrolysis of phosphodiester bonds of nucleic acids to produce nucleotide units or mononucleotides but not inorganic phosphates. Phospholipases hydrolyse phospholipids. We shall also discuss inorganic pyrophosphatase (pyrophosphate phosphohydrolases, EC 3.6.1.1), the enzyme that hydrolyses pyrophosphate to inorganic P, because pyrophosphate can be used as a fertilizer (Dick and Tabatabai 1978).

The aim of this review is to discuss the role of phosphatases in P mineralisation in soil and the response of these enzyme activities to changes in environmental factors, agricultural management and pollution. Particular attention will be given to phosphomonoesterase activities, which have been studied most among soil phosphatases. The meaning of measuring phosphatase activities and the drawbacks of the current protocols for enzyme assays have a central role because we think that a better understanding of the role of phosphatases (like that of any enzyme activity in soil) depends on improvement of the present enzyme assays by separating the contribution of extracellular stabilised phosphatase activities from the contribution of activities of phosphatases associated with active microbial cells. The effects of organic amendments, fertilizers and pollutants will be discussed by considering the present drawbacks of the currently used enzyme assays rather than listing all reports on the subject and underlining the contradictory data. We suggest that the reader also considers the review by Speir and Ross (1978) because it discusses the first reports and includes an extensive bibliography of the 1950s, 1960s and part of the 1970s on the effects of sterilisation, air drying, storage of soil samples before measurements, pH, temperature, soil properties, soil depth, fertilizers, trace elements, activators and inhibitors. We also refer to the reviews by Malcom (1983) and Tabatabai (1994) for a more detailed discussion of the analytical problems and for the state-of-the-art on kinetic properties and the effects of inhibitors, activators and soil properties on phosphatase activities.

9.2 Determination of Soil Phosphatase Activities

Activities of soil phosphomonoesterases have been the most studied, although phospholipids and nucleic acids, whose degradation is catalysed by phosphodiesterases, are among the major sources of fresh organic P inputs to soil (Cosgrove 1967). Before the advent of the simple, accurate and rapid enzyme assay based on the use of *p*-nitrophenyl phosphate (pNPP) by Tabatabai and Bremner (1969), phosphatase assays used natural substrates such as β-glycerolphosphate and nucleic acids (Speir and Ross 1978; Malcom 1983; Tabatabai 1994). The use of artificial substrates began in the early 1960s with phenyl phosphate (Hofmann 1963), phenolphthalein phosphate (Dubovenko 1964; Geller and Ginzburg 1979), pNPP (Bertrand and de Wolf 1968), α-naphthyl phosphate (Hochstein 1962) and β-naphthyl phosphate (Ramirez-Martinez and McLaren 1966). The choice of artificial substrates eliminated the determination of released phosphate, which is easily adsorbed by soil particles (Tabatabai 1994). The success of the pNPP assay stems also from the fact that hydrolysis of pNPP is much more rapid than that of natural substrates such as nucleic acids. The pNPP is hydrolysed to *p*-nitrophenol (pNP), which is usually determined spectrophotometrically at 400 nm under alkaline conditions. Soluble organic compounds can interfere with the quantification of the pNP (Vuorinen and Saharinen 1996). For this reason, Gerritse and van Dijk (1978) suggested the separation of pNP from pNPP and other soluble organic compounds, extracted from organic soils or animal wastes, by high pressure liquid chromatography on a cellulose column. They also observed a marked reduction of both acid and alkaline phosphomonoesterase activity by phosphate concentrations greater than 0.1 mM and therefore suggested using pNPP concentrations of 0.01–0.1 mM in the enzyme assay.

Because the phosphate group in pNPP is attached to the aromatic chromophore, hydrolysis may not reflect the activity of alkyl phosphomonoesterases. To determine this enzyme activity in soil, Avidov et al. (1993) proposed an assay based on the hydrolyis of 4-(*p*-nitrophenoxy)-1,2-butanediol phosphate with successive oxidation of the reaction product 4-(*p*-nitrophenoxy)-1,2-butanediol to pNP.

The hydrolysis of 4-methy-umbelliferyl phosphate (MUP) to 4-methylumbelliferone (MU) has also been used to assay phosphomonoesterase activity in soil by determining the fluorescence of the MU. This assay circumvents interferences by soluble organic compounds because fluorescence is measured for emission wavelengths after specific excitation (Marx et al. 2001). The MUP assay gave higher values than the pNPP assay, but the two enzyme activities (measured in modified universal buffer adjusted to the soil pH value) were significantly correlated ($P < 0.001$) when expressed on the basis of C content but not when expressed on the basis of dry soil weight (Drouillon and Merckx 2005). The MUP assay gave lower K_m values than the pNPP assay (Table 9.1) (Marx et al. 2001) and this may suggest that the former substrate mimics the hydrolysis of naturally occurring soil organic phosphate esters more closely (Freeman et al. 1995).

The phosphodiesterase assay is similar to the phosphomonoesterase assays because it is based on the release of pNP from bis-*p*-nitrophenyl phosphate (bpNPP) when the soil slurry is incubated with the substrate at pH 8.0 for 1 h (Browman and Tabatabai 1978). The bpNPP was first used by Ishii and Hayano (1974). Ohmura and Hayano (1986) showed that the optimum pH of phosphodiesterase activity of 15 soils ranged from 4.5 to 9.5, a broader pH optimum than that suggested in the assay by Browman and Tabatabai (1978). In addition, the enzyme activity was significantly correlated with soil pH.

Phosphotriesterase activity of soil has been determined by hydrolysis of tris-*p*-nitrophenyl phosphate, which is insoluble in water, to pNP (Eivazi and Tabatabai 1977).

According to Turner et al. (2002a, b), soil phytase has been poorly studied because it has been determined by the release of phosphate from phytate and not by using suitable artificial substrates (Yadav and Tarafdar 2003). Berry et al. (2007) proposed measuring the phytase activity of soil by using a chromophoric substrate

Table 9.1 Some K_m values of phosphatases

Enzyme	K_m (mM)	Substrate concentration (mM)	Temperature (°C)	Buffer	pH	References
Acid phosphomonoesterase	0.94–1.75	1–20[a]	37	MUB	6.5	Tabatabai and Bremner (1971)
Acid phosphomonoesterase	0.35–5.40[b]	–	37	Acetate	4.7	Cervelli et al. (1973)
Acid phosphomonoesterase	1.11–3.40	1–20[a]	37	MUB	6.5	Eivazi and Tabatabai (1977)
Acid phosphomonoesterase	0.1	0.05–0.50	30	Acetate	5.0	Gerritse and van Dijk (1978)
Acid phosphomonoesterase	1.71–6.99[b]	3.2–23.0[b]	30	MUB	5–6	Trasar-Cepeda and Gil-Sotres (1988)
Alkaline phosphomonoesterase	0.7	0.05–0.50	30	Tris	8.0	Gerritse and van Dijk (1978)
Alkaline phosphomonoesterase	0.44–4.94	1–20[a]	37	MUB	11	Eivazi and Tabatabai (1977)
Phosphodiesterase	0.25–1.25	1–15[a]	37	MUB	10	Eivazi and Tabatabai (1977)
Pyrophosphatase	21–51[c]	10–60[a]	37	MUB	–	Dick and Tabatabai (1978)

MUB **modified universal buffer**

Substrates were *p*-nitrophenyl phosphate for phosphomonoesterases, bis-*p*-nitrophenyl phosphate for phosphodiesereases, and pyrophosphate for pyrophosphatase

[a]Soil solution bases

[b]Corrected for the adsorption of the substrate

[c]Lineweaver–Burk plot

analogue of phytic acid whose disappearance can be monitored by high-performance liquid chromatography with UV detection. However, the method has not yet been set up for determining enzyme activity in soil.

Dick and Tabatabai (1977, 1978) set up an accurate method for determining inorganic pyrophosphatase activity of soil at pH 8.0 using pyrophosphate as the substrate and an improved determination of the released phosphate. This enzyme activity can be important from an agricultural point of view because pyrophosphate is a fertilizer P. According to Dick and Tabatabai (1978), the previous assays presented various drawbacks such as the adsorption of enzymatically released inorganic P by soil particles, hydrolysis of pyrophosphate to inorganic P after extraction from soil due to other reactions than that catalysed by pyrophosphatase, and interference of pyroposphate on the determination of inorganic P.

The hydrolysis of polyphosphates in soil has been determined by Dick and Tabatabai (1986). One of the polyphosphates used in agriculture is trimetaphosphate, a cyclic polyphosphate (Busman and Tabatabai 1985). The assay for determining trimetaphosphatase (trimetaphosphate hydrolase, EC 3.6.1.2) activity was set up by Busman and Tabatabai (1985). It involves the incubation of soil with trimetaphosphate at pH 8.0 for 5 h, followed by precipitation of residual trimetaphosphate, pyrophosphate and triphosphate. Phosphate is not precipitated and can then be determined. Trimetaphosphate is hydrolysed by trimetaphosphatase to triphosphate, which is then hydrolysed by triphosphatases to pyrophosphate and phosphate (Tabatabai 1994). Finally, pyrophosphate is hydrolysed to phosphate by pyrophosphatase. Therefore, the interpretation of the data obtained by this assay is complicated by the fact that the enzyme assay measures the activity of three enzymes, trimetaphosphatase, triphosphatase and pyrophosphatase (Tabatabai 1994).

9.3 Range and Kinetic Properties

Table 9.2 shows the range of phosphatase activities measured in soil with current assay procedures. Acid phosphomonoesterase activities in soil have been frequently measured at pH 6.5; however, at this pH the measured enzyme activity may include acid and alkaline phosphomonoesterase activity (Malcom 1983). Acid phosphomonoesterase activity generally prevails in acidic soils, whereas alkaline phosphomonoesterase activity prevails in alkaline soils, and for this reason the activities of the two enzymes are negatively correlated (Juma and Tabatabai 1978). Pang and Kolenko (1986) found a pH optimum of 7.0 for phosphomonoesterase activity in two forest soils. In comparing phosphatase activities (as for any other enzyme activity in soil) it is important to consider the period of the year in which soil sampling is done because enzyme activities of soil can change throughout the year (Schneider et al. 2001). Grierson and Adams (2000) observed that acid phosphomonoesterase activity of Jarrah (*Eucalyptus marginata* Donn ex Sm) forest soils

Table 9.2 Range of measured phosphatase activities

Enzyme	Substrate	Substrate concentration (mM)	Temperature (°C)	Duration of assay (h)	Buffer	pH	Range of enzyme activity (μmol product g^{-1} h^{-1})	References
Phosphomonoesterase	pNNP	50	20	1	MUB	At soil pH of 3.2–8.1	0.05–5.22	Drouillon and Merckx (2005)
Phosphomonoesterase	MUP	16 and 25	20	0.25	MUB	At soil pH of 3.2–8.1	0.48–10.41	Drouillon and Merckx (2005)
Acid Phosphomonoesterase	pNPP	10	37	0.5	0.5 M Tris maleate	6.5	10.4–307	Turner et al. (2002b)
Acid Phosphomonoesterase	pNPP	5	37	1	0.5 M Tris maleate	6.5	2.62–12.19	Turner and Haygarth (2005)
Acid Phosphomonoesterase	–	25	37	1	MUB	6.5	1.03–10.38 in air dried soils; 2.11–27.07 in moist soils	Baligar et al. (1988)
Acid phosphomonoesterase	MUP	0.01–0.40	–	–	Water	At soil pH of 4.0–4.3	0.57–1.08	Santruckova et al. (2004)
Acid phosphomonoesterase	pNNP	50	37	1	MUB	6.5	0.85–14.9	Dick et al. (1988)
Acid phosphomonoesterase	pNNP	50	37	1	MUB	6.5	0.31–3.15	Zornoza et al. (2009)
Acid phosphomonoesterase	pNNP	5	37	1	MUB	6.5	0.35–0.88	Eivazi and Tabatabai (1977)
Acid phosphomonoesterase	–	115	37	1	MUB	6.5	0.06–0.13	Ho (1979)
Alkaline phosphomonoesterase	pNNP	5	37	1	MUB	11	0.06–1.60	Eivazi and Tabatabai (1977)
Alkaline phosphomonoesterase	pNNP	50	37	1	MUB	11	0.34–5.50	Dick et al. (1988)
Phosphodiesterase	bpNPP	5	37	1	MUB	9–11	0.10–0.55	Eivazi and Tabatabai (1977)
Phosphotriesterase	tpNPP (insoluble)	5	37	1	MUB	10	0.01–0.08	Eivazi and Tabatabai (1977)

pNPP *p*-nitrophenyl phosphate, *MUP* 4-methyl umbelliferyl phosphate, *MUB* modified universal buffer, *bpNPP* bis-*p*-nitrophenyl phosphate, *tpNPP* tris-*p*-nitrophenyl phosphate ranged from 30 to 40 μmol pNP g^{-1} h^{-1} in winter and spring when soil was moist, whereas it was below 10 μmol pNP g^{-1} h^{-1} in the dry summer.

Activities of phosphodiesterases are lower than acid and alkaline phosphomonoesterase activities (Criquet et al. 2007) because the production of P monoesters from P diesters may stimulate the microbial synthesis of phosphomonoesterases (Turner and Haygarth 2005). It is reasonable to hypothesise that phosphodiesterase and phosphomonoesterase activities act sequentially in soil (Fig. 9.1). Phosphotriesterase activity was also lower than acid and alkaline phosphomonoesterase activities of soil (Eivazi and Tabatabai 1977).

Table 9.1 shows the K_m (the Michaelis–Menten constant) values of phosphatases in soil. Although phosphatases, like other hydrolases in soil, can derive from different sources and thus have different kinetic constants, the K_m value of phosphatase activity of a soil can be calculated. As discussed by Nannipieri and Gianfreda (1998), the calculated values probably represent a weighted average of the various constants of enzymes involved in the measured enzyme activity, with an unknown weighting factor. However, in the case of acid phosphomonoesterases, at least two enzymes with markedly different K_m values were found in pyrophosphate extracts from two soils by applying the Eadie–Scatchard plot (rate of reaction V vs. the substrate concentration S) (Nannipieri et al. 1982).

Brams and McLaren (1974) observed a marked deviation from linearity at higher substrate concentration for soil phosphomonoesterase (pH 6.90), and Irving and Cosgrove (1976) suggested that diffusional effects and adsorption of substrate by soil colloids were responsible for the fact that acid phosphomonoesterase of a Krasnozem did not follow Michaelis–Menten kinetics. Cervelli et al. (1973) proposed calculating the K_m value by considering the adsorption of the substrate (pNPP) by the Freundlich law; the corrected K_m value of acid phosphomonoesterase was lower than

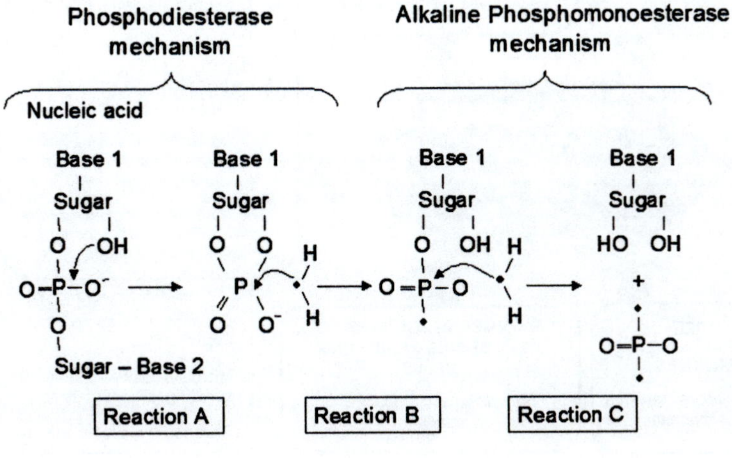

Fig. 9.1 Mechanisms of phosphodiesterase and alkaline phosphomonoesterase reaction involving molecular rearrangement (*Reaction A*) and incorporation of oxygen atoms (•) in the phosphate molecule (*Reactions B* and *C*). Redrawn from Blake et al. (2005)

the uncorrected value. The same was observed for the K_m values of acid phosphomonoesterase in acid and organic soils from Galicia, Spain (Trasar-Cepeda and Gil-Sotres 1988). The shaking of soil slurries can accelerate the diffusion of the substrate towards the enzymes. Indeed, K_m values of phosphomonoesterases measured at pH 6.9 were 2.5 times greater when measured in soil columns than when measured in a batch-type system with shaking (Brams and McLaren 1974).

9.4 Limitations of the Present Enzyme Assays

Although the drawbacks of the currently used enzyme assays have been extensively discussed (Skujins 1978; Burns 1978, 1982; Nannipieri 1994; Tabatabai 1994; Nannipieri et al. 2002; Gianfreda and Ruggiero 2006), they are still frequently neglected when soil enzyme activities are interpreted. Firstly, the present enzyme assays measure potential and not real enzyme activities (Fig. 9.2) because the assay conditions are different (optimal pH, optimal temperature, substrate present at saturating concentration, presence of buffer to control pH during the assay, soil slurry, shaking) from those occurring in situ (fluctuations of temperature and moisture of the soil; pH and substrate concentration are rarely at the optimum for enzyme activity, etc.). Secondly, we do not know which enzymes contributed to the

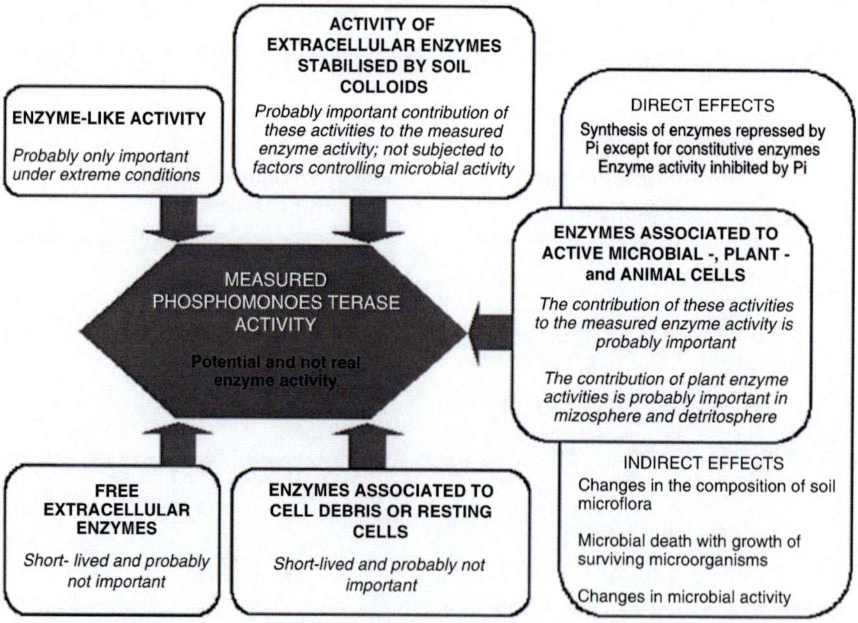

Fig. 9.2 Contribution of activities of phosphomonoesterase differently located in the soil matrix to the measured enzyme activity, and drawbacks of the currently used assays

measured enzyme activity. According to Burns (1982), enzymes catalysing the measured reaction can be:

1. Associated with active microbial cells, either intracellular or attached to the outer cell surface
2. Associated with cell debris or dead cells
3. Associated with resting cells, such as bacterial spores
4. Released as truly extracellular enzymes to degrade high molecular weight or insoluble substrates
5. Present as extracellular enzymes of enzyme–substrates complexes
6. Present as free extracellular enzymes
7. Present as extracellular enzymes stabilised by their association with surface-reactive particles (e.g. clay minerals, iron oxides and hydroxides)
8. Entrapped by humic matter (the humus–enzyme complexes)

The activity of enzyme-like catalysts is probably significant under extreme environmental conditions where these catalysts are present, whereas activities of free extracellular enzymes (6), enzymes associated with substrates (5), and enzymes of cell debris and dead cells (2) are probably short-lived because they can be rapidly degraded unless they are adsorbed by soil particles. The contribution of enzymes associated with resting cells is probably insignificant (Nannipieri et al. 2002). Therefore, it is reasonable to hypothesise that the measured enzyme activity depends on the activity of enzymes associated with active microbial cells, including enzyme activities of plant cells in the rhizosphere and detritosphere soil, and on the activity of extracellular enzymes stabilised by soil colloids (Fig. 9.2) (Nannipieri 1994; Nannipieri et al. 2002). The extracellular stabilised enzyme activity is not affected by changes in composition, abundance or activity of the soil microflora; thus the intracelluler enzyme activity of active cells should be used as an indicator of nutrient dynamics and changes in soil functioning due to agricultural management and ecological factors because it is well established that microbial activities of soil are more sensitive to these changes than other soil properties (Nannipieri et al. 2003). However, the separation of stabilised extracellular enzyme activity and enzyme activity associated with active microbial cells and plant cells is not possible with the present enzyme assays (Nannipieri et al. 2002). Most reports on soil enzymes assume that the present short assays only determine extracellular and stabilised enzyme activity. Despite this assumption, the measured enzyme activities are often taken as indicators of soil quality, which is strictly related to microbial activity and thus to intracellular enzyme activity. In addition, it is often assumed that changes in enzyme activities only reflect the response of microbiota to environmental factors, neglecting the fact that the measured enzyme activity also depends on the activity of stabilised extracellular enzymes.

Microbial inhibitors such as toluene have been used to inhibit the enzyme activity associated with active microbial cells, but this can create artefacts; for example, toluene can increase the permeability of cell membranes and thus the access of the urea substrate to intracellular ureases (Nannipieri et al. 2002). In addition, microorganisms can also use toluene as a substrate (Kaplan and

Hartenstein 1979). Toluene did not affect phosphodiesterase or acid and alkaline phosphomonoesterase activities but increased the phosphotriesterase activity of soil (Eivazi and Tabatabai 1977).

The so-called physiological response method is based on the measurements of enzyme activities and microbial biomass of soil during the period when microbial growth is stimulated by adding glucose and a nitrogen source to soil (Nannipieri et al. 2002). If enzyme activity is plotted against biomass, there is generally a significant and positive correlation between the enzyme activity (*y*-axis) and microbial biomass (*x*-axis). The extrapolation to zero of microbial biomass gives a positive intercept of the plot on the *y*-axis, which is the extracellular stabilised enzyme activity. This approach has been used to calculate extracellular acid phosphomonoesterase activity of a moist soil treated with different rates of sewage sludges, with measurement of microbial biomass by measuring ATP. Extracellular phosphomonoesterase activity of soil was 14.9, 5.3 and 4.3 μmol pNPP g^{-1} h^{-1} after addition of 0, 50 and 100 tons of sewage sludge per hectare, respectively (Nannipieri et al. 1996a). This approach can only work for constitutive but not for inducible or repressible enzymes, such as phosphomonoesterase, whose synthesis is generally repressed by inorganic phosphate (Nannipieri 1994). Indeed, changes in inducible or repressible enzyme activities are not related to changes in microbial biomass. In addition, the percentage of glucose-utilising microorganisms depends on soil type, management and pollution (Nannipieri et al. 2002). Acid phosphomonoesterase activity of two eucalypt forest soils was significantly correlated with ergosterol content and microbial P when all these properties were measured throughout the year, the relative plots giving a positive intercept on the *y*-axis (Grierson and Adams 2000). Obviously, the intercept obtained with the first correlation cannot represent the extracellular acid phosphomonoesterase activity because ergosterol content only determines the fungal biomass. However, the approach by Grierson and Adams (2000) does not involve the stimulation of microbial growth by adding easily degradable organic compounds to soil, and thus does not present the above-mentioned drawbacks (Nannipieri et al. 2002).

Chloroform fumigation has also been used to distinguish enzyme activity associated with active microbial and plant cells from the extracellular enzyme activity stabilised in soil (Klose and Tabatabai 1999). This method assumes that the present short-term enzyme assays measure the stabilised extracellular enzyme activity and that the increase in enzyme activity after $CHCl_3$ fumigation is due to the intracellular enzyme activity. Therefore, the intracellular enzyme activity of soil can be calculated by subtracting the enzyme activity before fumigation from that after fumigation. As discussed by Nannipieri et al. (2002), this approach presents the following problems: (1) $CHCl_3$ fumigation does not kill all microbial cells and the efficiency of cell lysis depends on soil structure (Arnebrant and Schnurer 1990), and (2) the assumption that the present short-term enzyme assays determine only the extracellular enzyme activity has never been proven. The fact that enzyme activities, including acid and alkaline phosphomonoesterase activities, can increase with microbial biomass when easily degradable organic compounds, such as glucose, are added to soil (Nannipieri et al. 1978, 1979, 1983; Renella et al. 2006a, b, 2007b) suggests that the present enzyme assays also measure the contribution of enzyme activities associated with active microbial cells of soil, and (3) proteases are active during the $CHCl_3$ fumigation period and degrade urease and both phosphomonoesterase enzymes (Renella et al. 2002); thus, protease activity needs to be inhibited during soil fumigation.

The use of sonication can increase the enzyme activity of soil. Indeed, the activity of acid phosphomonoesterase was 156% higher with soil sonication than without it (De Cesare et al. 2000). The increase probably depended on the release of extracellular enzymes stabilised by soil colloids and not on cell lysis, because the release of enzymes by sonication was not related to the release of ATP.

9.5 Role of Phosphatase in Organic P Mineralisation in Soil and the Effect of Inorganic P

As already mentioned, phosphodiesterase and phosphomonoesterase activities may act sequentially (Fig. 9.1). Pant and Warman (2000) observed that acid phosphomonoesterase (from wheat germ), alkaline phosphomonoesterase (from calf intestinal mucosa), phospholipase (from *Clostridium perfingrens*) and nuclease (from *Staphylococcus aureus*), all immobilised on positively charged supports, were able to mineralise (at pH 7.0) organic P extracted from different soils by water or NaOH. The activities of both phosphomonoesterases were generally increased when these enzymes were used with one of the two phosphodiesterases.

Soil acid phosphomonoesterase activity was higher at low inorganic P content of soil than at high content, and the enzyme activity of the low-P soil was significantly correlated with herbage yield, probably due to the importance of organic P mineralisation for plant P nutrition (Speir and Cowling 1991). Santruckova et al. (2004) found that higher enzymatic hydrolysis of organic P depended on the higher microbial P immobilisation but not on the higher mineralisation of organic P compounds.

Application of inorganic P can repress the synthesis of phosphomonoesterases in soil because it inhibits the expression of *PHO* genes (Oshima et al. 1996) and, indeed, phosphate inhibits the phosphatase activities of soil (Halstead 1964; Juma and Tabatabai 1977, 1978; Lima et al. 1996; Moscatelli et al. 2005; Nannipieri et al. 1978; Olander and Vitousek 2000; Spiers and McGill 1979). However, the absence of a response of phosphatase activities to P addition has also been reported. For example, the application of triple superphosphate to an oak soil in 1992 did not affect acid phosphomonoesterase activity of soil samples taken in 1993 and 1994 (Schneider et al. 2001). Addition of phosphate with glucose and inorganic N did not stimulate the phosphomonoesterase activity (pH 6.5) of soil, whereas the stimulation occurred in the respective soil treated only with glucose and inorganic N (Nannipieri et al. 1978). Presumably, the enzyme activity was not decreased by phosphate due to the presence of extracellular phosphomonoesterases stabilised by soil colloids or due to the presence of constitutive microbial phosphomonoesterase in soil. Enzyme assays discriminating the activities of extracellular stabilised enzymes from activities of enzymes associated with soil microorganisms would permit an understanding of the underlying mechanisms (Fig. 9.2).

9.6 Phosphatase Activities of Bulk and Rhizosphere Soil and the Origin of Phosphatases in Soil

It is well established that enzyme activities are higher in rhizosphere than bulk soil (Skujins 1978; Tarafdar and Chhonkar 1978; Dinkelaker and Marschner 1992). Both acid and alkaline phosphomonoesterase activites of soil were increased near the rhizoplane of *Brassica oleracea*, *Allium cepa*, *Triticum aestivum* and *Trifolium alexandrium* and such an increase depended on plant species, soil type and plant age (Tarafdar and Jungk 1987). Probably, the increase with plant age was due to the gradual formation of the rhizosphere microflora and to the release of plant phosphomonoesterases. The distance from the rhizoplane at which the rhizosphere effect on enzyme activities was observed was higher for acid (from 2 to 3.1 mm) than for alkaline (from 1.2 to 1.6 mm) phosphomonoesterase. There was an inverse and significant correlation between the acid or the alkaline phosphomonoesterase activity and the content of organic P of the rhizosphere soil sampled from *Triticum aestivum* and *Trifolium alexandrium,* whereas the content of inorganic P increased towards the rhizoplane. Increases in both acid and alkaline phosphomonoesterase activities near the rhizoplane of maize were accompanied by changes in the composition of bacterial communities as determined by PCR-DGGE (polymerase chain reaction-denaturing gradient gel electrophoresis) (Kandeler et al. 2002). Some agroforestry species (*Tithonia diversifolia, Tephrosia vogelii* and *Crotalaria grahmiana*) stimulated acid phosphomonoesterase activity of rhizosphere soil, whereas maize stimulated alkaline phosphomonoesterase activity of rhizosphere soil (George et al. 2002). Higher soil phosphomonoesterase activities were found under invader plant species than under grass and forbs, presumably due to the higher P uptake of the invading plants (Neal 1973). Izaguirre-Mayoral et al. (2002) found that nodulated legumes species growing in acid savanna soils stimulated the acid phosphomonoesterase activity of rhizosphere soil.

Interactions between soil microorganisms and plant species can also affect phosphomonoesterase activities of rhizosphere soil. It has been suggested that arbuscular mycorrhizal (AM) fungi stimulate the release of acid phosphomonoesterase from roots of subterranean clover (*Trifolium subterraneum* L.) (Joner and Jakobsen 1995). However, mycorrhizal infection of cucumber (*Cucumus sativus* L.) did not affect acid and alkaline phosphomonoesterase activities of soil (Joner et al. 1995). More information on the relationship between phosphatases and mycorrhizae is given by Jansa et al. (2011).

Obviously, it is difficult to interpret the measurement of phosphomonoesterase activities of rhizosphere soil if the contribution of plant and microbial phosphatases to the measured enzyme activity are not separated. Colvan et al. (2001) suggested that acid phosphomonoesterase activity of hay meadow soils was due to enzyme released by plants, because the enzyme activity was high and microbial P was low in soils never treated with fertilizer or treated with N or K fertilizer for 100 years. However, the measured acid phosphomonoesterase activities could also have been at least partly due to enzymes synthesised by the soil microflora in response to P-deficient conditions (Nannipieri 1994). A significant correlation between acid phosphomonoesterase activity of an oak soil and fine root length density of oak has been found (Schneider et al. 2001). The response of activities of hydrolases, including acid and alkaline phosphomonoesterase and phosphodiesterase, in rhizosphere soil depends on the type of root exudate stimulating microbial activity (Renella et al. 2006b, 2007a), which suggests that active microbial phosphomonoesterases are probably the major contributing enzymes to the measured enzyme activity of the rhizosphere soil.

Phosphorus nutrition of transgenic *Nicotiana tabacum* expressing a chimeric phytase gene (*ex::ph A*) from soil fungus *Aspergillus niger* was better in neutral than in acid soils, because the adsorption by soil of phytase released from roots was lower at neutral than at acid pH values (George et al. 2005a, b). The expression of phytase in the transgenic plant had no effect on the structure of microbial communities of rhizosphere soil compared to the wild type (George et al. 2009).

No correlations have been found between acid phosphomonoesterase activity and acid phosphomonoesterase-active bacterial colonies, and between alkaline phosphomonoesterase activity and alkaline phosphomonoesterase-active bacterial colonies of forest soils (Hysek and Sarapatka 1998), but this is not surprising since culturable bacteria only represent 1–10% of the bacteria inhabiting soil (Torsvik et al. 1996). Acid phosphomonoesterase was secreted by roots of three cereals (wheat, pearl millet and sorghum), three legumes (mung bean, moth bean and cluster bean), and three oil seed crops (groundnut, sesame and mustard) when these plants were grown in culture solution under P-deficient conditions (Yadav and Tarafdar 2001). The secretion pattern depended on the plant species, and the amount of root exudates increased with plant age and was higher with phytin than with leucithin and glycerolphosphate, each present as the sole source of P. Because acid phosphomonoesterases (like any enzyme molecule) diffuse poorly into the soil matrix, plant-released enzymes probably mineralise organic P from sloughed off or damaged cells rather than from native soil organic P (Lefebvre et al. 1990; Yadav and Tarafdar 2001).

Alkaline phosphomonoesterase activity has not been detected in plants (Dick et al. 1983; Juma and Tabatabai 1988a, b, c) and for this reason microbial cells supposedly synthesise most of the soil alkaline phosphomonoesterases (Tabatabai 1994). Both soil bacteria and soil microorganisms other than AM fungi (Joner and Jakobsen 1995) are thought to contribute to the measured soil alkaline phosphomonoesterase activity. Indeed, alkaline phosphatase activity of both rhizosphere and bulk soil depend on the composition of bacterial communities harbouring alkaline phosphatase genes, as determined by DGGE after amplification of extracted DNA by polymerase chain reaction with specific primers (Sakurai et al. 2008). Increases in acid and alkaline phosphomonoesterase activities due to the addition of ryegrass residues to soils were related to changes in the composition of bacterial communities, as determined by DGGE (Renella et al. 2006a). Changes in the ergosterol content of Jarrah forest soils explained 50% of the changes in acid phosphomonoesterase activity in each season, whereas the ergosterol content of soil sampled under *Banksia grandis*, an understorey dominant plant growing in dense thickets in the absence of fire, explained 74% of the changes in enzyme activity during the dry season (summer) but only 10% in moist soils (Grierson and Adams 2000). Both phytase- and phosphomonoesterase-producing fungi isolated from arid and semiarid soils of India belonged to the genera *Aspergillus*, *Emmericella* and *Penicillum* (Yadav and Tarafdar 2003). In conclusion, the present evidence confirms that bacteria are the main source of alkaline phosphomonoesterase activity in soil, whereas acid phosphomonoesterase and phytase can derive from plants, fungi and bacteria. However, studies such as that by Sakurai et al. (2008) involving detection of genes codifying these enzymes are necessary.

Renella et al. (2007b) estimated the production and persistence of acid and alkaline phosphomonoesterase and phosphodiesterase activities in soils with a wide range of properties, by stimulating microbial growth through the addition of glucose and inorganic N to soil. Phosphatase activities of the soil increased, with microbial biomass reaching a peak value, but then both declined on prolonging the incubation time. Enzyme production (Pr) was calculated by the relationship $Pr = H/t_H$, where H is the peak of the enzyme activity and t_H is the time of the peak after adding glucose plus N to soil. Enzyme persistence (Pe) was calculated by the relationship $Pe = (r/H)\Delta t$, where r indicates the residual enzyme activity at the end of the incubation time and Δt is the time interval between the peak value (H) and the residual activity (r). The Pr values of acid phosphomonoesterase activity were highest in soils under forest and set-aside management, whereas Pr values of alkaline phosphomonoesterase and phosphodiesterase activities were highest in alkaline and neutral soils. The Pe values of acid phosphomonoesterase activity were highest in acid soils, whereas no relationship was found between alkaline phosphomonoesterase or phosphodiesterase activity and the soil pH or management (Renella et al. 2007b).

Both phosphodiesterases and pyrophosphatases are ubiquitous in animal, plant and microbial cells because they are involved in the degradation of nucleic acids and in several basic metabolic pathways of cells, respectively (Browman and Tabatabai 1978; Cooperman et al. 1992; Tabatabai 1994).

9.7 Effects of Soil Handling, Soil Properties, Agricultural Management and Pollutants on Soil Phosphatase Activities

Here, we shall only discuss reports after the late 1970s because Speir and Ross (1978) have extensively reviewed the effects of soil sampling, handling and storage, soil properties, different agricultural managements, forest practices and pollutants on phosphatase activities.

9.7.1 *Effects of Soil Handling and Soil Properties on Phosphatase Activities*

Both air-drying and freeze-drying often decrease acid and alkaline phosphomonoesterase activity of soil (Gerritse and van Dijk 1978; Baligar et al. 1988; Adams 1992). However, Eivazi and Tabatabai (1977) found an increase in acid phosphomonoesterase and phosphotriesterase activities and a decrease in alkaline phosphomonoesterase and phosphodiesterase activites after air-drying of soil. Acid phosphomonoesterase activities in moist soils stored at 4°C and in the respective air-dried soils were significantly correlated (Baligar et al. 1988). Air-drying also decreased pyrophosphatase activity of soil, and the best storage conditions were to keep field-moist soils at 5°C (Tabatabai and Dick 1979). Probably the best strategy is to keep moist soils at 4°C and measure the enzyme activity as soon as possible. Kandeler (2007) suggests that if the determination of the enzyme activity requires storage periods longer than 3 weeks at 4°C, it is better to store the samples at −20°C than at 4°C. At the end of the storage period, soil samples are allowed to thaw at 4°C for about 2 days before the determination of the enzyme activity.

Steam sterilisation at 121°C for 1 h completely inactivated alkaline phosphomonoesterase, phosphodiesterase and phosphotriesterase activity, but increased acid phosphomonoesterase activity (Eivazi and Tabatabai 1977). Heating above 60°C inactivated the pyrophosphatase activity of soil (Tabatabai and Dick 1979).

It is well established that phosphatase activities are correlated with the content of organic matter and decrease with soil depth (Speir and Ross 1978; Tabatabai and Dick 1979; Prado et al. 1982; Pang and Kolenko 1986; Tabatabai 1994). Factors affecting phosphatase activity, measured by using sodium phenyl phosphate as a substrate in different woodland soils, could be ranked as rock type = vegetation type > soil type = season > soil depth (Harrison 1983). Some 20% of the variation of both acid and alkaline phosphomonoesterase activities of semiarid woodland soils depended on soil microclimate and surface depth (0–10 cm); soil temperature together with soil water potential was a better predictor of phosphatase activities than either factor alone (Kramer and Green 2000).

Humic acids competitively inhibited wheat phytase activity measured at 55°C and pH 5.15 (Pereira 1971). For forest soils, phosphomonoesterase activity of litter (pH in water 3.6–3.7), humus (pH in water 3.4–3.5) and mineral layers (pH in water 4.0–4.3), measured with water and MUP as the substrate, were correlated with the contents of organic C, microbial N and microbial P, and with soil respiration (Joergensen and Scheu 1999). Santruckova et al. (2004) observed that changes in soil pH and contents of total organic C, total N, total P, oxalate-soluble reactive and organic P, as well as oxalate-soluble Al and Fe can affect phosphomonoesterase activity and microbial biomass P in various ways, leading either to a surplus or a deficiency in available soil phosphate. The relationship between phosphomonoesterase activities and the distribution of P forms is not clear. Acid phosphomonoesterase and phosphodiesterase activities of Karri forest soils were not related to soil P fractions (non-occluded Fe- and Al-bound P, P sorbed by carbonates, occluded P, and Ca-bound P) or total P (Adams 1992).

Alkaline phosphomonoesterase activity, measured using disodium phenylphosphate as substrate, was mainly associated with silt and clay fractions of a Haplic Chernozem (Kandeler et al. 1999). By contrast, in a calcareous and in an acid soil, both alkaline and acid phosphomonoesterase activities were associated with larger soil fractions (100–2,000 μm particle diameter) containing plant debris and less humified organic matter and were characterised by the highest mineralisation of organic P among the soil fractions (Rojo et al. 1990).

9.7.2 Effect of Agricultural Management, Forest Practices and Fire on Soil Phosphatases

Hay meadow soils treated with farmyard manure for about 100 years had higher acid and alkaline phosphomonoesterase and phosphodiesterase activities and higher NH_4F–HCl-extractable P than those receiving mineral P (Colvan et al. 2001). However, both phosphomonoesterases and phosphodiesterase activities were positively ($P < 0.1$) correlated with extractable P in soils treated with farmyard manure or with phosphate, whereas acid phospomonoesterase activity was negatively ($P < 0.05$) correlated with extractable P. Acid phosphomonoesterase activity was negatively correlated with alkaline phosphomonoesterase ($P < 0.05$) and with phosphodiesterase ($P < 0.05$) activities when considering all treatments, but was always higher than the alkaline phosphomonoesterase activity because all soils were acidic.

Straw burning instead of straw soil incorporation decreased acid but not alkaline phosphomonoesterase activity of the soil (Dick et al. 1988). Sewage sludges applied to soils increased phosphodiesterase and acid and alkaline phosphomonoesterase activities, which subsequently decreased with time (Criquet et al. 2007). Long-term experiments have shown that repeated applications of manure increases both acid and alkaline phosphomonoesterase activities, particularly immediately after manure addition to the soil (Dick et al. 1988; Colvan et al. 2001), due to the stimulation of microbial growth. When the monitoring period is prolonged, stimulation of microbial synthesis of enzymes by easily degradable organic substrates decreases (Garcia et al. 1993; Nannipieri 1994). However, the interpretation of changes in enzyme activity of soils treated with organic materials is difficult because these materials add exogeneous enzymes associated with microorganisms and extracellularly stabilised enzymes to the soil.

No-till systems usually have higher enzyme activities in the surface soils than tilled soils because of the increase in soil organic matter content (Nannipieri 1994). However, this did not occur in an organic soil (Bergstrom et al. 1998).

In forest soils, fertilisation with urea or phosphate reduced acid phosphomonoesterase activity (Pang and Kolenko 1986). Fumigation of forest soils with methylbromide and chloropicrin for 24 h decreased phosphomonoesterase activity (pH 7.0), but there was a recovery when the fumigant was removed and the moist soils were incubated under controlled conditions (Pang and Kolenko 1986).

Controlled fire did not affect acid phosphomonoesterase activity of forest surface (0–5 cm) soil because the temperature never exceeded 50°C, whereas a marked reduction occurred with uncontrolled wildfires (Saa et al. 1993). Also, logging and/or burning operations almost eliminated acid phosphomonoesterase and phosphodiesterase activities of Karri forest soils (Adams 1992). Acid phosphomonoesterase activity did not recover after incubation of a wildfire-affected soil for 11 weeks under controlled conditions, probably because the high inorganic P repressed enzyme synthesis by soil microorganisms (Saa et al. 1998).

9.7.3 Effects of Pollutants on Soil Phosphatase Activities

Both acid and alkaline phosphomonoesterase activities have been monitored to evaluate the effects of several pollutants on organic P mineralisation in soil. The ecological dose (ED_{50}), i.e. the concentration of the pollutant that reduces the enzyme activity by 50%, has been calculated to quantify some of these effects.

Acid phosphomonoesterase activity of a blanket peat, an organic grassland soil and a calcareous grassland soil were high due to the P limitation induced by long-term atmospheric nitrogen deposition (Turner et al. 2002b), whereas sulfur pollution decreased the acid phosphomonoesterase activity of ectomycorrhizal roots in Norway spruce (Rejsek 1991).

The acid phosphomonoesterase activity of either moist or air-dried acid mesic fibrisols and histosols decreased after addition of copper to soil (Mathur and Rayment 1977; Mathur and Sanderson 1978). Juma and Tabatabai (1977) observed (depending on the soil type) the highest inhibition by Hg(II), As(V), W(VI) and Mo (VI) in the case of the acid phosphomonoesterase activity, and by Ag(I), Cd(II), V(IV) and As(V) in the case of the phosphomonoesterase activity. A negative and significant correlation was found between the sum of Cu and Zn total concentration in the soil and phosphomonoesterase activity determined using phenyl phosphate as the substrate (Tyler 1976).

Short-term laboratory incubations might not reflect the toxic effects in long-term heavy metal polluted soils. Alkaline phosphomonoesterase activity was still reduced in soils contaminated with Cd (concentration ranging from 0 to 0.36 nmol Cd kg^{-1}) in 1988–1990 and sampled in 2001, despite very low Cd availability, as determined either by water extraction or by the BIOMET bacterial biosensor system (Renella et al. 2004). In contrast, acid phosphomonoesterase activity and the composition of the bacterial community, determined either by plate counts or by DGGE, were unaffected, probably because the Cd pollution caused physiological adaptations rather than the selection of metal-resistant culturable bacteria. Addition of dry milled ryegrass to these long-term Cd-contaminated soils increased both microbial biomass and acid and alkaline phosphomonoesterase activities (Renella et al. 2005a). However, the ratio of alkaline phosphomonoesterase activity to ATP decreased while that of acid phosphomonoesterase activity to ATP was unaffected compared to the respective uncontaminated soils. Both acid and alkaline phosphomonoesterase activities and the respective ratios of hydrolase to microbial biomass C were reduced in soils contaminated on the long term with Ni and Cd, but not in those with a Mn and Zn contamination; the former contamination also changed the composition of the bacterial community, as determined by DGGE (Renella et al. 2005b).

Dose–response curves can combine the effects of pollutants and soil physicochemical properties on soil microflora (Babich et al. 1983) and can be used to calculate the effective ecological dose (ED_{50}) of enzyme activity of soil in response to pollution by heavy metals (Doelman and Haastra 1989). The ED_{50} values for Cd, Cu and Zn were 2.6 mmol kg^{-1} soil in sandy soils and 45 mmol kg^{-1} soil in clayey soils. Generally, the Cd toxicity was higher when observed 1.5 years after addition of heavy metal salts to soils than after 6 weeks. The presence of Cu or Zn increased the toxicity of Cd on acid and alkaline phosphomonoesterase activity of contrasting forest soils, as observed by comparing the relative ED_{50} values determined by the kinetic model (Renella et al. 2003). The toxicity was higher in sandy than in finer textured soils. The sensitivity of acid phosphomonoesterase activity was higher in alkaline than in acid or neutral soils, and the sensitivity of alkaline phosphomonoesterase activity showed an opposite behaviour (Renella et al. 2003).

Recovery of both acid and alkaline phosphomonoesterase activities occurred 7 years after the in situ remediation of soils contaminated with sludge-borne metals. Inorganic amendments were used, such as 5% (w/w) beringite (a coal fly ash) or 1% (w/w) zerovalent iron grid; both treatments reduced the heavy metal availability, whereas the composition of the bacterial community as determined by DGGE was not affected (Mench et al. 2006). The treatment of soils that were vegetated with a grass and herb mixture with alkaline fly ash and peat reduced leaching of Cu and Pb and increased the phosphodiesterase activity and the acid and alkaline phosphomonoesterase activites of the soil (Kumpiene et al. 2009). The treatment of an As-contaminated loamy sand soil with beringite (with or without zerovalent iron grid) reduced the extractable As, uptake of As by lettuce, and acid phosphomonoesterase activity because the treatment increased soil pH, whereas both alkaline phosphomonoesterase and phosphodiesterase activities were increased (Ascher et al. 2009). The composition of bacterial and fungal communities determined by DGGE both changed, with a decrease in microbial diversity induced by the treatments.

The interpretation of the effect of any pollutant on soil phosphatase activities (as for any soil enzyme activity) is problematic because of the limits of the enzyme assays currently used (as already discussed) and the presence of direct and indirect effects on the target enzyme (Nannipieri et al. 2002). For example, inhibition of enzyme activity by a pollutant may be masked by the growth of surviving microorganisms with expression of genes codifying the enzyme; the microbial growth can be caused by the use of microbial debris (derived from microbial cells killed by the pollutant) by the surviving microorganisms.

9.8 Stabilisation of Extracellular Phosphatases in Soil by Interaction with Surface-Reactive Particles or by Entrapment Within Humic Molecules

Three approaches have been followed to study the stabilisation of phosphatases in soil: (1) the use of pure enzymes to create model enzyme complexes, either with inorganic minerals such as clay, or with humus like materials, (2) the extraction and characterisation of phosphatases from soil, and (3) the visualisation of extracellular phosphatases in the soil matrix.

Enzymes, like any protein, are rapidly (within a few hours) adsorbed to clays and can be partially desorbed by washing the clay–protein complex; the molecules that cannot be desorbed by washing are referred to as "bound" proteins (Stotzky 1986; Nielsen et al. 2006). Protein adsorption depends on clay properties such as surface area, cation exchange capacity, charge density, type of saturating cation and degree of clay swelling (Stotzky 1986; Nannipieri et al. 1996b; Nielsen et al. 2006). The type of protein is also important because adsorption is generally maximal at pH values within the range of the protein's isoelectric point and thus involves an ion-exchange mechanism. However, hydrogen bonding, van der Waals forces and hydrophobic effects are also involved in the adsorption of proteins by clay minerals.

Reduced enzyme activity after the adsorption can occur due to modification in the tertiary structure of the protein or due to reduced accessibility of the substrate to the active site. Adsorption of alkaline phosphomonoesterase by illite reduced the enzyme activity more than adsorption of the enzyme by montmorillonite or kaolinite (Makboul and Ottow 1979). Inhibition of enzyme upon adsorption may depend on its function. Intracellular phytase was completely inhibited when adsorbed by clays, whereas extracellular phytase retained its catalytic activity (Quiquampoix and Mousain 2005). Usually, the adsorption of the enzyme by clay minerals increases $V_{\mathbf{max}}$ and K_{m} values (Nannipieri and Gianfreda 1998). Both values were increased when alkaline phosphomonoesterase was adsorbed by Ca-montmorillonite whereas they were decreased when the enzyme was adsorbed by Ca-illite (Makboul and Ottow 1979). The adsorption of acid phosphomonoesterase and pyrophosphatase by Ca-illite and Ca-montmorillonite did not affect the kinetic constants of enzymes (Dick and Tabatabai 1987). The protein adsorption by clays may also improve stability against thermal denaturation, wetting and drying cycles, and proteolysis; the resistance against proteolysis occurs if the protein penetrates the interlayer space of montmorillonite (Stotzky 1986; Nannipieri et al. 1996b; Nielsen et al. 2006). Indeed, acid phosphomonoesterase adsorbed on kaolinite was less resistant to proteolysis than that adsorbed on montmorillonite against proteolysis and thermal denaturation (Sarkar et al. 1989) and alkaline, whereas alkaline phosphomonoesterase adsorbed on Ca-illite was more resistant to proteolysis than the free enzyme (Makboul and Ottow 1979). Both acid and alkaline phosphomonoesterase bound to homo-ionic clays were degraded by soil microorganisms but degradation rates were higher with kaolinite complexes than with bentonite and vermiculate complexes, and higher with Ca-clays than with Al-clays (Chhonkar and Tarafdar 1985).

Acid phosphomonoesterase from sweet potato was more inhibited by tannic acid than urease and invertase (Rao et al. 1998).

Humus–enzyme complexes have been prepared by oxidative coupling of phenols in the presence of the enzyme to be immobilised (Nannipieri et al. 1996b). Acid phosphomonoesterase was immobilised in a resorcinol polymer synthesised by peroxidase; the enzyme was not linked to the resorcinol moiety by covalent bonds and it was more resistant to denaturation by pH, temperature and proteolysis than the free enzyme (Garzillo et al. 1996).

Acid phosphomonoesterase from potato adsorbed on Ca-polygalacturonate (a polymer simulating mucigel of the root–soil interface) by electrostatic interactions showed increased stability, but a decreased resistance against proteolytic and thermal denaturation (Marzadori et al. 1998).

Phosphatases have been extracted from soil using different solutions (Tabatabai and Fu 1992; Nannipieri et al. 1996b). Both acid and alkaline phosphomonoesterase were extracted by shaking litter with 1 M $CaCl_2$, 0.05% Tween 80 and polyvinylpolypyrrolidone, and the enzyme activities were determined after dialysis and concentration of the extract (Criquet et al. 2004). Soil moisture was the most important factor affecting the production of acid phosphomonoesterase. However, principal component analysis and multiple regressions showed that both temperature and the number of culturable heterotrophic bacteria also affected the dynamics of acid and alkaline phosphomonoesterase activities and organic P mineralisation.

Free extracellular alkaline phosphomonoesterase extracted from two soils by water was less resistant (i.e. had a lower inactivation temperature) and showed lower K_m values than the respective humic–enzyme complexes extracted by the chelating resin Chelex (Kandeler 1990).

Mayaudon (1986) suggested that several enzymes, including phosphomonoesterases and phosphodiesterases extracted from soil by phosphate-EDTA at pH 7–8, are fungal glycoenzymes associated with bacterial lipopolysaccharides, which are linked to humic compounds by Ca bridges. Humus–phosphomonoesterase complexes have been extracted from soil by pyrophosphate (Nannipieri et al. 1996b). **Successive exhaustive ultrafiltration divided the soil extract into** two fractions: one with molecular weights higher than 100,000 (A_I), and the other with molecular weights between 10,000 and 100,000 (A_{II}). Gel chromatography of the A_I fraction gave three peaks of enzyme activity, whereas gel chromatography of the A_{II} fraction gave only one peak. The kinetic behaviour of some of the fractions showed the existence of two enzymes (or two forms of the same enzyme) catalysing the same reaction with markedly different kinetic properties. In addition, humus–phosphomonoesterase complexes with higher molecular weight were more resistant to thermal and proteolytic denaturation than those with lower molecular weight (Nannipieri et al. 1996b). It was suggested that humus–enzyme complexes of higher molecular weight are likely to possess the molecular arrangement proposed by Burns et al. (1972), in which enzymes are surrounded by a network of humic molecules with pores large enough to permit the passage of substrates and products of the enzyme reaction, but not that of proteolytic enzymes.

The yield of phosphodiesterases extracted from a forest soil using a 0.1 M phosphate buffer was increased when KCl and EDTA were added to the buffer, probably because the extracting solution desorbed extracellular enzymes adsorbed on the surface of soil colloids by ionic bonding (Hayano 1977). The treatment of the soil extract with protamine sulfate removed brown-coloured substances (probably humic molecules), and the partially purified enzyme showed a pH optimum in the range of 5.2–6.0 and hydrolysed either the 3′- or the 5′-phosphodiester bond of deoxythymidine *p*-nitrophenyl phosphate. Two phosphodiesterase fractions extracted from an A horizon of a larch forest Andosol showed high affinity to adenosine 3′- and 5′-mononucleotides (Hayano 1988), whereas a third fraction was 2′,3′-cyclic nucleotide 2′-phosphodiesterase (EC 3.1.4.16) or 2′,3′-cyclic nucleotide 3′-phosphodiesterase (EC 3.1.4.37), with a pH optimum of 5.0 (Hayano 1987). All three phosphodiesterase active fractions were inhibited by Hg^{2+}.

Approaches based on the study of clay–enzyme or humic–enzyme complexes, either prepared with pure enzymes or based on the extraction of humus–enzyme complexes from soil, present several drawbacks. In soil, neither enzymes nor surface-reactive particles, such as clays, are pure; for example, enzymes are released after cell death and lysis together with other cellular remains, which can affect adsorption of proteins to soil particles and the resistance of the formed complexes to thermal denaturation and proteolysis (Nannipieri et al. 1996b; Nielsen et al. 2006). Cell lysis with release of intracellular enzymes and artifacts due to interaction between enzymes and co-extracted soil components can occur during soil extraction.

The visualisation of the stabilised extracellular enzymes in the soil matrix by electron microscopy can give insights into the formation of stabilised enzyme complexes. Ultracytochemical tests have detected acid phosphomonoesterase activity in soil microbial cells and in fragments of microbial membranes as small as 7 × 20 nm, using electron microscopy. However, these tests were not able to locate enzymes in electron-dense minerals such as clays, or in soil components such as humic materials, reacting with counterstains such as OsO_4 (Ladd et al. 1996).

Techniques such as those based on the use of enzyme-labelled fluorescence (ELF)-97 phosphate seem to be promising for localising active phosphatases in the soil matrix because fluorescence is emitted after the enzymatic hydrolysis of the substrate (Wasaki et al. 2008; Wasaki and Maruyama 2011). This technique has detected acid phosphomonoesterase activity in roots of plants grown under P-deficient conditions, but its use in soil needs to be tested.

9.9 Conclusions and Future Research Needs

Probably, phosphodiesterase and phosphomonoesterase activities act sequentially in the mineralisation of organic P to inorganic P, which can be taken up by plant roots or by soil microorganisms (Fig. 9.1). These enzyme activities are higher in rhizosphere than in bulk soil and this suggests an important role for these enzyme activities in plant P nutrition. Acid phosphomonoesterase activity is more important in acid soils than is alkaline phosphomonoesterase activity, and vice versa in alkaline soils. Effects of soil properties, agrochemicals, soil tillage, forest practices and pollutants have been extensively studied. However, a better understanding of the underlying mechanisms as well as the role of phosphatases (as that of any other enzyme in soil) requires setting up assays that discriminate the activities of enzymes associated with soil microorganisms from those of extracellular stabilised enzymes. A better understanding of the role of phosphatases in soil can be gained from studies on oxygen isotopes (isotope fractionation) of the phosphate group because alkaline phosphomonoesterase, unlike pyrophosphatase, catalyses a unidirectional reaction, producing kinetic isotope effects (Blake et al. 2005; Frossard et al. 2011).

The problem of measuring real rather than soil potential phosphatase activities may be solved by comparing the activities obtained from currently used enzyme assays with the organic P mineralisation (using isotope dilution methods as described by Frossard et al. 2011) in soils with a broad spectrum of properties, and by using the quantitatively most important soil organic P esters as substrates.

Extracellular enzyme activity of soil has been considered in the model describing decomposition of organic matter (Schimel and Weintraub 2003). However, the validation of this model requires determination of soil extracellular enzyme activity.

Phosphatases originate mainly from soil microorganisms, but in the rhizosphere and detritosphere they can also originate from plant cells. Changes in phosphatase activities have been related to changes in the composition of microbial communities, as determined by molecular techniques, in order to better understand the origin of phosphatases in soil. However, further insights into the origin of phosphatases in soil require relating the phosphatase activities to the expression of genes codifying these enzymes. The research carried out by Sakurai et al. (2008) involving the detection of alkaline phosphomonoesterase genes should be extended to soils with a broad spectrum of properties. Detection of the other genes expressing the various phosphatases is needed together with a relative comparison of the activity of these enzymes with the target gene.

References

Adams MA (1992) Phosphatase activity and phosphorus fractions in karri (*Eucalyptus diversicolor* F. Muell.) forest soils. Biol Fertil Soils 14:200–204

Arnebrant K, Schnurer J (1990) Changes in ATP content during and after chloroform fumigation. Soil Biol Biochem 22:875–877

Ascher J, Ceccherini MT, Landi L, Mench M, Pietramellara G, Nannipieri P, Renella G (2009) **Composition, biomass and activity of microflora, and leaf yield and foliar elemental concen-**trations of lettuce, after in situ stabilization of an arsenic-contaminated soil. Appl Soil Ecol 41:351–359

Avidov E, Dick WA, Racke KD (1993) Proposed substrate for evaluating alkyl phosphomonoesterase activity in soil. Soil Biol Biochem 25:763–768

Babich H, Bewley RJF, Stotzky G (1983) Application of "ecological dose" concept to the impact of heavy metals in some microbe mediated ecological processes in soil. Arch Environ Contam Toxicol 12:421–426

Baligar VC, Wright RJ, Smedley MD (1988) Acid phosphatase activity in soil of the Appalachian Region. Soil Sci Soc Am J 52:1612–1616

Bergstrom DW, Monreal CM, King DJ (1998) Sensitivity of soil enzyme activities to conservation practices. Soil Sci Soc Am J 62:1286–1295

Berry DF, Shang C, Waltham Sajdak CA, Zelazny LW (2007) Measurement of phytase activity using tethered phytic acid as an artificial substrate: methods development. Soil Biol Biochem 39:361–367

Bertrand D, de Wolf A (1968) Effect of microelement applied as complementary fertilizers on the soil microflora. C R Acad Agric Fr 54:1130–1133

Blake RE, O'Neil JRO, Surkov AV (2005) Biogeochemical cycling of phosphorus: insights from oxygen isotope effects of phosphoenzymes. Am J Sci 305:596–620

Brams WH, McLaren AD (1974) Phosphatase reactions in columns of soil. Soil Biol Biochem 6:183–189

Browman MG, Tabatabai MA (1978) Phosphodiesetrase activity of soils. Soil Sci Soc Am J 42:284–290

Burns RG (1978) Enzyme activity in soil: some theoretical and practical considerations. In: Burns RG (ed) Soil enzymes. Academic, New York, pp 295–340

Burns RG (1982) Enzyme activity in soil: location and possible role in microbial ecology. Soil Biol Biochem 14:423–427

Burns RG, Pukite AH, McLaren AD (1972) Concerning the location and persistence of soil urease. Soil Sci Soc Am Proc 36:308–311

Busman LM, Tabatabai MA (1985) Hydrolysis of trimetaphosphate in soils. Soil Sci Soc Am J 49:630–636

Cervelli S, Nannipieri P, Sequi P (1973) Michaelis constant of soil acid phosphatase. Soil Biol Biochem 5:841–845

Chhonkar PK, Tarafdar JC (1985) Degradation of clay-enzyme complexes by soil microorganisms. Zbl Mikrobiol 140:471–474

Colvan SR, Syers JK, O'Donnel AG (2001) Effect of long-term fertiliser use on acid and alkaline phosphomonoesterase and phosphodiesterase activities in managed grassland. Biol Fertil Soils 34:258–263

Cooperman BS, Baykov AA, Lahti R (1992) Evolutionary conservation of the active site of soluble inorganic pyrophosphatase. Trends Biochem Sci 7:262–266

Cosgrove DJ (1967) Metabolism of organic phosphates in soil. In: McLaren AD, Peterson GH (eds) Soil biochemistry, vol 1. Marcel Dekker, New York, pp 216–228

Cosgrove DJ (1980) Inositol phosphates. Their chemistry, biochemistry and physiology. Elsevier, Amsterdam

Criquet S, Ferre E, Farnet AM, Le Petit J (2004) Annual dynamics of phosphatase activities in an evergreen oak litter: influence of biotic and abiotic factors. Soil Biol Biochem 36:1111–1118

Criquet S, Braud A, Neble S (2007) Short-term effects of sewage sludge application on phosphatase activities and available P fractions in mediterranean soil. Soil Biol Biochem 39:921–929

De Cesare F, Garzillo AM, Buonocore V, Badalucco L (2000) Use of sonication for measuring acid phosphatase activity in soil. Soil Biol Biochem 32:825–832

Dick WA, Tabatabai MA (1977) Determination of orthophosphate in aqueous solutions containing labile organic and inorganic phosphorus compounds. J Environ Qual 6:82–85

Dick WA, Tabatabai MA (1978) Inorganic pyrophosphatase activity of soil. Soil Biol Biochem 10:59–65

Dick RP, Tabatabai MA (1986) Hydrolysis of polyphosphate in soils. Soil Sci 142:132–140

Dick WA, Tabatabai MA (1987) Kinetic and activities of phosphatase-clay complexes. Soil Sci 143:5–15

Dick WA, Juma NG, Tabatabai MA (1983) Effects of soils on acid phosphatse and inorganic pyrophosphatase of corn roots. Soil Sci 136:19–25

Dick RP, Rasmussen PE, Kerle EA (1988) Influence of long-term residue management on soil enzyme activities in relation to soil chemical properties of a wheat-fallow system. Biol Fertil Soils 6:159–164

Dinkelaker B, Marschner H (1992) In vivo demonstration of acid phosphatase activity in the rhizosphere of soil-grown plants. Plant Soil 144:199–205

Doelman P, Haastra L (1989) Short- and long-term effects of heavy metals on phosphatise activity in soil: an ecological dose-response model approach. Biol Fertil Soils 8:235–241

Drouillon M, Merckx R (2005) Performance of *para*-nitrophenyl phosphate and 4-methylumbelliferyl phosphate as substrate analogues for phosphomonoesterase in soil with different organic matter content. Soil Biol Biochem 37:1527–1534

Dubovenko EK (1964) Phosphatase activity of different soils. Chem Abstr 62:13794f

Eivazi F, Tabatabai MA (1977) Phosphatases in soils. Soil Biol Biochem 9:167–177

Freeman C, Liska G, Ostle NJ, Jones SE, Lock MA (1995) The sue of fluorogenic substrates for measuring enzyme activity in peatlands. Plant Soil 175:147–152

Frossard E, Achat DL, Bernasconi SM, Bünemann EK, Fardeau J-C, Jansa J, Morel C, Rabeharisoa L, Randriamanantsoa L, Sinaj S, Tamburini F, Oberson A (2011) The use of tracers to investigate phosphate cycling in soil–plant systems. In: Bünemann EK, Oberson A, Frossard E (eds) Phosphorus in action: biological processes in soil phosphorus cycling. Soil biology, vol 26. Springer, Heidelberg. doi:10.1007/978-3-642-15271-9_3

Garcia C, Hernandez T, Costa C, Ceccanti B, Masciandaro G, Ciardi C (1993) A study of biochemical parameters of composted and fresh municipal wastes. Bioresour Technol 44: 17–23

Garzillo AMV, Badalucco L, De Cesare F, Grego S, Buonocore V (1996) Synthesis and characterization of an acid phosphatase-polyresorcinol complex. Soil Biol Biochem 9: 1155–1161

Geller IT, Ginzburg KY (1979) Determination of the phosphatase activity of various soil groups. Pochvovedeniye 1:97–105

George TS, Gregory PJ, Wood M, Read D, Buresh RJ (2002) Phosphatase activity and organic acids in the rhizosphere of potential agroforestry species and maize. Soil Biol Biochem 34:1487–1494

George TS, Richardson AE, Simpson RJ (2005a) Behaviour of plant-derived extracellular phytase upon addition to soil. Soil Biol Biochem 37:977–978

George TS, Richardson AE, Hadobas PA, Simpson RJ (2005b) Expression of fungal phytase gene in *Nicotiana tabacum* improves phosphorus nutrition of plants grown in amended soils. Plant Biotechnol J 3:129–140

George TS, Richardson AE, Li SS, Gregory PJ, Daniell TJ (2009) Extracellular release of heterologus phytase from root of transgenic plants: does manipulation of rhizosphere biochemistry impact microbial community structure. FEMS Microbiol Ecol 70:433–445

Gerritse RG, van Dijk H (1978) Determination of phosphatase activities of soils and animal wastes. Soil Biol Biochem 10:545–551

Gianfreda L, Ruggiero P (2006) Enzyme activities in soil. In: Nannipieri P, Smalla K (eds) Nucleic acids and proteins in soil. Springer, Heidelberg, pp 257–311

Grierson PF, Adams MA (2000) Plant species affect acid phosphatase, ergosterol and microbial P in a Jarrah (*Eucalyptus marginata* n ex SM) forest in south-western Australia. Soil Biol Biochem 32:1817–1827

Halstead RL (1964) Phosphatase activity of soils as influences by lime and other treatments. Can J Soil Sci 44:137–144

Halstead RL, McKercher RB (1975) Biochemsitry and cycling of phosphorus. In: Paul EA, McLaren AD (eds) Soil biochemistry, vol 4. Marcel Dekker, New York, pp 31–63

Harrison AF (1983) Relationship between intensity of phosphatase activity and physico-chemical properties in woodland soils. Soil Biol Biochem 15:93–99

Hayano K (1977) Extraction and properties of phosphodiesterase from a forest soil. Soil Biol Biochem 9:221–223

Hayano K (1987) Characterization of two phosphodiesterase components in an extract of a larch forest soil. Soil Sci Plant Nutr 34:393–403

Hayano K (1988) Characterization of a phosphodiesterase component in a forest soil extract. Biol Fertil Soils 3:159–164

Ho I (1979) Acid phosphatase activity in forest soil. Forest Sci 25:567–568

Hochstein L (1962) The fluorometric assay of soil enzymes. Chem Abstr 63:15164g

Hofmann E (1963) The analysis of enzymes in soil. In: Linskens HF, Tracey MV (eds) Moderne **methoden der pflanzenanalyse, vol VI. Springer, Berlin, pp 416–423**

Hysek J, Sarapatka B (1998) Relationship between phosphatase active bacteria and phosphatise activities in forest soils. Biol Fertil Soils 26:112–115

Irving GCJ, Cosgrove DJ (1976) The kinetics of soil acid phosphatase. Soil Biol Biochem 8:335–349

Ishii T, Hayano K (1974) Method for the estimation of phosphodiesterase activity of soil. Chem Abstr 83:8023u

Izaguirre-Mayoral ML, Flores S, Carballo O (2002) Determination of acod phosphatase and dehydrogenase activities in the rhizosphere of nodulated legume species native to two contrasting savanna sites in Venezuela. Biol Fertil Soils 35:470–472

Jansa J, Finlay R, Wallander H, Smith FA, Smith SE (2011) Role of mycorrhizal symbioses in phosphorus cycling. In: Bünemann EK, Oberson A, Frossard E (eds) Phosphorus in action: biological processes in soil phosphorus cycling. Soil biology, vol 26. Springer, Heidelberg. doi:10.1007/978-3-642-15271-9_6

Joergensen RG, Scheu S (1999) Response of soil microorganisms to the addition of carbon, nitrogen and phosphorus in a forest Rendzina. Soil Biol Biochem 31:859–866

Joner EJ, Jakobsen I (1995) Growth and extracellular phosphatase activity of arbuscular mycor-**rhizal hyphae as influenced by soil organic matter. Soil Biol Biochem 9:1153–1159**

Joner EJ, Magid J, Gahoonia TS, Jakobsen I (1995) P depletion and activity of phosphatases in the rhizosphere of mycorrhizal and non-mycorrhizal cucumber (*Cucumis sativus* L.). Soil Biol Biochem 27:1145–1151

Juma NG, Tabatabai MA (1977) Effects of trace elements on phosphase activity in soils. Soil Sci Soc Am J 41:343–346

Juma NG, Tabatabai MA (1978) Distribution of phosphomonoesterases in soils. Soil Sci 126: 101–108

Juma NG, Tabatabai MA (1988a) Hydrolysis of organic phosphates by corn and soybean roots. Plant Soil 107:31–38

Juma NG, Tabatabai MA (1988b) Phosphatase activity in corn and soybean roots: conditions for assay and effects of metals. Plant Soil 107:39–47

Juma NG, Tabatabai MA (1988c) Comparison of kinetic and thermodynamic parameters of phosphomonoesterases of soils and of corn and soybean roots. Soil Biol Biochem 20: 533–539

Kandeler E (1990) Characterization of free and adsorbed phosphatases in soils. Biol Fertil Soils 9:199–202

Kandeler E (2007) Physiological and biochemical methods for studying soil biota and their function. In: Paul EA (ed) Soil microbiology, ecology and biochemistry. Academic, Amsterdam, pp 53–83

Kandeler E, Palli S, Stemmer M, Gerzabeck MH (1999) Tillage changes in microbial biomass and enzyme activities in particle-size fractions of a Haplic Chernozem. Soil Biol Biochem 31:1253–1264

Kandeler E, Marschner P, Tscherko D, Gahoonia TS, Nielsen NE (2002) Microbial community composition and functional diversity in the rhizosphere of maize. Plant Soil 238:310–312

Kaplan DL, Hartenstein R (1979) Problems with toluene and the determination of extracellular enzyme activity in soils. Soil Biol Biochem 11:335–338

Klose S, Tabatabai MA (1999) Urease activity of microbial biomass in soils. Soil Biol Biochem 31:205–211

Kramer S, Green DM (2000) Acid and alkaline phosphatase dynamics and their relationship to soil microclimate in a semiarid woodland. Soil Biol Biochem 32:179–188

Kumpiene J, Guerri G, Landi L, Pietramellara G, Nannipieri P, Renella G (2009) Microbial biomass, respiration and enzyme activites after in situ aided phytostabilization of a Pb- and Cu-contaminated soil. Ecotoxicol Environ Saf 72:115–119

Ladd JN, Foster R, Nannipieri P, Oades JM (1996) Soil structure and biological activity. In: Stotzky G, Bollag J-M (eds) Soil biochemistry, vol 9. Marcel Dekker, New York, pp 23–78

Lefebvre DD, Duff MG, Fife CA, Julien-Inalsigh C, Plaxton WC (1990) Response of phosphate deprivation in *Brassica nigra* suspension cells. Plant Physiol 93:504–511

Lima JA, Nahas E, Gomes AC (1996) Microbial populations and activities in sewage sludge and phosphate fertilizer-amended soil. Appl Soil Ecol 4:75–82

Makboul HE, Ottow JCG (1979) Alkaline phosphatase and Michaelis constant in the presence of different clay minerals. Soil Sci 128:129–135

Malcom RE (1983) Assessment of phosphatase activity in soils. Soil Biol Biochem 15:403–408

Marx MC, Wood M, Jarvis SC (2001) A microplate fluorimetric assay for the study of enzyme diversity in soils. Soil Biol Biochem 33:1633–1640

Marzadori C, Gessa C, Ciurli S (1998) Kinetic properties and stability of potato phosphatase immobilized on Ca-polygalacturonate. Soil Biol Biochem 27:97–103

Mathur SP, Rayment AF (1977) Influence of trace element fertilization on the decomposition rate and phosphatase activity of a mesic fibrisol. Can J Soil Sci 57:397–408

Mathur SP, Sanderson RB (1978) Relationship between copper contents, rates of soil respiration and phosphatase activities of some histosols in an area of southwestern Quebec in the summer and the fall. Can J Soil Sci 58:125–134

Mayaudon J (1986) The role of carbohydrates in the free enzymes in soil. In: Fuchsman CH (ed) Peat and water. Elsevier, New York, pp 263–308

Mench M, Renella G, Gelsomino A, Landi L, Nannipieri P (2006) Biochemical parameters and bacterial species richness in soils contaminated by sludge-borne metals and remediated with inorganic soil amendments. Environ Pollut 14:24–31

Moscatelli MC, Lagomarsino A, De Angelis O, Grego S (2005) Seasonality of soil biological properties in a poplar plantation growing under elevated atmospheric CO_2. Appl Soil Ecol 30:162–173

Nannipieri P (1994) The potential use of soil enzymes as indicators of productivity, sustainability and pollution. In: Pankhurst CE, Doube BM, Gupta VVSR, Grace PR (eds) soil biota: management in sustainable farming systems. CSIRO, East Melbourne, pp 238–244

Nannipieri P, Gianfreda L (1998) Kinetics of enzyme reactions in soil environments. In: Huang PM, Buffle SN, Senesi N, Buffle J (eds) Structure and surface reactions of soil particles. IUPAC series on analytical and physical chemistry of environmental systems, vol 4. Wiley, New York, pp 449–479

Nannipieri P, Johanson RL, Paul EA (1978) Criteria for measurement of microbial growth and activity in soil. Soil Biol Biochem 10:223–229

Nannipieri P, Pedrazzini F, Arcara PG, Piovanelli C (1979) Changes in amino acids, enzyme activities and biomasses during soil microbial growth. Soil Sci 127:26–34

Nannipieri P, Ceccanti B, Cervelli S, Conti C (1982) Hydrolases extracted from soil: kinetic parameters of several enzymes catalysing the same reaction. Soil Biol Biochem 14:429–432

Nannipieri P, Muccini L, Ciardi C (1983) Microbial biomass and enzyme activities: production and persistence. Soil Biol Biochem 15:679–685

Nannipieri P, Sastre J, Landi L, Lobo MC, Pietramellara G (1996a) Determination of extracellular neutral phosphomonoeterase activity in soil. Soil Biol Biochem 28:107–112

Nannipieri P, Sequi P, Fusi P (1996b) Humus and enzyme activity. In: Piccolo A (ed) Humic substances in terrestrial ecosystems. Elsevier, Amsterdam, pp 293–328

Nannipieri P, Kandeler E, Ruggiero P (2002) Enzyme activities and microbiological and biochemical processes. In: Burns RG, Dick RP (eds) Enzymes in the environment. Activity, ecology and applications. Marcel Dekker, New York, pp 1–3

Nannipieri P, Ascher J, Ceccherini MT, Landi L, Pietramellara G, Renella G (2003) Microbial diversity and soil functions. Eur J Soil Sci 54:655–670

Neal JL (1973) Influence of selected grasses and forbs on soil phosphatase activity. Can J Soil Sci 53:119–121

Nielsen KM, Calamai L, Pietramellara G (2006) Stabilization of extracellular DNA and proteins by transient binding to various soil components. In: Nannipieri P, Smalla K (eds) Nucleic acids and proteins in soil. Springer, Heidelberg, pp 141–188

Ohmura H, Hayano K (1986) Effect of pH on phosphomonoesterase activity in soil. Tochigi Agric Exp Stn Natl Inst Agro-Environ Sci: 55–61 (Japanese article with a summary in English)

Olander LP, Vitousek PM (2000) Regulation of soil phosphatase and chitinase activity by N and P availability. Biogeochemistry 49:175–190

Oshima Y, Ogawa N, Harashima S (1996) Regulation of phosphatase synthesis in *Saccharomyces cerevisae* – a review. Gene 179:171–177

Pang PCK, Kolenko H (1986) Phosphomonoesterase activity in forest soils. Soil Biol Biochem 18:35–40

Pant HK, Warman PR (2000) Enzymatic hydrolysis of soil organic phosphorus by immobilized phosphatases. Biol Fertil Soils 30:306–311

Pereira M (1971) The effect of humic acids on the activity of phytase. Humus Planta 357–364

Prado R, Tena M, Pinilla J-A (1982) Relationship between soil phosphatase activity and organic matter content. Agronomie 2:539–544

Quiquampoix H, Mousain D (2005) Enzymatic hydrolysis of organic phosphorus. In: Turner BL, Frossard E, Baldwin DS (eds) Organic phosphorous in the environment. CABI, Wallingford, pp 89–112

Ramirez-Martinez JR (1968) Organic phosphorus mineralization and phosphatase activity in soils. Folia Microbiol 13:161–174

Ramirez-Martinez JR, McLaren AD (1966) Determination of soil phosphatase activity by a fluorimetric technique. Enzymologia 30:243–253

Rao MA, Violante MA, Gianfreda L (1998) Interactions between tannic acid and acid phosphatase. Soil Biol Biochem 30:11–112

Rejsek K (1991) Acid phosphomonoesterase activity of ectomycorrhizal roots in Norway spruce pure stands exposed to pollution. Soil Biol Biochem 23:667–671

Renella G, Landi L, Nannipieri P (2002) Hydrolase activities during and after the chloroform fumigation of soil as affected by protease activity. Soil Biol Biochem 34:51–60

Renella G, Ortigoza ALR, Landi L, Nannipieri P (2003) Additive effects of copper and zinc on cadmium toxicity on phosphatise activities and ATP content of soil as estimated by the ecological dose (ED_{50}). Soil Biol Biochem 35:1203–1210

Renella G, Munch M, van del Lilie D, Pietramellara G, Ascher J, Ceccherini MT, Landi L, Nannipieri P (2004) Hydrolase activity, microbial biomass and community structure in long-term Cd-contaminated soils. Soil Biol Biochem 36:443–451

Renella G, Mench M, Landi L, Nannipieri P (2005a) Microbial activity and hydrolase synthesis in long-term Cd-contaminated soils. Soil Biol Biochem 37:133–139

Renella G, Mench M, Gelsomino A, Landi L, Nannipieri P (2005b) Functional activity and microbial community structure in soils amended with bimetallic sludges. Soil Biol Biochem 37:1498–1506

Renella G, Landi L, Ascher MT, Ceccherini MT, Pietramellara G, Nannipieri P (2006a) Phosphomonoesterase production and persistence and composition of bacterial communities during plant material decomposition in soils in soil with different pH values. Soil Biol Biochem 38:795–802

Renella G, Egamberdiyeva D, Landi L, Mench M, Nannipieri P (2006b) Microbial activity and hydrolase activities durino decomposition of root exudates released by an artificial root surface in Cd-contaminated soils. Soil Biol Biochem 38:702–708

Renella G, Landi L, Valori F, Nannipieri P (2007a) Microbial and hydrolase activity after release of low molecular weight organic compounds by a model root surface in a clayey and a sandy soil. Appl Soil Ecol 36:124–129

Renella G, Szukcis U, Landi L, Nannipieri P (2007b) Quantitative assessment of hydrolase production and persistence in soil. Biol Fertil Soils 44:321–329
Rojo MJ, Carcedo SG, Mateos MP (1990) Distribution and characterization of phosphatase and organic phosphorus in soil fractions. Soil Biol Biochem 22:169–174
Saa A, Trasar-Cepeda MC, Gil-Sotres F, Carballas T (1993) Change in soil phosphorus and acid phosphatase activity immediately following forest fires. Soil Biol Biochem 25:1223–1230
Saa A, Trasar-Cepeda MC, Gil-Sotres F, Carballas T (1998) Soil P status and phosphomonoesterase activity of recently burnt and unburnt soil following laboratory incubation. Soil Biol Biochem 30:419–428
Sakurai M, Wasaki J, Tomizawa Y, Shinano T, Osaki M (2008) Analysis of bacterial communities on alkaline phosphatase gene in organic matter applied soil. Soil Biol Biochem 54:62–71
Santruckova H, Vrba J, Picek T, Kopacek J (2004) Soil biochemical activity and phosphorus transformations and losses from acidified forest soils. Soil Biol Biochem 36:1569–1576
Sarkar JM, Leonowicz A, Bollag JM (1989) Immobilization of enzymes on clay and soils. Soil Biol Biochem 21:223–230
Schimel JP, Weintraub MN (2003) The implications of exoenzyme activity on microbial carbon and nitrogen limitation in soil: a theoretical model. Soil Biol Biochem 35:549–563
Schmidt G, Laskowski M Sr (1961) Phosphatase ester cleavage (survey). In: Boyer PD, Lardy H, Myrback K (eds) The enzymes, 2nd edn. Academic, New York, pp 3–35
Schneider K, Turrion M-B, Grierson PF, Gallardo JF (2001) Phosphatase activity, microbial phosphorus, and fine root growth in forest soils in the Sierra de Gata, western Spain. Biol Fertil Soils 34:151–155
Skujins J (1978) History of abiontic soil enzyme research. In: Burns RG (ed) Soil enzymes. Academic, London, pp 1–49
Speir TW, Cowling JC (1991) Phosphatse activities of pasture plants and soils: relationship with plant productivity and P fertility indices. Biol Fertil Soils 12:189–194
Speir TW, Ross DJ (1978) Soil phosphatase and sulphatase. In: Burns RG (ed) Soil enzymes. Academic, London, pp 197–250
Spiers GA, McGill WB (1979) Effects of phosphorus addition and energy supply on acid phosphatase production and activity in soils. Soil Biol Biochem 11:3–8
Stotzky G (1986) Influence of soil mineral colloids and metabolic processes, growth adhesion, and ecology of microbes and viruses. In: Huang M, Schnitzer M (eds) Interactions of soil minerals with natural organics and microbes, special publication 17. Soil Science Society of America, Madison, pp 305–428
Tabatabai MA (1994) Soil enzymes. In: Weaver RW, Angle S, Bottomley P, Bezdicek D, Smith S, Tabatabai A, Wollum A (eds) Methods of soil analysis. Part 2. Microbiological and biochemical properties. Soil Science Society of America, Madison, pp 775–833
Tabatabai MA, Bremner JM (1969) Use of *p*-nitrophenyl phosphate for assay of soil phosphatise activity. Soil Biol Biochem 1:301–307
Tabatabai MA, Bremner JM (1971) Michaelis constant of soil enzymes. Soil Biol Biochem 3:317–323
Tabatabai MA, Dick WA (1979) Distribution and stability of phosphatase in soil. Soil Biol Biochem 11:655–659
Tabatabai A, Fu M (1992) Extraction of enzymes from soil. In: Stotzky G, Bollag J-M (eds) Soil biochemistry, vol 7. Marcel Dekker, New York, pp 197–227
Tarafdar JC, Chhonkar PK (1978) Status of phosphatases in the root-soil interface of leguminous and non-leguminous crops. Z Pflanz Bodenkunde 141:347–351
Tarafdar JC, Jungk A (1987) Phosphatise activity in the rhizosphere and its relation to the depletion of soil organic phosphorus. Biol Fertil Soils 3:199–204
Torsvik V, Sørheim R, Gorksøyr J (1996) Total bacterial diversity in soil and sediment communities – a review. J Ind Microbiol 17:170–178
Trasar-Cepeda MC, Gil-Sotres F (1988) Kinetics of acid phosphatase activity in various soils of Galicia (NW Spain). Soil Biol Biochem 20:275–280

Turner BL, Haygarth PM (2005) Phosphatase activity in temperate pasture soils: potential regulation of labile organic phosphorus turnover by phosphodiesterase activity. Sci Total Environ 344:27–36

Turner BL, Paphazy MJ, Haygarth PM, McKelvie ID (2002a) Inositol phosphates in the environment. Philos Trans R Soc Lond B Biol Sci 357:449–469

Turner BL, Baxter R, Whitton BA (2002b) Seasonal phosphatse activity in three characteristic soils of the English uplands polluted by long-term atmospheric nitrogen deposition. Environ Pollut 120:313–317

Tyler G (1976) Heavy metal pollution, phosphatase activity and mineralization of organic phosporus in forest soils. Soil Biol Biochem 8:327–332

Vuorinen AH, Saharinen MH (1996) Effects of soil organic matter extracted from soil on acid phosphomonoesterase. Soil Biol Biochem 28:1477–1481

Wasaki J, Maruyama H (2011) Molecular approaches to the study of biological phosphorus cycling. In: Bünemann EK, Oberson A, Frossard E (eds) Phosphorus in action: biological processes in soil phosphorus cycling. Soil biology, vol 26. Springer, Heidelberg. doi: 10.1007/978-3-642-15271-9_4

Wasaki J, Kojima S, Maruyama H, Haase S, Osaki M, Kandeler E (2008) Localization of acid phosphatase activities in the roots of white lupin plants grown under phosphorus-deficient conditions. Soil Sci Plant Nutr 54:95–102

Yadav RS, Tarafdar JC (2001) Influence of organic and inorganic phosphorus supply on the maximum secretion of acid phosphatase by plants. Biol Fertil Soils 34:140–143

Yadav RS, Tarafdar JC (2003) Phytase of fungi in arid and semi-arid soils and their efficiency in hydrolysing different organic P compounds. Soil Biol Biochem 35:1–7

Zornoza R, Mataix-Solera J, Guerrero C, Arcenegui V, Mataix-Beneyto J (2009) Comparison of soil physical, chemical and biochemical properties among native forest, maintained and abandoned almond orchards in mountainous areas of eastern Spain. Arid Land Res Manag 23:267–282

Chapter 10
Phosphorus Nutrition: Rhizosphere Processes, Plant Response and Adaptations

Timothy S. George, Ann -Mari Fransson, John P. Hammond, and Philip J. White

10.1 Introduction

Phosphorus (P) is an essential element required for cellular function and when deficient has a significant impact on plant growth and fecundity. Poor availability of P in soil and consequent P deficiency represents a major constraint to crop production globally (Runge-Metzger 1995). Soil P status is also a key factor that controls the competitive dynamics and species composition in different natural ecosystems (McGill and Cole 1981; Attiwill and Adams 1993), and thus may have significant impact on biodiversity (Wassen et al. 2005). Many plant species have evolved in P-limited environments and, as a consequence, are known to possess a number of adaptive features that can enhance the acquisition of P from soil (Raghothama 1999; Vance et al. 2003; Richardson et al. 2007). Most plants have evolved to respond to P starvation by increasing the ability of their root systems to acquire P from the soil (White et al. 2005; Hammond and White 2008; Lynch and Brown 2008; White and Hammond 2008; Fang et al. 2009). Plant root cells take up P as orthophosphate ($H_2PO_4^-$, abbreviated here as Pi), whose concentrations in the soil solution is extremely low (<10 μM). The mass flow of Pi in the soil solution is insufficient to supply the P requirements of a plant (Kirkby and Johnston 2008). Hence, roots must proliferate throughout the soil to acquire sufficient P for plant nutrition. Although some soil P is present as labile Pi bound to soil particles, most is present as sparingly soluble inorganic salts, such as calcium (Ca) phosphate in

T.S. George(✉) and P.J. White
SCRI, Invergowrie, Dundee DD25DA, UK
e-mail: tim.george@scri.ac.uk, philip.white@scri.ac.uk

A.-M. Fransson
Plant Ecology and Systematics, Lund University, Ecology Building, 2236 Lund, Sweden
e-mail: Ann-Mari.Fransson@ekol.lu.se

J.P. Hammond,
Warwick HRI, University of Warwick, Wellesbourne, Warwick CV359EF, UK
e-mail: John.Hammond@warwick.ac.uk

E.K. Bünemann et al. (eds.), *Phosphorus in Action*, Soil Biology 26,
DOI 10.1007/978-3-642-15271-9_10,

alkaline soils or aluminium (Al) and iron (Fe) phosphates in acidic soils, or as complex organic compounds in soil organic material or soil organisms (Hinsinger 2001; Oberson and Joner 2005; Turner 2007; Kirkby and Johnston 2008). Organic P generally accounts for around 50% of soil P, and is largely comprised of monoesters with lesser amounts of diesters and phosphonates (Newman and Tate 1980; Hawkes et al. 1984; Condron et al. 1990). In order to be available to plants, inorganic P must be either desorbed or solubilised, and organic P must be mineralised to release Pi. Once in the soil solution, Pi is acquired rapidly by plant roots such that its concentration in close proximity to the root surface is estimated to be in the order of ~0.05 to 0.2 μM (Barber 1984), which is significantly less than elsewhere in the soil environment, where soil solution concentrations are typically in the range of 1–5 μM (Bieleski 1973). Slow Pi diffusion through soil to the roots is the ultimate limitation to P supply to the root surface, and can thus restrict P acquisition.

The conserved responses of plants to P starvation that increase P acquisition include:

1. Acidification of the rhizosphere and secretion of low molecular weight organic anions and phosphatase enzymes into the soil to mobilize Pi from inorganic and organic P sources (Marschner 1995; Hinsinger 2001; Jones et al. 2003; Delhaize et al. 2007; Jain et al. 2007b; George and Richardson 2008).
2. Investment of a greater proportion of plant biomass in the root system, and alterations in the morphology of the root system to enable greater exploration of the soil volume and the exploitation of localized patches of high Pi availability (White et al. 2005; Hermans et al. 2006; Hammond and White 2008; Lynch and Brown 2008).
3. Increasing the capacity of root cells to take up Pi, thereby reducing Pi concentrations in the rhizosphere, increasing the rate of diffusion of Pi towards the rhizosphere and stimulating the release of Pi from labile sources (Marschner 1995; Bucher 2007; Jain et al. 2007b).

In addition, most plants foster symbiotic relationships with mycorrhizal fungi to increase their ability to explore the soil volume and mobilize P from remote inorganic and organic sources (Bucher 2007; Smith and Read 2007; Jansa et al. 2011). All these responses are coordinated by a small number of regulatory systems controlled by both the P status of the shoot and Pi availability in the rhizosphere (White et al. 2005; Amtmann et al. 2006; White and Hammond 2008; Hammond and White 2008).

In this chapter, we will discuss the current state of knowledge regarding (1) how plants react to limited P availability by changing their physiological response in root growth and rhizosphere biochemistry traits, (2) how they coordinate this response to P limitation, and (3) how they respond to the re-supply of P once P becomes available again. Gaps in knowledge will be identified and priorities for future research will be discussed.

10.2 Root and Rhizosphere Responses of Plants to P Deficit

10.2.1 Morphological Adjustment of Roots to P Deficiency

Most species partition a greater proportion of their total dry matter into root growth when grown under P deficiency (Bradshaw et al. 1960; Hill et al. 2006). A number of studies indicate that the capacity to adjust root mass ratio in favour of root growth is expressed most effectively by species that have evolved in fertile soils (Christie and Moorby 1975; Boot and Mensink 1990), and consequently this adjustment is considered characteristic of plants that can compete effectively in high-nutrient environments (Chapin 1980). Many species adjusting to low P conditions concurrently increase specific root length (SRL) to achieve longer or more branched roots per unit of root dry matter (Christie 1975; Fitter 1985; Hill et al. 2006) and increase their root hair length and density (Itoh and Barber 1983). Increased SRL can be achieved by reducing root mass density (Fitter 1985; Fan et al. 2003) and/or by decreasing root diameter (Hill et al. 2006). In addition, a common physiological response is root agravitropism or topsoil foraging, putting roots where concentrations of P are relatively large (Lynch 2005). The formation of specialised root structures that increase P acquisition is an alternative means by which plants adjust to low soil P levels. Most notable of these are the cluster (or proteoid) roots (dense bottle-brush-like clusters of rootlets) formed on white lupin (*Lupinus albus*) (Gardner et al. 1981; Dinkelaker et al. 1995; Keerthisinghe et al. 1998; Neumann et al. 1999). The formation of cluster roots by white lupin is induced and regulated by the P status of the shoot rather than the P concentration of the root system or the soil solution (Keerthisinghe et al. 1998; Shane et al. 2003; Shen et al. 2005). In particular, cluster roots significantly increase root-surface area and thus soil contact, and are usually found on species that are either non-mycorrhizal or weakly mycorrhizal (Skene 1998).

10.2.2 Formation of Root Hairs in Response to P Deficit

Root hair development may be sparse in high-P conditions, but both density and length increase when plants are grown on a low-P supply (Bates and Lynch 1996; Gahoonia and Nielsen 1997), thus increasing the capacity for P acquisition (Itoh and Barber 1983; Föhse et al. 1991). The benefit of increased root hair density reaches a plateau when the P-depletion zones around each root hair begin to overlap, with the optimal relationship between root hair density and length dependent on the P-diffusion coefficient of the soil (Ma et al. 2001). In a study of ten grassland species, Hill et al. (2006) concluded that root-morphology adjustments helped plants to maintain root length under a range of low-P conditions, consequently improving the potential for P uptake from P-deficient soil. However, the intrinsic morphological characteristics of each species (particularly an extensive, fine root system), as opposed to the ability to adjust root morphology, was the most important determinant of whether a plant had a low P fertiliser requirement for maximum growth rate (Hill et al. 2006).

The use of root hair mutants has demonstrated that the presence of root hairs increases the root soil contact, evidenced by enhanced rhizosheath (soil adhering to roots) production (Haling et al. 2010; Brown et al. 2010) and this, in association with the consequent enhanced root surface area, has been demonstrated to enhance tolerance of Al-toxic (Haling et al. 2010) and P-deficient conditions (Brown et al. 2010). The major mechanism by which root hairs are beneficial to P acquisition is likely to be the greater volume of soil exploited by long root hair varieties, as evidenced by differential zones of depletion around roots of various barley root-hair mutants (Fig. 10.1) (Gahoonia and Nielsen 1997).

10.2.3 Release of Extracellular Organic Anions

Many studies have shown that organic anions modify the chemistry of the rhizosphere and mobilise various forms of inorganic and organic P. This is achieved by an increase in the dissolution of sparingly soluble P minerals, reduced sorption of P by alteration of the surface characteristics of soil particles, desorption of Pi from sorption sites (ligand exchange and ligand dissolution), and through the chelation of cations (e.g. Al and Fe in acidic soils or Ca in alkaline soils) that are commonly associated or complexed with Pi in soil (Bar-Yosef 1991; Jones and Darrah 1994; Lan et al. 1995; Jones 1998). Organic anions may also promote the growth of rhizosphere microorganisms that improve plant P acquisition. The importance of

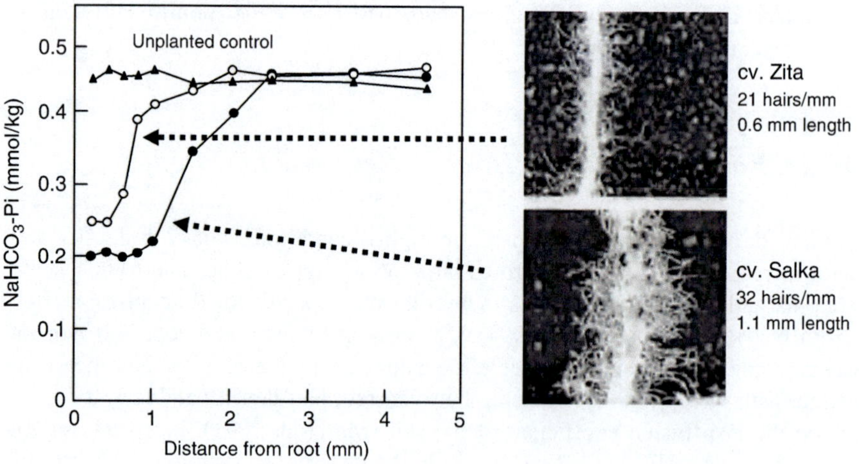

Fig. 10.1 Depletion of $NaHCO_3$-extractable inorganic P (mmol P/kg soil) from the rhizosphere of two barley cultivars (*Hordeum vulgare* cvs Zita and Salka) with different root hair morphologies. The cultivars are compared to an unplanted control soil (from Gahoonia and Nielsen 1997)

organic anions in increasing the availability of organic P, and its subsequent mineralisation by phosphatases, has also been identified recently (Jones 1998; Otani and Ae 1999; Hayes et al. 2000a; Hens et al. 2003; Li et al. 2003).

It is evident that exudation of organic anions from plant roots is facilitated by transport proteins (Neumann et al. 1999; Ryan et al. 2001). At concentrations commonly found in the rhizosphere (10–100 μM; Jones 1998) citrate and oxalate have a greater potential for P mobilisation than other organic anions. In fact, high rates of citrate exudation from cluster roots of white lupins are associated with a large capacity for P mobilisation in soil by this species (Vance et al. 2003; Richardson et al. 2007). Other plant species vary in the nature and amounts of organic anions they exude from roots (Veneklaas et al. 2003; Wouterlood et al. 2004; Pearse et al. 2006). But, generally increased organic anion efflux from roots stimulated by P-deficient conditions (Hedley et al. 1982; Lipton et al. 1987; Hoffland et al. 1989; Kirk et al. 1999), is a common phenomenon.

The heterologous expression of genes for enzymes involved in organic anion synthesis in roots has been investigated as a means to increase exudation of organic anions from roots. Overexpression of a bacterial gene encoding citrate synthase (CS) in tobacco (*Nicotiana tabacum*) has been reported to increase citrate efflux from roots of transgenic lines compared to control plants (de la Fuente-Martínez et al. 1997). However, using similar gene constructs and in some cases the same transgenic lines, Delhaize et al. (2001) could not confirm these results. Moreover, tobacco plants that overexpressed a tobacco CS, or were downregulated for isocitrate dehydrogenase expression, showed no significant increase in citrate efflux even though in some cases the plants had greater internal citrate concentration (Delhaize et al. 2003). Notwithstanding this, it is apparent that there is potential to enhance organic acid exudation by targeting the citrate synthesis biosynthetic pathway. Overexpression of a plant gene for mitochondrial CS in *Arabidopsis thaliana* enhanced citrate efflux, with an associated small improvement in P acquisition (Koyama et al. 2000).

Genes that encode channels involved in the transport of organic anions may be another target for a gene technology approach to improving tolerance to P deficit. Citrate-permeable channels in the plasma membrane of cluster roots of white lupin have been identified (Zhang et al. 2004), and a gene encoding a malate channel has been cloned from wheat (Sasaki et al. 2004). When expressed in transgenic barley (GP-ALMT1), this gene (*TaALMT1*) resulted in increased exudation of malate, albeit in an Al-activated manner (Delhaize et al. 2004), and has been demonstrated to be beneficial to the P nutrition of plants when grown in acidic soils (Fig. 10.2) (Delhaize et al. 2009).

10.2.4 Release of Extracellular Phosphatase

A number of studies have demonstrated significant rates of organic P mineralisation in proportion to soil phosphatase activity (Trasar-Cepeda and Carballas 1991;

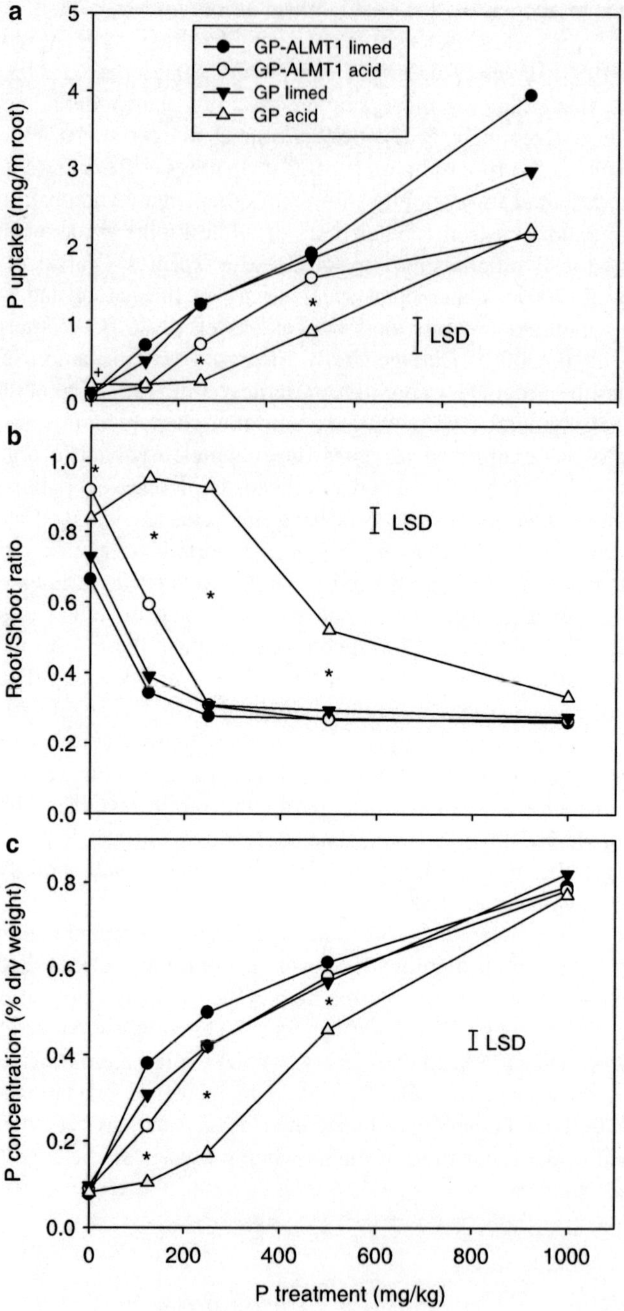

Fig. 10.2 Expression of the wheat aluminium resistance gene (*TaALMT1*) in transgenic barley enhances phosphorus uptake per unit root and reduces the root/shoot ratios of plants grown on an acid soil with a range of phosphorus supplies. Effect of phosphorus supply on phosphorus uptake

Lopez-Hernandez et al. 1998; Oehl et al. 2001; George et al. 2002). In natural ecosystems, mineralisation of soil organic P is thought to provide the major proportion of P to plants (Fox and Comerford 1992; Polglase et al. 1992). Similarly, in organic-based farming systems, and where green-manure crops are used for fertilisation, high rates of organic P cycling have been observed (Oberson et al. 1996, 2001; Nziguheba et al. 1998; Maroko et al. 1999; Oehl et al. 2004).

The hydrolysis of organic P is mediated by the action of phosphatase enzymes in the extracellular environment, a process that is necessary for the subsequent uptake of Pi by plant roots (see Nannipieri et al. 2011). At present there is no evidence for direct uptake of dissolved organic P compounds by plants, although organic P substrates may be hydrolysed within the root apoplast (Duff et al. 1994; George et al. 2008). Extracellular phosphatase activity of plant roots is induced under conditions of P deficiency and is associated with either root cell walls (McLachlan 1980; Dracup et al. 1984; Barrett-Lennard et al. 1993; Hayes et al. 1999; Hunter and McManus 1999) or is released directly into the rhizosphere (Tarafdar and Claassen 1988; Tadano et al. 1993; Li et al. 1997; Gaume et al. 2001). The cloning of genes encoding extracellular phosphatases from *A. thaliana* (Haran et al. 2000) and *L. albus* has provided direct evidence for extracellular secretion and regulation of phosphatase expression in response to P deficiency (Wasaki et al. 2000; Miller et al. 2001).

Extracellular secretion of phosphatases from roots is correlated with the ability of plants to obtain P from organic P sources when grown under sterile conditions (Tarafdar and Claassen 1988; Hayes et al. 2000b; Richardson et al. 2000; George et al. 2008). For example, wheat and a range of pasture species are able to utilise P from various monoester (e.g. glucose-6-phosphate) and diester (e.g. ribonucleic acid) forms, but show limited capacity to acquire P directly from *myo*-inositol hexakisphosphate (Richardson et al. 2000; George et al. 2008), despite inositol phosphates being an abundant form of organic P in many soils. It is likely that the biological importance of the different forms of organic P will be dictated by their turnover rates. Direct hydrolysis of organic P and subsequent utilisation of the mineralised Pi by roots has also been demonstrated in soil-grown plants. Depletion of various pools of extractable organic P from the rhizosphere has been linked with greater phosphatase activities around plant roots (Chen et al. 2002; George et al.

Fig. 10.2 (continued) by roots (**a**), root/shoot ratios (**b**), and shoot phosphorus concentrations (**c**) of transgenic barley (*GP*, *triangles*) and transgenic barley expressing *TaALMT1* (*GP-ALMT1*, *circles*) grown on an unamended acid ferrosol (*open symbols*) or on the same soil that had been limed (*closed symbols*). Phosphorus uptake per unit root was calculated from total shoot phosphorus only, because the phosphorus contents of roots could not be accurately determined because of adhering soil. Plants were harvested 26 days after sowing and the data for each show the treatment means ($n = 4$) and least-significant difference (*LSD*) ($P = 0.05$; untransformed data). *Asterisks* indicate where the means of the genotypes grown on the acid soil with a particular phosphorus treatment differed by more than the LSD (untransformed data), and *crosses* indicate where additional differences between genotypes were apparent using log10-transformed data (from Delhaize et al. 2009)

2002, 2006). However, the relative contribution of extracellular phosphatases derived from roots and from microorganisms in the utilisation of soil organic P is unclear because the numbers and activity of bacteria and fungi are greater within the rhizosphere than in the bulk soil (Chen et al. 2002; Richardson et al. 2005). In addition, there is some evidence that phosphatases derived from soil fungi have a greater affinity for organic P compounds compared to phosphatases derived from plant roots (Tarafdar et al. 2001). Either way, it is evident that the mineralisation of organic P occurs in the rhizosphere and could make an important contribution to the orthophosphate requirement of plants for growth.

Research efforts have been focussed on improving the ability of plants to acquire P directly from common forms of soil organic P, such as inositol phosphates. A number of studies have developed transgenic plants with heterologous expression of microbial phytases (Richardson et al. 2001; Zimmermann et al. 2003; Lung et al. 2005; Xiao et al. 2005). Transgenic plants that produce microbial phytase and release it from their roots have novel ability to hydrolyse P from *myo*-inositol hexakisphosphate and, when grown under controlled conditions, showed enhanced growth and P nutrition (Fig. 10.3) (Richardson et al. 2001; Mudge et al. 2003;

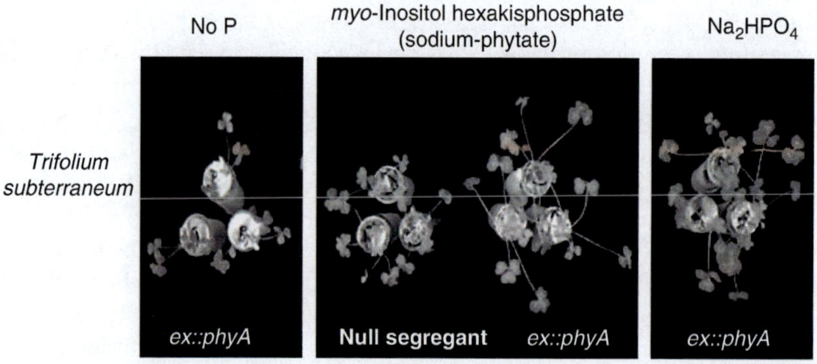

Shoot dry weight (mg/plant)	28.1	40.7	51.8	47.9
Shoot phosphorus (μg P/shoot)	48.3	103.5	299.0	305.3
Exuded root phytase activity[a] (nKat/g root dry wt)	–	1.3	107.9	–

[a]Activity for wild-type plants was 0.6 nKat/g root dry weight.

Fig. 10.3 Growth, phosphorus nutrition and activity of phytase exuded from the roots of transgenic *Trifolium subterraneum*. The images show plants that release the *Aspergillus niger* phytase (ex::phyA) as an extracellular enzyme and the corresponding null-segregant transgenic control line. Plants were grown for 28 days in sterile agar either without added phosphorus (no P) or with phosphorus supplied as sodium phytate (*myo*-inositol hexakisphosphate) or disodium phosphate (Na_2HPO_4) at 0.9 mM (with respect to phosphate) (taken from George et al. 2004)

Zimmermann et al. 2003; George et al. 2004). However, when grown in a range of soils, these plants have generally shown limited capacity to access additional P over that of control plants (George et al. 2004, 2005b). These results highlight the complexity inherent in attempting to improve multimechanistic tolerance traits by a single gene approach. Whilst potential exists for manipulating P-use efficiency at a genetic scale, success will often be limited by poor understanding of the control of the mechanisms imposed by different soil environments.

10.3 Coordinating Plant Responses to Variations in P Supply

In P-replete plants, small metabolites, nucleic acids and phospholipids contribute approximately equally to leaf P content (Marschner 1995; Dörmann and Benning 2002; White and Hammond 2008). When plants lack sufficient P, they restrict their use of P to essential cellular functions and improve the ability of their root systems to acquire P from the soil.

Many of the responses of plants to P starvation appear to be initiated, or modulated, by a decrease in the delivery of Pi to the shoot and the consequent reduction in the Pi available for shoot metabolism (Fig. 10.4, response 1). This has a direct effect on photosynthesis, glycolysis and respiration, which is reinforced by transcriptional reprogramming (Plaxton and Carswell 1999; Hammond et al. 2003,

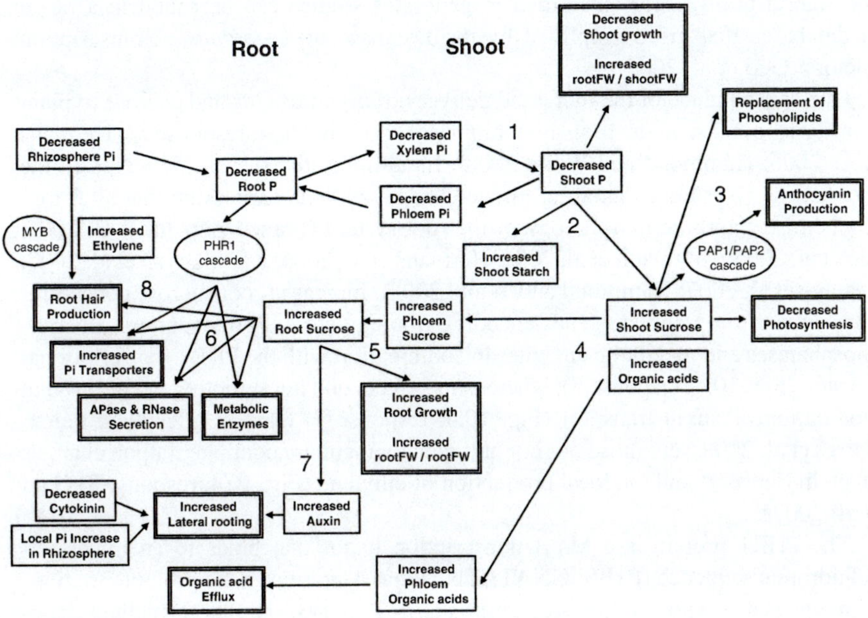

Fig. 10.4 Regulatory networks coordinating plant responses to variations in P supply. *Numbers* indicate different plant responses and are explained in the text (from White and Hammond 2008)

2005; Misson et al. 2005; Hermans et al. 2006; Wasaki et al. 2006; Morcuende et al. 2007; White and Hammond 2008). The changes in carbohydrate metabolism result in the accumulation of organic acids, starch and sucrose in leaves of P-starved plants (Fig. 10.4, response 2) (Hermans et al. 2006; Morcuende et al. 2007). Metabolism is rerouted by employing reactions that do not require Pi or adenylates and, under severe P starvation, intracellular phosphatases and nucleases are produced to remobilize P from cellular metabolites and nucleic acids (Plaxton and Carswell 1999; Hammond et al. 2003; Wasaki et al. 2006; Morcuende et al. 2007; Müller et al. 2007).

Increased leaf sucrose concentrations lead indirectly to (1) a reduction in photosynthesis through decreased expression of genes encoding many photosystem subunits and small subunits of RuBisCo (Lloyd and Zakhleniuk 2004; Amtmann et al. 2006; Hermans et al. 2006; Rook et al. 2006; Morcuende et al. 2007), (2) an increase in leaf sulfolipid and galactolipid concentrations through the upregulation of genes involved in their biosynthesis (Dörmann and Benning 2002; Hammond et al. 2003; Benning and Ohta 2005; Misson et al. 2005; Franco-Zorrilla et al. 2005; Gaude et al. 2008), and (3) the production of anthocyanins through a transcriptional cascade involving the transcription factors TTG1-TT8/EGL3-PAP1/PAP2 (Fig. 10.4, response 3) (Lloyd and Zakhleniuk 2004; Teng et al. 2005; Amtmann et al. 2006; Solfanelli et al. 2006). An increased leaf sucrose concentration also results in the upregulation of genes encoding transport proteins delivering organic acids and sucrose to the phloem, which facilitates the movement of these compounds to the root (Fig. 10.4, response 4) (Hermans et al. 2006). Details of the genes and transcription factors identified in genomics studies can be found in a range of databases that are exemplified by the Database of *Arabidopsis* Transcription Factors (Guo et al. 2005).

One consequence of the increased delivery of organic acids and sucrose to plant roots is an increase in the root/shoot biomass ratio (Fig. 10.4, response 5) (Hermans et al. 2006; Hammond and White 2008). In addition, the sucrose delivered to the root acts as a systemic signal to initiate changes in gene expression that alter root biochemistry and the morphology of the root system (Franco-Zorrilla et al. 2005; Liu et al. 2005; Amtmann et al. 2006; Hermans et al. 2006; Karthikeyan et al. 2007; Tesfaye et al. 2007; Hammond and White 2008). Increased root sucrose concentrations appear to upregulate genes encoding riboregulators, Pi transporters, RNases, phosphatases and metabolic enzymes in combination with the PHR1 transcriptional cascade (Fig. 10.4, response 6), whereas its effects on lateral rooting occur through modulation of auxin transport (Fig. 10.4, response 7) (Jain et al. 2007a; Pérez-Torres et al. 2008) and those on root hair development are contingent upon changes in auxin transport and the local production of ethylene (Fig. 10.4, response 8) (Jain et al. 2007a).

The PHR1 protein is a MYB transcription factor that binds to an imperfect-palindromic sequence (P1BS; GNATATNC) that is present in the promoter regions of many genes whose expression responds to P starvation. These include genes encoding transcription factors, protein kinases, Pi transporters, RNases, phosphatases, metabolic enzymes and enzymes involved in the synthesis of sulfolipids and

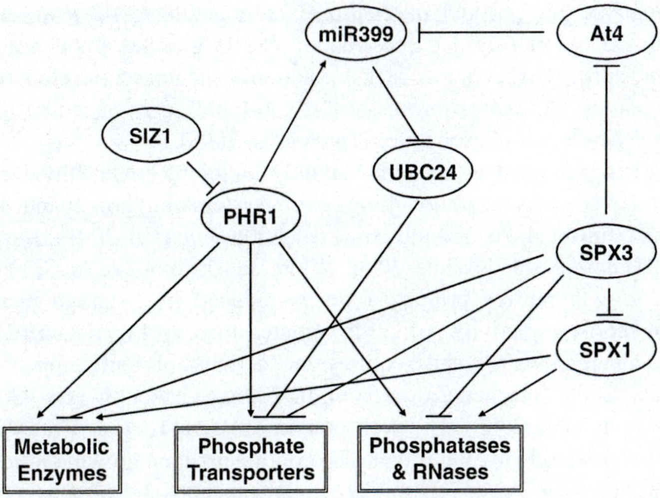

Fig. 10.5 Gene Regulatory networks impacting on PHR1-mediated acclimatory responses to P starvation. *Arrows* indicate positive regulation. *Blunt-ended lines* indicate negative regulation

galactolipids (Fig. 10.5) (Rubio et al. 2001; Hammond et al. 2004; Franco-Zorrilla et al. 2004; Misson et al. 2005; Jain et al. 2007b; Fang et al. 2009; Lin et al. 2009). The expression of *PHR1* appears to be constitutive, but the PHR1 protein is targeted by a small ubiquitin-like modifier (SUMO) E3 ligase (SIZ1), whose expression is increased by P starvation (Miura et al. 2005). The activity of SIZ1 acts as a negative regulator of plant responses to P starvation (Miura et al. 2005). The PHR1-mediated increase in Pi transport is contingent upon the activity of PHF1, an ER protein that facilitates the trafficking of PHT1-family Pi transporters, whose expression is upregulated upon P starvation (González et al. 2005). Amongst the targets of the PHR1 protein are members of the miR399 microRNA family and the SPX gene family (Bari et al. 2006; Franco-Zorrilla et al. 2007; Nilsson et al. 2007; Lin et al. 2009). The expression of *miR399s* is specifically and rapidly upregulated by P starvation (Chiou 2007). The target for miR399s is *AtUBC24*, which is downregulated during P starvation. This gene encodes the ubiquitin E2 conjugating enzyme responsible for the *pho2* mutant phenotype, which is thought to downregulate the transcription of a subset of P-starvation-responsive genes through intermediary transcription factors (Chiou 2007; Fang et al. 2009). Expression of *AtUBC24* in roots appears to be regulated systemically by shoot P status and the translocation of *miR399s* in the phloem (Buhtz et al. 2008; Lin et al. 2008; Pant et al. 2008). The rate of their translocation in the phloem is likely to be influenced indirectly by sucrose loading and unloading in the shoot and root, respectively. Members of the *TPSI1/Mt4/At4* family of non-coding transcripts, whose expression is rapidly and specifically induced in response to P starvation, appear to bind and sequester miR399s thereby attenuating miR399-mediated transcriptional responses to P starvation (Franco-Zorrilla et al. 2007). In *Arabidopsis*, the expression of *AtSPX1* and *AtSPX3* are greatly increased by P starvation (Duan et al. 2008). Increased expression of

AtSPX1 upregulates transcription of several genes, including *PAP2*, *RNS1* and *ACP5* (Duan et al. 2008). Increased expression of AtSPX3 occurs upon prolonged P starvation and appears to act in feedback regulation of plant responses to P starvation by downregulating the expression of *AtSPX1*, *At4* and genes encoding several Pi transporters, RNases and phosphatases (Duan et al. 2008).

Crosstalk between local and systemic signals (including auxin, ethylene, cytokinin and sucrose) controls the remodelling of root system morphology in response to P starvation (White et al. 2005; Amtmann et al. 2006; Jain et al. 2007a; Karthikeyan et al. 2007; Hammond and White 2008; White and Hammond 2008; Fang et al. 2009). The growth rate of primary roots is reduced in P-starved plants by a reduction of meristem activity, which is initiated directly by contact of the root cap with media lacking Pi and requires the activity of multicopper oxidases (Ticconi et al. 2004; Sánchez-Calderón et al. 2006; Svistoonoff et al. 2007; Jain et al. 2007a; Fang et al. 2009). The proliferation of lateral roots of P-starved plants in regions of increased Pi availability is also contingent upon growth of the primary root apex through these regions (Drew 1975), but appears to be initiated by changes in auxin transport and perception (Nacry et al. 2005; Sánchez-Calderón et al. 2006; Jain et al. 2007a; Hammond and White 2008; Pérez-Torres et al. 2008), with greater sucrose availability increasing the responsiveness to auxin (Nacry et al. 2005; Jain et al. 2007a). Specifically, the *TIR1* gene, which encodes the auxin receptor component of the ubiquitin protein ligase complex SCFTIR1, is upregulated by P starvation (Pérez-Torres et al. 2008). The upregulation of *TIR1* results in the degradation of AUX/IAA auxin response repressors, allowing the expression of ARF transcription factors, such as *ARF19*, to modulate the expression of genes that enable the initiation and emergence of lateral roots without increasing root auxin concentrations (Pérez-Torres et al. 2008). The initiation of lateral roots is also promoted by reduced cytokinin concentrations in roots of P-deficient plants, which appears to be a secondary consequence of the crosstalk between sugar and local P-signalling cascades (Franco-Zorrilla et al. 2005). This phenomenon is comparable to the proliferation of specialised cluster roots in regions of local Pi enrichment observed in diverse non-mycorrhizal plant species when they lack sufficient P (Lamont 2003; Lambers et al. 2006; Vance 2008). The initiation and elongation of root hairs are stimulated by locally elevated concentrations of auxin and ethylene, and both are stimulated when more sucrose is available to the roots (Jain et al. 2007a; Hammond and White 2008). Finally, the topsoil-foraging phenotype of P-deficient plants appears to be modulated primarily by the sensitivity of root gravitropism to ethylene, which increases with P starvation (Basu et al. 2007).

10.4 Response of Plants to P Re-supply

A root growing in soil is likely to find sites with a high concentration of potentially available P and sites with almost no P (Hodge 2009), though how the plant senses and reacts to this is yet unknown. There will be microsites with active microflora and/or microflora where competition for nutrients is great. An individual root will then experience local P competition, P-limiting and P-sufficient conditions at different times.

As highlighted above, the physiological state of a P-deficient plant is quite specific and the response is multigenic in nature with, for example, over 1,000 genes being differentially regulated under these conditions in *Arabidopsis* (Wu et al. 2003; Hammond et al. 2003; Morcuende et al. 2007). Under conditions of P starvation, plants have increased root/shoot biomass ratio (Lynch 1995), alteration of root architecture (Williamson et al. 2001; López-Bucio et al. 2000), many more lateral roots and long root hairs (Bates and Lynch 1996). Also high-affinity P transporters are more abundant (Mudge et al. 2002; Smith et al. 2003) and organic acids and phosphatases are synthesized and secreted (Raghothama 1999; del Pozo et al. 1999; Li et al. 2002). There are fewer P-containing metabolites (Zrenner et al. 2006), phospholipids are replaced in part by sulfolipids and galactolipids (Dörmann and Benning 2002; Kelly et al. 2003), and cells have a reduced level of RNA (Hewitt et al. 2005). Vacuolar Pi has been remobilised, carbohydrates such as starch and sugars have been accumulated and anthocyanin has been produced. In addition, some plants may have begun processes of senescence or flowering (Morcuende et al. 2007).

The hypothetical transcriptional response of plants to P re-supply or upon discovery of a P resource in a heterogeneous environment would be to reverse many of these changes and we consider that this response will take several forms:

1. Initially, there is an immediate non-specific response to perturbation of the system, which is likely to be rapid and transient and shows typical characteristics of other perturbation responses, involving increased expression of genes that are likely to protect plants against abiotic and biotic stresses (AbuQamar et al. 2009). It has previously been suggested that these may be cell-autonomous and related to changes in cell membrane potential, initiating cytosolic Ca^{2+} signalling cascades (Hammond et al. 2003; Amtmann et al. 2006). There is considerable crosstalk between abiotic and biotic signalling pathways, and these are often integrated in the cytosolic Ca^{2+} signature.
2. Sensing of altered P availability and initiation of regulatory cascades, which will not necessarily be a reversal of those cascades initiated upon P starvation. These responses are also rapid, occurring within 30 min of P re-supply (Amtmann et al. 2006). Although many of these responses are regulated systemically by sucrose concentration in P-deficient plants, upon re-supply of P the necessary gene cascades change regulation far in advance of any changes in sucrose content, suggesting other signalling pathways (Amtmann et al. 2006).
3. Reversal of tissue P economy. Morcuende et al. (2007) demonstrated that upon P re-supply there is rapid (<3 h) upregulation of nucleic acid synthesis to promote growth and re-optimise metabolic pathways for energy production and reversal of sulfolipid and galactolipid synthesis, to allow sulfur to become available for protein synthesis.

4. Reducing energetic investment in costly P-tolerance mechanisms. Morcuende et al. (2007) demonstrated that expression of genes regulating root growth, organic acid and phosphatases synthesis and efflux are also rapidly downregulated upon P re-supply.
5. Upregulation of mechanisms to prevent Pi toxicity such as sequestration and complexation, due to the persistence of Pi-transport proteins.

Results from our own c-DNA microarray experiments, in which *A. thaliana* was grown in P-deficient conditions then re-supplied with P, generally support these hypotheses of the response of P-starved plants to P re-supply. There was a distinct time-dependent response in the roots (Table 10.1) but this time dependence was not obvious in the leaves (data not shown). Most transcripts reacted after 3 h. Some transcripts increased initially and decreased within 24 h. In the first 45 min of P re-supply, the largest proportion of genes differentially regulated are associated with initiating cell production and include genes associated with cell function (3.5%), DNA synthesis (5.2%), protein degradation and synthesis (8.4%), RNA processing and regulation (2.7%) and amino acid metabolism (1.3%). After 3 h of re-supply of P, many of the same genes are still differentially regulated but a number of other processes have started to occur, notably cascades involved in cell wall development and lipid metabolism (3.1%), signalling cascades (8.8%) and sulfur assimilation (4.8%), most likely promoting protein synthesis. Beyond this, and between 0.5 and 2 days after re-supplying P, while the RNA processing, protein metabolism and cell division pathways are still differentially regulated, the DNA synthesis genes have equilibrated and differential regulation of signalling cascades has moderated. Interestingly, it is only at this stage that abiotic stress genes are differentially regulated, much later than anticipated in the hypotheses. It is also at this stage that we see genes involved in carbohydrate metabolism and transporters differentially regulated.

So, from this single study it is apparent that the earliest responses of plants to P re-supply is to upregulate cellular molecular machinery, which is quickly followed by cell division and lipid and protein biosynthesis. It is only later that the plant shows a generic stress response and upregulation of specific signalling cascades and P transport. The key result from this and other studies is that the vast majority (~35%) of differentially regulated genes at all time points are of unknown origin, which means that there are a lot of response mechanisms and regulatory cascades associated with P re-supply that are yet to be understood.

10.5 Can P Starvation and Re-supply Responses Be Genetically Manipulated for Agricultural Benefit?

Increased pressure on P fertiliser usage and costs due to the depletion of non-renewable natural resources (Heffer et al. 2006; Cordell et al. 2009; Gilbert 2009), their potential negative impacts on local environments and water quality (White and

Table 10.1 Differential regulation of genes in root and shoot tissue upon re-supply of P to roots in comparison to a salt-stress control

Cell process		Number of genes	Percentage of genes differentially regulated				Cell process		Number of genes	Percentage of genes differentially regulated			
			0.75 h	3 h	12 h	48 h				0.75 h	3 h	12 h	48 h
Cell	Cell	162	3.5	2.8	2.8	–	Nucleotide metabolism	Synthesis	12	–	0.2	–	–
	Division	24	0.5	–	–	–		Degradation	10	–	0.2	–	–
	Cycle	32	0.7	0.1		–	RNA	RNA	603	–	–	11.8	–
	Vesicle transport	34	–	0.7	0.6	–		Processing	51	1.1	–	–	–
	Late embryogenesis	5	0.1	–	–	–		Transcription	27	–	–	0.5	–
Cell wall	Cell wall	73	–	1.4	1.4	–		Regulation of transcription	477	1.6	0.9	9.3	0.9
	Precursor synthesis	2	–	–	–	–	DNA	DNA	150	3.0	2.9	–	–
	Cell wall proteins	14	–	0.1	0.3	–		Synthesis/chromatin structure	103	2.2	2.0	0.1	–
	Degradation	15	–	–	–	0.4		Unspecified	23	–	–	–	0.6
	Modification	13	–	0.2	–	0.4	Protein	Synthesis	129	1.6	2.5	2.3	2.5
Glycolysis	Glycolysis	–	–	–	0.3	–		Postranslational modification	163	0.2	2.9	3.2	3.4
	Mitochondrial e-transport/ATP synthesis	39	–	0.6	0.8	0.7		Degradation	367	6.8	7.0	6.7	0.2
Lipid	Lipid metabolism	–	–	1.7	–	–		Protein folding	21	–	–	0.4	–
	Exotics	18	–	0.3	–	–	Amino acid	Amino acid metabolism	69	1.3	1.3	1.2	–
	Phospholipase D	4	–	–	0.1	–		Synthesis	44	–	0.2	0.9	0.9
Carbohydrate (CHO) metabolism	Raffinose family	4	–	0.1	–	4.7		Degradation	28	0.5	0.5	–	0.4
	Trehalose	7	0.2	0.1	–	0.3	Hormones	Hormone metabolism	118	0.1	2.2	0.4	0.1
	Myo-inositol	2	–	–	–	–	Signalling	Signalling	248	–	4.7	–	–
	CHO metabolism	24	–	0.5	–	–		Receptor kinases	92	0.2	1.8	0.3	0.7
	Synthesis	10	–	0.2	–	–		Calcium	40	0.9	–	–	–

(*continued*)

Table 10.1 (continued)

Cell process		Number of genes	Percentage of genes differentially regulated				Cell process		Number of genes	Percentage of genes differentially regulated			
			0.75 h	3 h	12 h	48 h				0.75 h	3 h	12 h	48 h
Stress	Stress	–	–	–	3.1	–		Phosphoinositides	4	0.1	–	–	–
	PR-proteins	4	–	–	–	0.1		Miscellaneous	3	–	0.1	–	–
	Abiotic	98	–	–	1.9	–	Transport	Transport	191	0.2	0.9	3.7	4.0
	Heat	61	1.0	1.1	1.2	1.2	Nitrogen	Metabolism	5	–	–	0.1	–
	Cold	6	0.1	0.1	–	–	Sulfur	Assimilation	252	0.1	4.8	0.1	0.1
	Unspecified	22	0.4	0.4	0.3	0.4	Miscellaneous	Miscellaneous	252	5.4	0.7	1.1	1.2
Redox	Redox regulation	42	0.2	0.6	0.8	0.8	Not Assigned		1,833	33.9	33.9	35.8	35.0
Polyamine	Metabolism	3	–	–	0.1	–							

The plants were grown in a P-poor soil for 8 weeks and after that P was added to increase the P availability fourfold. At the same time, other plants were also treated with a salt stress to remove any general abiotic stress responses. Phosphorus addition resulted in an extractable P concentration increased by 20 times in the soils. Roots and above ground tissue were sampled initially, after 45 min, 3, 12, and 48 h. c-DNA microarray experiments were carried out to assess the percentage of genes differentially regulated. Missing values indicate non-detectable changes

Hammond 2008), and the energy required and carbon dioxide evolved in their production and use (Helsel 1992; Jenssen and Kongshaug 2003), have increased the need to manage P fertiliser input more carefully. Over 85% of P mined is used in food production (Heffer et al. 2006) and peak P production (akin to peak oil) is estimated to occur by 2033 (Raven 2008; Cordell et al. 2009). These pressures will be exacerbated by increasing demand on food production systems as the human population increases, and by fluctuation in oil prices (Cordell et al. 2009). The breeding of new crop varieties that yield well with reduced P fertiliser inputs is now a priority for sustainable agriculture in the future.

Breeding crops that acquire and/or use P more efficiently is one strategy to reduce the use of P fertilisers. Such crops could produce comparable yields with lower inputs of inorganic Pi fertilisers or have reduced physiological P requirements and tissue P concentrations, thus reducing the amount of P removed by the crop and, thereby, the amount of P needed to maintain the availability of Pi in the soil. New varieties can be bred conventionally, based on trait-focused screens of germplasm collections. However, a great deal of information is now available about how plants regulate P homeostasis and acquisition from the soil, particularly at the genetic level (Franco-Zorrilla et al. 2004; Jain et al. 2007b; Hammond and White 2008; White and Hammond 2008: Fang et al. 2009; Lin et al. 2009). Mutants with allelic variation and/or altered expression of genes affecting P acquisition or P use within the plant have been generated. Several of these mutants illustrate strategies for developing crop plants that acquire and/or use P more efficiently.

Mutations that improve P acquisition from the soil could improve crop growth when P availability in the soil is poor. Transgenic plants that secrete microbial phytases into the rhizosphere have the potential to release P from inositol phosphates and show enhanced growth and P nutrition when inositol hexaphosphate is the major source of P (Richardson et al. 2001; Mudge et al. 2003; Zimmermann et al. 2003; George et al. 2004, 2005a). However, when grown in most soils, these plants have comparable growth and P nutrition to control plants (George et al. 2004, 2005b). Similarly, overexpression of a bacterial gene encoding citrate synthase in tobacco has been reported to increase citrate efflux from roots and to increase the availability of P from Ca-P (de la Fuente-Martínez et al. 1997; López-Bucio et al. 2000), but an effect on plant growth and P acquisition is not always observed (Delhaize et al. 2001). The expression of a wheat malate transporter gene (*ALMT1*) in barley has been shown to be effective in increasing P uptake by transgenic plants, but only in severely acidic soil conditions (Delhaize et al. 2009). Increased expression of specific phosphate transporters has been shown to increase biomass accumulation in tobacco cell cultures under P-limiting conditions (Mitsukawa et al. 1997), but did not enhance P uptake rates or growth of transgenic barley in soil (Rae et al. 2004).

Mutations altering root morphology also have the potential to enable plants to acquire more P. For example, barley genotypes with long root hairs have higher yields than genotypes with no root hairs on soils with low P availability (Brown et al. 2010), and genotypes of bean, maize and brassica with larger root systems have better growth under P-limiting conditions (Rubio et al. 2003; Liu et al. 2004; Hammond et al. 2009). The overexpression of miR399, or the downregulation of

UBC24 (*pho2*) expression, results in greater accumulation of P (Delhaize and Randall 1995; Aung et al. 2006; Bari et al. 2006; Chiou et al. 2006). A T-DNA insertional knockout of AtSIZ1 caused *Arabidopsis* to exhibit exaggerated Pi starvation responses, including cessation of primary root growth, extensive lateral root and root hair development, increase in root/shoot biomass quotient, and greater anthocyanin accumulation, even though intracellular Pi levels in *siz1* plants were similar to those in the wild type. All three mutants exhibit constitutive P-deficiency symptoms, including increased P uptake, which might be beneficial in some agricultural systems.

Mutations that improve crop growth when soil P availability is low, through better physiological utilisation of P, may also be useful in breeding crops for reduced P inputs. For example, OsPTF1, a bHLH transcription factor from rice, whose expression increases in the roots of P-starved plants, has been shown to enhance tolerance to P starvation (Yi et al. 2005). Also, transgenic tobacco cells that lack an alternative oxidase had improved growth under P-limiting conditions (Parsons et al. 1999).

10.6 Concluding Remarks

Management of soil P remains a crucial issue for the economic and environmental sustainability of agriculture and natural ecosystems globally. It is therefore essential that we have appropriate understanding of the mechanisms by which plants are able to acquire P from soil. In this chapter, various processes and physiological traits of plants that facilitate the availability and acquisition of P from soil have been outlined and some possibilities for deploying these traits into agricultural germplasm discussed. Better understanding of these processes and development of improved germplasm may ultimately improve the P-use efficiency of agriculture systems and provide valuable information for wider-scale land and resource management. However, at present it is evident that the full extent of the complexity of the gene-by-gene, and gene-by-environment interactions that are associated with plant P nutrition are not well appreciated, and that our comprehension of the functional redundancy and compatibility of different mechanisms both within individual plants and between coexisting organisms is poor. It is therefore important that a systems approach to P management continues to be developed for a more sustainable agriculture.

References

AbuQamar S, Luo HL, Laluk K, Mickelbart MV, Mengiste T (2009) Crosstalk between biotic and abiotic stress responses in tomato is mediated by the *AIM1* transcription factor. Plant J 58:347–360

Amtmann A, Hammond JP, Armengaud P, White PJ (2006) Nutrient sensing and signalling in plants: potassium and phosphorus. Adv Bot Res 43:209–257

Attiwill PM, Adams MA (1993) Nutrient cycling in forests. New Phytol 124:561–582

Aung K, Lin SI, Wu CC, Huang YT, Su CL, Chiou TJ (2006) *pho2*, a phosphate overaccumulator, is caused by a nonsense mutation in a microRNA399 target gene. Plant Physiol 141:1000–1011
Barber SA (1984) Soil nutrient bioavailability: a mechanistic approach. Wiley, New York
Bari R, Pant BD, Stitt M, Scheible WR (2006) PHO2, microrna399, and PHR1 define a phosphate-signaling pathway in plants. Plant Physiol 141:988–999
Barrett-Lennard EG, Dracup M, Greenway H (1993) Role of extracellular phosphatase in the phosphorus-nutrition of clover. J Exp Bot 44:1595–1600
Bar-Yosef B (1991) Root excretions and their environmental effects. Influence on availability of phosphorus. In: Waisel Y, Eshel A, Kafkafi U (eds) Plant roots: the hidden half. Dekker, New York, pp 529–557
Basu P, Zhang YJ, Lynch JP, Brown KM (2007) Ethylene modulates genetic, positional, and nutritional regulation of root plagiogravitropism. Funct Plant Biol 34:41–51
Bates TR, Lynch JP (1996) Stimulation of root hair elongation in *Arabidopsis thaliana* by low phosphorus availability. Plant Cell Environ 19:529–538
Benning C, Ohta H (2005) Three enzyme systems for galactoglycerolipid biosynthesis are coordinately regulated in plants. J Biol Chem 280:2397–2400
Bieleski RL (1973) Phosphate pools, phosphate transport and phosphate availability. Annu Rev Plant Physiol 24:225–252
Boot GA, Mensink M (1990) Size and morphology of root systems of perennial grasses from contrasting habitats as affected by nitrogen supply. Plant Soil 129:291–299
Bradshaw AD, Chadwick MJ, Jowett D, Lodge RW, Snaydon RW (1960) Experimental investigations into the mineral nutrition of several grass species. III. Phosphate level. J Ecol 48:631–637
Brown LK, George TS, Thompson JA, Wright G, Lyon J, Hubbard SF, White PJ (2010) What are the implications of variation in root hair length on P-limited yield in barley (*Hordeum vulgare* L.)? Plant Cell Environ (in press)
Bucher M (2007) Functional biology of plant phosphate uptake at root and mycorrhiza interfaces. New Phytol 173:11–26
Buhtz A, Springer F, Chappell L, Baulcombe D, Kehr J (2008) Identification and characterization of small RNAs from the phloem of *Brassica napus*. Plant J 53:739–749
Chapin FS (1980) The mineral nutrition of wild plants. Annu Rev Ecol Syst 11:233–260
Chen CR, Condron LM, Davis MR, Sherlock RR (2002) Phosphorus dynamics in the rhizosphere of perennial ryegrass (*Lolium perenne* L.) and radiata pine (*Pinus radiata* D. Don). Soil Biol Biochem 34:487–499
Chiou TJ (2007) The role of microRNAs in sensing nutrient stress. Plant Cell Environ 30:323–332
Chiou TJ, Aung K, Lin SI, Wu CC, Chiang SF, Su CI (2006) Regulation of phosphate homeostasis by microRNA in *Arabidopsis*. Plant Cell 18:412–421
Christie EK (1975) Physiological responses of semiarid grasses. II. The pattern of root growth in relation to external phosphorus concentration. Aust J Agric Res 26:437–446
Christie EK, Moorby J (1975) Physiological responses of semiarid grasses. I. The influence of phosphorus supply on growth and phosphorus absorption. Aust J Agric Res 26:423–436
Condron LM, Frossard E, Tiessen H, Newman RH, Stewart JWB (1990) Chemical nature of organic phosphorus in cultivated and uncultivated soils under different environmental conditions. J Soil Sci 41:41–50
Cordell D, Drangert J-O, White S (2009) The story of phosphorus: global food security and food for thought. Glob Environ Change 19:292–305
de la Fuente-Martínez JM, Ramirez-Rodriguez V, Cabrera-Ponce JL, Herrera-Estrella L (1997) Aluminum tolerance in transgenic plants by alteration of citrate synthesis. Science 276: 1566–1588
del Pozo JC, Allona I, Rubio V, Leyva A, De la Pena A, Aragoncillo C, Paz-Ares J (1999) Type 5 acid phosphatase gene from *Arabidopsis thaliana* is induced by phosphate starvation and by some other types of phosphate mobilising/oxidative stress conditions. Plant J 19:579–589
Delhaize E, Randall PJ (1995) Characterisation of a phosphate-accumulator mutant of *Arabidopsis thaliana*. Plant Physiol 107:207–213

Delhaize E, Hebb DM, Ryan PR (2001) Expression of a *Pseudomonas aeruginosa* citrate synthase gene in tobacco is not associated with either enhanced citrate accumulation or efflux. Plant Physiol 125:2059–2067

Delhaize E, Ryan PR, Hocking PJ, Richardson AE (2003) Effects of altered citrate synthase and isocitrate dehydrogenase expression on internal citrate concentrations in tobacco (*Nicotiana tabacum* L.). Plant Soil 248:137–144

Delhaize E, Ryan PR, Hebb DM, Yamamoto Y, Sasaki T, Matsumoto H (2004) Engineering high level aluminum tolerance in barley with the ALMT1 gene. Proc Natl Acad Sci USA 101:15249–15254

Delhaize E, Gruber BD, Ryan PR (2007) The roles of organic anion permeases in aluminium resistance and mineral nutrition. FEBS Lett 581:2255–2262

Delhaize E, Taylor P, Hocking PJ, Simpson RJ, Ryan PR, Richardson AE (2009) Transgenic barley (*Hordeum vulgare* L.) expressing the wheat aluminium resistance gene (*TaALMT1*) shows enhanced phosphorus nutrition and grain production when grown on an acid soil. Plant Biotechnol J 7:391–400

Dinkelaker B, Hengeler C, Marschner H (1995) Distribution and function of proteoid and other root clusters. Bot Acta 108:183–200

Dörmann P, Benning C (2002) Galactolipids rule in seed plants. Trends Plant Sci 7:112–118

Dracup MNH, Barrett-Lennard EG, Greenway H, Robson AD (1984) Effect of phosphorus deficiency on phosphatase activity of cell walls from roots of subterranean clover. J Exp Bot 35:466–480

Drew MC (1975) Comparison of the effects of a localized supply of phosphate, nitrate, ammonium and potassium on the growth of the seminal root system, and the shoot, in barley. New Phytol 75:479–490

Duan K, Yi K, Dang L, Huang H, Wu W, Wu P (2008) Characterization of a sub-family of *Arabidopsis* genes with the SPX domain reveals their diverse functions in plant tolerance to phosphorus starvation. Plant J 54:965–975

Duff SMG, Sarath G, Plaxton WC (1994) The role of acid phosphatases in plant phosphorus metabolism. Physiol Plant 90:791–800

Fan M, Zhu J, Richards C, Brown KM, Lynch JP (2003) Physiological roles for aerenchyma in phosphorus-stressed roots. Funct Plant Biol 30:493–506

Fang Z, Shao C, Meng Y, Wua P, Chen M (2009) Phosphate signaling in *Arabidopsis* and *Oryza sativa*. Plant Sci 176:170–180

Fitter AH (1985) Functional significance of root morphology and root system architecture. In: Fitter AH, Atkinson D, Read DJ, Useher MB (eds) Ecological interactions in soil-plant, microbes and animals. Blackwell, London, pp 87–106

Föhse D, Claassen N, Jungk A (1991) Phosphorus efficiency of plants. II. Significance of root radius, root hairs and cation-anion balance for phosphorus influx in seven plant species. Plant Soil 132:261–272

Fox TR, Comerford NB (1992) Rhizosphere phosphatase activity and phosphatase hydrolyzable organic phosphorus in two forested spodosols. Soil Biol Biochem 24:579–583

Franco-Zorrilla JM, González E, Bustos R, Linhares F, Leyva A, Paz-Ares J (2004) The transcriptional control of plant responses to phosphate limitation. J Exp Bot 55:285–293

Franco-Zorrilla JM, Martín AC, Leyva A, Paz-Ares J (2005) Interaction between phosphate-starvation, sugar, and cytokinin signaling in *Arabidopsis* and the roles of cytokinin receptors CRE1/AHK4 and AHK3. Plant Physiol 138:847–857

Franco-Zorrilla JM, Valli A, Todesco M, Mateos I, Puga MI, Rubio-Somoza I, Leyva A, Weigel D, García JA, Paz-Ares J (2007) Target mimicry provides a new mechanism for regulation of microRNA activity. Nat Genet 39:1033–1037

Gahoonia TS, Nielsen NE (1997) Variation in root hairs of barley cultivars doubled soil phosphorus uptake. Euphytica 98:177–182

Gardner WK, Parbury DG, Barber DA (1981) Proteoid root morphology and function in *Lupinus albus*. Plant Soil 60:143–147

Gaude N, Nakamura Y, Scheible W, Ohta H, Dörmann P (2008) Phospholipase C5 (NPC5) is involved in galactolipid accumulation during phosphate limitation in leaves of *Arabidopsis*. Plant J 56:28–39

Gaume A, Machler F, Deleon C, Narro L, Frossard E (2001) Low-P tolerance by maize (*Zea mays* L.) genotypes: significance of root growth, and organic acids and acid phosphatase root exudation. Plant Soil 228:253–264

George TS, Richardson AE (2008) Potential and limitations to improving crops for enhanced phosphorus utilization. In: Hammond JP, White PJ (eds) The ecophysiology of plant–phosphorus interactions. Springer, Dordrecht, pp 247–270

George TS, Gregory PJ, Robinson JS, Buresh RJ (2002) Changes in phosphorus concentrations and pH in the rhizosphere of some agroforestry and crop species. Plant Soil 246:65–73

George TS, Richardson AE, Hadobas PA, Simpson RJ (2004) Characterisation of transgenic *Trifolium subterraneum* L. which expresses *phyA* and releases extracellular phytase: growth and P nutrition in laboratory media and soil. Plant Cell Environ 27:1351–1361

George TS, Simpson RJ, Hadobas PA, Richardson AE (2005a) Expression of a fungal phytase gene in *Nicotiana tabacum* improves phosphorus nutrition in plants grown in amended soil. Plant Biotechnol J 3:129–140

George TS, Richardson AE, Smith JB, Hadobas PA, Simpson RJ (2005b) Limitations to the potential of transgenic *Trifolium subterraneum* L. plants that exude phytase when grown in soils with a range of organic P content. Plant Soil 278:263–274

George TS, Turner BL, Gregory PJ, Richardson AE (2006) Depletion of organic phosphorus from oxisols in relation to phosphatase activities in the rhizosphere. Eur J Soil Sci 57:47–57

George TS, Gregory PJ, Hocking PJ, Richardson AE (2008) Variation in root-associated phosphatase activities in wheat contributes to the utilisation of organic P substrates in vitro, but does not effectively predict P-nutrition in different soils. Environ Exp Bot 64:239–249

Gilbert N (2009) The disappearing nutrient. Nature 461:716–718

González E, Solano R, Rubio V, Leyva A, Paz-Ares J (2005) PHOSPHATE TRANSPORTER TRAFFIC FACILITATOR1 is a plant-specific SEC12-related protein that enables the endoplasmic reticulum exit of a high-affinity phosphate transporter in *Arabidopsis*. Plant Cell 17: 3500–3512

Guo A, He K, Liu D, Bai S, Gu X, Wei L, Luo J (2005) DATF: a database of *Arabidopsis* transcription factors. Bioinformatics 21:2568–2569, http://datf.cbi.pku.edu.cn. Accessed 30 July 2010

Haling RE, Simpson RJ, Delhaize E, Hocking PJ, Richardson AE (2010) Effect of lime on root growth, morphology and the rhizosheath of cereals growing in an acid soil. Plant Soil 327:199–212

Hammond JP, White PJ (2008) Sucrose transport in the phloem: integrating root responses to phosphorus starvation. J Exp Bot 59:93–109

Hammond JP, Bennett MJ, Bowen HC, Broadley MR, Eastwood DC, May ST, Rahn C, Swarup R, Woolaway KE, White PJ (2003) Changes in gene expression in *Arabidopsis* shoots during phosphate starvation and the potential for developing smart plants. Plant Physiol 132:578–596

Hammond JP, Broadley MR, White PJ (2004) Genetic responses to phosphorus deficiency. Ann Bot 94:323–332

Hammond JP, Broadley MR, Craigon DJ, Higgins J, Emmerson Z, Townsend H, White PJ, May ST (2005) Using genomic DNA-based probe-selection to improve the sensitivity of high-density oligonucleotide arrays when applied to heterologous species. Plant Methods 1:10

Hammond JP, Broadley MR, White PJ, King GJ, Bowen HC, Hayden R, Meacham MC, Mead A, Overs T, Spracklen WP, Greenwood DJ (2009) Shoot yield drives phosphorus use efficiency in *Brassica oleracea* and correlates with root architectural traits. J Exp Bot 60:1953–1968

Haran S, Logendra S, Seskar M, Bratanova M, Raskin I (2000) Characterization of *Arabidopsis* acid phosphatase promoter and regulation of acid phosphatase expression. Plant Physiol 124:615–626

Hawkes GE, Powlson DS, Randall EW, Tate KR (1984) A ^{31}P nuclear magnetic resonance study of the phosphorus species in alkali extracts of soils from long-term field experiments. J Soil Sci 35:35–45

Hayes JE, Richardson AE, Simpson RJ (1999) Phytase and acid phosphatase activities in roots of temperate pasture grasses and legumes. Aust J Plant Physiol 26:801–809

Hayes JE, Richardson AE, Simpson RJ (2000a) Components of organic phosphorus in soil extracts that are hydrolysed by phytase and acid phosphatase. Biol Fertil Soils 32:279–286

Hayes JE, Simpson RJ, Richardson AE (2000b) The growth and phosphorus utilisation of plants in sterile media when supplied with inositol hexaphosphate, glucose 1-phosphate or inorganic phosphate. Plant Soil 220:165–174

Hedley MJ, White RE, Nye PH (1982) Plant-induced changes in the rhizosphere of rape (*Brassica napus* var. Emerald) seedlings. III. Changes in L value, soil phosphate fractions and phosphatase activity. New Phytol 91:45–56

Heffer P, Prud'homme MPR, Muirheid B, Isherwood KF (2006) Phosphorus fertilisation: issues and outlook. Proceedings 586. International Fertiliser Society, York

Helsel ZR (1992) Energy and alternatives for fertiliser and pesticide use. In: Fluck RC (ed) Energy in world agriculture, vol 6. Elsevier, Oxford, pp 177–210

Hens M, Turner BL, Hocking PJ (2003) Chemical nature and bioavailability of soil organic phosphorus mobilized by organic anions. In: Rengel Z (ed) Proceedings of the second international symposium on phosphorus dynamics in the soil-plant continuum. Uniprint: University of Western Australia, Perth, pp 16–17

Hermans C, Hammond JP, White PJ, Verbruggen N (2006) How do plants respond to nutrient shortage by biomass allocation? Trends Plant Sci 11:610–617

Hewitt MM, Carr JM, Williamson CL, Slocum RD (2005) Effects of phosphate limitation on expression of genes involved in pyrimidine synthesis and salvaging in *Arabidopsis*. Plant Physiol Biochem 43:91–99

Hill JO, Simpson RJ, Moore AD, Chapman DF (2006) Morphology and response of roots of pasture species to phosphorus and nitrogen nutrition. Plant Soil 286:7–19

Hinsinger P (2001) Bioavailability of soil inorganic P in the rhizosphere as affected by root induced chemical changes: a review. Plant Soil 237:173–195

Hodge A (2009) Root decisions. Plant Cell Environ 32:628–640

Hoffland E, Findenegg GR, Nelemans JA (1989) Solubilization of rock phosphorus by rape. II. Local root exudation of organic acids in response to P-starvation. Plant Soil 113:161–165

Hunter DA, McManus MT (1999) Comparison of acid phosphatase in two genotypes of white clover with different responses to applied phosphate. J Plant Nutr 22:679–692

Itoh S, Barber SA (1983) Phosphorus uptake by six plant species as related to root hairs. Agron J 75:457–461

Jain A, Poling MD, Karthikeyan AS, Blakeslee JJ, Peer WA, Titapiwatanakun B, Murphy AS, Raghothama KG (2007a) Differential effects of sucrose and auxin on localized phosphate deficiency-induced modulation of different traits of root system architecture in *Arabidopsis*. Plant Physiol 144:232–247

Jain A, Vasconcelos MJ, Raghothama KG, Sahi SV (2007b) Molecular mechanisms of plant adaptation to phosphate deficiency. Plant Breed Rev 29:359–419

Jansa J, Finlay R, Wallander H, Smith FA, Smith SE (2011) Role of mycorrhizal symbioses in phosphorus cycling. In: Bünemann EK, Oberson A, Frossard E (eds) Phosphorus in action: biological processes in soil phosphorus cycling. Soil biology, vol 26. Springer, Heidelberg. doi:10.1007/978-3-642-15271-9_6

Jenssen TK, Kongshaug G (2003) Energy consumption and greenhouse gas emissions in fertiliser production. Proceedings 509. International Fertiliser Society, York

Jones DL (1998) Organic acids in the rhizosphere – a critical review. Plant Soil 205:25–44

Jones DL, Darrah PR (1994) Role of root derived organic acids in the mobilization of nutrients in the rhizosphere. Plant Soil 166:247–257

Jones DL, Dennis PG, Owen AG, van Hees PAW (2003) Organic acid behavior in soils – misconceptions and knowledge gaps. Plant Soil 248:31–41

Karthikeyan AS, Varadarajan DK, Jain A, Held MA, Carpita NC, Raghothama KG (2007) Phosphate starvation responses are mediated by sugar signaling in *Arabidopsis.* Planta 225:907–918

Keerthisinghe G, Hocking PJ, Ryan PR, Delhaize E (1998) Effect of phosphorus supply on the formation and function of proteoid roots of white lupin (*Lupinus albus* L.). Plant Cell Environ 21:467–478

Kelly AA, Froehlich JE, Dörmann P (2003) Disruption of the two digalactosyldiacylglycerol synthase genes DGD1 and DGD2 in *Arabibopsis* reveals the existence of an additional enzyme of galactolipid synthase. Plant Cell 15:2694–2706

Kirk GJD, Santos EE, Findenegg GR (1999) Phosphate solubilization by organic anion secretion from rice (*Oryza sativa* L.) growing in aerobic soil. Plant Soil 211:11–18

Kirkby EA, Johnston AE (2008) Soil and fertilizer phosphorus in relation to crop nutrition. In: Hammond JP, White PJ (eds) The ecophysiology of plant–phosphorus interactions. Springer, Dordrecht, pp 177–223

Koyama H, Kanamura A, Kihara T, Hara T, Takita E, Shibata D (2000) Overexpression of mitochondrial citrate synthase in *Arabidopsis thaliana* improved growth on a phosphorus limited soil. Plant Cell Physiol 41:1030–1037

Lambers H, Shane MW, Cramer MD, Pearse SJ, Veneklaas EJ (2006) Root structure and functioning for efficient acquisition of phosphorus: matching morphological and physiological traits. Ann Bot 98:693–713

Lamont BB (2003) Structure, ecology and physiology of root clusters – a review. Plant Soil 248:1–19

Lan M, Comerford NB, Fox TR (1995) Organic anions effect on phosphorus release from spodic horizons. Soil Sci Soc Am J 59:1745–1749

Li M, Osaki M, Rao IM, Tadano T (1997) Secretion of phytase from the roots of several plant species under phosphorus-deficient conditions. Plant Soil 195:161–169

Li D, Zhu H, Liu K, Liu X, Leggewie G, Urdvardi M, Wang D (2002) Purple acid phosphatase of *Arabidopsis thaliana.* Comparative analysis and differential regulation by phosphate deprivation. J Biol Chem 277:27772–27781

Li L, Tang C, Rengel Z, Zhang F (2003) Chickpea facilitates phosphorus uptake by intercropped wheat from an organic phosphorus source. Plant Soil 248:297–303

Lin SI, Chiang SF, Lin WY, Chen JW, Tseng CY, Wu PC, Chiou TJ (2008) Regulatory network of microRNA399 and PHO2 by systemic signaling. Plant Physiol 147:732–746

Lin WY, Lin SI, Chiou TJ (2009) Molecular regulators of phosphate homeostasis in plants. J Exp Bot 60:1427–1438

Lipton DS, Blanchar RW, Blevins DG (1987) Citrate, malate, and succinate concentration in exudates from P-sufficient and P-stressed *Medicago sativa* L. seedlings. Plant Physiol 85:315–317

Liu Y, Mi G, Chen F, Zhang J, Zhang F (2004) Rhizosphere effect and root growth of two maize (*Zea mays* L.) genotypes with contrasting P efficiency at low P availability. Plant Sci 167:217–223

Liu JQ, Samac DA, Bucciarelli B, Allan DL, Vance CP (2005) Signaling of phosphorus deficiency-induced gene expression in white lupin requires sugar and phloem transport. Plant J 41:257–268

Lloyd JC, Zakhleniuk OV (2004) Responses of primary and secondary metabolism to sugar accumulation revealed by microarray expression analysis of the *Arabidopsis* mutant, *pho3.* J Exp Bot 55:1221–1230

López-Bucio J, de la Vega OM, Guevara-García A, Herrera-Estrella L (2000) Enhanced phosphorus uptake in transgenic tobacco plants that overproduce citrate. Nat Biotechnol 18:450–453

Lopez-Hernandez D, Brossard M, Frossard E (1998) P-isotopic exchange values in relation to P mineralization in soils with very low P-sorbing capacities. Soil Biol Biochem 30:1663–1670

Lung S-C, Chan W-L, Yip W, Wang L, Yeung EC, Lim BL (2005) Secretion of beta-propeller phytase from tobacco and *Arabidopsis* roots enhances phosphorus utilisation. Plant Sci 169:341–349

Lynch JP (1995) Root architecture and plant productivity. Plant Physiol 109:7–13

Lynch JP (2005) Root architecture and nutrient acquisition. In: BassiriRad H (ed) Nutrient acquisition by plants: an ecological perspective. Springer, Berlin, pp 147–183

Lynch JP, Brown KM (2008) Root strategies for phosphorus acquisition. In: White PJ, Hammond JP (eds) The ecophysiology of plant–phosphorus interactions. Springer, Dordrecht, pp 83–116

Ma Z, Walk TC, Marcus A, Lynch JP (2001) Morphological synergism in root hair length, density, initiation and geometry for phosphorus acquisition in *Arabidopsis thaliana*: a modeling approach. Plant Soil 236:221–235

Maroko JB, Buresh RJ, Smithson PC (1999) Soil phosphorus fractions in unfertilized fallow-maize systems on two tropical soils. Soil Sci Soc Am J 63:320–326

Marschner H (1995) Mineral nutrition of higher plants, 2nd edn. Academic, London

McGill WB, Cole CV (1981) Comparative aspects of cycling of organic C, N, S and P through soil organic matter. Geoderma 26:267–286

McLachlan KD (1980) Acid phosphatase activity of intact roots and phosphorus nutrition of plants. I. Assay conditions and phosphatase activity. Aust J Agric Res 31:429–440

Miller SS, Liu JQ, Allan DL, Menzhuber CJ, Fedorova M, Vance CP (2001) Molecular control of acid phosphatase secretion into the rhizosphere of proteoid roots from phosphorus-stressed white lupin. Plant Physiol 127:594–606

Misson J, Raghothama KG, Jain A, Jouhet J, Block MA, Bligny R, Ortet P, Creff A, Somerville S, Rolland N, Doumas P, Nacry P, Herrerra-Estrella L, Nussaume L, Thibaud M-C (2005) A genome-wide transcriptional analysis using *Arabidopsis thaliana* Affymetrix gene chips determined plant responses to phosphate deprivation. Proc Natl Acad Sci USA 102:11934–11939

Mitsukawa N, Okumura S, Shirano Y, Sato S, Kato T, Harashima S, Shibata D (1997) Overexpression of an *Arabidopsis thaliana* high affinity phosphate transporter gene in tobacco cultured cells enhances cell growth under phosphate limited conditions. Proc Natl Acad Sci USA 94:7098–7102

Miura K, Rus A, Sharkhuu A, Yokoi S, Karthikeyan AS, Raghothama KG, Baek D, Koo YD, Jin JB, Bressan RA, Yun D-J, Hasegawa PM (2005) The *Arabidopsis* SUMO E3 ligase SIZ1 controls phosphate deficiency responses. Proc Natl Acad Sci USA 102:7760–7765

Morcuende R, Bari R, Gibon Y, Zheng WM, Pant BD, Bläsing O, Usadel B, Czechowski T, Udvardi MK, Stitt M, Scheible WR (2007) Genome-wide reprogramming of metabolism and regulatory networks of *Arabidopsis* in response to phosphorus. Plant Cell Environ 30:85–112

Mudge SR, Rae AL, Diatloff E, Smith FW (2002) Expression analysis suggests novel roles for members of the Pht1 family of phosphate transporters in *Arabidopsis*. Plant J 31:341–353

Mudge SR, Smith FW, Richardson AE (2003) Root-specific and phosphate-regulated expression of phytase under the control of a phosphate transporter promoter enables *Arabidopsis* to grow on phytate as a sole phosphorus source. Plant Sci 165:871–878

Müller R, Morant M, Jarmer H, Nilsson L, Nielsen TH (2007) Genome-wide analysis of the *Arabidopsis* leaf transcriptome reveals interaction of phosphate and sugar metabolism. Plant Physiol 143:156–171

Nacry P, Canivenc G, Muller B, Azmi A, Van Onckelen H, Rossignol M, Doumas P (2005) A role for auxin redistribution in the responses of the root system architecture to phosphate starvation in *Arabidopsis*. Plant Physiol 138:2061–2074

Nannipieri P, Giagnoni L, Landi L, Renella G (2011) Role of phosphatase enzymes in soil. In: Bünemann EK, Oberson A, Frossard E (eds) Phosphorus in action: biological processes in soil phosphorus cycling. Soil biology, vol 26. Springer, Heidelberg. doi:10.1007/978-3-642-15271-9_9

Neumann G, Massonneau A, Martinoia A, Römheld V (1999) Physiological adaptations to phosphorus deficiency during proteiod root development in white lupin. Planta 208:373–382

Newman RH, Tate KR (1980) Soil phosphorus characterization by ^{31}P-nuclear magnetic resonance. Commun Soil Sci Plant Anal 11:835–842

Nilsson L, Müller R, Nielsen TH (2007) Increased expression of the MYB-related transcription factor, PHR1, leads to enhanced phosphate uptake in *Arabidopsis thaliana*. Plant Cell Environ 30:1499–1512

Nziguheba G, Palm CA, Buresh RJ, Smithson PC (1998) Soil phosphorus fractions and adsorption as affected by organic and inorganic sources. Plant Soil 198:159–168

Oberson A, Joner EJ (2005) Microbial turnover of phosphorus in soil. In: Turner BL, Frossard E, Baldwin DS (eds) Organic phosphorus in the environment. CABI, Wallingford, pp 133–164

Oberson A, Besson JM, Maire N, Sticher H (1996) Microbial processes in soil organic transformations in conventional and biological cropping systems. Biol Fertil Soils 21:138–148

Oberson A, Friesen DK, Rao IM, Bühler S, Frossard E (2001) Phosphorus transformations in an oxisol under contrasting land-use systems: the role of the microbial biomass. Plant Soil 237:197–210

Oehl F, Oberson A, Sinaj S, Frossard E (2001) Organic phosphorus mineralization studies using isotopic dilution techniques. Soil Sci Soc Am J 65:780–787

Oehl F, Frossard E, Fliessbach A, Dubois D, Oberson A (2004) Basal organic phosphorus mineralization in soils under different farming systems. Soil Biol Biochem 36:667–675

Otani T, Ae N (1999) Extraction of organic phosphorus in andosols by various methods. Soil Sci Plant Nutr 45:151–161

Pant BD, Buhtz A, Kehr J, Scheible W-R (2008) MicroRNA399 is a long-distance signal for the regulation of plant phosphate homeostasis. Plant J 53:731–738

Parsons HL, Yip JYH, Vanlerberghe GC (1999) Increased respiratory restriction during phosphate-limited growth in transgenic tobacco cells lacking alternative oxidase. Plant Physiol 121:1309–1320

Pearse SJ, Venaklaas EJ, Cawthray G, Bolland MDA, Lambers H (2006) Carboxylate composition of root exudates does not relate consistently to a crop species' ability to use phosphorus from aluminium, iron or calcium phosphate sources. New Phytol 173:181–190

Pérez-Torres C-A, López-Bucio J, Cruz-Ramírez A, Ibarra-Laclette E, Dharmasiri S, Estelle M, Herrera-Estrella L (2008) Phosphate availability alters lateral root development in *Arabidopsis* by modulating auxin sensitivity via a mechanism involving the TIR1 auxin receptor. Plant Cell 20:3258–3272

Plaxton WC, Carswell MC (1999) Metabolic aspects of the phosphate starvation response in plants. In: Lerner HR (ed) Plant responses to environmental stresses: from phytohormones to genome reorganisation. Dekker, New York, pp 349–372

Polglase PJ, Attiwill PM, Adams MA (1992) Nitrogen and phosphorus cycling in relation to stand age of *Eucalyptus regnans* F. Muell. III. Labile inorganic and organic P, phosphatase activity and P availability. Plant Soil 142:177–185

Rae AL, Jarmey JM, Mudge SR, Smith FW (2004) Over-expression of a high-affinity phosphate transporter in transgenic barley plants does not enhance phosphate uptake rates. Funct Plant Biol 31:141–148

Raghothama KG (1999) Phosphate acquisition. Annu Rev Plant Physiol Plant Mol Biol 50: 665–693

Raven JA (2008) Phosphorus and the future. In: White PJ, Hammond JP (eds) The ecophysiology of plant–phosphorus interactions. Springer, Dordrecht, pp 271–283

Richardson AE, Hadobas PA, Hayes JE (2000) Phosphomonoesterase and phytase activities of wheat (*Triticum aestivum* L.) roots and utilisation of organic phosphorus substrates by seedlings grown in sterile culture. Plant Cell Environ 23:397–405

Richardson AE, Hadobas PA, Hayes JE (2001) Extracellular secretion of *Aspergillus* phytase from *Arabidopsis* roots enables plants to obtain phosphorus from phytate. Plant J 25:641–649

Richardson AE, George TS, Hens M, Simpson RJ (2005) Utilisation of soil organic phosphorus by higher plants. In: Turner BL, Frossard E, Baldwin DS (eds) Organic phosphorus in the environment. CABI, Wallingford, pp 165–184

Richardson AE, George TS, Jakobsen I, Simpson RJ (2007) Plant utilization of inositol phosphates. In: Turner BL, Richardson AE, Mullaney EJ (eds) Inositol phosphates: linking agriculture and the environment. CABI, Wallingford, pp 242–260

Rook F, Hadingham SA, Li Y, Bevan MW (2006) Sugar and ABA response pathways and the control of gene expression. Plant Cell Environ 29:426–434

Rubio V, Linhares F, Solano R, Martín AC, Iglesias J, Leyva A, Paz-Ares J (2001) A conserved MYB transcription factor involved in phosphate starvation signaling both in vascular plants and in unicellular algae. Genes Dev 15:2122–2133

Rubio G, Liao H, Yan XL, Lynch JP (2003) Topsoil foraging and its role in plant competitiveness for phosphorus in common bean. Crop Sci 43:598–607

Runge-Metzger A (1995) Closing the cycle: obstacles to efficient P management for improved global security. In: Tiessen H (ed) Phosphorus in the global environment: transfers, cycles and management. Wiley, Chichester, pp 27–42

Ryan PR, Delhaize E, Jones DL (2001) Function and mechanism of organic anion exudation from plant roots. Annu Rev Plant Physiol Plant Mol Biol 52:527–560

Sánchez-Calderón L, López-Bucio J, Chacón-López A, Gutiérrez-Ortega A, Hernández-Abreu E, Herrera-Estrella L (2006) Characterization of low phosphorus insensitive mutants reveals a crosstalk between low P-induced determinate root development and the activation of genes involved in the adaptation of *Arabidopsis* to P deficiency. Plant Physiol 140:879–889

Sasaki T, Yamomoto Y, Ezaki B, Katsuhara M, Ahn SJ, Ryan PR, Delhaize E, Matsumoyo H (2004) A wheat gene encoding an aluminium-activated malate transporter. Plant J 37:645–653

Shane MW, de Vos M, De Roock S, Lambers H (2003) Shoot P status regulates cluster-root growth and citrate exudation in *Lupinus albus* grown with a divided root system. Plant Cell Environ 26:265–273

Shen J, Li H, Neumann G, Zhang F (2005) Nutrient uptake, cluster root formation and exudation of protons and citrate in *Lupinus albus* as affected by localized supply of phosphorus in a split-root system. Plant Sci 168:837–845

Skene KR (1998) Cluster roots: some ecological considerations. J Ecol 86:1060–1064

Smith SE, Read DJ (2007) Mycorrhizal symbiosis, 3rd edn. Elsevier, London

Smith FW, Mudge SR, Rae AL, Glassop D (2003) Phosphate transport in plants. Plant Soil 248:71–83

Solfanelli C, Poggi A, Loreti E, Alpi A, Perata P (2006) Sucrose-specific induction of the anthocyanin biosynthetic pathway in *Arabidopsis*. Plant Physiol 140:637–646

Svistoonoff S, Creff A, Reymond M, Sigoillot-Claude C, Ricaud L, Blanchet A, Nussaume L, Desnos T (2007) Root tip contact with low-phosphate media reprograms plant root architecture. Nat Genet 39:792–796

Tadano T, Ozawa K, Sakai H, Osaki M, Matsui H (1993) Secretion of acid phosphatase by the roots of crop plants under phosphorus-deficient conditions and some properties of the enzyme secreted by lupin roots. Plant Soil 155(156):95–98

Tarafdar JC, Claassen N (1988) Organic phosphorus compounds as a phosphorus source for higher plants through the activity of phosphatases produced by plant roots and microorganisms. Biol Fertil Soils 5:308–312

Tarafdar JC, Yadav RS, Meena SC (2001) Comparative efficiency of acid phosphatase originated from plant and fungal sources. J Plant Nutr Soil Sci 164:279–282

Teng S, Keurentjes J, Bentsink L, Koornneef M, Smeekens S (2005) Sucrose-specific induction of anthocyanin biosynthesis in *Arabidopsis* requires the MYB75/PAP1 gene. Plant Physiol 139:1840–1852

Tesfaye M, Liu J, Allan DL, Vance CP (2007) Genomic and genetic control of phosphate stress in legumes. Plant Physiol 144:594–603

Ticconi CA, Delatorre CA, Lahner B, Salt DE, Abel S (2004) *Arabidopsis pdr2* reveals a phosphate-sensitive checkpoint in root development. Plant J 37:801–814

Trasar-Cepeda MC, Carballas T (1991) Liming and the phosphatase activity and mineralization of phosphorus in an acidic soil. Soil Biol Biochem 23:209–215

Turner BL (2007) Inositol phosphates in soil: amounts, forms and significance of the phosphorylated inositol stereoisomers. In: Turner BL, Richardson AE, Mullaney EJ (eds) Inositol phosphates: linking agriculture and the environment. CABI, Wallingford, pp 186–206

Vance CP (2008) Plants without arbuscular mycorrhizae. In: Hammond JP, White PJ (eds) The ecophysiology of plant–phosphorus interactions. Springer, Dordrecht, pp 117–142

Vance CP, Ehde-Stone C, Allan DL (2003) Phosphorus acquisition and use: critical adaptations by plants for securing a nonrenewable resource. New Phytol 157:423–447

Veneklaas EJ, Stevens J, Cawthray GR, Turner S, Grigg AM, Lambers H (2003) Chickpea and white lupin rhizosphere carboxylates vary with soil properties and enhance phosphorus uptake. Plant Soil 248:187–197

Wasaki J, Omura M, Ando M, Dateki H, Shinano T, Osaki M, Ito H, Matsui H, Tadano T (2000) Molecular cloning and root specific expression of secretory acid phosphatase from phosphate deficient lupin (*Lupinus albus* L.). Soil Sci Plant Nutr 46:427–437

Wasaki J, Shinano T, Onishi K, Yonetani R, Yazaki J, Fujii F, Shimbo K, Ishikawa M, Shimatani Z, Nagata Y, Hashimoto A, Ohta T, Sato Y, Miyamoto C, Honda S, Kojima K, Sasaki T, Kishimoto N, Kikuchi S, Osaki M (2006) Transcriptomic analysis indicates putative metabolic changes caused by manipulation of phosphorus availability in rice leaves. J Exp Bot 57: 2049–2059

Wassen MJ, Venterink HO, Lapshina ED, Tanneberger L (2005) Endangered plants persist under phosphorus limitation. Nature 437:547–550

White PJ, Hammond JP (2008) Phosphorus nutrition of terrestrial plants. In: White PJ, Hammond JP (eds) The ecophysiology of plant–phosphorus interactions. Springer, Dordrecht, pp 51–81

White PJ, Broadley MR, Greenwood DJ, Hammond JP (2005) Genetic modifications to improve phosphorus acquisition by roots. Proceedings 568. International Fertiliser Society, York

Williamson LC, Ribrioux SPCP, Fitter AH, Leyser HMO (2001) Phosphate availability regulates root system architecture in *Arabidopsis*. Plant Physiol 126:875–890

Wouterlood M, Cawthray GR, Scanlon TT, Lambers H, Veneklaas EJ (2004) Carboxylate concentrations in the rhizosphere of lateral roots of chickpea (*Cicer arietinum*) increase during plant development, but are not correlated with phosphorus status of soil or plants. New Phytol 162:745–753

Wu P, Ma L, Hou X, Wang M, Wu Y, Liu F, Deng XW (2003) Phosphate starvation triggers distinct alterations if genome expression in *Arabidopsis* roots and leaves. Plant Physiol 132:1260–1271

Xiao K, Harrison MJ, Wang ZY (2005) Transgenic expression of a novel *Medicago truncatula* phytase gene results in improved acquisition of organic phosphorus by *Arabidopsis*. Planta 222:27–36

Yi K, Wu Z, Zhou J, Du L, Guo L, Wu Y, Wu P (2005) OsPTF1, a novel transcription factor involved in tolerance to phosphate starvation in rice. Plant Physiol 138:2087–2096

Zhang WH, Ryan PR, Tyerman SD (2004) Citrate-permeable channels in the plasma membrane of cluster roots from white lupin. Plant Physiol 136:3771–3783

Zimmermann P, Zardi G, Lehmann M, Zeder C, Amrhein N, Frossard E, Bucher M (2003) Engineering the root-soil interface via targeted expression of a synthetic phytase gene in trichoblasts. Plant Biotechnol J 1:353–360

Zrenner R, Stitt M, Schmidt R, Sonnewald U, Boldt R (2006) Pyramidine and purine biosynthesis and degradation in plants. Annu Rev Plant Biol 57:805–836

Part III
Ecosystems and Management

Chapter 11
Biological Phosphorus Cycling in Grasslands: Interactions with Nitrogen

Claire Jouany, Pablo Cruz, Tanguy Daufresne, and Michel Duru

11.1 Introduction

Grassland ecosystems offer the opportunity to exploit several pathways in the phosphorus (P) cycle because biological as well as geochemical processes occur on different scales in the soil–plant system. We still need to improve our knowledge of these processes because accurate management of mineral resources becomes a priority when developing sustainable herbage production systems where economic and environmental issues are the main considerations (Ehlert et al. 2003; Abbot and Murphy 2003; Marriot et al. 2004). In this context, accurate P management within farming systems based on grass production is an important goal because P is often growth-limiting (Elser et al. 2007).

For cultivated grasslands, where appropriate P nutrition is required for optimizing forage production, the aim is to meet crop needs, which are largely determined by the level of N supplied from soil reserves and/or organic or mineral fertilizers. This amount varies greatly according to the site and the function of the grassland in the system (Bélanger et al. 1989; Schellberg et al. 1999; **Griffin et al.** 2002).

In natural grasslands in the absence of deliberate anthropic inputs, biological cycling is a major process to consider in order to understand the relationships between grassland P fertility and species richness. Grassland management for nature conservation purposes requires the maintenance of P-limited ecosystems because long-term high P availability is an obstacle to the restoration of natural grassland (Janssens et al. 1998; Critchley et al. 2002; McCrea et al. 2004; Wassen et al. 2005). This is important in places where the effect of atmospheric N deposition on species diversity is directly related to P availability (Limpens et al. 2004). Conservation practices recommend soil fertility reduction, principally through soil phosphate mining, in order to promote and sustain species-rich vegetation.

C. Jouany (✉), P. Cruz, and M. Duru
INRA-UMR AGIR, Chemin de Borde Rouge, BP 52627, 31326 Castanet-Tolosan Cedex, France
e-mail: cjouany@toulouse.inra.fr

T. Daufresne
INRA-CEFS, Chemin de Borde Rouge, BP 52627, 31326 Castanet-Tolosan Cedex, France

E.K. Bünemann et al. (eds.), *Phosphorus in Action*, Soil Biology 26,
DOI 10.1007/978-3-642-15271-9_11, © Springer-Verlag Berlin Heidelberg 2011

Exhaustion of P reserves is achieved by forcing the grassland to produce biomass by means of a high N supply (Olde Venterink et al. 2001; Fagan et al. 2008).

11.1.1 P Cycle in Grassland Ecosystems

A schematic representation of the P cycle is given in Fig. 11.1. In grassland soils, the total concentration of P ranges between 200 and 1,100 mg kg^{-1} soil according to soil age (Walker and Adams 1958). Larger stocks are found in poorly developed young soils whereas the lowest concentrations occur in highly weathered ones (Lambers et al. 2008). As expected, P is present in both mineral and organic forms, whose relative contribution to the total stock varies between 20% and 90% (Fig. 11.1). P cycling is controlled simultaneously by geochemical and biological processes (Frossard et al. 1995), unlike the N cycle, which is largely controlled by biological processes. The total organic P concentration in grassland soils is not related to the total stock, but is a direct function of total organic carbon and N concentration (Walker and Adams 1958).

P concentration in aboveground biomass is commonly reported to vary between 1 and 5 mg g^{-1} dry matter according to the growth stage: major nutrient concentrations decrease with biomass accumulation as a consequence of dilution with growth (Lemaire and Gastal 1997; Duru and Ducrocq 1997). For grass swards, the

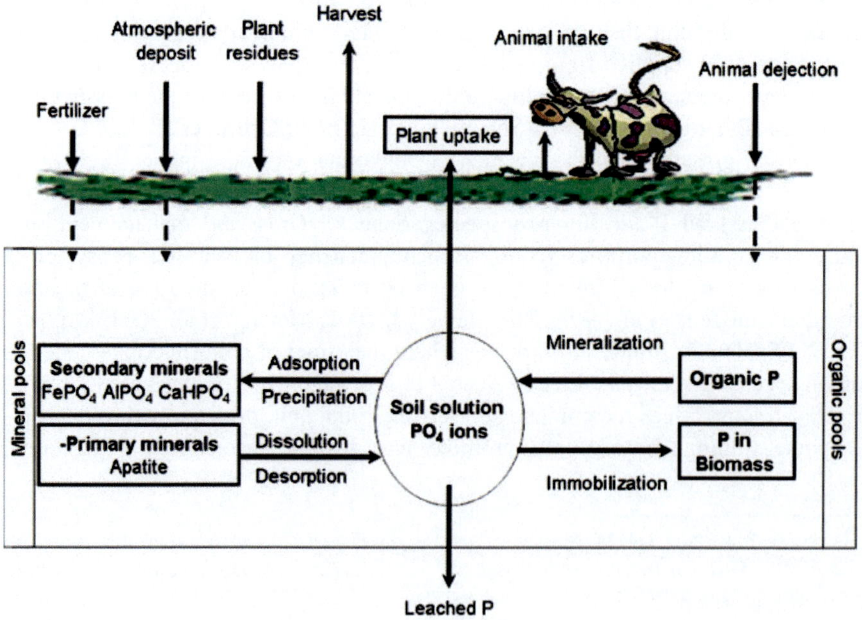

Fig. 11.1 Schematic presentation of the P cycle in a grassland ecosystem

P concentration decrease is a linear function of the sward N concentration (Duru and Thélier-Huché 1997). In cultivated grasslands, the intensity of the P flux leaving the system (output) is a direct function of the N management regime (Stroia et al. 2007; Watson and Matthews 2008). According to the methods of defoliation (cutting for hay or grazing) and fertilization, P balances on a field scale can differ considerably, from a negative balance where large P exports are not counterbalanced by fertilization, to a large surplus in over-fertilized grasslands (Stroia et al. 2007), which can represent important sources of P to surface runoff waters (Schärer et al. 2007). In natural grassland, fluxes of P are lower and are directly controlled by plant and animal residues, which decompose on the soil surface. Organic matter inputs result in very uneven nutrient availability (Güsewell et al. 2005) and, consequently, vegetation dynamics (Gillet et al. 2010).

In both fertilized and natural grassland ecosystems these processes lead to P accumulation in surface horizons due to the absence of ploughing and the low mobility of P ions. This results in a vertical gradient in the distribution of plant-available P down the soil profile and, consequently, a preferential localization of roots in the upper soil layers (Jobbágy and Jackson 2001).

11.1.2 Aims of the Chapter

Section 11.1.1 explains the complexity of soil–plant interactions in grassland ecosystems and highlights the importance of studying biological P cycling in relation with that of N. In order to discuss this, the chapter relies on several case studies that take advantage of original approaches developed by both agronomists and ecologists. They illustrate the interactions existing between N and P and their effects on ecosystem functioning. First, the nutrition index approach, based on nutrient dilution in the process of biomass accumulation, is presented (Sect. 11.2.2). This formalism, developed by agronomists, makes it possible to evaluate the relative response of biomass production in grassland in relation to changes in nutritional status (N and P). In parallel, the functional characterization of grassland vegetation from plant functional type (PFT) definition will be presented (Sect. 11.3.1.1). This relies on the fact that grassland communities may contain a wide diversity of species that influence nutrient biological cycling and regulate the biogeochemical cycle of nutrients. Finally, the effect of grazing herbivores on the biogeochemical cycle of major elements is addressed (Sect. 11.4).

11.2 Direct N–P Interactions in Grasslands

11.2.1 Influence of N–P Interaction on Grassland Production

N–P interaction is a major process that controls photosynthetic production in ecosystems. In a recent meta-analysis based on an exhaustive survey, Elser et al.

(2007) demonstrated that combined N and P enrichment produce similarly strong synergistic effects in all habitats, whether terrestrial, freshwater or marine. In agro-ecosystems, similar results are frequently reported for crops as well as grasslands (Aulack and Malhi 2005; Bélanger et al. 1989; Loeppky et al. 1999), or rangelands (Snyman 2002). These interactions do not always occur; they depend on annual weather conditions, because soil nutrient availability is controlled by water supply. N nutrition can be limited due to excessive rain and loss by leaching or low soil water availability, which limits N absorption (Mills et al. 2009). For P, low water supply limits P diffusion at the root interface and absorption by plants (Hinsinger 1998). Fertilization experiments conducted since 1999 on grassland with low soil P availability in Angladure (Ercé, French Pyrenees) revealed a significant positive N–P interaction (Stroia 2007). Four treatments were tested: control, +N, +P and +NP. N and P were added at the beginning of the vegetation period; extra N was supplied for the second cycle. An interaction for total annual dry matter yield (DMY) was observed in 10 years out of 11; when it occurred, its magnitude varied as shown in Table 11.1. The interaction appears to be more frequent for the first growth cycle and is occasionally found for the second or the autumn regrowth in wet years; however, it is difficult to identify the exact role of N and P in that interaction.

11.2.2 Analysis of N–P Interaction Using the Nutrient Index Approach

11.2.2.1 The Plant Nutrient Index Approach

This approach, developed by agronomists, relies on the nutrient dilution that takes place during biomass accumulation. The ratio of structural (low N concentration) to non-structural (metabolic tissue of high N concentration) tissue increases as biomass accumulates throughout regrowth (Lemaire and Gastal 1997). For grass-dominated swards, the decline in N concentration with biomass accumulation follows a curve with the following form (Fig. 11.2): $N_{\mathrm{critical}}\ (\%) = 4.8 \times \mathrm{DM}^{-0.32}$, where DM is accumulated dry matter (Lemaire et al. 1989). This critical N curve gives the minimum N concentration for maximum growth for different levels of biomass accumulation in swards, and can be used for diagnosing nutrient status. For a given sward biomass, the N nutrition index (Ni) is defined as the ratio (expressed as a percentage) of measured N concentration (N_{act}) to its corresponding optimum value obtained for identical biomass from the critical N curve (N_{c}) (Lemaire et al. 1989). This approach has been successfully applied to diagnose the N nutritional status of the different components in mixed cropping systems (Cruz and Soussana 1997) and grassland (Duru et al. 1997). A similar approach was later developed for major nutrients: Duru and Thélier-Huché (1997) demonstrated that for P there is no single critical P curve as found for N, since the variation in optimum P

Table 11.1 Average total annual dry matter yield response to N, P and NP

Treatments	Year										
	1999	2000	2001	2002	2003	2004	2005	2006	2007	2008	2009
Control	11.6 ± 1.2	11.3 ± 0.4	10.1 ± 1.9	8 ± 0.9	7.7 ± 2	4.9 ± 1	7.4 ± 1.8	7.3 ± 1.7	7.9 ± 1.2	6.3 ± 0.6	6.9 ± 0.9
+P	11.2 ± 2.6	11.2 ± 0.5	10.3 ± 0.5	9.5 ± 1.18	10 ± 0.6	6.6 ± 0.7	6.7 ± 0.6	11.3 ± 0.7	8.4 ± 0.6	8 ± 1.8	7.4 ± 0.7
+N	13.8 ± 1.6	16.7 ± 1.0	15.7 ± 0.8	8.6 ± 0.4	11 ± 1.3	8.9 ± 0.8	10.3 ± 0.7	10.7 ± 1.4	9.6 ± 0.7	8.9 ± 1.5	7.2 ± 0.4
+NP	14.4 ± 1.7	17.2 ± 0.9	16.5 ± 1.5	12.6 ± 0.5	15.3 ± 3.7	10.3 ± 2.4	10.9 ± 1.5	15.2 ± 2.9	11.8 ± 1.7	11.1 ± 1.3	9.8 ± 1.4
Interaction (%)[a]	56	11	10	119	36	−5	59	7	77	12	283

Values are given in t DMY ha^{-1} together with standard deviation ($n = 4$) measured on the Angladure long-term field experiment (Ercé, French Pyrenees) from 1999 to 2009

[a]Interaction (%) = 100 × [(yield response to N and P) − (yield response to N + yield response to P)]/(yield response to N + yield response to P)

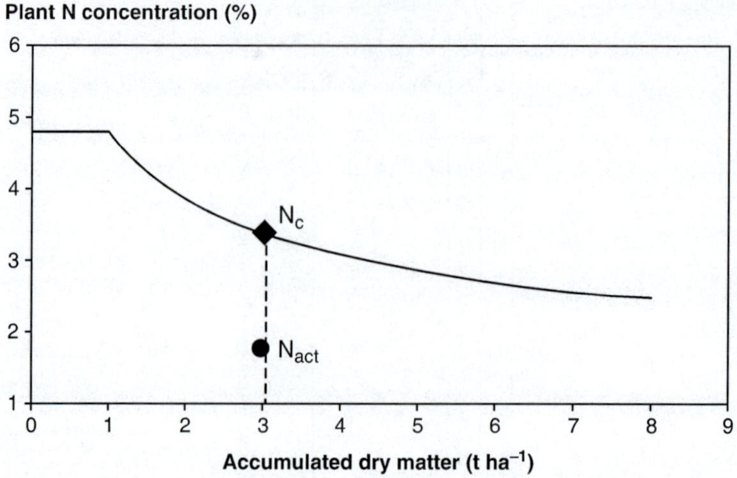

Fig. 11.2 Determination of the nitrogen nutrition index (Ni); Ni $= N_{act}/N_c$ where N_{act} is the measured concentration and N_c is the optimum value obtained from the critical N curve (adapted from Lemaire and Gastal 1997)

concentration during sward growth is dependent on the N application rate. Consequently, they proposed the following relationship: $P\%_{\mathbf{critical}} = 0.15 + 0.065\ N\%_{\mathbf{measured}}$, which provides, at a given time, the minimal P concentration in a sward needed to produce maximum dry matter. This relationship remains linear within the range of N and P concentrations observed in grasslands. Similarly to Ni, they proposed calculation of a P nutrition index (Pi) as the ratio of $P\%_{\mathbf{measured}}$ in the given sward to $P\%_{\mathbf{critical}}$ obtained with the critical P curve, expressed as a percentage. Whichever nutrient is considered (N or P), the nutrition index indicates the extent to which the sward nutrient requirements for maximum growth have been fulfilled by the fertilizer inputs and/or the soil nutrient reserves. In practice, indices vary between values above 100 (signifying an adequate or excessive nutrition level) to below 40 (indicating severe deficiency). Duru and Ducrocq (1997) applied this method to diagnose both a set of natural swards with severe P deficiency and a legume fraction contributing to less than 20% of the total harvested biomass. They showed that this index gives an accurate diagnosis of the P nutrition level during growth, albeit after the event. More recently, the approach has been extended to corn (Ziadi et al. 2007) and spring wheat (Ziadi et al. 2008). In practice, index values are used in France by the extension services to make fertilizer recommendations for farmers (Thélier-Huché et al. 1999). This approach appears to be more reliable than those based either on a single critical concentration (Pinkerton and Randall 1994) or a nutrient concentration ratio (Walworth et al. 1986; Bailey et al. 2000). In the field of ecology, N:P ratios are used in order to determine whether biomass production is N- or P-limited at community level. The review by Güsewell (2004) reports that terrestrial plant communities with N:P < 10 are N-limited and those with N:P > 20 are P-limited; as a consequence biomass production should be enhanced by N or P fertilization, respectively.

11.2.2.2 Relationships Between Nutrition Indices and Growth: Analysis of N–P Interaction

The nutrient index approach has been used by agronomists to establish relationships between N nutrient status of swards and biomass production. Lemaire and Gastal (1997), for sown grasslands, and Duru and Ducrocq (1997) for permanent grasslands demonstrated that with non-limiting P and water supply there is a direct relationship between Ni and the relative yield, i.e. the ratio of DMY measured to maximum DMY, expressed as a percentage. From this, the Ni was used to quantify the DMY response of grassland swards to N and P supply (Duru and Ducrocq 1997) and their interactions, and to analyse more precisely the interactions between mineral nutrition and water supply (Duru et al. 2000; Gonzales-Dugo et al. 2005; Mills et al. 2009). As an example, we show in Fig. 11.3 the relationship between Ni and the relative response to N and P for the Angladure grassland established for the first growth cycle in 2007. For this first harvest, the biomass production measured represents 39%, 47%, 38% and 40% of the total annual biomass production (Table 11.1) for control, +P, +N and +NP treatments (N0P0, N0P1, N1P0 and N1P1 in Fig. 11.3), respectively. The relative response was calculated as the ratio of

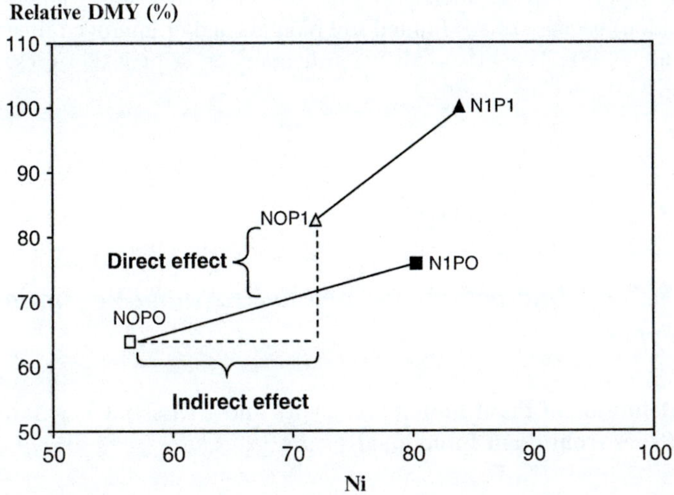

Fig. 11.3 Relationship between the relative response ($DMY_{treatment}/DMY_{optimum}$) and N nutrition index (Ni) for the first growth cycle measured in 2007 on the Angladure experiment grassland (Ercé, French Pyrenees). The direct effect on growth represents the increase of the sward efficiency for N conversion in biomass for +P treatments; the indirect effect on growth is a consequence of the increase in N nutrition status of the sward for +P treatments. *N0P0* control, *N1P0* + N treatment, *N0P1* + P treatment, *N1P1* + NP treatment

the DMY measured on a given treatment to the maximum DMY (on N1P1 plots). From Fig. 11.3, it is possible to identify graphically the response in terms of nutrient status. Comparison of the N0P1 treatment with the control (N0P0) quantifies the increase in growth in response to P supply, which results from the indirect effect of P and increases the N nutrition status (Ni passes from 55 to 72), and the direct effect of P on growth, which increases the sward efficiency for N conversion in biomass. The sward is more efficient in converting N into biomass under non-limiting P supply than with limiting P supply; the slope for the P1 treatments (upper line in Fig. 11.3) being higher than the slope for the P0 treatments (lower line in Fig. 11.3) (Duru and Ducrocq 1997).

Several hypotheses are proposed to explain the increase in N nutrition level of the sward following P supply (indirect effect). The first is that improved P nutrition status increases the growth of aerial parts and the root system and thus improves soil exploration capacity by the root system and, hence, nutrient interception (Duru 1992). This process is particularly efficient in the drainage phase when improved root colonization of the soil can lead to better interception of mobile nutrients such as nitrate-N. As an example, the spectacular interaction observed in 2002 (Table 11.1) can be explained by these processes; heavy rainfall in spring led to a positive water balance, the amount of drained water being much more significant than usual, i.e. 900 mm instead of 400 mm on average. As a consequence, N losses were greater in N1P0 than in N1P1 plots, where adequate plant P nutrition allowed better root growth and, consequently, better interception of nitrate in the soils. Another hypothesis concerns the effect of P on organic matter mineralization: increased P supply leads to increased soil N internal recycling (mineralization and nitrification) because of a P-limited soil biomass and/or improved litter quality (Parfitt et al. 2005). The effect of P fertilization on legumes is presented in Sect. 11.3.2.

11.3 Indirect N–P Interactions

11.3.1 Indirect Interactions Related to Grassland Vegetation Types

11.3.1.1 Definition of Plant Functional Types and Grassland Vegetation Types from Plant Functional Traits

In the field of plant ecology, the development of a functional approach has provided new methods based on an analysis of the functional traits developed by plants growing in communities. The determination of the functional composition of the vegetation can be based on the identification of a PFT, which defines a group of species that fulfil similar functions in the ecosystem, without necessarily presenting phylogenetic relationships (Gitay and Noble 1997). In order to better understand ecosystem behaviour, one needs to define the functional groups associated with the main processes taking place in the ecosystem, and the species most characteristic of these groups (Hooper and Vitousek 1997). These species correspond to the dominant ones present in the community because they represent the largest pool of matter and energy, and control its structure and behaviour (Goldberg 1997). A group brings together species that share common values of one or several functional traits, i.e. biological characteristics (morphological, physiological, phenological, demographic etc.) that are the expression of similar behaviour or strategies. One distinguishes response traits, whose values change in response to environmental factors or agricultural practices (fertilization and/or defoliation regime), and effect traits, which act on the processes of the ecosystem (productivity and nutrient cycling among others) (Lavorel and Garnier 2002). This functional approach has been used by Ansquer et al. (2004) to identify different PFTs based on the measurement of the leaf dry matter content (LDMC) for grasses in a community (Table 11.2). These plant types have been used to develop a simplified method to determine the functional composition of the grassland vegetation at the community scale, and to define grassland vegetation types (GVTs). In their recent work, Ansquer et al. (2009a, c) demonstrated that the identification of PFTs is an operational approach for assessing agricultural services defined in terms of productivity, phenology and quality.

11.3.1.2 Relationships Between Response Traits and Fertility Gradients

The value of traits varies in response to both nutrient availability and defoliation regime. Thus it becomes possible to group species according to the strategies they employ to adapt to different habitats; grassland species can be ranked according to their aptitude for acquisition, or conservation of nutrients and carbon. For instance, species adapted to nutrient-rich environments, such as *Lolium perenne*, will display resource capture strategies. The corresponding foliar traits are a high specific leaf area (SLA) and low LDMC. Conversely, species adapted to low fertility environments that display nutrient conservation strategies, such as *Briza media* and *Festuca ovina,* have low SLA and high LDMC values. Besides these traits, which are relatively easy to measure, others should be considered, either because their

Table 11.2 Plant functional types (PFTs) of grass species defined on the basis of a hierarchical analysis applied to dry matter contents of leaf blades (LDMC) (Ansquer et al. 2004)

Type A	Type B	Type C	Type D
Holcus lanatus, *Lolium perenne*	*Anthoxanthum odoratum*, *Arrhenatherum elatius*, *Dactylis glomerata*, *Festuca arundinacea*, *Poa trivialis*	*Agrostis capillaris*, *Avena pubescens*, *Festuca rubra*, *Phleum pratense*, *Trisetum flavescens*	*Brachypodium pinnatum*, *Briza media*, *Cynosurus cristatus*, *Deschampsia caespitosa*, *Festuca ovina*

agronomic impact is important or because they better account for the species' capacity to succeed in the competition for nutrients.

For instance, species that display resource conservation strategies have a leaf lifespan (LLS) longer than that of species with acquisition strategies, and consequently are more efficient in the use of mineral resources. This is evaluated by measuring their nutrient use efficiency (NUE), i.e. the crop yield per unit of nutrient absorbed (Ryser 1996). This efficiency results from two components: (1) the productivity associated with a given nutrient (the amount needed for one unit of biomass production) and (2) its mean residence time (MRT) in leaves (Berendse and Aerts 1987).

As a case study for P and N NUE, Table 11.3 shows an experiment conducted on *Dactylis glomerata* and *Festuca rubra* transplanted into grassland that initially displayed low N and P fertility levels. In order to differentiate N and P nutrition levels, three treatments were applied: +N, +P and +NP. N and P were applied at 0 and 150 kg ha^{-1} as NH_4NO_3 and triple superphosphate. N and P supply provided non-limiting nutrition conditions, whereas the unfertilized control displayed N- and/or P-limiting nutritional status.

Under N-limited growth conditions, DMY was increased by 33% by P, but this increase was not significant. Under non-limiting N growth conditions, there is a significant increase in DMY following P supply. Under both N levels, there was no difference in DMY between the two species.

Under N-limiting growth conditions there is a significant effect of the species on the NUE for N, which increases in response to P limitation: by 4 points (from 51 to 55) for type B species (*Dactylis glomerata*) and 7 points (from 60 to 67) for type C (*Festuca rubra*). Under non-limiting P supply, NUE for P is significantly different between species, NUE for *Festuca rubra* being 25% higher than for *Dactylis*. For both species, increasing P stress on control plots leads to an increase in NUE for P, but the values for the two species are not significantly different. Under non-limiting N supply, we do not observe any significant change in NUE for N and P in response to P stress. It is likely that when N nutrition is non-limiting, both species are better able to adapt to P stress than when N is limiting. This experiment demonstrates that both species increase their NUE for N and for P when their availability decreases, as expected (Lajtha and Harrison 1995). However, the size of the response varies according to the species, type C (*Festuca rubra*) always displaying higher NUE than type B (*Dactylis glomerata*) for the same amount of forage production. As an example, a *Dactylis glomerata* sward would export more P and more N than a *Festuca rubra* one (25% and 15% more on average in our case). We need to keep in mind that some treatments forced both species to grow under fertility levels different from those usually met, i.e. limiting N and P for *Dactylis* and ample N and P supply for *Festuca*. The NUE values measured in their natural habitats would be different, close to that of the N- and P-limiting treatment for *Festuca* and N and P non-limiting treatment for *Dactylis*. In natural habitats, *Festuca* is most likely to display high NUE values (67 and 757 for N and P, respectively) and *Dactylis* low values (36 and 233 for N and P, respectively).

Table 11.3 Comparison of *Dactylis glomerata* and *Festuca rubra* responses to P, N and NP

	N − P−			N − P+			N + P−			N + P+		
	NUE_N (kg DM kg^{-1} N)	NUE_P (kg DM kg^{-1} P)	DMY (t ha^{-1})	NUE_N (kg DM kg^{-1} N)	NUE_P (kg DM kg^{-1} P)	DMY (t ha^{-1})	NUE_N (kg DM kg^{-1} N)	NUE_P (kg DM kg^{-1} P)	DMY (t ha^{-1})	NUE_N (kg DM kg^{-1} N)	NUE_P (kg DM kg^{-1} P)	DMY (t ha^{-1})
Dactylis glomerata	55^a	717	2.4	51^a	375^a	3.2	26	496	2.8	36	233	4.5
Festuca rubra	67^b	757	2.4	60^b	459^b	3.2	28	510	2.9	28	209	4.5

DMY average dry matter yield, NUE_N nutrient use efficiency for N, NUE_P nutrient use efficiency for P. All measurements were made at the first cut ($n = 3$)
Figures in a column followed by different letters are significantly different ($P < 0.001$)

11.3.1.3 Relationships Between Plant Functional Types and Nutrient Cycling

In order to illustrate the relationships existing between PFT and nutrient cycling, we rely on a survey conducted in 2001 on a garden collection of 16 grasses representing the four types described in Sect. 11.3.1. The grasses were grown in Auzeville (France) with non-limiting mineral and water supply. The nutrient concentrations were measured for the spring growth cycle on the youngest adult leaves, and then for senescent leaves in July. Under similar growing conditions, P concentrations for green tissues vary between 0.21% and 0.41% according to species and type; meanwhile N concentrations vary between 2.1% and 3.8%. Species with conservation strategies (types C and D) tend to have lower N and P concentrations than species with acquisition strategies (types A and B). In parallel, large variations in N and P concentrations are observed for the senescent leaves. When we plot the N:P ratios of the green tissue as a function of N:P ratios for senescent ones (Fig. 11.4), we observe an increase in N:P ratio from type A to type D, varying from 6.6 to 14.8. We also note that for all species, N:P ratios are lower for senescent leaves than for green ones; they vary between 2.7 and 9.7, indicating that for all these species and under these growing conditions, resorption efficiency is higher for N than for P.

These examples demonstrate that differences in species strategies for resource acquisition lead to wide ranges in mineral concentrations for residues and litter returned to the soil, with positive or negative feedback on the rate of P cycling: according to N:P and or C:P ratios, either mineralization or immobilization pathways will control P cycling and its availability for plants (White and Tayoub 1983; Güsewell et al. 2005). Similarly, comparing ten contrasting sites in Europe, Fortunel et al. (2009) demonstrated that species with conservation strategies have a

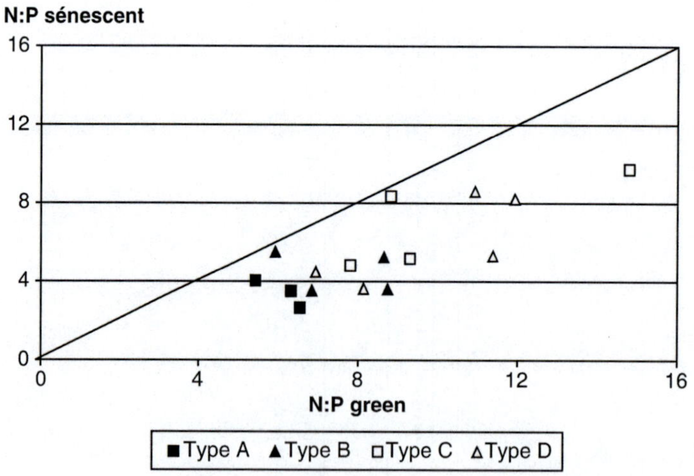

Fig. 11.4 Relationship between N:P ratio measured on green leaves and senescent leaves for 16 grasses representative of the four plant functional types defined by Ansquer et al. (2004) (see Table 11.2 for examples)

chemical composition that results in slower decomposition rates of litter and residues, lengthening the time until nutrients will again become available for plants.

11.3.2 *Effects of P on Biological Nitrogen Fixation*

There is a general consensus that N application will often favour grasses, which are more competitive than legumes under non-limiting N, and thus result in a decline in the proportion of legumes in grassland communities (Laidlaw and Withers 1998; Loiseau et al. 2001). In contrast, for low residual heights, a frequent cutting regime and absence of N fertilization are practices frequently reported to maintain or increase clover content in mixed swards (Barthram et al. 1992). Conversely, P fertilization alone, associated with intense defoliation, will often favour the development of legumes. These trends are acknowledged for grazed grasslands and rangelands (Aydin and Uzun 2005; Martiniello and Berardo 2007) as well as grasslands cut for hay (Nevens and Rehuel 2003; Jouany et al. 2004). This increased contribution of legumes to the sward biomass in response to P fertilization leads to an increase in the sward N concentration because the N concentration of the legume fraction is generally higher than that of the grass fraction (Mackay et al. 1995; Henkin et al. 1996). As a consequence, the N:P ratio for pure legume swards or associations remains generally higher than that of grass swards (Jouany et al. 2004). Results in the literature show that when P is limiting, nodule growth and N_2-fixation activity are limited (Haynes and Ludecke 1981). Høgh-Jensen et al. (2002) demonstrated that under low P availability, white clover displays whole-plant adaptative responses by modifying the relative growth of shoots, roots and nodules. From an agronomic point of view, one should keep in mind that optimization of P nutrition might help in maximizing N inputs into grasslands by symbiotic N fixation.

11.4 Analyzing the Effect of Herbivores on N and P Cycles in Grassland Ecosystems

Most of the grasslands in the world are grazed by large herbivores, mostly domestic herds of ruminants grazing at low intensity (Allard et al. 2003). This being so, the biogeochemistry of major elements (e.g. N or P) in grasslands cannot be thoroughly addressed without considering the role of grazers. The effect of grazing on the N and P cycles has been highlighted by numerous empirical and experimental studies (Hobbs 1996). In short, terrestrial grazers have been shown to promote soil nutrient heterogeneity through the concentration of faeces and urine in specific parts of the landscape (Augustine 2003; Augustine and Frank 2001), to increase soil nutrient availability through faeces and urine deposition (Carline et al. 2005), to decrease soil nutrient availability through the selective grazing of plants producing good nutrient recycling litter (Pastor et al. 1993), and to affect the coupling between the N and P cycles (Frank 2008), with important consequences for the intensity of, and limitations on, primary production.

The combined direct (grass consumption) and indirect (recycling through urine and/or faeces deposition) effects of herbivores have been studied using various experimental setups. For instance, controlled grazing in enclosures (Allard et al. 2003) versus deposition of faeces in exclosures to prevent grazing can, hence, disentangle the direct and indirect effects of the herbivores (Van der Wal et al. 2004). Artificial clipping and deposition of synthetic urine in mesocosms (Attard et al. 2008) have been performed. From these studies, it appears that the net effect of grazers on nutrient cycles and its consequences for primary production seem not to follow a general rule. Instead, they appear to vary across locations and to be dependent on other factors, such as the climate, the nature of the bedrock etc. (Augustine and McNaughton 2006; Frank and Groffman 1998; Frank 2008; Van der Wal et al. 2004). Yet the effect of grazers is generally significant and, in some cases, the combined effect of grazing and other factors (e.g. the increase in atmospheric N deposition) on biogeochemical cycles can even lead to serious habitat degradation (Van der Wal et al. 2003). For some specific cases, the mechanistic interpretation of the herbivore effect on nutrient cycles is trivial. For instance, by preferentially consuming legumes herbivores can decrease N fixation and, hence, decrease the soil N budget (Ritchie et al. 1998). Yet, in most cases the mechanisms are less obvious, and most authors mention the complexity of the soil processes and stress the indirect effect of grazers on the community structure of soil microorganisms (Carline et al. 2005). For instance, artificial defoliation in grassland has been shown to reduce root biomass and affect the structure of soil food webs, with consequences for the soil inorganic N budget (Mikola et al. 2001). Other experiments have concluded that a decrease of soil microbial respiration follows artificial defoliation, probably due to the fact that defoliation reduces the labile carbon available to soil microbes (Stark and Kytöviita 2006). On the other hand, urine deposition has been shown to affect the community structure of soil bacteria, especially ammonia-oxidizing and nitrite-reducing bacterial communities, which in turn affect the N cycle and budget (Orwin et al. 2010). Again, the magnitude of these effects is greatly variable and seems to depend on other factors such as soil moisture, trampling etc. (Sørensen et al. 2009).

On the other hand, modelling approaches have stressed the effect of grazers on the speed of recycling and on nutrient budgets in the soil–vegetation compartment (e.g., De Mazancourt et al. 1998, 1999). This kind of insight can help a lot with the interpretation of data. However these modelling approaches have focussed on the cycling of only one nutrient and, to our knowledge, the case of coupled cycles of two or more nutrients has not been thoroughly addressed. Some insights could come from limnology, where the effect of planktonic grazers on the coupling between the cycles of N and P has long been studied (Frank 2008). In pelagic aquatic ecosystems, planktonic biomass represents the main stock of nutrients and planktonic grazers can represent a significant fraction of this stock (Andersen 1997). In this context, models have shown that the ratio of nutrients in planktonic grazers can affect the ratios of nutrients available to algae, with important consequences for algal community structure and nutrient limitation status (Sterner 1990; Andersen 1997; Daufresne and Loreau 2001). These results suggest that one cannot understand the effect of grazers on the P cycle without considering the N cycle. Yet, these results are not readily applicable to terrestrial ecosystems, where the nutrient fraction stored in the herbivore biomass is usually negligible compared to that stored in the soil–vegetation system. Hence, the study of grassland ecosystems requires the derivation of specific soil–grass–grazer models incorporating both the N and P cycles. Because the soil N and P represent the main stocks of nutrients in grassland, it is expected that the effect of the grazers on the loss rates of N and P from the soil will have a key effect on the N:P ratio in these systems. In particular, the higher sensitivity of N to leaching and to volatilization (Ambus et al. 2007), in comparison to P, should play an essential role. These models should provide valuable theoretical tools for predicting the effect of grazing on the N and P cycles in grassland, and should shed some light on the interpretation of empirical patterns.

11.5 Conclusion

The approaches and case studies introduced in this chapter highlight the importance of studying biological P cycling in relation to that of N. A comprehensive approach to nutrient cycling in grassland ecosystems, whether cultivated or natural, needs to be considered when developing sustainable production systems with environmental and economic constraints. Experimental approaches developed in recent years by agronomists and ecologists have led to a better understanding of the interactions between N and P and their effects on ecosystems. Long-term field experiments are essential for analysis and evaluation of these interactions. There still remain some aspects that deserve more attention in the future:

1. Although studies conducted on a field scale have demonstrated that there is similarity between grasses and dicots for plant phenology (Ansquer et al. 2009b) and that grasses and dicotyledonous (especially rosette) plant life forms behave very similarly (Ansquer et al. 2009a), little is known about similarities and differences in nutrient use and recycling processes between different life forms. This point is important when considering the role of legumes in N–P interactions in grassland.
2. We now need to improve our knowledge of the effects of global climate change on P biological cycling. Direct effects as a consequence of decreased water availability or indirect ones as a consequence of dramatic changes in vegetation can be expected from global warming.
3. The derivation of specific soil–grass–grazer models incorporating both the N and P cycles should provide valuable theoretical tools to predict the effect of grazing on the N and P cycles.

4. Finally, one aspect of the question remains unexplored, it concerns the soil microbial community and its function in regulating nutrient fluxes. Improved knowledge of the interactions between soil biota and plants and grazing animals should help in developing more efficient and sustainable grassland systems.

References

Abbot LK, Murphy DV (2003) What is soil biological fertility? In: Abbot LK, Murphy DV (eds) Soil biological fertility. A key to sustainable land use in agriculture. Kluwer, Dordrecht, pp 1–15

Allard V, Newton PC, Lieffering M, Clark H, Matthew C, Soussana JF, Gray IS (2003) Nitrogen cycling in grazed pastures at elevated CO_2: N returns by ruminants. Glob Change Biol 9:1731–1742

Ambus P, Petersen SO, Soussana JF (2007) Short-term carbon and nitrogen cycling in urine patches assessed by combined carbon-13 and nitrogen-15 labeling. Agric Ecosyst Environ 121:84–92

Andersen T (1997) Pelagic nutrient cycles, herbivores as sources and sinks. Ecological studies, vol 129. Springer, Berlin

Ansquer P, Theau JP, Cruz P, Viegas J, Al Haj Khaled R, Duru M (2004) Caractérisation de la diversité fonctionnelle des prairies à flore complexe, vers la construction d'outils de gestion. Fourrages 179:353–368

Ansquer P, Al Haj Khaled R, Cruz P, Theau JP, Therond O, Duru M (2009a) Characterizing and predicting plant phenology in species-rich grasslands. Grass Forage Sci 64:57–70

Ansquer P, Duru M, Theau JP, Cruz P (2009b) Functional traits as indicators of fodder provision over a short time scale in species-rich grasslands. Ann Bot 103:117–126

Ansquer P, Duru M, Theau JP, Cruz P (2009c) Convergence in plant traits between species with grassland communities simplifies their monitoring. Ecol Indic 9:1020–1029

Attard E, Degrange V, Klumpp K, Richaume A, Soussana JF, Le Roux X (2008) How do grassland management history and bacterial micro-localization affect the response of bacterial community structure to changes in aboveground grazing regime? Soil Biol Biochem 40:1244–1252

Augustine DJ (2003) Long-term, livestock-mediated redistribution of nitrogen and phosphorus in a East African savanna. J Appl Ecol 40:137–149

Augustine DJ, Frank DA (2001) Effects of migratory grazers on spatial heterogeneity of soil nitrogen properties in a grassland ecosystem. Ecology 82:3149–3162

Augustine DJ, McNaughton SJ (2006) Interactive effects of ungulate herbivores, soil fertility, and variable rainfall on ecosystem processes in a semi-arid savanna. Ecosystems 9:1242–1256

Aulack MS, Malhi SS (2005) Interactions of nitrogen with other nutrients and water: effect on crop yield and quality, nutrient use efficiency, carbon sequestration, and environmental pollution. Adv Agron 86:341–409

Aydin I, Uzun F (2005) Nitrogen and phosphorus fertilization of rangeland affects yield, forage quality and botanical composition. Eur J Agron 23:8–14

Bailey JS, Dills RA, Foy RH, Patterson D (2000) The diagnosis and recommendation system (DRIS) for diagnosing the nutrient status of grassland swards. III. Practical applications. Plant Soil 222:255–262

Barthram GT, Grant SA, Elston DA (1992) The effects of sward height and nitrogen fertilizer application on changes in sward composition, white clover growth and the stock carrying capacity of an upland perennial ryegrass/white clover sward grazed by sheep for four years. Grass Forage Sci 47:326–341

Bélanger G, Richards JE, Walton RB (1989) Effects of 25 years of N, P and K fertilization on yield, persistence and nutritive value of timothy sward. Can J Plant Sci 69:501–512

Berendse F, Aerts R (1987) Nitrogen-use-efficiency: a biologically meaningful definition? Funct Ecol 1:293–296

Carline KA, Jones HE, Bardgett RD (2005) Large herbivores affect the stoichiometry of nutrients in a regenerating woodland ecosystem. Oikos 110:453–460

Critchley CNR, Chambers BJ, Fowbert JA, Bhogal A, Rose SC, Sanderson RA (2002) Plant species richness, functional type and soil properties of grasslands and allied vegetations in English environmentally sensitive areas. Grass Forage Sci 57:82–92

Cruz P, Soussana JF (1997) Mixed crops. In: Lemaire G (ed) Diagnosis of the nitrogen status in crops. Springer, Berlin, pp 130–144

Daufresne T, Loreau M (2001) Plant-herbivore interactions and ecological stoichiometry: when do herbivores determine plant nutrient limitation? Ecol Lett 4:196–206

De Mazancourt C, Loreau M, Abbadie L (1998) Grazing optimization and nutrient cycling: when do herbivores enhance plant production? Ecology 79:2242–2252

De Mazancourt C, Loreau M, Abbadie L (1999) Grazing optimization and nutrient cycling: potential impact of large herbivores in a savanna system. Ecol Appl 9:784–797

Duru M (1992) Diagnostic de la nutrition minérale de prairies permanentes au printemps. I. Etablissement de références. Agronomie 12:219–233

Duru M, Ducrocq H (1997) A nitrogen and phosphorus herbage nutrient index as a tool for assessing the effect of N and P supply on the dry matter yield for permanent pastures. Nutr Cycl Agroecosys 47:59–69

Duru M, Thélier-Huché L (1997) N and PK status of herbage: use for diagnosis of grasslands. Diagnostic procedures for crop N management, Les Colloques, n° 82, éd. INRA, Paris, 125–138

Duru M, Thélier-Huché L (1997) N and PK status of herbage: use for diagnosis of grasslands. In: Lemaire G, Burns IG (eds) Diagnostic procedure for crop N management. Les colloques 82. INRA, Paris, pp 125–138

Duru M, Lemaire G, Cruz P (1997) Grasslands, Lemairen G, Burnsed IG éd, Diagnosis of the nitrogen status in crops, Springer-Verlag pp 59–72

Duru M, Lemaire G, Cruz P (1997) Grasslands. In: Lemaire G, Burns IG (eds) Diagnostic procedures for crop N management. Les colloques 82. INRA, Paris, pp 130–144

Duru M, Cruz P, Jouany C, Theau JP (2000) Intérêt pour le conseil, du diagnostique de nutrition azotée de prairies de graminées par analyse de plante. Fourrages 164:381–395

Ehlert P, Morel C, Fotyma M, Destain JP (2003) Potential role of phosphate buffering capacity of soils in fertilizer management strategies fitted to environmental goals. J Plant Nutr Soil Sci 166:409–415

Elser J, Bracken MES, Cleland EE, Gruner DS, Harpole WS, Hillebrand H, Ngai JT, Seabloom EW, Shurin JB, Smith E (2007) Global analysis of nitrogen and phosphorus limitation of primary producers in freshwater, marine and terrestrial ecosystems. Ecol Lett 10:1135–1142

Fagan KC, Pywell RF, Bullock JM, Marrs RH (2008) Do restored calcareous grasslands on former arable fields resemble ancient targets? The effect of time, methods and environment on outcomes. J Appl Ecol 45:1293–1303

Fortunel C, Garnier E, Joffre R, Kazakou E, Quested E, Grigulis K, Lavorel S, Ansquer P, Castro H, Cruz P, Dolezal J, Eriksson O, Freitas H, Golodets C, Jouany C, Kigel J, Kleyer M, Lehstein V, Leps J, Meir T, Pakeman R, Papadimitriou M, Papanastasis VP, Quetier F, Robson M, Sternberg M, Theau JP, Thebault A, Zarovali M (2009) Leaf traits capture the effects of land use changes and climate on litter decomposability of grasslands across Europe. Ecology 90:598–611

Frank DA (2008) Ungulate and topographic control of nitrogen:phosphorus stoichiometry in temperate grassland; soils, plants and mineralization rates. Oikos 117:591–601

Frank DA, Groffman PM (1998) Ungulate vs. landscape control of soil C and N processes in grasslands of Yellowstone National Park. Ecology 79:2229–2241

Frossard E, Brossard M, Hedley MJ, Metherell A (1995) Reaction controlling the cycling of P in soil. In: Tiessen H (ed) Phosphorus in the global environment. Wiley, Chichester, pp 107–138

Gillet F, Kohler F, Vandenberghe C, Buttler A (2010) Effect of dung deposition on small-scale patch structure and seasonal vegetation dynamics in mountain pastures. Agric Ecosyst Environ 135:34–41

Gitay H, Noble IR (1997) What are plant functional types and how should we seek them? In: Smith TM, Shugart HH, Woodward FI (eds) Plant functional types: their relevance to ecosystem properties and global change. Cambridge University Press, Cambridge, pp 3–19

Goldberg DE (1997) Competitive ability: definitions, contingency and related traits. In: Silverstown J, Franco M, Harper JL (eds) Plant life histories. Cambridge University Press, Cambridge, pp 283–306

Gonzales-Dugo V, Durand JL, Gastal F, Picon-Cochard C (2005) Short term response of the nitrogen nutrition status of tall fescue and Italian ryegrass swards under water deficit. Aust J Agric Res 56:1269–1276

Griffin T, Giberson E, Wiedenhoeft M (2002) Yield response of long-term mixed grassland swards and nutrient cycling under different nutrient sources and management regimes. Grass Forage Sci 57:268–278

Güsewell S (2004) N:P ratios in terrestrial plants: variation and functional signifiance. New Phytol 164:243–266

Güsewell S, Jewell PJ, Edwards PJ (2005) Effects of heterogeneous habitat use by cattle on nutrient availability and litter decomposition in soils of an Alpine pasture. Plant Soil 268: 135–149

Haynes RJ, Ludecke TE (1981) Yield, root morphology and chemical composition of two pasture legumes as affected by lime and phosphorus applications to an acid soil. Plant Soil 62:241–254

Henkin Z, Noy-Meir I, Kafkafi U, Seligman N (1996) Phosphate fertilization primes production of rangeland on brown rendzina soils in the Galilee, Israel. Agric Ecosyst Environ 59:43–53

Hinsinger P (1998) How do plant acquire mineral nutrients? Chemical processes involved in the rhizosphere. Adv Agron 64:225–265

Hobbs NT (1996) Modification of ecosystems by ungulates. J Wildl Manage 60(4):695–713

Høgh-Jensen H, Schjoerring JK, Soussana JF (2002) The influence of phosphorus deficiency on growth and nitrogen fixation of white clover plants. Ann Bot 90:745–753

Hooper DU, Vitousek PM (1997) The effects of plant composition and diversity on ecosystem processes. Science 272:1302–1305

Janssens F, Peeters A, Tallowin JRB, Bakker JP, Bekker RM, Fillat F, Oomes MJM (1998) Relationship between soil chemical factors and grassland diversity. Plant Soil 202:69–78

Jobbágy EG, Jackson RB (2001) The distribution of soil nutrients with depth: global patterns and the imprint of plants. Biogeochemistry 53:51–77

Jouany C, Cruz P, Petibon P, Duru M (2004) Diagnosing phosphorus status of natural grassland in the presence of white clover. Eur J Agron 1:273–285

Laidlaw AS, Withers JA (1998) Changes in contribution of white clover to canopy structure in perennial ryegrass/white clover swards in response to N fertilizer. Grass Forage Sci 53:287–291

Lajtha K, Harrison AF (1995) Strategies of phosphorus acquisition and conservation by plant species and communities. In: Tiessen H (ed) Phosphorus in the global environment. Wiley, Chichester, pp 139–147

Lambers H, Raven JA, Shaver GR, Smith GE (2008) Plant nutrient-acquisition strategies change with soil age. Trends Ecol Evol 23:95–103

Lavorel S, Garnier E (2002) Predicting changes in community composition and ecosystem functioning from plant traits: revisiting the Holy Grail. Funct Ecol 16:545–556

Lemaire G, Gastal F (1997) N uptake and distribution in plant canopies. In: Lemaire G (ed) Diagnosis of the nitrogen status in crops. Springer, Berlin, pp 3–43

Lemaire G, Gastal F, Salette J (1989) Analysis of the effect of N nutrition on dry matter yield of a sward by reference to potential yield and optimum N content. In: Proceedings XVI International Grassland Congress, Nice, pp 179–180

Limpens J, Berendse F, Klees H (2004) How phosphorus availability affects the impact of nitrogen deposition on *Sphagnum* and vascular plants in bogs. Ecosystems 7:793–804

Loeppky HA, Horton PR, Bittman S, Townley-Smith L, Wright T, Nuttall WF (1999) Forage seeds yield response to N and P fertilizers and soil nutrients in Northeastern Saskatchewan. Can J Soil Sci 79:265–271

Loiseau P, Soussana JF, Louault F, Delpy R (2001) Soil N contributes to the oscillations of the white clover content in mixed swards of perennial ryegrass under conditions that simulate grazing over years. Grass Forage Sci 56:205–217

Marriot CA, Fothergill M, Jeangros B, Scotton M, Louault F (2004) Long term impacts of extensification of grassland management on biodiversity and productivity in upland areas. Agronomie 24:447–462

Mackay AD, Saggar S, Trolove SN, Lambert MG (1995) Use of an unsorted pasture sample in herbage testing for sulphur, phosphorus and nitrogen. New Zeal J Agr Res 38:483–493

Martiniello P, Berardo N (2007) Residual fertilizer effects on dry-matter yield and nutritive value of Mediterranean pastures. Grass Forage Sci 62:87–99

McCrea AR, Trueman IC, Fullen MA (2004) Factors relating to soil fertility and species diversity in both semi-natural and created meadows in the West Midlands of England. Eur J Soil Sci 55:335–348

Mikola J, Yeates GW, Barker GM, Wardle DA, Bonner KI (2001) Effects of defoliation intensity on soil food-web properties in an experimental grassland community. Oikos 92:333–343

Mills A, Moot DJ, Jamieson PD (2009) Quantifying the effect of nitrogen on productivity of cocksfoot (*Dactylis glomerata* L.) pastures. Eur J Agron 30:63–69

Nevens F, Rehuel D (2003) Effects of cutting or grazing grass swards on herbage yield, nitrogen uptake and residual soil nitrate at different levels of N fertilization. Grass Forage Sci 58: 431–449

Olde Venterink H, van der Vliet RE, Wassen MJ (2001) Nutrient limitation along a productivity gradient in wet meadows. Plant Soil 234:171–179

Orwin KH, Bertram JE, Clough TJ, Condron LM, Sherlock RR, O'Callaghan M, Ray J, Baird DB (2010) Impact of bovine urine deposition on soil microbial activity, biomass, and community structure. Appl Soil Ecol 44:89–100

Parfitt RL, Yeates GW, Ross DJ, Mackay AD, Budding PJ (2005) Relationships between biota, nitrogen and phosphorus availability, and pasture growth under organic and conventional management. Appl Soil Ecol 28:1–13

Pastor J, Dewey B, Caiman RJ, McInnes PF, Cohen Y (1993) Moose browsing and soil fertility in the boreal forests of Isle Royale national park. Ecology 74(2):467–480

Pinkerton A, Randall PJ (1994) Internal phosphorus requirement of six legumes and two grasses. Aust J Exp Agric 34:373–379

Ritchie ME, Tilman D, Knops JMH (1998) Herbivore effects on plant and nitrogen dynamics in oak savanna. Ecology 79:165–177

Ryser P (1996) The importance of tissue density for growth and life span of leaves and roots: a comparison of five ecologically contrasting grasses. Funct Ecol 10:717–723

Schärer M, Stamm C, Vollmer T, Frossard E, Oberson A, Flühler H, Sinaj S (2007) Reducing phosphorus losses from over-fertilized grassland soils proves difficult in the short term. Soil Use Manage 23:154–164

Schellberg J, Möseler BM, Kühbauch W, Rademacher IF (1999) Long-term effects of fertilizer on soil nutrient concentration, yield, forage quality and floristic composition of a hay meadow in the Eifel mountains, Germany. Grass Forage Sci 54:195–207

Snyman HA (2002) Short term response of rangeland botanical composition and productivity to fertilization (N and P) in a semi-arid climate of South Africa. J Arid Environ 50:167–183

Sørensen LI, Mikola J, Kytöviita MM, Olofsson J (2009) Trampling and spatial heterogeneity explain decomposer abundances in a sub-arctic grassland subjected to simulated reindeer grazing. Ecosystems 12:830–842

Stark S, Kytöviita MM (2006) Simulated grazer effects on microbial respiration in a subarctic meadow: implications for nutrient competition between plants and soil microorganisms. Appl Soil Ecol 31:20–31

Sterner RW (1990) The ratio of nitrogen to phosphorus resupplied by herbivores: zooplankton and the algal competitive arena. Am Nat 136:209–229

Stroia C (2007) Etude de fonctionnement de l'écosystème prairial en conditions de nutrition N et P sub limitantes. Application au diagnostic de nutrition. PhD thesis, Institut National Polytechnique, Toulouse. http://ethesis.inp-toulouse.fr/archive/00000489/. Accessed 3 Aug 2010

Stroia C, Morel C, Jouany C (2007) Dynamics of diffusive soil phosphorus in two grassland experiments determined both in field and laboratory conditions. Agric Ecosyst Environ 19:60–74

Thélier-Huché L, Farrugia A, Castillon P (1999) L'analyse d'herbe: un outil pour le pilotage de la fertilisation phosphatée et potassique des prairies naturelles et temporaires. Institut de l'Elevage, Paris

Van der Wal R, Pearce I, Brooker R, Scott D, Welch D, Woodin S (2003) Interplay between nitrogen deposition and grazing causes habitat degradation. Ecol Lett 6:141–146

Van der Wal R, Bardgett RD, Harrison KA, Stien A (2004) Vertebrate herbivores and ecosystem control: cascading effects of faeces on tundra ecosystems. Ecography 27:242–252

Walker TW, Adams AFR (1958) Studies on soil organic matter: 1. Influence of phosphorus content of parent material on accumulations of carbon, nitrogen, sulphur and organic phosphorus in grassland soils. Soil Sci 85:307–318

Walworth JL, Sumner ME, Isaac RA, Plank CO (1986) Preliminary DRIS norms for alphalpha in the southeastern United States and a comparison with Midwestern norms. Agron J 78:1046–1052

Wassen MJ, Olde Venterink H, Lapshina ED, Tanneberger F (2005) Endangered plant species persist under phosphorus limitation. Nature 437:547–549

Watson CJ, Matthews DI (2008) A 10-years study of phosphorus balances and the impact of grazed grassland on total P redistribution within the soil profile. Eur J Soil Sci 59:1171–1176

White RE, Tayoub AT (1983) Decomposition of plant residues of variable C:P ratio and the effect on soil phosphate availability. Plant Soil 74:163–173

Ziadi N, Bélanger G, Cambouris AN, Tremblay N, Nolin MC, Claessens A (2007) Relationship between P and N concentrations in corn. Agron J 99:833–841

Ziadi N, Bélanger G, Cambouris AN, Tremblay N, Nolin MC, Claessens A (2008) Relationship between phosphorus and nitrogen concentrations in spring wheat. Agron J 100:80–86

Chapter 12
Biological Phosphorus Cycling in Arctic and Alpine Soils

Michael N. Weintraub

12.1 Introduction

Climate change is having disproportionately large effects on arctic and alpine ecosystems (McGuire 1995; McGuire and Hobbie 1997; Melillo et al. 1995). Furthermore, soils from arctic environments play important roles in the global carbon (C) cycle because they contain a disproportionately large reservoir of soil C (Post et al. 1982). Because both primary production and decomposition have the potential to be limited by phosphorus (P) availability in arctic and alpine environments, understanding the biological controls on P cycling and availability is necessary to understand the impacts of environmental changes in these regions. The objective of this review is to draw together current knowledge on biological P cycling in arctic and alpine soils. The main topics considered are: the unique aspects of these environments and how they might influence biological P cycling; the dominant controls on P availability in these systems; interactions between seasonal dynamics and P cycling; studies on natural P availability; and responses of plants and soil microorganisms to P fertilization experiments.

The relative lack of literature on P dynamics, particularly biological P cycling, in arctic and alpine ecosystems is notable, especially for the high arctic. A disproportionate number of the studies that are available have been conducted in northern Alaska, USA (especially near Toolik Field Station), in the area near Abisko, Sweden, and in the Rocky Mountains of Colorado, USA. Thus, information from large areas of the arctic in North America and Europe is lacking, and the same is true for many alpine areas around the world.

M.N. Weintraub
Department of Environmental Sciences, University of Toledo, 2801 W. Bancroft Street, Mail Stop 604, Toledo OH 43606, USA
e-mail: michael.weintraub@utoledo.edu

E.K. Bünemann et al. (eds.), *Phosphorus in Action*, Soil Biology 26,
DOI 10.1007/978-3-642-15271-9_12,

12.2 Characteristics of Arctic and Alpine Soils

There are many similarities between the cold climates of arctic and alpine regions, as evidenced by the similarities between their plant communities. For example, 37% of the alpine plant species in the mountains of Colorado, USA, also occur in the arctic (Bliss 1956). Environmental conditions in both regions may vary markedly, with precipitation, angle of insolation, photoperiod, fluctuations in daily temperature, atmospheric thickness, partial pressure of atmospheric gases, and the duration of the plant growing season all depending upon altitude, latitude, and geographical position (Richardson et al. 2003). Even though there is greater seasonal variability in air temperatures in arctic than in alpine regions, the presence of permafrost in arctic soils tends to buffer soil temperatures from fluctuations, and results in smaller seasonal fluctuations in soil temperatures in arctic regions than in mid-latitude alpine regions (Richardson et al. 2003). There are gradients of environmental severity in both arctic and alpine environments, with conditions becoming increasingly harsh with latitude and elevation, respectively (Bliss 1956).

To a limited extent, it is possible to generalize about large, highly variable regions; for example, water is more often limiting in alpine soils than in low arctic soils (Bliss 1956; Richardson et al. 2003), although both environments encompass a wide range of soil moistures. This is because relative humidity tends to decrease with increasing elevation, which can exacerbate moisture stress in alpine environments (Richardson et al. 2003). In arctic soils, permafrost often isolates the active layer, preventing the deep drainage of water (Bliss 1956). However, drier polar desert and semidesert become predominant in the high arctic. Alpine soils less frequently have continuous permafrost to trap soil moisture, and therefore lack the ability to retain water that arctic soils possess (Richardson et al. 2003). Alpine regions tend to be characterized by topographic gradients in moisture availability that are caused by the high topographic relief (Billings 1973). Steep slopes serve to enhance runoff (Richardson et al. 2003), and ridge tops or plateaus often experience strong winds that blow snow onto leeward slopes, leading to differences in snowpack depth and high spatial variation.

In alpine environments, topographic gradients in moisture availability result in steep gradients in plant community composition and primary productivity (Billings 1973). Although alpine floras may differ among regions, there are often similar patterns of vegetation across the topographic gradients: cushion and rosette plants on ridges and between rocks; herbaceous plants and graminoids along the slopes; dwarf shrubs with herbaceous dicots and graminoids below the melting snowdrifts; and sedges, grasses, low shrubs, and mosses in poorly drained areas (Billings 1973). Higher moisture availability and primary productivity in areas of snow accumulation result in increased soil organic matter content and biological activity. For example, of the alpine plant communities on Niwot Ridge in the Rocky Mountains of Colorado, USA, wet meadows growing in relatively flat, low lying areas have the highest soil moisture, plant productivity, soil organic matter, microbial biomass, and microbial activity (Costello and Schmidt 2006; Fisk and Schmidt 1995; Fisk et al. 1998). In locations where the snow accumulation is so high that it inhibits plant growth, however, organic matter accumulation may be inhibited, as has been found to be the case on some leeward slopes at Niwot Ridge (Litaor et al. 2005).

In many arctic soils, the wet, cold conditions tend to inhibit decomposition, and as a result, organic matter often accumulates, especially in flat, poorly drained areas (Schimel et al. 1996). In the alpine tundra, the bog and wet meadow soils have comparable moisture availability to arctic soils, whereas other alpine soils tend to be well drained after snow melt (Bliss 1956). Thus, although organic soils can predominate in poorly drained alpine areas, these are the only alpine soils that tend to be similar to their arctic counterparts (Bliss 1956). Otherwise, alpine soils are often drier and tend to have less accumulated organic matter than arctic soils.

In soils where the lack of moisture inhibits soil development, the total P pool may be relatively higher than in wetter soils of similar age and parent material, but a greater proportion of the total P pool is likely to be bound in primary minerals (Cross and Schlesinger 1995; Walker and Syers 1976). As soil organic matter content decreases, soil P availability to plants and soil microorganisms becomes increasingly controlled by geochemical, rather than biochemical, reactions (Cross and Schlesinger 1995).

12.3 P Availability and Uptake in Arctic and Alpine Soils

The widely acknowledged controls on soil P availability include soil organic matter content, age, parent material, pH, and concentrations of soluble aluminum, iron, and calcium cations (Brady and Weil 2008). Tundra soils often contain significant amounts of soil organic matter, especially where they are not well drained. Soils rich in organic matter are often low in plant-available P due to the occlusion of P in organic matter (Walker and Syers 1976). In order for soil microorganisms and plant roots to acquire P from organic sources, they must first mineralize the organic P with phosphatase enzymes, which mineralize phosphate by hydrolyzing it from organic P compounds such as nucleic acids (for more detail on phosphatases see Nannipieri et al. 2011). Even if the organic matter contains enough total P to meet plant and microbial demand, energy and resources are required for organisms to produce the phosphatase enzymes that are necessary to liberate P from soil organic matter, limiting the rate of P mineralization (Allison et al. 2011; Sinsabaugh and Moorhead 1994). Thus, sites high in organic matter may exhibit P limitation on plant growth.

12.3.1 P Forms and Distribution in Arctic Soils

Giblin et al. (1991) conducted a study of nutrient dynamics across a toposequence of Alaskan arctic tundra plant communities in northern Alaska, USA (Table 12.1),

Table 12.1 Descriptions of the arctic and alpine vegetation types most frequently discussed in this review (Fisk et al. 1998; Jonasson et al. 1996; Seastedt and Vaccaro 2001; Weintraub and Schimel 2003)

Vegetation types	Location	Description
Arctic shrub	Northern Alaska, USA	A moist tundra community that generally occurs on moderately hilly topography with silty to gravelly soils. Dominated by relatively high-stature (>1–2 m) *Betula nana* and *Salix pulchra* shrubs with several other shrub species as lesser components (e.g., *Vaccinium vitis-idaea*)
Arctic tussock	Northern Alaska, USA	Found on well-drained, often gravelly soils in some upland sites and along water tracks. Dominated by the sedge *Eriophorum vaginatum*. Also includes *Carex bigelowii*, feather mosses (e.g., *Hylocomium splendens*, *Dicranum elongatum*), *Sphagnum rubellum*, and a mix of dwarf deciduous and evergreen shrubs (e.g., *Salix* spp., *Betula nana*, *Vaccinium vitis-idaea*, *Empetrum nigrum*)
Arctic wet sedge meadow	Northern Alaska, USA	Occurs in low-lying areas and is dominated by low-stature (<20–30 cm) rhizomatous sedges such as *Carex aquatilis* and *Eriophorum angustifolium*, with some *Eriophorum scheuchzeri*
Arctic heath	Northern Sweden	Dominated by ericaceous dwarf shrubs such as *Vaccinium uliginosum*, *Empetrum hermaphroditum*, *Rhododendron lapponicure*, *Dryas octopetata*, and prostrate *Betula nana*. Contains a scattered mix of forbs and graminoids, with abundant mosses covering most of the ground
Arctic fellfield	Northern Sweden	Has a thin, discontinuous organic horizon, and uneven vegetation cover dominated by *Empetrum hermaphroditum*, *Cassiope tetragona*, and *Vaccinium vitis-idaea*, with scattered forbs and a discontinuous moss cover
Alpine dry meadow	Rocky Mountains, Colorado, USA	This community occurs where relatively little snowpack accumulates, and the soils are dry throughout the growing season relative to adjacent plant communities. Dominated by *Kobresia myosuroides* (50–70% of total cover), with *Acomastylis rossii*, *Polygonum viviparum*, and one or more species of *Trifolium*
Alpine wet meadow	Rocky Mountains, Colorado, USA	Occurs in low-lying areas that melt out relatively late (mid-June to mid-July). Comprised mostly of the sedge *Carex scopulorum*

and concluded that most of these arctic plant communities occur on highly weathered soils with low concentrations of primary mineral P and relatively high proportions of the mineral P bound to Al and Fe (samples were sequentially extracted with 1 M NH_4Cl to remove P loosely bound to carbonates; then with citrate–dithionate–bicarbonate followed by 1 M NaOH to remove strongly adsorbed P plus P bound in Fe or Al oxides; then with 0.5 M HCl to remove P from apatite minerals). This distribution of P results in low P inputs from weathering, with decomposition responsible for replenishing most or all of the soil solution P (Chapin et al. 1978; Giblin et al. 1991). Chapin et al. (1978) found that 64% of the P in the upper 20 cm of a wet sedge meadow soil in Barrow, Alaska, USA, is organic (dissolved inorganic and organic P were measured on soil solution collected with a pressure membrane apparatus; labile P was extracted from soil with an anion exchange resin, and exchangeable P was determined by ^{32}P exchange with soil; organic P was calculated by subtracting inorganic P before and after ashing). Furthermore, they found that 84% of the inorganic P is non-exchangeable, and that most of the small pool of exchangeable P is tightly complexed by iron. They estimated that annual plant P demand in the wet sedge meadow on the arctic coastal plain in Alaska, USA, is 150 times larger than the pool of dissolved inorganic P and 6.5 times the labile P pool. Given the distribution of soil P they observed, the authors concluded that the replenishment of the dissolved P pool and P demand must be met by decomposition in this environment, and that biologically mediated P mineralization is the primary driver of P cycling in these organic-matter-rich soils.

Giblin et al. (1991) found dissolved P concentrations in soil water and in soil potassium chloride and calcium chloride extractions to be variable, but consistently low (frequently below detection) during the growing season (June to August), with substantially higher, but still low, concentrations in organic horizons relative to the mineral soil. These results are consistent with other observations of low concentrations of dissolved and labile P in arctic soils (Chapin et al. 1978; Chapin and Shaver 1981; Jonasson et al. 1993). Furthermore, Giblin et al. (1991) observed greater seasonal and interannual variability in extractable P than in extractable inorganic N.

Several studies have concluded that one of the largest reservoirs of potentially available soil P in arctic soils is the microbial biomass. Giblin et al. (1991) hypothesized that low arctic soil P availability in the summer could be at least partly due to microbial P immobilization during the growing season because they observed 1–2 orders of magnitude more P in microbial biomass (estimated by comparing P extracted from soils treated with a biocide, such as chloroform, or hexanol in this case, with P extracted from untreated soils) than in the extractable P pool across the toposequence they sampled in the arctic tundra of Alaska, USA. In a study of arctic heath and fellfield soils in northern Sweden, Jonasson et al. (1996) found that microbial biomass accounted for ~35% of total P, compared to ~3.5% of total N. Inorganic extractable P content was less than 1% of the total soil pool in the same soils.

12.3.2 P Forms and Distribution in Alpine Soils

As discussed above, alpine soils may be similar to tundra soils, but can be drier and contain less organic matter, depending on topographic position. Where organic matter contents are low, soil P availability may be more dependent on geochemical, rather than biochemical, reactions.

In a study of alpine Spodosols and Inceptisols collected at 2,000–2,400 m in the eastern Pyrenees of France, Cassagne et al. (2000) found that despite significant differences in P distribution among horizons between the two soil types, both soils from their higher altitude sites, at or above treeline, contained up to 10% of total P as soluble, exchangeable inorganic P [anion exchange resin P as measured by the Hedley et al. (1982) P sequential fractionation scheme]. The proportion of resin P in these soils is high compared to soils from other ecosystems (Cassagne et al. 2000), and suggests that P may be relatively available in the higher altitude soils of this alpine ecosystem. It is important to note, however, that this labile P pool is subject to significant temporal variability in response to plant and microbial uptake, and the date of collection was not reported in this study. NaOH-extractable organic P comprised over 45% of total P in all depths of the Inceptisols. NaOH-extractable organic P was generally lower in the Spodosols, particularly in the eluvial horizons, but tended to increase with depth. Overall, these soils were found to have relatively high total and organic P contents. The authors suggest that low mineralization rates associated with low temperatures are probably responsible for the accumulation of organic matter.

Makarov et al. (1997) found that organic P amounted to 92–99% of NaOH-extractable P (P adsorbed to Fe and Al oxides and carbonates or associated with humic acids; Cross and Schlesinger 1995) across a toposequence of alpine soils in the northern Caucasus. Cassagne et al. (2000) also found that NaOH-extractable organic P was a major component of the alpine Spodosols and Inceptisols in their study. Moreover, they observed little difference in NaOH-extractable organic P between depths in the Inceptisols. The authors interpreted this finding as an arrest of biological P transformation or translocation, possibly due to low temperatures. However, microbial activity and phosphatase production may continue even at low temperatures (Brooks et al. 1998; Lipson et al. 1999, 2002; Schmidt et al. 2004; see discussion below). An alternative explanation is that an accumulation of readily mineralized organic P is the result of relatively low plant and microbial P demand.

Turner et al. (2004) characterized soil P chemistry across a latitudinal alpine (and arctic) gradient consisting of three tundra heath/birch forest sites on Spodosols in the Fennoscandian mountains: Dovrefjell (Sør-Tröndelag, Norway; subarctic), Abisko (Norrbotten, Sweden; arctic) and Joatka (Finnmark, Norway; arctic). They used NaOH–EDTA extraction and solution ^{31}P nuclear magnetic resonance spectroscopy to determine the distribution of P compounds in these subarctic–arctic alpine soils. They observed a diverse mixture of inorganic and organic P compounds, with similar composition to peat. High concentrations of inorganic polyphosphate and orthophosphate diesters were present, which the authors suggest are at least partly derived from microbial biomass and probably reflect high microbial biomass P (e.g., Jonasson et al. 1996). At the same time, these compounds are relatively labile, indicating that these soils contain a large pool of potentially bioavailable soil organic P. The authors concluded that the high concentrations of easily degraded P compounds reflect the accumulation of organic matter resulting from slow decomposition in these cold, acidic soils (Turner et al. 2004).

In a study of P availability across a topographic and snow-depth gradient in the Rocky Mountains of Colorado, USA, Litaor et al. (2005) found a larger percentage of organic P (36–79% of total P) than inorganic P (10–45% of total P), based on the Hedley et al. (1982) P fractionation scheme. They also found that the relatively low-lying level sites, where runoff from snowmelt accumulated, contained more organic P than the windward or leeward slopes, and had the highest concentrations of soluble P. This provides additional evidence that organic P accumulates in areas with relatively high moisture availability, and that this results in higher P availability. Furthermore, the authors observed significant correlations between P availability (soil solution P and resin and sodium bicarbonate-extractable P) and aboveground biomass of graminoids, the dominant plant group. However, they did not observe significant correlations with aboveground plant biomass as a whole, or with herbaceous or forb plant groupings. These results suggest that some alpine species may be P-limited even though primary productivity as a whole may not appear to be.

12.3.3 P Mineralization Dynamics in Arctic and Alpine Soils

Differences in plant litter quality may cause relatively large differences in net P mineralization between soils and litter from contrasting plant communities, even in soils in close proximity to one another (Giblin et al. 1991; Jonasson et al. 1993; Schmidt et al. 1999). However, although few studies have been conducted, net P mineralization (measured as dissolved inorganic P accumulation over time on anion exchange resins) in arctic soils has generally been found to be low, or even negative, during the growing season (Giblin et al. 1991; Jonasson et al. 1993; Nadelhoffer et al. 1991).

Although data are limited for arctic and alpine environments, studies in other environments have observed an accumulation of labile soil P at the onset of winter. For example, Fabre et al. (1996) observed significantly increased concentrations of sodium bicarbonate-extractable inorganic and organic P (representative of labile P adsorbed on soil particles) and NaOH-extractable inorganic P beginning in November and continuing through March in a temperate floodplain forest soil in southern France. They hypothesize that this increase in extractable P was caused by P leaching from fresh litter inputs in the fall, a limitation of P mineralization resulting from low temperatures, and low plant and microbial uptake. Chen et al. (2003) also observed substantial winter increases in bicarbonate-extractable organic P concentrations in forest and grassland soils on the south island of New Zealand. They hypothesized that reduced microbial P mineralization resulting from low temperatures, as well as the leaching of dissolved organic P from fresh litter fall in the forest soil, leads to the accumulation of labile soil organic P.

Thus, it has been hypothesized that low temperatures prohibit microbial activity, and therefore P mineralization, during the winter (Cassagne et al. 2000; Chen et al. 2003; Fabre et al. 1996). Recent studies, however, demonstrate that microbial biomass and enzyme activities are relatively high in snow-covered soils (Brooks et al. 1998; Lipson et al. 1999, 2002; Schmidt et al. 2004), and that a significant proportion of yearly decomposition can occur beneath the snowpack (Schmidt and Lipson 2004). A deep snowpack can create a thermal buffer and sustain soil temperatures at or around 0°C, maintaining the presence of liquid water and protecting the microbial community from harsh winter conditions.

Additionally, a laboratory incubation of soils from several arctic plant communities in northern Alaska, USA, found that dissolved inorganic P release from tussock tundra and wet sedge meadow organic soils was 5–10 times greater at 3°C than at either 9°C or 15°C (Nadelhoffer et al. 1991). This could be because P immobilization increases more rapidly with temperature than gross P mineralization in these soils, or because of an inhibition of cold-adapted phosphatases with increasing temperature.

The results described above indicate that winter is actually an extremely active time for tundra microbial communities. For example, the highest levels of microbial biomass in an alpine soil were observed under the snowpack (Lipson et al. 1999). Although microbial activity may continue beneath the snow, snowmelt and its effects on microorganisms (such as lysis resulting from freeze–thaw effects and changes in osmolarity with snowmelt) can have significant effects on P cycling. Lipson et al. (1999) observed a crash in the microbial biomass and a pulse of nutrients in response to spring snowmelt, which they hypothesized to be derived from lysed microbial cells. These nutrients may be flushed from the system with snowmelt or taken up by plants as the growing season begins. In tundra soils, P is also rapidly released after the soil thaws in the spring (Chapin et al. 1978; Schimel et al. 1996). Chapin et al. (1978) estimate that the spring pulse of dissolved inorganic P from a crash in microbial biomass at snowmelt represents as much as 30% of annual plant P uptake in a wet sedge meadow on the arctic coastal plain of Alaska, USA. This P may be taken up immediately, or may enter the exchangeable P pool, which can serve as a buffer for large pulses of phosphate. This pulse of nutrients represents a large proportion of the annual flux (i.e., up to 50% of the microbial biomass), and is derived from nutrient mineralization in the soil, rather than nutrient release from the melting snowpack (Schimel et al. 1996).

Thus, it has been hypothesized that high winter P mineralization could be the result of microbial turnover in the winter (Giblin et al. 1991). In a test of this hypothesis in two arctic soils collected from a tree-line heath and a high-altitude fellfield near Abisko Scientific Research Station in northern Sweden, Schmidt et al. (1999) found that this pattern held only for the fellfield soil. Thus, while the pattern of high levels of P release in winter has been observed in both arctic and alpine soils, it does not appear to be a universal feature of cold soils. It is currently unclear, however, why this occurs in some soils but not others. Some studies have found that some cold-adapted microbial communities may be relatively resistant to freeze–thaw cycles (Lipson and Monson 1998; Lipson et al. 2000; Grogan et al. 2004). Soil temperature, moisture content at the time of freezing, variability in the timing of snowfall and the depth of the snowpack, and the rate of soil thaw are additional factors that have the potential to influence microbial P release at snowmelt. There is also the possibility for relatively high rates of P loss when large pulses of P are released at snowmelt (Larsen et al. 2007).

12.3.4 Seasonal Dynamics of Arctic and Alpine Plant P Uptake

Arctic and alpine systems have relatively short, snow-free growing seasons, when plants are active and taking up nutrients from the soil (Lipson and Monson 1998; Weintraub and Schimel 2005). In order to understand the biological controls on soil P cycling in arctic and alpine ecosystems, it is necessary to consider the seasonal dynamics of nutrient uptake and the predominant role of winter.

Differences in soil moisture availability between arctic and alpine tundra environments often result in alpine plants having deeper penetrating roots than their arctic counterparts (Bliss 1956), which may cause differences in the distribution and timing of root C inputs to, and nutrient uptake from, the soil. Root C inputs to the soil are one of the principal sources of labile C to mycorrhizae and microbial decomposers during the growing season (Bertin et al. 2003; Kuzyakov 2002). Differences in the distribution and timing of root growth and exudation among plant species and across plant communities have the potential to result in significant differences in soil microbial P cycling across both time and space by influencing microbial C uptake, nitrogen (N) and P acquisition (including phosphatase activity) and P immobilization (e.g., Weintraub and Schimel 2005; Weintraub et al. 2007). This influence has the potential to vary between arctic and alpine regions, and among plant communities within these regions, based on the timing and depth of root growth.

Previous research indicates that arctic and alpine plant growth early in the growing season often does not depend on nutrient uptake from the soil, but instead depends on nutrients stored in plant tissues (Chapin et al. 1980, 1986; Mullen and Schmidt 1993; Shaver and Kummerow 1992). This adaptation is necessary for growth in cold climates with a short growing season because these soils may remain frozen for several weeks after air temperatures are above freezing, at the time of year when solar radiation is highest. The most intensive period of root growth in deciduous tundra plants generally does not start until after leaf expansion is well underway (Shaver and Kummerow 1992). On the other hand, high rates of spring N uptake in alpine plants have also been observed (Bilbrough et al. 2000; Jaeger et al. 1999; Mullen et al. 1998). This may be because winter and spring arctic air and soil surface temperatures are often much lower than those in alpine regions, possibly extending dormancy (Bilbrough et al. 2000), and because moisture availability often becomes limiting after snowmelt in alpine soils. However, the limited evidence available also suggests that P uptake in arctic and alpine plants occurs after N uptake. For example, the Rocky Mountain alpine herb *Ranunculus adoneus* takes up N during snowmelt (Mullen et al. 1998), but takes up P much later in the growing season (Mullen and Schmidt 1993). This evidence, along with data on the timing of plant N and P accumulation in response to fertilization of arctic tussock and wet sedge meadow tundra (Shaver and Chapin 1995), and the seasonal dynamics of soil N and P availability in these two communities and shrub tundra (Weintraub and Schimel 2005), indicate that deciduous plant N uptake generally occurs earlier in the growing season than P acquisition in the tundra communities for which data are available.

12.3.5 The Role of Phosphatases in P Acquisition

A study of arctic and temperate strains of mycorrhizal fungi in the genus *Hebeloma* found that phosphatase activity (measured by using an artificial substrate that releases a colored or fluorescent reaction product when acted upon by the enzyme) in the arctic strains was highest at 2°C relative to 6°C, 12°C, or 22°C, and was only significantly higher than the temperate strains at 2°C (Tibbett et al. 1998). On average, the arctic strains also grew more slowly at all temperatures than the temperate strains. This suggests that organisms in cold soils can respond to the kinetic constraints imposed on enzyme activity by low temperatures, either by increasing enzyme production or by producing isozymes that are more active at low temperatures. Furthermore, phosphatase activity and microbial metabolism have been detected at soil temperatures as low as −20°C (Bremner and Zantua 1975; Christner 2002; Mikan et al. 2002), providing additional evidence for soil microbial activity and nutrient cycling in cold soils. Löffler et al. (2008) also observed significant phosphatase activity in soils collected frozen in March from an alpine altitudinal gradient in the Norwegian arctic. Because their samples were analyzed within an hour of thawing, they concluded that the enzyme activity they observed was not the result of in vitro production, but rather reflected the extant pool of soil enzymes. Bremner and Zantua (1975) attribute the occurrence of enzyme activity in soils at subzero temperatures to enzyme–substrate interactions in unfrozen water at the surfaces of soil particles (enzyme activities were measured at subzero temperatures by combining soils and enzyme assay substrates prior to freezing, incubating at subzero temperatures, and subsequent analysis of the reacted substrate). A similar conclusion was reached by Mikan et al. (2002) in a study of soil respiration at subzero temperatures.

In a study of seasonal dynamics of soil nutrient availability in arctic tundra (tussock, shrub, and wet sedge meadow plant communities), Weintraub and Schimel (2005) observed rapid increases in potassium sulfate-extractable phosphate in tussock and wet sedge meadow soils (Fig. 12.1). Extractable phosphate concentrations in shrub increased gradually in late July, and continued to increase until the end of August. By the end of the growing season, shrub had the highest extractable phosphate concentrations of any soil in this study. Nadelhoffer et al. (1991) concluded that wet sedge meadow soils may be P-limited, but Weintraub and Schimel (2005) determined that P availability was strongly seasonal, and was actually relatively high in wet sedge meadow soils in late July. Weintraub and Schimel (2005) attributed the elevation in arctic tundra soil phosphate availability they observed late in the

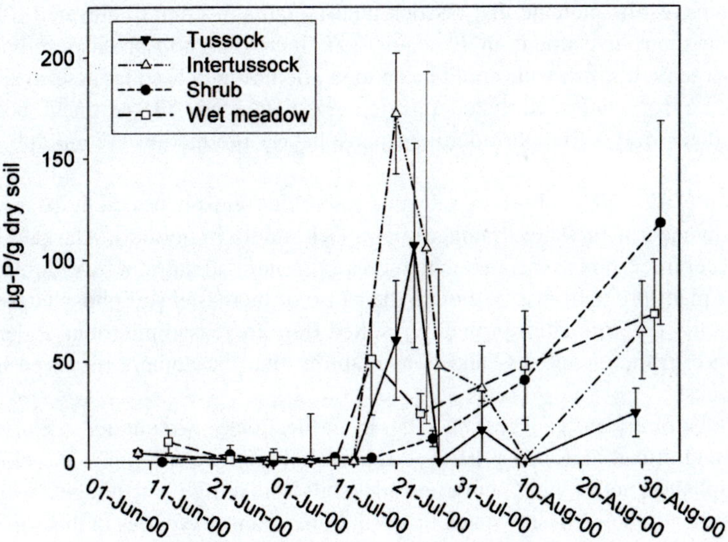

Fig. 12.1 Seasonal dynamics of 0.5 M potassium sulfate-extractable soil phosphate from three different plant communities (intertussock samples were collected between *Eriophorum vaginatum* tussocks in the tussock tundra community) at Toolik Field Station in northern Alaska in the summer of 2000. Reprinted with permission from Weintraub and Schimel (2005)

growing season to increases in root and microbial phosphatase activity. This conclusion was based on previous research demonstrating significant phosphatase activity on, and organic P utilization by, the roots of *Eriophorum vaginatum*, the predominant plant in the tussock tundra community (Kroehler and Linkins 1988, 1991; Moorhead et al. 1993). Moorhead et al. (1993) estimated that, on an annual basis, *E. vaginatum* root surface phosphatases mineralize almost twice as much P as is required for plant growth. Thus, the authors concluded that *E. vaginatum* may meet much of its P demand from the activities of root phosphatases, and estimated that approximately 28% of total annual tussock phosphatase activity (plants and soil combined) occurs during a brief late season pulse in soil P availability (Moorhead et al. 1993). As a result, they suggested that the majority of P uptake in *E. vaginatum* occurs late in the growing season, after the period of growth, and that this P is then stored in the rhizomes for use during the following growing season.

Because organisms generally only produce phosphatase in response to P demand, there is often an inverse relationship between phosphatase activity and inorganic P availability (Sinsabaugh and Moorhead 1994; Tibbett et al. 1998). However, the findings of Weintraub and Schimel (2005) indicate that P availability and phosphatase activity can also be positively correlated when phosphatase activity mineralizes phosphate in excess of organisms' demand.

Further, Moorhead and Linkins (1997) found that after a 3-year exposure of tussock tundra to elevated CO_2 (680 μmol mol^{-1}), phosphatase activity was higher on *E. vaginatum* root surfaces, ectomycorrhizal rhizomorphs and mantles associated with *Betula nana* roots, and in the organic soil horizons associated with plant roots. Their results indicate that tussock tundra plants respond to elevated CO_2 by increasing their investment in P acquisition. Increased phosphatase activity in tussock organic horizon soils could be an indication of increased labile C availability resulting from increased rhizodeposition in the elevated CO_2 treatment, because root exudates may stimulate microbial phosphatase production (Weintraub et al. 2007).

Chapin et al. (1988) observed elevated microbial phosphatase activity in water tracks (channels with water drainage) in tussock tundra in northern Alaska (USA). These water tracks are associated with increased water and nutrient flux to roots and to higher plant productivity. Although the cause of increased soil phosphatase was not explicitly determined, it probably resulted from increased microbial P demand in response to the elevated C and N availability that the authors observed in the water tracks.

In a subalpine conifer forest at 3,050 m in the Rocky Mountains of Colorado, USA, Weintraub et al. (2007) observed relatively high levels of microbial biomass and phosphatase activity in O horizon soils (soils with a high percentage of organic matter) beneath the ~1 m snowpack in April. Phosphatase activities in this soil were higher beneath the snow in April 2005 than they were at any time until August 2005 (when the dataset ends), due to increasingly dry conditions as the growing season progressed. Weintraub et al. (2007) also observed a trend toward increased organic horizon microbial phosphatase activity in response to increases in rhizodeposition associated with the spring initiation of photosynthesis by the trees in this forest. These results exemplify how the seasonal dynamics of soil moisture availability and root growth and exudation can influence microbial P acquisition from the soil.

12.4 P Limitation in Arctic and Alpine Soils

P is often limiting to plant growth in arctic and alpine soils (Billings and Mooney 1968). Nutrient addition studies conducted in the arctic tundra at Toolik Field Station in northern Alaska, Abisko Research Station in northern Sweden, and in the alpine tundra at Niwot Ridge LTER in the Rocky Mountains of Colorado, indicate that plant growth in both arctic and alpine environments is typically N-limited, but P may be co- or solely limiting at times (Jonasson et al. 1999; Shaver and Chapin 1986; Theodose and Bowman 1997).

12.4.1 Case Studies from Arctic P Addition Experiments

Chapin et al. (1975) found that the addition of N, P, and potassium (NPK fertilizer) consistently increased production of the wet sedge meadow tundra in Barrow, Alaska, USA (Table 12.2). This increase was particularly evident when a high-P fertilizer was applied, and a high-P fertilizer stimulated production more than a

Table 12.2 Arctic nutrient addition studies

Study	Location	Vegetation type(s)	Treatment
Chapin et al. (1975)	Northern Alaska, USA	Wet sedge meadow	NPK commercial fertilizer addition 45 g m^{-2}, 8–32–16 in one plot and 20–10–10 in the other
Chapin and Shaver (1985)	Northern Alaska, USA	Tussock, wet sedge meadow	NPK commercial fertilizer addition: 25 g m^{-2} N, 25 g m^{-2} P, and 31 g m^{-2} K applied in July 1978. Sampling occurred during the 1979–1981 growing seasons
Kielland and Chapin (1994)	Northern Alaska, USA	Tussock, shrub, wet sedge meadow, dry heath	10 g m^{-2} P as triple superphosphate. Sampling occurred 2 years after fertilizer application
Shaver and Chapin (1995)	Northern Alaska, USA	Tussock, wet sedge meadow	N (25 g m^{-2}), P (25 g m^{-2}), and K (31.6 g m^{-2}) added singly and in all possible combinations. Different plots were set up in 1976, 1977, and 1978, with fertilizer applied only in the year the experiment was set up. Sampling occurred during 1976–1979 growing seasons
Jonasson et al. (1996)	Northern Sweden	Arctic heath and fellfield	NPK fertilizer (5:1.25:3.75 g m^{-2}) followed by a second addition later in the growing season with twice this amount
Jonasson et al. (1999)	Northern Sweden	Arctic heath and fellfield	NPK fertilization: 4.9 g m^{-2} N, 1.3 g m^{-2} P, 6.0 g m^{-2} K, and 0.36 g m^{-2} Mg in 1989. From 1990 to 1992, additions were 10.0, 2.6, 9.0, and 0.8 g m^{-2}, respectively. No fertilization in 1993, when sampling occurred
Nadelhoffer et al. (2002)	Northern Alaska, USA	Tussock, wet sedge meadow	N and P additions once per year starting in 1988 and continuing through sampling in 1994 and 1995. Annual additions were 10 g m^{-2} N, and 5 g m^{-2} P, except in 1988 when 10 g m^{-2} P was added
Madan et al. (2007)	Svalbard	High arctic polar semidesert	Factorial combination of N (0.5 g m^{-2} $year^{-1}$ in low N plots, and 5 g m^{-2} $year^{-1}$ in high N plots) and P (1 g m^{-2} $year^{-1}$) added five times during 2000–2002 the growing seasons

Fertilizer was applied once at the beginning of the growing season in the year of sampling unless noted high-N fertilizer. Nadelhoffer et al. (2002) measured changes in fine root N and P concentrations in response to fertilization in tussock and wet sedge meadow tundra at Toolik Field Station in northern Alaska, USA. They found that root N and P increased in both communities, but that P concentrations increased more than N concentrations in wet sedge tundra, whereas relative increases in N and P concentrations in roots from tussock tundra plant were similar. These results suggest that wet sedge meadow tundra is P-limited, consistent with the findings of Chapin et al. (1975), whereas tussock tundra is likely co-limited by N and P.

A factorial N and P addition study in high arctic mixed heath in Svalbard by Gordon et al. (2001) found that this tundra community is also co-limited by N and P. Bryophytes, in particular, were found to respond strongly to P addition with increased biomass. This result is contrary to the findings of other studies at lower latitude tundra sites, where increased shrub and graminoid growth in response to fertilization resulted in most bryophytes being shaded out (Gordon et al. 2001).

A study of P uptake in Alaskan (USA) arctic tundra plants by Kielland and Chapin (1994) found that P uptake significantly increased in response to fertilization for plants from tussock, deciduous shrub, and dry heath tundra communities, but not wet sedge meadow. They also observed a significant correlation between plant P accumulation and 0.1 N sulfuric acid-extractable soil P (a measure of exchangeable P). Furthermore, P uptake capacity was correlated with soil P availability across all growth-forms. These results indicate that plant P uptake is typically dependent upon its availability in this environment, and suggest an unmet plant P demand. Thus, Kielland and Chapin (1994) conclude that the soil processes that govern P availability largely control the P absorption of the tundra species in their study.

A long-term (3–10 years) large scale fertilization experiment across multiple tussock and wet sedge meadow sites in northern Alaska, USA, by Shaver and Chapin (1995) found that the same plant communities were limited by different nutrients in different locations, and that it is not possible to conclude that a particular tundra community is always N- or P-limited. An examination of leaf N: P ratios suggested that N limitation was approximately three times more frequent than P limitation in tussock tundra, whereas P limitation was at least four times as common as N limitation in wet sedge meadow (Shaver and Chapin 1995). In accordance with the finding that tundra communities do not always consistently respond to increases in nutrient availability, Chapin and Shaver (1985) found that plant species in tussock and wet sedge meadow tundra in northern Alaska (USA) respond individualistically to nutrient additions, and that species that are more typical of nutrient-rich sites tended to respond more strongly to fertilization than species associated with nutrient-poor sites.

Jonasson et al. (1996) measured the changes in microbial biomass C, N, and P pools in response to NPK fertilization in heath and fellfield tundra soils in northern Sweden. They found that microbial biomass did not increase in response to fertilization, but they observed significant increases in microbial N and P. They also observed increased microbial respiration, suggesting increased microbial activity. Jonasson et al. (1999) found that the addition of fertilizer to Swedish arctic heath and fellfield soils led to greater increases in microbial P than N. The high proportion of soil P in microbial biomass (Jonasson et al. 1996) suggests that there may be intense competition between plants and microbes for P, and that P release from microbial biomass has the potential to be a significant P source to plants (Schimel et al. 1996). However, Jonasson et al. (1999) found that microbial biomass P increased only when soil inorganic P availability increased, suggesting that microorganisms acquired additional P only in cases of declining P-sink strength in plants. This in turn suggests that, at least in some tundra communities, plants can successfully compete against soil microorganisms for P.

Madan et al. (2007) conducted a factorial N and P addition study at a high arctic semidesert in Svalbard. They observed relatively few effects on the plant community using an N deposition rate of 0.5 g N m^{-2} $year^{-1}$ (five times ambient deposition). However, they observed less bare soils, and a trend toward increased plant species richness and diversity when they also added 1.0 g P m^{-2} $year^{-1}$. Although the N deposition rate of 0.5 g N m^{-2} $year^{-1}$ is somewhat realistic for the future at five times ambient deposition, their P application rate of 1.0 g P m^{-2} $year^{-1}$ was three orders of magnitude higher than ambient deposition levels, and is therefore unrealistic for the foreseeable future. On the basis of these results, the authors concluded that low P availability is likely to limit the response of this polar semidesert vegetation to N deposition.

12.4.2 *Case Studies from Alpine P Addition Experiments*

A study of aboveground production responses to N and P additions in wet and dry meadow alpine tundra communities at Niwot Ridge in the Rocky Mountains of Colorado, USA (Table 12.3) found dry meadow to be N-limited and wet meadow to be co-limited by N and P (Bowman et al. 1993). Wet meadow forbs increased tissue P concentrations after P fertilization nearly three times more than wet meadow graminoids, and two times more than dry meadow forbs and graminoids (Bowman 1994; Bowman et al. 1993). Following up on this research, a 4-year nutrient addition study in dry and moist alpine tundra communities at Niwot Ridge concluded that there was clear evidence of P limitation affecting the overall production of both of these alpine tundra communities (Seastedt and Vaccaro 2001). This disparity with earlier findings may be because the study plots used by Seastedt and Vaccaro (2001) contained a higher proportion of species sensitive to P limitations, had higher N availability, or both. The Seastedt and Vaccaro (2001) study occurred on old, organic-matter-rich soils that were not affected by the most recent glaciation, and the combination of high organic matter content and soil age may have exacerbated P limitation.

Another study at Niwot Ridge found that within both wet and dry meadow community types, different plant functional groups responded individualistically to N and P fertilization, based on assessment of growth form and mycorrhizal

Table 12.3 Alpine nutrient addition studies

Study	Location	Vegetation type(s)	Treatment
Bowman et al. (1993)	Rocky Mountains, Colorado, USA	Alpine wet and dry meadows	N and P factorial fertilization, with N applied as 25 g m^{-2} urea osmocote pellets, and P applied as 25 g m^{-2} P_2O_5 osmocote pellets in mid-June 1990 and early July 1991, with sampling in 1991
Bowman (1994)	Rocky Mountains, Colorado, USA	Alpine wet and dry meadows	Same plots as Bowman et al. (1993), sampled in 1991
Theodose and Bowman (1997)	Rocky Mountains, Colorado, USA	Alpine wet and dry meadows	Same plots as Bowman et al. (1993), sampled from 1990 to 1994. Fertilizer application continued through 1994, with the exception of 1992, when no fertilizer was applied
Seastedt and Vaccaro (2001)	Rocky Mountains, Colorado, USA	Alpine wet and dry meadows	N and P factorial fertilization. In September 1993 and August 1994, 20 g m^{-2} N and 2 g m^{-2} P were added. No fertilizer was added to any plots in 1995, and in 1996 and 1997 10 g m^{-2} N as $(NH_4)_2SO_4$ and 1 g m^{-2} P as P_2O_5 was applied in July of each year. Sampling occurred in 1997
Soudzilovskaia and Onipchenko (2005)	Northwestern Caucasus, Russia	Alpine heath	N and P factorial fertilization. N was added as urea (9 g N m^{-2} $year^{-1}$); P was added as double superphosphate ($CaH_4(PO_4)_2$) (2.5 g P m^{-2} $year^{-1}$). Fertilizer was applied at the beginning of every growing season from 1999 to 2002, the period when the study was conducted

Fertilizer was applied once in the year of sampling unless noted

associations (Theodose and Bowman 1997). In particular, N_2 fixing and non-mycorrhizal forbs in the dry meadow community were found to be P-limited. Sedges, the dominant functional group in the dry meadow community, were unaffected by fertilization, however. A long-term increase in grasses was also observed in both the wet and dry meadow alpine tundra communities in response to P fertilization.

Taken together, these studies from Niwot Ridge Colorado, USA, indicate that the areas with the deepest snowpack (wet meadow) are likely to have relatively high N availability, and are more likely to be P-limited. This is because accumulated N deposition in snow is redistributed along with the snowpack and its meltwater (Bowman 1994; Bowman et al. 1993; Theodose and Bowman 1997; Walker et al. 1993). These results exemplify how the landscape heterogeneity, climate, and snow distribution interact to influence the relative degree of P limitation across the landscape. Alpine communities on Niwot Ridge with high inputs of water and N from snow are co-limited by N and P availability, whereas communities with low snow cover and drier soils are limited by N and/or water availability (Bowman et al. 1993; Theodose and Bowman 1997).

Additionally, a factorial nutrient addition study conducted by Soudzilovskaia and Onipchenko (2005) in an alpine heath community in northwestern Caucasus, Russia, concluded that community plant density and flowering was co-limited by N and P. However, they also found that different plant species had different responses to the nutrient addition treatments. In response to P addition, they observed increases in *Festuca ovina* and decreases in *Carex* spp., suggesting that an increase in P availability altered the competitive dynamics between these plants, which were the most abundant at their site, with *Carex* spp. being more N-limited and *Festuca ovina* more P-limited. These results add to the growing body of evidence that P availability may constrain vegetation responses to increased N availability in alpine communities, and that changes in the relative availabilities of soil N and P are likely to alter plant competitive dynamics and, ultimately, community composition. This conclusion is supported by the findings of Arnesen et al. (2007), who analyzed the relationships between plant community composition and bedrock-derived soil nutrients and pH in rocky alpine habitats in northern Norway. They found that after soil pH (correlated with carbonate content at their sites), P availability was the soil factor that best explained the floristic variation among their sites. They also noted that in exposed alpine habitats plant litter is likely to be transported off site, increasing the dependence on bedrock-derived nutrients.

12.5 Conclusions

Organic P has generally been found to predominate in the arctic soils studied, which are mostly from the low arctic, where there is a lower proportion of polar desert than in the high arctic, which has not been well studied. The combination of relatively high moisture availability, often from snow accumulation and redistribution, along with low temperatures, can result in the accumulation of soil organic matter, including organic P. Poorly drained alpine soils where snowmelt water accumulates may be similar to these arctic soils in that both tend to have relatively high organic matter content, with P cycles dominated by biological P cycling. Alpine soils are often drier than low arctic soils, however, with less accumulated organic material. As soil organic matter content decreases, soil P availability to plants and soil microorganisms may become increasingly controlled by geochemical, rather than biochemical, reactions. As has been found to be the case in the wet meadow tundra at Niwot Ridge in the Rocky Mountains of Colorado, USA, redistribution of the snowpack can result in the redistribution of deposited N along with snow, alleviating N limitation and potentially exacerbating P limitation where snowmelt water accumulates (Theodose and Bowman 1997). Given the increasing intensity of N deposition, the potential for elevated N availability to exacerbate P limitation warrants additional investigation.

It has been hypothesized that low temperatures and frozen soil prohibit microbial activity, and therefore P mineralization, especially during the winter, and that this is the ultimate cause of organic P accumulation in arctic and alpine soils. However, phosphatase activity and microbial metabolism have been detected at soil temperatures as low as −20°C, and winter has actually been found to be an extremely active time for alpine tundra microbial communities. Further study will be required to resolve this apparent contradiction.

Although microbial activity may continue beneath the snow, snowmelt and its effects on microorganisms can have significant effects on P cycling, resulting in a rapid release of P when the soil thaws in the spring. The microbial biomass is one of the largest reservoirs of potentially available soil P in arctic and alpine soils, and a release of microbial P in the spring can represent a large proportion of the annual flux. The release of P from microbial biomass at snowmelt has not been found to occur consistently, however, and the controls on this process and the extent to which it may be responsible for soil P losses are not well understood.

Net P mineralization and available P in arctic and alpine soils have typically been found to be low during the growing season. However, elevated soil phosphate availability late in the growing season has been observed in low arctic tundra communities, and has been attributed to late summer increases in root and microbial phosphatase activity, and may be associated with late growing season plant nutrient accumulation to support growth the following spring (Weintraub and Schimel 2005). There is little data available on the activity of arctic and alpine root phosphatases, however, and further investigation into their dynamics, their interaction with plant phenology, and their influence on soil P availability and the possibility of organic P uptake is warranted.

The results of P fertilization studies in arctic and alpine ecosystems have been mixed, with some plant communities, such as alpine wet meadow, showing signs of P limitation to primary productivity, whereas other communities are either N-limited or co-limited by N and P. A cross-site comparison of fertilization experiments in the Alaskan (USA) arctic concluded that the same plant communities may be limited by different nutrients in different locations, presumably as a result of site-specific differences in edaphic factors such as parent material and soil age.

Because both primary production and decomposition have the potential to be limited by P availability in arctic and alpine environments, which may change if decomposition rates, plant phenology and community composition are altered by climate change, understanding the biological controls on P cycling and availability is necessary to understand the impacts of environmental changes in these regions. Our current understanding of P limitation in arctic and alpine ecosystems comes largely from a few intensively studied areas, however, and often lacks mechanistic detail. In order to predict how P limitation may affect primary productivity and decomposition as the climate changes, more mechanistic studies across a more representative range of arctic and alpine environments will be required.

References

Allison SD, Weintraub MN, Gartner TB, Waldrop MP (2011) Evolutionary-economic principles as regulators of soil enzyme production and ecosystem function. In: Shukla GC, Varma A (eds) Soil enzymology. Soil biology, vol 22. Springer, New York

Arnesen G, Beck PSA, Engelskjøn T (2007) Soil acidity, content of carbonates, and available phosphorus are the soil factors best correlated with alpine vegetation: evidence from Troms, north Norway. Arct Antarct Alp Res 39:189–199

Bertin C, Yang X, Weston LA (2003) The role of root exudates and allelochemicals in the rhizosphere. Plant Soil 256:67–83

Bilbrough CJ, Welker JM, Bowman WD (2000) Early spring nitrogen uptake by snow-covered plants: a comparison of arctic and alpine plant function under the snowpack. Arct Antarct Alp Res 32:404–411

Billings WD (1973) Arctic and alpine vegetations: similarities, differences, and susceptibility to disturbance. Bioscience 23:697–704

Billings WD, Mooney HA (1968) The ecology of arctic and alpine plants. Biol Rev 43:481–529

Bliss LC (1956) A comparison of plant development in microenvironments of arctic and alpine tundras. Ecol Monogr 26:303–337

Bowman WD (1994) Accumulation and use of nitrogen and phosphorus following fertilization in two alpine tundra communities. Oikos 70:261–270

Bowman WD, Theodose TA, Schardt JC, Conant RT (1993) Constraints of nutrient availability on primary production in two alpine tundra communities. Ecology 74:2085–2097

Brady NC, Weil RR (2008) The nature and properties of soils, 14th edn. Pearson/Prentice Hall, Upper Saddle River

Bremner JM, Zantua MI (1975) Enzyme activity in soils at subzero temperatures. Soil Biol Biochem 7:383–387

Brooks PD, Williams MW, Schmidt SK (1998) Inorganic nitrogen and microbial biomass dynamics before and during spring snowmelt. Biogeochemistry 43:1–15

Cassagne N, Remaury M, Gauquelin T, Fabre A (2000) Forms and profile distribution of soil phosphorus in alpine Inceptisols and Spodosols (Pyrenees, France). Geoderma 95:161–172

Chapin FS III, Shaver GR (1981) Changes in soil properties and vegetation following disturbance of alaskan arctic tundra. J Appl Ecol 18:605–617

Chapin FS III, Shaver GR (1985) Individualistic growth response of tundra plant species to environmental manipulations in the field. Ecology 66:564–576

Chapin FS III, Cleve KV, Tieszen LL (1975) Seasonal nutrient dynamics of tundra vegetation at Barrow, Alaska. Arct Antarct Alp Res 7:209–226

Chapin FS III, Barsdate RJ, Barel D (1978) Phosphorus cycling in Alaskan coastal tundra: a hypothesis for the regulation of nutrient cycling. Oikos 31:189–199

Chapin FS III, Johnson DA, McKendrick JD (1980) Seasonal movement of nutrients in plants of differing growth form in an Alaskan tundra ecosystem: implications for herbivory. J Ecol 68:189–209

Chapin FS III, Shaver GR, Kedrowski RA (1986) Environmental controls over carbon, nitrogen and phosphorus fractions in *Eriophorum vaginatum* in Alaskan tussock tundra. J Ecol 74:167–196

Chapin FS III, Fetcher N, Kielland K, Everett KR, Linkins AE (1988) Productivity and nutrient cycling of Alaskan tundra: enhancement by flowing soil water. Ecology 69:693–702

Chen CR, Condron LM, Davis MR, Sherlock RR (2003) Seasonal changes in soil phosphorus and associated microbial properties under adjacent grassland and forest in New Zealand. For Ecol Manage 177:539–557

Christner BC (2002) Incorporation of DNA and protein precursors into macromolecules by bacteria at -15°C. Appl Environ Microbiol 68:6435–6438

Costello EK, Schmidt SK (2006) Microbial diversity in alpine tundra wet meadow soil: novel Chloroflexi from a cold, water-saturated environment. Environ Microbiol 8:1471–1486

Cross AF, Schlesinger WH (1995) A literature review and evaluation of the Hedley fractionation: applications to the biogeochemical cycle of soil phosphorus in natural ecosystems. Geoderma 64:197–214

Fabre A, Pinay G, Ruffinoni C (1996) Seasonal changes in inorganic and organic phosphorus in the soil of a riparian forest. Biogeochemistry 35:419–432

Fisk MC, Schmidt SK (1995) Nitrogen mineralization and microbial biomass nitrogen dynamics in three alpine tundra communities. Soil Sci Soc Am J 59:1036–1043

Fisk MC, Schmidt SK, Seastedt TR (1998) Topographic patterns of above- and belowground production and nitrogen cycling in alpine tundra. Ecology 79:2253–2266

Giblin AE, Nadelhoffer KJ, Shaver GR, Laundre JA, McKerrow AJ (1991) Biogeochemical diversity along a riverside toposequence in arctic Alaska. Ecol Monogr 61:415–436

Gordon C, Wynn JM, Woodin SJ (2001) Impacts of increased nitrogen supply on high arctic heath: the importance of bryophytes and phosphorus availability. New Phytol 149:461–471

Grogan P, Michelsen A, Ambus P, Jonasson S (2004) Freeze–thaw regime effects on carbon and nitrogen dynamics in sub-arctic heath tundra mesocosms. Soil Biol Biochem 36:641–654

Hedley MJ, Stewart JWB, Chauhan BS (1982) Changes in inorganic and organic soil phosphorus fractions induced by cultivation practices and by laboratory incubations. Soil Sci Soc Am J 46:970–976

Jaeger CH, Monson RK, Fisk MC, Schmidt SK (1999) Seasonal partitioning of nitrogen by plants and soil microorganisms in an alpine ecosystem. Ecology 80:1883–1891

Jonasson S, Havström M, Jensen M, Callaghan TV (1993) In situ mineralization of nitrogen and phosphorus of arctic soils after perturbations simulating climate change. Oecologia 95:179–186

Jonasson S, Michelsen A, Schmidt IK, Nielsen EV, Callaghan TV (1996) Microbial biomass C, N and P in two arctic soils and responses to addition of NPK fertilizer and sugar: implications for plant nutrient uptake. Oecologia 106:507–515

Jonasson S, Michelsen A, Schmidt IK, Nielsen EV (1999) Responses in microbes and plants to changed temperature, nutrient, and light regimes in the arctic. Ecology 80:1828–1843

Kielland K, Chapin FS (1994) Phosphate uptake in arctic plants in relation to phosphate supply: the role of spatial and temporal variability. Oikos 70:443–448

Kroehler C, Linkins A (1988) The root surface phosphatases of *Eriophorum vaginatum*: effects of temperature, pH, substrate concentration and inorganic phosphorus. Plant Soil 105:3–10

Kroehler CJ, Linkins AE (1991) The absorption of inorganic phosphate from P-labeled inositol hexaphosphate by *Eriophorum vaginatum*. Oecologia 85:424–428

Kuzyakov Y (2002) Review: factors affecting rhizosphere priming effects. J Plant Nutr Soil Sci 165:382–396

Larsen KS, Grogan P, Jonasson S, Michelsen A (2007) Dynamics and microbial dynamics in two subarctic ecosystems during winter and spring thaw: effects of increased snow depth. Arct Antarct Alp Res 39:268–276

Lipson DA, Monson RK (1998) Plant–microbe competition for soil amino acids in the alpine tundra: effects of freeze–thaw and dry–rewet events. Oecologia 113:406–414

Lipson DA, Schmidt SK, Monson RK (1999) Links between microbial population dynamics and nitrogen availability in an alpine ecosystem. Ecology 80:1623–1631

Lipson DA, Schmidt SK, Monson RK (2000) Carbon availability and temperature control the post-snowmelt decline in alpine soil microbial biomass. Soil Biol Biochem 32:441–448

Lipson DA, Schadt CW, Schmidt SK (2002) Changes in soil microbial community structure and function in an alpine dry meadow following spring snow melt. Microb Ecol 43:307–314

Litaor MI, Seastedt TR, Walker MD, Carbone M, Townsend A (2005) The biogeochemistry of phosphorus across an alpine topographic/snow gradient. Geoderma 124:49–61

Löffler UCM, Cypionka H, Löffler J (2008) Soil microbial activity along an arctic-alpine altitudinal gradient from a seasonal perspective. Eur J Soil Sci 59:842–854

Madan NJ, Deacon LJ, Robinson CH (2007) Greater nitrogen and/or phosphorus availability increase plant species' cover and diversity at a high arctic polar semidesert. Polar Biol 30:559–570

Makarov MI, Malysheva TI, Haumaier L, Alt HG, Zech W (1997) The forms of phosphorus in humic and fulvic acids of a toposequence of alpine soils in the northern Caucasus. Geoderma 80:61–73

McGuire AD (1995) The responses of net primary production (NPP) and total carbon storage for the continental United States to changes in atmospheric CO_2, climate, and vegetation. Bull Ecol Soc Am 76:177

McGuire AD, Hobbie JE (1997) Global climate change and the equilibrium responses of carbon storage in arctic and subarctic regions. The Arctic Research Consortium of the United States, Fairbanks

Melillo JM, Kicklighter DW, McGuire AD, Peterjohn WT, Newkirk KM, Zepp RG, Sontaff CH (1995) Global change and its effects on soil organic carbon stocks. Wiley, New York

Mikan CJ, Schimel JP, Doyle AP (2002) Temperature controls of microbial respiration in arctic tundra soils above and below freezing. Soil Biol Biochem 34:1785–1795

Moorhead DL, Linkins AE (1997) Elevated CO_2 alters belowground exoenzyme activities in tussock tundra. Plant Soil 189:321–329

Moorhead DL, Kroehler CJ, Linkins AE, Reynolds JF (1993) Extracellular acid phosphatase activities in *Eriophorum vaginatum* tussocks: a modeling synthesis. Arct Antarct Alp Res 25:50–55

Mullen RB, Schmidt SK (1993) Mycorrhizal infection, phosphorus uptake, and phenology in *Ranunculus adoneus*: implications for the functioning of mycorrhizae in alpine systems. Oecologia 94:229–234

Mullen RB, Schmidt SK, Jaeger CH (1998) Nitrogen uptake during snowmelt by the snow buttercup, *Ranunculus adoneus*. Arct Antarct Alp Res 30:121–125

Nadelhoffer KJ, Giblin AE, Shaver GR, Laundre JA (1991) Effects of temperature and substrate quality on element mineralization in six arctic soils. Ecology 72:242–253

Nadelhoffer KJ, Johnson L, Laundre J, Giblin AE, Shaver GR (2002) Fine root production and nutrient content in wet and moist arctic tundras as influenced by chronic fertilization. Plant Soil 242:107–113

Nannipieri P, Giagnoni L, Landi L, Renella G (2011) Role of phosphatase enzymes in soil. In: Bünemann E, Oberson A, Frossard E (eds) Phosphorus in action: biological processes in soil phosphorus cycling. Soil biology, vol 26. Springer, Heidelberg. doi:10.1007/978-3-642-15271-9_9

Post WM, Emanuel WR, Zinke PJ, Stangenberger AG (1982) Soil carbon pools and world life zones. Nature 298:156–159

Richardson MI, Moore IT, Soma KK, Lei FM, Wingfield JC (2003) How similar are high latitude and high altitude habitats? A review and a preliminary study of the adrenocortical response to stress in birds of the Qinghai-Tibetan Plateau. Acta Zool Sin 49:1–19

Schimel JP, Kielland K, Chapin FS (1996) Nutrient availability and uptake by tundra plants. In: Reynolds JF, Tenhunen JD (eds) Ecological studies. Springer, Berlin, pp 203–221

Schmidt SK, Lipson DA (2004) Microbial growth under the snow: implications for nutrient and allelochemical availability in temperate soils. Plant Soil 259:1–7

Schmidt IK, Jonasson S, Michelsen A (1999) Mineralization and microbial immobilization of N and P in arctic soils in relation to season, temperature and nutrient amendment. Appl Soil Ecol 11:147–160

Schmidt SK, Lipson DA, Ley RE, Fisk MC, West AE (2004) Impacts of chronic nitrogen additions vary seasonally and by microbial functional group in tundra soils. Biogeochemistry 69:1–17

Seastedt TR, Vaccaro L (2001) Plant species richness, productivity, and nitrogen and phosphorus limitations across a snowpack gradient in alpine tundra, Colorado, USA. Arct Antarct Alp Res 33:100–106

Shaver GR, Chapin FS (1986) Effect of fertilizer on production and biomass of tussock tundra, Alaska, USA. Arct Antarct Alp Res 18:261–268

Shaver GR, Chapin FS (1995) Long-term responses to factorial, NPK fertilizer treatment by Alaskan wet and moist tundra sedge species. Ecography 18:259–275

Shaver GR, Kummerow J (1992) Phenology, resource allocation, and growth of arctic vascular plants. In: Chapin FS, Jefferies RL, Reynolds JF, Shaver GR, Svoboda J (eds) Arctic ecosystems in a changing climate. Academic, New York, pp 193–211

Sinsabaugh RL, Moorhead DL (1994) Resource allocation to extracellular enzyme production: a model for nitrogen and phosphorus control of litter decomposition. Soil Biol Biochem 26:1305–1311

Soudzilovskaia NA, Onipchenko VG (2005) Experimental investigation of fertilization and irrigation effects on an alpine heath, northwestern Caucasus, Russia. Arct Antarct Alp Res 37:602–610

Theodose TA, Bowman WD (1997) Nutrient availability, plant abundance, and species diversity in two alpine tundra communities. Ecology 78:1861–1872

Tibbett M, Sanders FE, Cairney JWG (1998) The effect of temperature and inorganic phosphorus supply on growth and acid phosphatase production in arctic and temperate strains of ectomycorrhizal *Hebeloma* spp. in axenic culture. Mycol Res 102:129–135

Turner BL, Baxter R, Mahieu N, Sjogersten S, Whitton BA (2004) Phosphorus compounds in subarctic Fennoscandian soils at the mountain birch (*Betula pubescens*)–tundra ecotone. Soil Biol Biochem 36:815–823

Walker TW, Syers JK (1976) The fate of phosphorus during pedogenesis. Geoderma 15:1–19

Walker DA, Halfpenny JC, Walker MD, Wessman CA (1993) Long-term studies of snow–vegetation interactions. Bioscience 43:287–301

Weintraub MN, Schimel JP (2003) Interactions between carbon and nitrogen mineralization and soil organic matter chemistry in arctic tundra soils. Ecosystems 6:129–143

Weintraub MN, Schimel JP (2005) The seasonal dynamics of amino acids and other nutrients in Alaskan arctic tundra soils. Biogeochemistry 73:359–380

Weintraub MN, Scott-Denton LE, Schmidt SK, Monson RK (2007) The effects of tree rhizodeposition on soil exoenzyme activity, dissolved organic carbon, and nutrient availability in a subalpine forest ecosystem. Oecologia 154:327–338

Chapter 13
Phosphorus Nutrition of Forest Plantations: The Role of Inorganic and Organic Phosphorus

Thomas R. Fox, Bradley W. Miller, Rafael Rubilar, Jose L. Stape, and Timothy J. Albaugh

13.1 Introduction

Forests are an important land use throughout the world. Forests covered 3,952,925,000 ha in the world in 2005, which is approximately 30% of the earth's total land surface area (FAO 2009). Forests are vital to the world's ecological, social, and economic health. Forests produce a large portion of the earth's oxygen and sequester a substantial portion of its carbon, and thus play a major role in regulating climate change. Forests preserve biodiversity and provide habitats for much of the world's plants, animals, and microorganisms. Today, wood from forests is a major economic commodity, serving as the raw material for building materials, paper, packaging, and fuelwood. Fuelwood remains the single largest use

T.R. Fox (✉)
Department of Forest Resources and Environmental Conservation, Virginia Polytechnic Institute and State University, 228 Cheatham Hall, Blacksburg, VA 24061, USA
e-mail: trfox@vt.edu

B.W. Miller
U. S. Environmental Protection Agency, National Risk Management Research Laboratory, Land Remediation and Pollution Control Division, 5995 Center Hill Avenue, Cincinnati, OH 45224-1702, USA
e-mail: Miller.BradleyW@epamail.epa.gov

R. Rubilar
Departamento de Silvicultura, Facultad de Ciencias Forestales, Universidad de Concepción, Concepción, Chile
e-mail: rrubila@unity.ncsu.edu

J.L. Stape
Department of Forestry and Environmental Resources, North Carolina State University, 3108 Jordan Hall, Raleigh, NC 27695, USA
e-mail: jlstape@ncsu.edu

T.J. Albaugh
Department of Forestry and Environmental Resources, North Carolina State University, 3108 Jordan Hall, Raleigh, NC 27695, USA
e-mail: talbaugh@ncsfnc.cfr.ncsu.edu

E.K. Bünemann et al. (eds.), *Phosphorus in Action*, Soil Biology 26,
DOI 10.1007/978-3-642-15271-9_13, © Springer-Verlag Berlin Heidelberg 2011

of wood in the world today (FAO 2009). In the near future, wood from forests will provide much of the raw material for a wide range of novel bio-based fuels and materials (Kimbrel et al. 2009).

Because of the continued increase in the worlds' population and the expanding use of wood for both traditional and novel products, the demand for wood will increase rapidly in the future (FAO 2009). However, in both temperate and tropical regions large amounts of forest land continue to be lost as they are converted to other land uses such as agricultural crops, pasture, and urban development (Brown et al. 2006; Jauregui 2007; Weir and Gries 2002). Total forest area in the world declined by 7,317,000 ha between 2000 and 2005 (FAO 2009). In addition, because of increased emphasis on noncommodity values of forests such as biodiversity, recreation, watershed protection, and wildlife habitat, the timber harvests from native forests are increasingly being restricted in many parts of the world (Sedjo and Botkin 1997).

In response to the increasing demand for wood from a shrinking forested land-base, the paradigm for forest management in many regions of the world has been shifting from a hunt and gather approach, whereby mature timber is harvested from extensive areas with little management, to an agronomic model whereby limited areas of forests best suited to production are planted and managed intensively for commodity production (Binkley 1997; Fox 2000; Fox et al. 2007c; Sedjo and Botkin 1997). Biodiversity, wildlife habitat, and other nontimber management goals can be integrated into plantation management systems, particularly when a landscape perspective is employed (Brown et al. 2006).

The use of intensive management in forest plantations to increase wood production and decrease rotation lengths can help meet the increasing demand for wood and fiber in the world, while still preserving large areas of native forests for other uses because more wood can be grown on less land (Sedjo and Botkin 1997). As an example, productivity of pine plantations in the southern USA more than doubled and the rotation length was cut in half between 1950 and 2000 as a result of improved silvicultural practices (Fox et al. 2007c). Planted forests currently produce approximately 1.4 million m^3 per year, which is approximately one-third of the world's industrial wood. This could increase to more than 2.1 million m^3 per year, which would be about one-half of the worlds' industrial timber by 2030 if more intensive management regimes were implemented (FAO 2009). Theoretical models (Battaglia et al. 2004; Landsberg et al. 2001; Stape et al. 2004b) and empirical field trials (Allen et al. 2005; Birk and Turner 1992; Borders and Bailey 2001; Ferreira and Stape 2009; Jokela et al. 2000; Gonçalves et al. 2008; Rubilar et al. 2008b; Stape et al. 2008; Trichet et al. 2009) show that these growth increases are possible in plantation forests in many regions of the world. Most forests occur on soils that are less fertile than soils used for agronomic crops and, consequently, deficiencies of P, either alone or in combination with N, limit growth of most forests. Therefore, fertilization with phosphorus alone or in combination with other nutrients is needed in many forests to achieve high levels of productivity (Fox et al. 2007a; Gonçalves et al. 2008; Rubilar et al. 2008a; Stape et al. 2004a, 2006; Trichet et al. 2009).

The objective of this chapter is to review the impact of phosphorus fertilization on growth of forest plantations and to discuss phosphorus dynamics in forest ecosystems. This review highlights the importance of both inorganic P (P_i) and organic P (P_o) in tree nutrition, and the transfer of P between these two pools. We discuss how trees and their associated mycorrhizas can modify the soil and increase the solubility of both P_i and P_o, increase phosphatase activity and mineralization of P_o, and increase uptake of both P_i and P_o. There are several reviews that also discuss aspects of P nutrition and fertilization in forest trees (Ballard 1984; Comerford and de Barros 2005; Fox et al. 2007a; Gonçalves and Benedetti 2004; May et al. 2009; Trichet et al. 2009) and organic P in soils (Condron et al. 2005; Chen et al. 2008).

13.2 Phosphorus Fertilization of Forest Plantations

Chronically low levels of available soil nutrients, principally phosphorus and nitrogen, are the most important factors limiting forest growth in many areas of the world (Fox et al. 2007a; Gonçalves and Benedetti 2004; May et al. 2009; Rubilar et al. 2008b; Trichet et al. 2009). On an annual basis, over 100 kg ha^{-1} of N and 10 kg ha^{-1} of P must be available for fully stocked *Pinus* stands to maintain the high leaf area index levels necessary for maximum volume production (Battaglia et al. 2004; Ducey and Allen 2001). Greater amounts of nutrients are needed in faster growing species such as eucalyptus (Barros et al. 2004). However, most forest soils are unable to provide the levels of available nutrients required to maintain rapid growth of plantations (Comerford and de Barros 2005; Fox et al. 2007a; Miller 1981). Reduced leaf area and decreased growth efficiency occur in many forest plantations because of low soil nutrient availability, which results in poor growth and productivity (Albaugh et al. 1998; Vose and Allen 1988). In addition, nutrient limitations may develop in intensively managed forest plantations that might not develop in natural stands when other silvicultural treatments (e.g., tree breeding, vegetation control, and tillage) are used to increase crop tree growth, which also subsequently increases nutrient demand (Fox et al. 2007a).

Remarkably, little fertilization is conducted in naturally regenerated forests, even on soils where nutrient availability severely limits tree growth (FAO 2006). In contrast, forest fertilization is a widespread silvicultural practice in forest plantations in many regions of the world (Fox et al. 2007a; Gonçalves and Benedetti 2004; May et al. 2009). In the southern USA over 486,000 ha of pine plantations were fertilized with P or N+P in 2004 (Albaugh et al. 2007). In the Pacific Northwest of the USA, operational fertilization is also a common treatment with about 40,000 ha of forest land fertilized annually (Fox et al. 2007b). The main tree species fertilized in the southern USA are *Pinus taeda* and *P. elliottii*, and in the Pacific Northwest the main species is *Pseudotsuga menziesii*. Extensive plantations of *Pinus pinaster* occur in the Landes de Gascogne region of southwest France and P fertilization is a common practice in this region (Trichet et al. 2009). In Australia, approximately 173,000 ha of eucalyptus plantations including *Eucalyptus globulus*,

E. nitens, E. dunni and 84,000 ha of pine plantations including *Pinus radiata, P. pinaster, P. caribaeae* × *P.elliottii* are fertilized annually (May et al. 2009). In South America, P fertilization of eucalyptus plantations is also widely used (Gonçalves and Benedetti 2004; Rubilar et al. 2008b). In the USA, P is most commonly applied as diammonium phosphate (DAP), although in the past triple superphosphate and ground-rock phosphate were commonly used P sources in forestry (Albaugh et al. 2007). In other regions of the world, other sources of P such as ground rock phosphate or blends of nitrogen, P, and potassium (NPK) such as 6:30:6 are more commonly used. For example, in Australia and South America, blends of NPKS (NPK plus sulfur) are more widely used than DAP (Rubilar et al. 2008b; May et al. 2009). In several reports, the source of the P applied seems to be less important than the rate applied (Trichet et al. 2009).

The benefits of P fertilization on forest soils that have severe P deficiencies have long been recognized (Pritchett et al. 1961). Volume growth gains in plantations of *Pinus radiata* (Ballard 1978a), eucalyptus (Barros et al. 2004), *Pinus taeda* (Gent et al. 1986), *Pinus elliottii* (Pritchett and Comerford 1982), and *Pinus pinaster* (Trichet et al. 2009) ranging from 20% to more than 100% are common on severely P-deficient soils following P fertilization near the time of planting. The magnitude of the growth response can vary widely depending on the species, soil type, understory competition, and climate (Comerford and de Barros 2005). More remarkable than the magnitude of the response is the longevity of the response to P fertilization in many forest plantations on P-deficient soils. The response to a single application of 56 kg ha^{-1} of P may last for 20 or more years (Pritchett and Comerford 1982; Comerford and de Barros 2005). There are several reports where a single application of P in one rotation continued to increase growth in subsequent rotations (Ballard 1978a; Comerford et al. 2002; Crous et al. 2007; Everett and Palm-Leis 2009; Gentle et al. 1986). Turner et al. (2002) have found that the response to P fertilization in *Pinus radiata* in Australia may last more than 50 years. The efficient cycling of nutrients in forest ecosystems contributes to these long-term P fertilization responses (Attiwill and Leeper 1987; Binkley 1986; Jordan 1985).

In intermediate-aged stands of both pine and eucalyptus, little response is typically observed to additions of P alone, except on the P-deficient sites described above where P was not added at the time of planting (Barros and Novais 1996; Fox et al. 2007a; Rubilar et al. 2008b). However, in many pine plantations in the southern USA, by the time of crown closure, the tree's potential to use both P and N is typically greater than the available soil supply, resulting in restricted leaf area development and growth (Allen et al. 1990; Fox et al. 2007a). These stands are generally very responsive to additions of P and N at this time in the rotation. In the majority of stands, both P and N are deficient, and the growth response following fertilization is much greater when both P and N are applied (Fox et al. 2007a). Results from an extensive series of intermediate-aged fertilizer trials in *Pinus taeda* stands established throughout the southern USA indicate that growth gains averaging 3.5 m^3 ha^{-1} per year over an 8-year period occur following a one-time application of 224 kg ha^{-1} N and 28 kg ha^{-1} P (Fox et al. 2007a). The growth response was less than half of this when either N or P were added alone. Over 85% of the stands fertilized were responsive to additions of N+P during this stage of stand development. The need for balanced fertilization with both P and N has also been shown in *Pinus radiata* in Chile (Rubilar et al. 2008b) and Australia (May et al. 2009).

Identification of forest plantations in need of P fertilization can be based on soil parent material, soil type, soil analysis, and foliage analysis (Comerford and de Barros 2005; May et al. 2009). In the Lower Coastal Plain of the Atlantic and Gulf Coasts of the USA, poorly drained, clayey Ultisols tend to be severely P deficient (Fox et al. 2007a). Along the Gulf Coast, well-drained clayey to loamy soils on the Citronelle and associated geologic formations have also been found to be P deficient (Allen and Lein 1998). The likelihood of response to P fertilization in New South Wales, Australia, can be determined on the basis of parent material (Turner et al. 1996). In some regions, P deficiencies are so widespread that all soils are considered P deficient, and P fertilization is a routine practice in all plantations. This is the case for eucalyptus plantations in South America and pine plantations in South Africa, where almost all upland soils tend to be P deficient (Attiwill and Adams 1996; Donald et al. 1987). Where soil P deficiencies are less widespread, soil analysis for P has been found to be a reliable tool for diagnosis of nutrient deficiencies in plantation forests (Ballard 1974; Wells et al. 1986). However, the critical concentration of available P varies between soils, extracting solutions, and plant age (Barros and Novais 1996). The critical value for extractable soil P in the A horizon, below which a fertilizer response is expected in *Pinus taeda* in the southern USA, is 4–6 μg g^{-1} based on the Mehlich-3 extraction procedure (Wells et al. 1986). The critical concentration of Bray-1-extractable P in the A horizon for *Eucalyptus grandis* in Brazil varies from 17 to 89 μg g^{-1}, depending on the clay content and P sorption maximum (Barros and Novais 1996). $CaCl_2$-extractable P has been used to identify sites supporting *Eucalyptus globulus* and *E. nitens* in Australia that require P fertilization (Mendham et al. 2002). Foliage analysis is also widely used to evaluate the need for P fertilization in forests (Colbert and Allen 1996; Lambert and Turner 1998). Critical values for foliar P concentrations vary by species and range from 0.09 mg g^{-1} for *Pinus elliottii*, 0.11 mg g^{-1} for *Pinus taeda* (Allen 1987; Everett and Palm-Leis 2009; Wells et al. 1986), and 0.14 mg g^{-1} in *Pinus radiata* (Will 1985). Critical values of foliar P in more demanding species such as eucalyptus are generally much higher, ranging up to 0.4 or 0.5 mg g^{-1} (Attiwill and Adams 1996; Dell et al. 2001). Total P concentrations in litter have also been used to diagnose P deficiencies in *Pinus radiata* stands in southeast South Australia (May et al. 2009).

Stand attributes such as basal area and site index have also been used to determine whether nutrient deficiencies exist in forests (Duzan et al. 1982). Linkages among nutrient availability, leaf area, and forest productivity (Albaugh et al. 1998) also permit the use of leaf area as a diagnostic tool (Fox et al. 2007a; May et al. 2009). Differences between a stand's current leaf area and its potential leaf area can be used to estimate responsiveness to nutrient additions. For example, the leaf area index of a fully stocked *Pinus taeda* stand in the southern USA should be

3.5 or greater. If the leaf area index is less than this, the stand is probably in need of N+P fertilization unless other obvious problems have altered leaf area (e.g., fire, ice damage, insect attack, disease). The probability and magnitude of response will be greater in stands with lower leaf area. Remote sensing techniques using Landsat satellite imagery have been developed that can accurately determine leaf area in southern pine stands (Flores et al. 2006). Leaf area is used in a similar manner to diagnose stands of both pine and eucalyptus in Australia that are nutrient deficient (May et al. 2009).

13.3 Phosphorus Dynamics in Forest Ecosystems

It is somewhat ironic that tree growth is commonly limited by P because the absolute quantities of P in forests soils are usually large and appear sufficient to support robust tree growth. However, the pools of labile P in the soil solution are small and typically growth-limiting. The concepts of soil quantity, intensity, and buffer capacity help to elucidate the processes in forest soils that determine plant-available P present in soil solution (Fig. 13.1). In forest soils, both chemical reactions and biological processes determine the buffer capacity for P and include dissolution/precipitation, mineralization/immobilization, and ligand exchange reactions (Fox 1995; Pierzynski et al. 2005).

Trees have evolved a variety of mechanisms to increase P availability and uptake in low-P soil environments, including changes in root morphology and architecture, mycorrhizal symbiosis, preferential root growth into more fertile zones in the soil, higher phosphatase activity in the rhizosphere, and the secretion of low molecular weight organic acids (Vance et al. 2003). The ability of trees to increase P availability in soils was well documented in a series of studies in which *Pinus radiata* was planted on pastures in New Zealand (Chen et al. 2008). Afforestation of grassland with *Pinus radiata* improved P availability in the topsoil due to enhanced

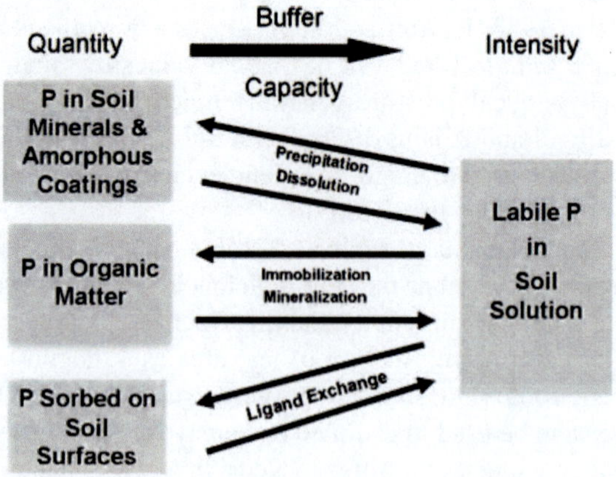

Fig. 13.1 Representation of quantity, intensity, and buffer capacity of pools of P that exist in forest soils and the processes that convert P into a labile form

mineralization of organic P, improved solubility of P caused by root and microbial exudates, effects of mycorrhizas, and changes in soil moisture and temperature regimes (Chen et al. 2008).

13.3.1 Phosphorus Pools and Cycling in the Forest Floor

Forest soils differ from agricultural soils in several significant ways that affect P availability and cycling in forest ecosystems (Comerford and de Barros 2005). However, the most obvious and important difference is the presence of a well-developed forest floor (O horizon) in forests that is typically absent in agricultural soils (Pritchett and Fisher 1987). The forest floor is a major pool of nutrients, including P, in forest ecosystems. Phosphorus content in the forest floor varies from less than 10 kg ha^{-1} to more than 300 kg ha^{-1} depending on the climate, soil type, species, and age of the forest (Pritchett and Fisher 1987). The importance of the forest floor to tree nutrition in forest ecosystems is well documented (Berg and Laskowski 2006). For example, repeated removal of litter from the forest floor to use as animal bedding in Germany during the 1800s led to well-documented site degradation in many forests (Evers 1991). There are examples in several parts of the world where windrowing or piling during site preparation, which removed the forest floor, along with slash and logging debris to facilitate planting caused severe deficiencies of N, P and possibly other nutrients, which decreased forest growth (Ballard 1978b; Fox et al. 1989; Jonard et al. 2009; Merino et al. 2003). The importance of the forest floor is highlighted by the fact that it has been shown that whole-tree harvesting that removes all the aboveground tree biomass, but leaves the forest floor intact, will have little impact of growth of the subsequent stand (Johnson and Todd 1998; Fox 2000).

The mass and nutrient content of the forest floor is determined by the competing process of litterfall input and decomposition (Berg and Laskowski 2006). In most forests, decomposition rates are less than litterfall and consequently the forest floor builds up over time (Richter et al. 2006). In general, hardwood litter decomposes and releases P faster than pine litter (Fisher and Binkley 2000). Well-drained soils in tropical regions supporting hardwood forests tend to have smaller forest floors with less P accumulation, whereas coniferous forest in temperate regions with poorly drained soils tend to accumulate large forest floors with more P. Forest management practices such as fertilization that increase the growth rate of the forest often increases the mass of the forest floor because of increased litterfall from stands with higher leaf area. Harding and Jokela (1994) reported that P fertilization significantly increased the size of the forest floor and the nutrient pools in *Pinus elliottii* stands. Similar results were observed in a study with *Pinus taeda* in North Carolina (Sanchez 2001).

There are substantial temporal variations in the mass and P content of the forest floor due to disturbances in the forest. Fire can consume the forest floor and oxidize the organic matter (Flinn et al. 1979). However, even in intense fires, much of the P remains in the ecosystem and its release during the fire can increase P availability, whereas N can be lost even in relatively low intensity fires (Fisher and Binkley 2000; Flinn et al. 1979). Harvesting also affects P dynamics in the forest floor. Increased temperature and moisture in the forest floor following tree harvest accelerate decomposition and mineralization of organic P. The higher decomposition rate coupled with the lack of litterfall inputs generally results in a significant decrease in the mass of the forest floor within a few years of harvest. This accelerated decomposition of the forest floor and mineralization of organically bound nutrients causes a substantial flush of nutrients into the soil known as the Assart Effect (Fisher and Binkley 2000). This results in a substantial increase in growth of vegetation for the first few years after a major disturbance such as harvest or fire. After the readily decomposed and mineralized portion of the forest floor is gone, available nutrients decline and growth rates drop. Over time, the forest floor increases again as the stand develops and the reserves of nutrients are built up again (Richter et al. 2006). This temporal variation in forest floor dynamics and associated pools of labile nutrients in the soil is an important aspect of nutrient cycling in forest ecosystems that is different from most agricultural systems.

13.3.2 Organic P in Forest Soils

A large portion of the P present in forest soils is an organic form (Condron et al. 2005). Labile P in forest soil is incorporated into microbial and plant biomass (Achat et al. 2009; Walbridge 1991), which will gradually shift soil P into organic forms in forests. Organic P pools can represent 20–90% of the total P present in soils (Achat et al. 2009; Condron et al. 2005; Turner and Lambert 1985). Organic P pools in the soil generally increase over time, which suggests that biological cycling becomes more important as forest stands develop (Richter et al. 2006; Wells and Jorgensen 1975). Understanding the P_o species present and how soil biota influence P_o pools may be an important tool for predicting the quantity of labile P pools, especially in forest soils with a high P fixation capacity.

Several fractionation schemes have been developed that use sequential chemical extractions to partition soil P into inorganic and organic P pools that may be related to bioavailability (Chang and Jackson 1957; Hedley et al. 1982; Tiessen and Moir 1993). Each of these fractionation procedures uses a variety of extractions of increasing strength to divide soil P into pools of decreasing biological availability. For example, in the Tiessen and Moir (1993) modifications of the Hedley sequential fractionation procedure, the P_i held in soil solution is first removed by anion exchange membranes. The second P fraction is extracted with 0.5 M $NaHCO_3$. These two fractions represent labile P that is thought to cycle over a short time period, such as a single growing season (Bowman and Cole 1978; Cross and Schlesinger 1995; Johnson et al. 2003). The remaining P fractions are extracted with 0.1 M NaOH before and after sonification to break up soil aggregates, followed by an extraction with 1 M HCl, hot concentrated HCl, and a final extraction with H_2O_2 and concentrated H_2SO_4. The P pools extracted by these solutions represent P that is moderately labile to highly recalcitrant (Tiessen and Moir 1993).

The rationale behind these fractionation procedures is that they represent various forms of P that vary in their availability to plants. Unfortunately, these various fractions have not been well correlated with P availability to forest trees or the growth response of trees following P fertilization (Miller 2008). The P_o pools collected from the soil fractionation systems do not provide useful information on the species of P_o present and are not able to reliably quantify labile P_o pools (Condron et al. 2005; Richter et al. 2006). For example, the P_o extracted in the NaOH fraction of the Hedley fractionation is considered to be only moderately labile and only contributes to long-term P cycling. However, Liu et al. (2004) found that the NaOH-extractable P_o pool decreased in the rhizosphere of *Pinus radiata* planted on an allophanic soil, which was accompanied by an increase in the NaOH-extractable P_i pool. There were no effects on the resin-extracted P_i content of the bulk soil or rhizosphere, which suggests that biological activity in the rhizosphere caused these short-term changes in the NaOH pools, indicating that these fractions are more labile than the fractionation scheme would indicate. Similar conclusions were reached by Richter et al. (2006) in their study of long-term changes in soil P in an aggrading *Pinus taeda* plantation established on an abandoned agricultural field. Likewise, other research has show that the residual P pools can be influenced by the soil biota (Chen et al. 2002).

Liquid state ^{31}P nuclear magnetic resonance (NMR) of soil extracts can identify and quantify the P_o in soils (Condron et al. 2005; Newman and Tate 1980; Turner et al. 2003; Turner and Richardson 2004). Using NMR it is possible to separate soil P into various forms such as inorganic P, orthophosphate monoester and diester phosphates, phosphonates, polyphosphates, and pyrophosphates, which vary in availability to plants (Condron et al. 1997). Although it is difficult to make comparisons of P_o pools between different sites because of site specific conditions, Cade-Menun et al. (2000) found that the P_o spectra from the forest floor (O_e and O_a horizons) of a 10-year-old burned clear-cut and an old growth forest were very similar. Although total P concentrations remained unchanged between sites, there was a decrease in the P_o concentration of the spodic (Bhf2) horizon of the clear cut and burned site (Cade-Menun et al. 2000). The change in the Bhf2 was attributed to a decrease in the illuviation of P_o through the soil profile.

Monoester and diester phosphates are among the most common forms of P_o in soils. Chen et al. (2004) measured P_o in a variety of soils planted with *Pinus radiata* pine in a greenhouse study and found that orthophosphate monoesters were the most common forms of P_o in a variety of soils. Orthophosphate esters may be monoesters with one C moiety per P, or diesters with two C moieties per P. The most abundant monoester in soils are inositol phosphates, which occur in various stereoisomeric forms, *myo*, *scyllo*, D-*chiro*, and *neo* (Turner et al. 2002, 2005). The monoesters also include sugar phosphates, phosphoproteins, and mononucleotides (Condron et al. 2005). Although most of the monoesters can be readily hydrolyzed by phosphatases present in the soil (Alvarez et al. 2004; He et al. 2004), monoesters adsorb tightly to soil colloids and consequently often become resistant to enzyme hydrolysis (Chen et al. 2004).

The diester phosphates, including nucleic acids and phospholipids, are frequently found in larger quantities in acid forest soils than in agricultural soils (Cade-Menun et al. 2000; Cade-Menun 2005). Although diesters only constituted approximately 10% of the P_o in agricultural soils, Cade-Menun et al. (2000) found that they accounted for more than 50% of the extractable P_o in the acid forest soils of British Columbia. Turner et al. (2007) measured P_o in a glacial chronosequence and found that diester P continued to accumulate as the age of the soil increased and made up a significant proportion of total soil P in the older soils. Diester phosphates are generally less tightly sorbed to soil colloids than are monoesters and therefore may be more susceptible to hydrolysis by phosphatases present in the soil (Condron et al. 2005; Magid et al. 1996).

Phosphonates are organic orthophosphate compounds with C–P bonds and were first reported in soils by Newman and Tate (1980). The C–P bond is more resistant to hydrolysis and oxidation than C–O–P bonds, making their functions as a P storage compound in biotic systems less likely (Quin and Quin 2001). The phosphonates accumulate in soils that are wet (Tate and Newman 1982), cold, and acidic (Cade-Menun et al. 2000; Dai et al. 1996; Gil-Sotres et al. 1990). These conditions exist in many forest soils (Fisher and Binkley 2000), which suggest that phosphonates may be present in greater quantities in forest soils than in well-drained agricultural soils.

13.3.3 Availability of Organic P in Forest Ecosystems

Conventional views of nutrient cycling and tree nutrition hold that organic P must be converted to inorganic P before it can be utilized by plants (Adams and Pate 1992; Hayes et al. 2000; Tate 1984). The mineralization of organic P is mediated by phosphatases that hydrolyze C–O–P ester bonds (Condron et al. 2005). For example, phytase activity is needed to hydrolyze inositol phosphate, which often comprises a large portion of the organic P in soil (Richardson 2001; Tang et al. 2006). Acid phosphatases dominate in most acid forest soils (Tabatabai 1982). Tree roots, fungi, and bacteria produce phosphatases in soil (Chen et al. 2008; Fox and Comerford 1992b; Speir and Ross 1978). Fox and Comerford (1992b) found higher acid phosphatase activity in the rhizosphere of *Pinus elliotti* roots.

In spite of the fact that P limits productivity in many forest ecosystems and that the forest floor contains large pools of organic P that seem to influence ecosystem productivity, much less attention has been paid to P mineralization in the forest floor compared to N mineralization (Comerford and de Barros 2005). In many forest ecosystems, it appears that forest floors immobilize N in undisturbed forests and thus are a sink for N during most of the rotation (Miller 1981; Piatek and Allen

2001). Berg and Laskowski (2006) compared the results from a large number of forest ecosystems in northern temperate climates and found that in the majority of them, N accumulated in the forest floor through the rotation. It is less certain whether the forest floor is a sink or source of P in forests. Piatek and Allen (2001) found that the forest floor was a sink for P in *Pinus taeda*, whereas Polglase et al. (1992b), using both laboratory and field studies with *Pinus taeda*, concluded that P was mineralized from the forest floor. Studies with other species including several species of pine also found that P was also released from the forest floor (Comerford and de Barros 2005).

Fertilization with P may accelerate cycling of P in forest ecosystems by increasing P mineralization from the forest floor (Comerford and de Barros 2005; Polglase et al. 1992b). The P content in litterfall from fertilized forest is greater than in unfertilized forests. (Harding and Jokela 1994). Fertilization can increase total P returns in litterfall from 150 to 400% (Dalla-Tea and Jokela 1991; Piatek and Allen 2001). Release of P from decomposing needles is strongly and positively related to the needle concentration of inorganic P (Polglase et al. 1992a), which increases with fertilization (Polglase et al. 1992c). The net mineralization of organic P is probably much greater following fertilization. The importance of P fertilization on the mineralization of P was clearly documented in the 6-year-old *Pinus taeda* plantations growing on Spodosols in Florida studied by Polglase et al. (1992c). Higher P mineralization in laboratory incubations showed that P fertilization increased organic matter quality. In the field, mineralization in the surface 5 cm of soil supplied 0 and 25% of the annual P requirements in unfertilized and fertilized plots, respectively. Higher laboratory and field mineralization rates with fertilization were attributed to rapid recycling of P in readily mineralizable compounds. In contrast, slow growth in nutrient-deficient control plots was attributed to slower biological cycling of phosphorus.

Unfortunately, the relationship between measured amounts of phosphatase in the soil and the mineralization of organic P is not well understood (Condron and Tiessen 2005). Increased mineralization of organic P in the rhizosphere of *Pinus radiata* was attributed to higher concentrations of acid and alkaline phosphatases and phosphodiesterase (Liu et al. 2004, 2005). However, in other studies soil phosphatase activity was not related to organic P mineralization (Adams 1992; Chen et al. 2008).

Mycorrhizae associated with tree roots may have the ability to produce phosphomonoesterase that can hydrolyze P_o (Alexander and Hardy 1981; Read and Perez-Moreno 2003; Smith and Read 1997; Williamson and Alexander 1975). Dinkelaker and Marschner (1992) found that phosphomonoesterase activity was greater in the mycorrhizal roots of *Picea* and rhizomorphs of *Thelephora terrestris* than in nonmycorrhizal roots. Bartlett and Lewis (1973) measured phosphatase activity on ectomycorrhizae of *Fagus spp.* and found that more than one phosphatase was present. They suggest that the activity of phosphatases on the surface of ectomycorrhizaea may lead to the immediate recycling of the organic P present in the forest floor back into the mycorrhizal root systems of trees. Although ecotmycorrhizaea can obtain P from the enzymatic breakdown of complex organic compounds (Read and Perez-Moreno 2003), recent work suggests that mineralization of organic P may be less important that previously thought in coniferous forests (Achat et al. 2009; Johnson and Gehring 2007). Lindahl et al. (2002) proposed that ectomycorrhizal fungi can directly acquire organic forms of nutrients in the forest floor and transport them to the host trees. In this hypothesis, ectomycorrhizae can acquire P_o directly from the soil and litter and may also capture organic nutrients from other soil organisms such as saprotrophic fungi (Johnson and Gehring 2007). Because mycorrhizal fungi are common in both the forest floor and the surface mineral soil, this may be an important mechanism of uptake of P_o in forest ecosystems.

13.3.4 Impact of Organic Acids on Phosphorus Dynamics in Forest Soils

Because tree growth is typically limited by the quantity of labile P present in the soil environment, the simple and economically viable solution in plantation forestry has been to apply inorganic P fertilizers to meet plant growth requirements (Fox et al. 2007a). Applied inorganic P is rapidly converted from labile to nonlabile P in the soil. These transformations are regulated by plants, soil microbes, and the P sorption capacity of the soil. Phosphorus sorption is rapid and there is a large sink for P in most forest soils (Beauchemin et al. 1996; Kelly and Kelly 2001; Comerford and de Barros 2005). In acid forest soils, P sorption occurs primarily at surfaces of Fe- and Al-oxide and hydrous oxide (Delgado and Torrent 2000; Pierzynski et al. 2005; Sample et al. 1980). Phosphate anions act as a Lewis base and form inner-sphere complexes with hydroxyl groups on the oxide surface, which acts as a Lewis acid. The strong sorption of P in soils is responsible for the substantial decrease in inorganic P in soil solution over time. With age and weathering, the importance of Al- and Fe-sorbed P increases (Parfitt et al. 1975).

Classical concepts involving dissolution of P minerals based on the thermodynamics of mineral equilibrium provide a theoretical basis for long-term trends soil nutrient availability in forest soils (Lindsay 1979). However, this approach has limited utility in most forest soils because P is sorbed to Fe and Al oxide surfaces and is often occluded within amorphous Fe and Al oxides that are not in thermodynamic equilibrium with soil solution. Most of these amorphous coatings are complex mixtures of Fe, Al, and organic matter rather than distinct minerals such as variscite. The heterogeneity of the soil environment affecting P sorption reactions adds to the complexity of P cycling in forest soils (Huang and Schnitzer 1986).

The desorption of surface complexed P and the dissolution of amorphous coatings containing occluded P can be greatly accelerated in forest soils by the presence of organic acids such as malate, citrate, and oxalate (Fox 1995; Jones and Darrah 1994). Solution P concentrations can be 10–1,000 times higher following the addition of organic acids to the soil (Jones 1998). The organic acids function as organic ligands that increase P release into soil solution by (1) ligand exchange with P held at metal oxide surfaces, (2) complexation of Al, Fe, and other metals in soil solution, and (3) dissolution of amorphous metal–organic matter coatings that contain occluded P (Fox 1995). Based on the kinetics of P and Al release, ligand exchange reactions can be separated from dissolution reactions (Fox et al. 1990b). Sato and Comerford (2006) suggested that P released by desorption can be separated from P released by dissolution of amorphous organic matter–metal oxide surfaces by using anion exchange membranes and soil extractions with various concentrations of oxalate to estimate disequilibria-desorbable and ligand-desorbable P.

A wide variety of naturally occurring organic acids have been identified in forest soils (Fox and Comerford 1990; Jones 1998; Pohlman and McColl 1988; Stevenson 1967). However, not all organic acids can exchange P from oxide surfaces or increase dissolution of these surfaces (Fox et al. 1990b). In acid forest soils with large amounts of Al and Fe, P desorption and dissolution reactions are mediated by organic ligands that form stable complexes with Al and Fe (Fox 1995; Jones 1998). For example, in forest soils with large amounts of Al such as Spodosols of the southern USA, those organic acids with an Al stability constant ($\log K_{\mathrm{Al}}$) greater than 3.5–4.0 had a much greater impact on P release than those that have a lower stability constant (Fox et al. 1990a; Lan et al. 1995). The Al stability constant is determined by the type and arrangement of functional groups on the organic acid. Those with multiple carboxylic acid groups have higher stability constants because they can form stable five- and six-membered ring structures with Al and Fe (Fox 1995). Consequently, organic acids such as oxalate, citrate and malate, which have $\log K_{\mathrm{Al}}$ values greater than 4, increase P release from soils whereas organic acids such as formate, acetate, lactate with $\log K_{\mathrm{Al}}$ values less than 4 have little impact on P release (Fox et al. 1990a; Lan et al. 1995; Jones 1998).

P release from soils generally increases as concentrations of organic ligands increase (Fox and Comerford 1992a; Gerke 1992). The concentration of low molecular weight organic acids is generally quite low in forest soils, typically less than 0.01–0.1 mM (Fox and Comerford 1990; Pohlman and McColl 1988). However, even at very low concentrations of $<1.0\ \mu\mathrm{mol\ g^{-1}}$ soil continuous release of low molecular weight organic acids may promote desorption of P and dissolution of amorphous mineral coatings containing P over the course of weeks to months. Fox and Comerford (1992a) showed that, over time, the cumulative amount of P released from forest soils was similar when oxalate was added sequentially over time at very low concentrations compared to a single addition of oxalate at higher concentrations.

Tree roots and their associated mycorrhizae can modify the rhizosphere and, through the release of organic acids, can increase P availability (Fox 1995; Fox and Comerford 1990; Grierson 1992; Hinsinger 2001; Jones 1998; Raghothama 1999, 2005; Smith and Read 1997). It is well documented that ectomycorrhizal roots can exude enough oxalate to dissolve amorphous minerals (Cromack et al. 1979; Sato and Comerford 2006). Phosphorus desorption and dissolution reactions in the rhizosphere are often greater than in the bulk soil because of higher concentrations of organic acids (Cardon and Whitbeck 2007; Jones 1998; Smith and Read 1997). The close proximity to the source of exudation and reduced volume of solution in contact with the surrounding soil can lead to increased concentrations in the rhizosphere (Jones 1998; Cardon and Whitbeck 2007). Most low molecular weight organic acids are rapidly degraded in forest soils (van Hees et al. 2002). Consequently, organic acids must constantly be replenished in the soil solution through root exudation or other means to maintain the observed concentrations (Fox 1995; Jones 1998). Large quantities of organic acids such as citrate are released by proteoid roots of shrubs such as *Banksia integrifiolia* (Grierson 1992). Ectomycorrhizal production of oxalate in the soil can be significant. For example, very high oxalate concentrations have been observed in forest soils associated with fungal maps in *Pseudotsuga menziesii* forests (Cromack et al. 1979). Fungi and soil microbes such as *Penicillium* and *Pseudomonas spp.* have been proven to be a source of oxalate in both laboratory and field experiments (Arvieu et al. 2003; Casarin et al. 2003; van Hees et al. 2000, 2003; Illmer and Schinner 1992).

The majority of organic P in soils is not soluble and thus is not susceptible to mineralization or uptake as P_o by mycorrhizas (Chen et al. 2008). As discussed above, low molecular weight organic acids have a significant impact on the release of inorganic P in many soils. However, little is known about the impact of low molecular weight organic acids on release of organic P (Bar-Yosef 1996; Jones 1998; Ström et al. 2002). Previous work has shown that low molecular weight organic acids such as oxalate can significantly increase the release of P_o in at least some soils (Fox and Comerford 1992a). For example, in forested Spodosols in the southern USA the release of organic P in the presence of low molecular weight organic acids was equal to or greater than the release of inorganic P in surface and spodic horizons (Fox and Comerford 1992a; Fox et al. 1990a, b). Through time, the release of organic P increased more than the release of inorganic P in these soils following addition of organic acids such as oxalate (Fox et al. 1990a, b). The release of organic P may be associated with the dissolution of the amorphous Al– and Fe–organic matter coatings that contain organic P (Fox et al. 1990b). Once released into solution, the organic P may be subject to enzyme hydrolysis by phosphatases in the soil (Adams and Pate 1992; Hayes et al. 2000). Fox and Comerford (1992b) found higher acid phosphatase activity in the rhizosphere of *Pinus elliotti* roots. It thus seems that the presence of organic acids in the rhizosphere of trees will increase the release of both inorganic and organic P. Because the solubility of organic P and not soil phosphatase activity may limit mineralization of organic P in soils (Adams and Pate 1992; Chen et al. 2008), the increased release of organic P mediated by organic acids may significantly impact P dynamics and long-term productivity in forest ecosystems (Turner and Lambert 1985; Condron et al. 2005). This effect would be even more important if the P_o released by the elevated concentrations of organic acids in the rhizosphere was also directly taken up by the ectomycorrhizas and transported to the trees, as proposed by Lindahl et al. (2002).

13.4 Summary and Implications to Tree P Nutrition

This review highlights the complexity of P dynamics in forest soils. Large amounts of both P_i and P_o are found in forest soils, including the forest floor and mineral horizons. However, most of this P may not be in a form that is readily available to trees. Trees and their associated mycorrhizas have evolved a number of mechanisms to modify the soil to increase P availability and uptake. Many of these mechanisms appear to operate on several levels and influence availability and uptake of both P_i and P_o. Concentrations of low molecular weight organic acids and phosphatases are greater in the rhizosphere of trees, which increases the solubility of both P_i and P_o and mineralization of P_o. Ectomycorrhizas of trees present in the forest floor and the mineral soil may be able to acquire P_o directly. The ability of mycorrhizas to directly utilize P_o may be a significant aspect of tree nutrition and dramatically alter our view of nutrient cycling in forest ecosystems. A comprehensive understanding of tree nutrition and forest productivity is not possible without a better understanding of both P_i and P_o in forest ecosystems. Additional research is clearly needed to elucidate the complex dynamics and transformations of P in forest ecosystems.

References

Achat DL, Bakker MR, Morel C (2009) Process-based assessment of phosphorus availability in low phosphorus sorbing forest soil using isotopic dilution methods. Soil Sci Soc Am J 73:2131–2142

Adams MA (1992) Phophatase activity and phosphorus fractions in Karri (*Eucalyptus diversicolor* F Muell) forest soils. Biol Fertil Soils 14:200–204

Adams MA, Pate JS (1992) Availability of organic and inorganic forms of phosphorus. Plant Soil 145:107–113

Albaugh TJ, Allen HL, Dougherty PM, Kress LW, King JS (1998) Leaf area and above- and belowground growth responses of loblolly pine to nutrient and water additions. For Sci 44: 1–12

Albaugh TA, Allen HL, Fox TR (2007) Historical patterns of forest fertilization in the southern United States from 1969 to 2004. South J Appl For 31:129–137

Alexander IJ, Hardy K (1981) Surface phosphatase activity of Sitka spruce mycorrhizas from a serpentine site. Soil Biol Biochem 13:301–305

Allen HL (1987) Forest fertilizers: nutrient amendment, stand productivity and environmental impacts. J For 85:37–46

Allen HL, Lein S (1998) Effects of site preparation, early fertilization, and weed control on 14-year old loblolly pine. Proc South Weed Sci Soc 51:104–110

Allen HL, Dougherty PM, Campbell RG (1990) Manipulation of water and nutrients – practice and opportunity in Southern U.S. pine forests. For Ecol Manage 30:437–453

Allen HL, Fox TR, Campbell RG (2005) What is ahead for intensive pine plantation silviculture in the South? South J Appl For 29:62–69

Alvarez M, Godoy R, Heyser W, Hartel S (2004) Surface-bound phosphatase activity in living hyphae of ectomycorrhizal fungi of *Nothofagus obliqua*. Mycologia 96:479–487

Arvieu JC, Leprince F, Plassard C (2003) Release of oxalate and protons by ectomycorrhizal fungi in response to P-deficiency and calcium carbonate in nutrient solution. Ann Forest Sci 60:815–821

Attiwill PM, Adams MA (eds) (1996) Nutrition of eucalypts. CSIRO, Collingwood, VIC, Australia

Attiwill PM, Leeper GW (1987) Forest soils and nutrient cycles. Melbourne University Press, Melbourne

Ballard R (1974) Use of soil testing for predicting phosphate fertilizer requirements of radiata pine at time of planting. NZ J For Sci 4:27–34

Ballard R (1978a) Effect of first rotation phosphorus applications on fertilizer requirements of second rotation pine. NZ J For Sci 8:135–145

Ballard R (1978b) Effects of slash and soil removal on the productivity of second rotation radiata pine on a pumice soil. NZ J For Sci 8:248–258

Ballard R (1984) Fertilization of plantations. In: Bown GD, Nambiar EKS (eds) Nutrition of plantation forest. Academic, London, pp 327–360

Barros NF, Novais RF (1996) Eucalyptus nutrition and fertilizer regimes in Brazil. In: Attiwill PM, Adams MA (eds) Nutrition of eucalypts. CSIRO, Collingwood, VIC, Australia, pp 335–356

Barros NF, Neves JCL, Novais RF (2004) Mineral fertilizer recommendations for eucalypt plantations. In: Gonçalves JL, Benedetti V (eds) Forest nutrition and fertilization. Institute of Forest Research and Study, Piracicaba, São Paulo, Brazil, pp 271–284

Bartlett EM, Lewis DH (1973) Surface phosphatase activity of mycorrhizal roots of beech. Soil Biol Biochem 5:249–257

Bar-Yosef B (1996) Root excretions and their environmental effects: influence on availability of phosphorus. In: Waisel Y, Eshel A, Kafkafi U (eds) Plant roots: the hidden half. Marcel Dekker, New York, pp 581–605

Battaglia M, Sands P, White D, Mummery D (2004) Cabala: a linked carbon, water and nutrient model of forest growth fro silvicultural decision support. For Ecol Manage 193: 251–282

Beauchemin S, Simard RR, Cluis D (1996) Phosphorus sorption-desorption kinetics of soil under contrasting land uses. J Environ Qual 25:1317–1325

Berg B, Laskowski R (2006) Litter decomposition: a guide to carbon and nutrient turnover. Academic, Amsterdam

Binkley D (1986) Forest nutrition management. Wiley, New York

Binkley C (1997) Preserving nature through intensive plantation forestry: the case for forestland allocation with illustrations from British Colombia. For Chron 73:553–559

Birk EM, Turner J (1992) Response of flooded gum (*Eucalyptus grandis*) to intensive cultural treatments – biomass and nutrient content of eucalyptus plantations and native forests. For Ecol Manage 47:1–28

Borders BE, Bailey RL (2001) Loblolly pine – pushing the limits of growth. South J Appl For 25:69–74

Bowman RA, Cole CV (1978) An exploratory method for fractionation of organic phosphorus from grassland soils. Soil Sci 125:95–101

Brown S, Palola S, Lorenzo M (2006) The possibility of plantations: integrating ecological forestry into plantation systems. National Wildlife Federation, Reston, VA

Cade-Menun BJ (2005) Characterizing phosphorus in environmental and agricultural samples by ^{31}P nuclear magnetic resonance spectroscopy. Talanta 66:359–371

Cade-Menun BM, Berch SM, Preston CM, Lavkulich LM (2000) Phosphorus forms and related soil chemistry of podzolic soils on northern Vancouver Island. I. A comparison of two forest types. Can J For Res 30:1714–1725

Cardon ZG, Whitbeck JL (2007) The rhizosphere: an ecological perspective. Elsevier, Amsterdam

Casarin V, Plassard C, Souche G, Arvieu JC (2003) Quantification of oxalate ions and protons released by ectomycorrhizal fungi in rhizosphere soil. Agronomie 23:461–469

Chang SC, Jackson ML (1957) Fractionation of soil phosphorus. Soil Sci 84:133–144

Chen CR, Condron LM, Davis MR, Sherlock RR (2002) Phosphorus dynamics in the rhizosphere of perennial ryegrass (*Lolium perenne* L.) and radiata pine (*Pinus radiata* D. Don.). Soil Biol Biochem 34:487–499

Chen CR, Condron LM, Turner BL, Mahieu N, Davis MR, Xu ZH, Sherlock RR (2004) Mineralisation of soil orthophosphate monoesters under pine seedlings and ryegrass. Aust J Soil Res 42:189–196

Chen CR, Condron LM, Xu ZH (2008) Impacts of grassland afforestation with coniferous trees on soil phosphorus dynamics and associated microbial processes: a review. For Ecol Manage 255:396–409

Colbert SR, Allen HL (1996) Factors contributing to variability in loblolly pine foliar nutrient concentrations. South J Appl For 20:45–52

Comerford NB, de Barros NF (2005) Phosphorus nutrition of forest trees. In: Sims TJ, Sharpley AN (eds) Phosphorus: agriculture and the environment. Agronomy monograph 46. American Society of Agronomy, Madison, WI, pp 541–558

Comerford NB, McLeod M, Skinner M (2002) Phosphorus form and bioavailability in the pine rotation following fertilization – P fertilization influences P form and potential bioavailability to pine in the subsequent rotation. For Ecol Manage 169:203–211

Condron LM, Tiessen H (2005) Interactions of organic phosphorus in terrestrial ecosystems. In: Turner BL, Frossard E, Baldwin D (eds) Organic phosphorus in the environment. CABI, Wallingford, UK, pp 295–307

Condron L, Frossard E, Newman RH, Tekely P, Morel JL (1997) Use of ^{31}P NMR in the study of soils and the environment. In: Nannny MA (ed) Nuclear magnetic resonance spectroscopy in environmental chemistry. Oxford University Press, New York, pp 247–271

Condron LM, Turner BL, Cade-Menum BJ (2005) Chemistry and dynamics of soil organic phosphorus. In: Sims TJ, Sharpley AN (eds) Phosphorus: agriculture and the environment. Agronomy monograph 46. American Society of Agronomy, Madison, WI, pp 87–121

Cromack KR, Sollins P, Graustein WC, Speidel K, Todd AW, Spycher G, Li CY, Todd RL (1979) Calcium oxalate accumulation and soil weathering in mats of the hypogeous fungus *Hysterangium crassum*. Soil Biol Biochem 11:463–468

Cross AF, Schlesinger WH (1995) A literature review and evaluation of the Hedley fractionation: applications to the biogeochemical cycle of soil phosphorus in natural ecosystems. Geoderma 64:197–214

Crous JW, Morris AR, Schole MC (2007) The significance of residual phosphorus and potassium fertilization in countering yield decline in a fourth rotation of *Pinus patula* in Swaziland. South Hemisph For J 69:1–8

Dai KH, David MB, Vance GF, Krzyszowska AJ (1996) Characterization of phosphorus in a spruce-fir spodosol by phosphorus-31 nuclear magnetic resonance spectroscopy. Soil Sci Soc Am J 60:1943–1950

Dalla-Tea F, Jokela EJ (1991) Needlefall, canopy light interception and productivity of young intensively managed slash and loblolly pine stands. For Sci 37:1298–1313

Delgado A, Torrent J (2000) Phosphorus forms and desorption patterns in heavily fertilized calcareous and limed acid soils. Soil Sci Soc Am J 64:2031–2037

Dell B, Malajczuk N, Xu D, Grove TS (2001) Nutrient disorders in plantation eucalyptus, 2nd edn. ACIAR monograph 74. Australian Centre for International Agricultural Research, Canberra, Australia

Dinkelaker B, Marschner H (1992) In vivo demonstration of acid phosphatase activity in the rhizosphere of soil grown plants. Plant Soil 144:199–205

Donald DGM, Lange PW, Schutz CJ, Morris AR (1987) The application of fertilizers to pines in South Africa. S Afr For J 141:53–62

Ducey MJ, Allen HL (2001) Nutrient supply and fertilization efficiency in midrotation loblolly pine plantations: a modeling analysis. For Sci 47:96–102

Duzan HW, Allen HL, Ballard R (1982) Predicting fertilizer response in established loblolly pine plantations using basal area and site index. South J Appl For 6:15–19

Everett CJ, Palm-Leis H (2009) Availability of residual phosphorus fertilizer for loblolly pine. For Ecol Manage 258:2207–2213

Evers FH (1991) Forest fertilization – present state and history with special reference to south German conditions. Fertil Res 27:71–86

FAO (2006) Global forest resources assessment (2005) FAO forestry paper 147. Food and Agriculture Organization of the United Nations, Rome, Italy

FAO (2009) State of the world's forests 2009. Food and Agriculture Organization of the United Nations, Rome, Italy

Ferreira JM, Stape JL (2009) Productivity gains by fertilization in *Eucalyptus urophylla* clonal plantations across gradients in site and stand conditions. South For 71:253–258

Fisher RF, Binkley D (2000) Ecology and management of forest soils, 3rd edn. Wiley, New York

Flinn DW, Hopmans P, Farrell PW, James JM (1979) Nutrient loss from burning of *Pinus radiata* logging residue. Aust For Res 9:17–23

Flores FJ, Allen HL, Chesire HM, Davis JM, Fuentes M, Kelting D (2006) Using multispectral satellite imagery to estimate leaf area and response to silvicultural treatments in loblolly pine stands. Can J For Res 36:1587–1596

Fox TR (1995) The influence of low molecular weight organic acids on properties and processes in forest soils. In: McFee WW, Kelly JM (eds) Carbon forms and functions in forest soils. Soil Science Society of America, Madison, WI, pp 43–61

Fox TR (2000) Sustained productivity of intensively managed forest plantations. For Ecol Manage 138:187–202

Fox TR, Comerford NB (1990) Low-molecular-weight organic acids in selected forest soils of the southeastern USA. Soil Sci Soc Am J 54:1139–1144

Fox TR, Comerford NB (1992a) Influence of oxalate loading on phosphorus and aluminum solubility in Spodosols. Soil Sci Soc Am J 56:290–294

Fox TR, Comerford NB (1992b) Rhizosphere phosphatase activity and phosphatase hydrolysable organic P in two forested Spodosols. Soil Biol Biochem 24:579–583

Fox TR, Morris LA, Maimone RA (1989) The impact of windrowing on the productivity of a rotation age loblolly pine plantation in the North Carolina Piedmont. In: Proceedings fifth southern silvicultural research conference. General technical report SO-74. USDA Forest Service, New Orleans, LA, pp 133–140

Fox TR, Comerford NB, McFee WW (1990a) Phosphorus and aluminum release from a spodic horizon mediated by organic acids. Soil Sci Soc Am J 54:1763–1767

Fox TR, Comerford NB, McFee WW (1990b) Kinetics of phosphorus release from Spodosols: effects of oxalate and formate. Soil Sci Soc Am J 54:1441–1447

Fox TR, Allen HL, Albaugh TJ, Rubilar R, Carlson CA (2007a) Tree nutrition and forest fertilization of pine plantations in the southern United States. South J Appl For 31:5–11

Fox TR, Allen HL, Albaugh TJ, Rubilar R, Carlson C (2007b) Forest fertilization and water quality in the United States. Better Crops 91:7–9

Fox TR, Jokela EJ, Allen HL (2007c) The development of pine plantation silviculture in the southern United States. J For 105:337–347

Gent JA, Allen HL, Campbell RG, Wells CG (1986) Magnitude, duration, and economic analysis of loblolly pine growth response following bedding and phosphorus fertilization. South J Appl For 10:124–128

Gentle SW, Humphreys FR, Lambert MJ (1986) Continuing response of *Pinus radiata* to phosphate fertilizers over two rotations. For Sci 32:822–829

Gerke J (1992) Phosphate, aluminium and iron in the soil solution of three different soils in relation to varying concentrations of citric acid. Z Pflanzenernahrung Bodenk 155:339–343

Gil-Sotres F, Zech W, Alt HG (1990) Characterization of phosphorus fractions in surface horizons of soils from Galicia (N.W. Spain) by ^{31}P NMR spectroscopy. Soil Biol Biochem 22:75–79

Gonçalves JL, Benedetti V (eds) (2004) Forest nutrition and fertilization. Institute of Forest Research and Study, Piracicaba, São Paulo, Brazil

Gonçalves JLM, Stape JL, Laclau JP, Bouillet JP, Ranger J (2008) Assessing the effects of early silvicultural management on long-term site productivity in fast-growing eucalypt plantations: the Brazilian experience. South For 70:105–118

Grierson PF (1992) Organic acids in the rhizosphere of *Banksia integrifolia* L. Plant Soil 144:259–265

Harding RB, Jokela EJ (1994) Long-term effects of forest fertilization on site organic matter nutrients. Soil Sci Soc Am J 58:216–221

Hayes JE, Richardson AE, Simpson RJ (2000) Components of organic phosphorus in soil extracts that are hydrolyzed by phytase and acid phosphatases. Biol Fertil Soils 8:128–133

He Z, Griffin TS, Honeycutt CW (2004) Enzymatic hydrolysis of organic phosphorus in swine manure and soil. J Environ Qual 33:367–372

Hedley MJ, Stewart JWB, Chauhan BS (1982) Changes in inorganic and organic soil phosphorus fractions induced by cultivation practices and by laboratory incubations. Soil Sci Soc Am J 46:970–976

Hinsinger P (2001) Bioavailability of soil inorganic P in the rhizosphere as affected by root-induced chemical changes: a review. Plant Soil 237:173–195

Huang PM, Schnitzer M (eds) (1986) Interactions of soil mineral with natural organics and microbes. SSSA special publication 17. Soil Science Society of America, Madison, WI

Illmer P, Schinner F (1992) Solubilization of inorganic phosphates by microorganisms isolated from forest soils. Soil Biol Biochem 24:389–395

Jauregui S (2007) International forest policy and options for climate change forest policy in developing countries. In: Free-Smith PH, Broadmeadow MSJ, Lynch JM (eds) Forestry and climate change. CABI, Cambridge, MA, pp 184–196

Johnson NC, Gehring CA (2007) Mycorrizas: symbiotic mediators of rhizosphere and ecosystem process. In: Cardon ZG, Whitebeck JL (eds) The rhizosphere: an ecological perspective. Academic, New York, pp 73–100

Johnson DW, Todd DE (1998) Harvesting effects on long-term changes in nutrient pools of mixed oak forest. Soil Sci Soc Am J 62:1725–1735

Johnson AH, Frizano J, Vann DR (2003) Biogeochemical implications of labile phosphorus in forest soils determined by the Hedley fractionation procedure. Oecologia 135:487–499

Jokela EJ, Wilson DS, Allen JE (2000) Early growth responses of slash and loblolly pines following fertilization and herbaceous weed control treatments at establishment. South J Appl For 24:23–30

Jonard M, Augusto L, Morel C, Achat DL, Saur E (2009) Forest floor contribution to phosphorus nutrition: experimental data. Ann For Sci 66:510

Jones DL (1998) Organic acids in the rhizosphere – a critical review. Plant Soil 205:25–44

Jones DL, Darrah PH (1994) Role of root derived organic acids in the mobilization of nutrients from the rhizosphere. Plant Soil 166:247–257

Jordan CF (1985) Nutrient cycling in tropical forest ecosystems. Wiley, Chichester

Kelly JM, Kelly JK (2001) Phosphorus and potassium uptake kinetics in red maple seedlings. For Sci 47:397–402

Kimbrel AR, Maness T, Brown H (2009) More energy from wood: what are the prospects. J For 107:267–270

Lambert MJ, Turner JL (1998) Interpretation of nutrient concentrations in *Pinus radiata* foliage at Belanglo state forest. Plant Soil 108:237–244

Lan M, Comerford NB, Fox TR (1995) Effects of organic anions with different Al complexation constants on P release from Spodic horizons. Soil Sci Soc Am J 59:1745–1749

Landsberg JJ, Johnsen KH, Albaugh TJ, Allen HL, McKeand SE (2001) Applying 3-PG, a simple process-based model designed to produce practical results, to data from loblolly pine experiments. For Sci 47:43–51

Lindahl BO, Taylor AFS, Finlay RD (2002) Defining nutrition contraints on carbon cycling in boreal forests – towards a less 'phytocentric' perspective. Plant Soil 242:123–135

Lindsay WL (1979) Chemical equilibrium in soil. Wiley, New York

Liu Q, Laganathan P, Hedley MJ, Skinner MF (2004) The mobilization and fate of soil and rock phosphate in the rhizosphere of ectomycorrhizal *Pinus radiata* seedlings in an allophanic soil. Plant Soil 264:219–229

Liu Q, Laganathan P, Hedley MJ (2005) Influence of ectomycorrhizal hyphae on phosphate fractions and dissolution of phosphate rock in rhizosphere soils of *Pinus radiata*. J Plant Nutr 28:1525–1540

Magid J, Tiessen H, Condron LM (1996) Dynamics of organic phosphorus in soils under natural and agricultural ecosystems. In: Piccolo G (ed) Humic substances in terrestrial ecosystems. Elsevier, Amsterdam, pp 429–466

May B, Smethurst P, Carlyle C, Mendham D, Bruce J, Baillie C (2009) Review of fertilizer use in Australian forestry. Forest and Wood Products Australia, Melbourne VIC, Australia. http://www.fwpa.com.au. Accessed 02 Aug 2010

Mendham D, Smethurst P, Holz G, Menary R, Grove T, Baker T (2002) Soil analysis as indicators of P status in young *Eucalyptus nitens* and *E. globulus* plantations. Soil Sci Soc Am J 66:959–968

Merino A, López AR, Brañas J, Rodríquez-Soalleiro R (2003) Nutrition and growth of newly established plantations of *Eucalyptus globulus* in northwestern Spain. Ann For Sci 60:509–517

Miller HG (1981) Forest fertilization: some guiding concepts. Forestry 54:157–167

Miller BW (2008) Long term effects of fertilization on phosphorus biogeochemistry in forest soils. Ph.D. dissertation, Virginia Polytechnic Institute and State University, Blacksburg, VA

Newman R, Tate KR (1980) Soil characterized by ^{31}P nuclear magnetic resonance. Commun Soil Sci Plant Anal 11:835–842

Parfitt R, Atkinson RJ, Smart RC (1975) The mechanism of phosphate fixation on iron oxides. Soil Sci Soc Am J 39:837–841

Piatek KB, Allen HL (2001) Are forest floors in mid-rotation stands of loblolly pine (*Pinus taeda* L.) a sink for nitrogen and phosphorus? Can J For Res 31:1164–1174

Pierzynski G, McDowell RW, Sims JT (2005) Chemistry, cycling, and potential movement of inorganic phosphorus in soils. In: Sims JT, Sharpley AN (eds) Phosphorus: agriculture and the environment. American Society of Agronomy, Madison, WI, pp 53–86

Pohlman AA, McColl JG (1988) Soluble organics from forest litter and their role in metal dissolution. Soil Sci Soc Am J 52:265–271

Polglase PJ, Jokela EJ, Comerford NB (1992a) Phosphorus, nitrogen, and carbon fractions in litter and soil of southern pine plantations. Soil Sci Soc Am J 56:566–573

Polglase PJ, Comerford NB, Jokela EJ (1992b) Mineralization of nitrogen and phosphorus from soil organic matter in southern pine plantations. Soil Sci Soc Am J 56:921–927

Polglase PJ, Jokela EJ, Comerford NB (1992c) Nitrogen and phosphorus release from decomposing needles of southern pine plantations. Soil Sci Soc Am J 56:914–920

Pritchett WL, Comerford NB (1982) Long-term response to phosphorus fertilization on selected southeastern coastal plain soils. Soil Sci Soc Am J 46:640–644

Pritchett WL, Fisher RF (1987) Properties and management of forest soils, 2nd edn. Wiley, New York

Pritchett WL, Llewellyn WR, Swinford KR (1961) Response of slash pine to collodial phosphate fertilization. Soil Sci Soc Am Proc 25:397–400

Quin LD, Quin GS (2001) Screening for carbon- bound phosphorus in marine animals by high resolution ^{31}P-NMR spectroscopy: coastal and hydrothermal vent invertebrates. Comp Biochem Physiol 128:173–185

Raghothama KG (1999) Phosphate acquisition. Annu Rev Plant Physiol Plant Mol Biol 50:665–693

Raghothama KG (2005) Phosphorus and plant nutrition: an overview. In: Sims, JT, Sharpley AN (eds) Phosphorus: agriculture. Agronomy monograph 46. American Society of Agronomy, Madison, WI, pp 355–378

Read DJ, Perez-Moreno J (2003) Mycorrhizas and nutrient cycling in ecosystems – a journey toward relevancy. New Phytol 157:475–492

Richardson AE (2001) Prospects for using soil micro-organisms to improve the acquisition of phosphorus by plants. Aust J Plant Physiol 28:897–906

Richter DB, Allen HL, Li J, Markewitz D, Raikes J (2006) Bioavailability of slowly cycling soil phosphorus: major restructuring of soil P-fractions over four decades in an aggrading forest. Oecologia 150:259–271

Rubilar R, Blevins L, Toro J, Vita A, Muñoz F (2008a) Early response of *Pinus radiata* plantations to weed control and fertilization on metamorphic soils of the coastal range, Maule Region, Chile. Bosque 29:74–84

Rubilar R, Fox T, Allen L, Albaugh T, Carlson C (2008b) Manejo intensive al estableciento de plantaciones forestales de *pinus sp*. y *eucalyptus sp* en Chile y Argentina. Info Agro Cono Sur 40:1–6

Sample EC, Soper RJ, Racz GJ (1980) Reactions of phosphate fertilizers in soils. In: Khasawneh FE, Sample EC, Kamprath EJ (eds) The role of phosphorus in agriculture. American Society of Agronomy, Madison, WI, pp 263–310

Sanchez FG (2001) Loblolly pine needle decomposition as affected by irrigation, fertilization, and substrate quality. For Ecol Manage 152:85–96

Sato S, Comerford NB (2006) Organic anions and phosphorus desorption and bioavailability in a humid Brazilian Ultisol. Soil Sci 171:695–705

Sedjo RA, Botkin D (1997) Using plantations to spare natural forests. Environment 39:14–20

Smith SE, Read DJ (1997) Mycorrhizal symbiosis, 2nd edn. Academic, Amsterdam

Speir TW, Ross DJ (1978) Soil phosphatases and sulphatases. In: Burns RG (ed) Soil enzymes. Academic, New York, pp 197–249

Stape JL, Binkley D, Ryan MG (2004a) Eucalyptus production and the supply, use and efficiency of use of water, light and nitrogen across a geographic gradient in Brazil. For Ecol Manage 193:17–31

Stape JL, Ryan MG, Binkley D (2004b) Testing the utility of the 3-PG model for growth of *Eucalyptus grandis* × *urophylla* with natural and manipulated supplies of water and nutrients. For Ecol Manage 193:219–234

Stape JL, Binkley D, Jacob WS, Takahashi EN (2006) A twin-plot approach to determine nutrient limitation and potential productivity in *Eucalyptus* plantations at landscape scales in Brazil. For Ecol Manage 223:358–362

Stape JL, Binkley D, Ryan MG (2008) Production and carbon allocation in a clonal *Eucalyptus* plantation with water and nutrient manipulations. For Ecol Manage 255:920–930

Stevenson FJ (1967) Organic acids in soils. In: McLaren AD, Peterson GH (eds) Soil biochemistry. Marcel Dekker, New York, pp 119–146

Ström L, Owen AG, Godbold DL, Jones DL (2002) Organic acid mediate P mobilization in the rhizosphere and uptake by maize roots. Soil Biol Biochem 34:703–710

Tabatabai MA (1982) Soil enzymes. In: Page AL, Miller RH, Keeney DR (eds) Methods of soil analysis. Part 2. Chemical and microbiological properties, 2nd edn. American Society of Agronomy, Madison, WI, pp 903–948

Tang J, Leung A, Leung C, Lim BL (2006) Hydrolysis of precipitated phytate by three distinct families of phytases. Soil Biol Biochem 38:1316–1324

Tate KR (1984) The biological transformation of P in soil. Plant Soil 76:245–256

Tate KR, Newman RH (1982) Phosphorus fractions of a clinosequence of soils in New Zealand tussock grassland. Soil Biol Biochem 14:191–196

Tiessen H, Moir JO (1993) Characterization of available P by sequential extraction. In: Carter MR (ed) Soil sampling and methods of analysis. Lewis and CRC, Boca Raton, FL, pp 75–86

Trichet P, Bakker MR, Augusto L, Alazard P, Merzeau D, Saur E (2009) Fifty years of fertilization experiments on *Pinus pinaster* in southwest France: the importance of phosphorus as a fertilizer. Forest Sci 55:390–402

Turner J, Lambert MJ (1985) Soil phosphorus forms and related tree growth in a long term *Pinus radiata* phosphate fertilizer trial. Commun Soil Sci Plant Anal 16:275–288

Turner BL, Richardson AE (2004) Identification of scyllo-inositol phosphates in soil by solution phosphorus-31 nuclear magnetic resonance spectroscopy. Soil Sci Soc Am J 68:802–808

Turner J, Knott JH, Lambert M (1996) Fertilization of *Pinus radiata* after thinning. I. Productivity gains. Aust For 59:7–21

Turner J, Lambert MJ, Humphreys FR (2002) Continuing growth response to phosphate fertilizers by a *Pinus radiata* plantation over fifty years. For Sci 48:556–568

Turner BL, Mahieu N, Condron LM (2003) Phosphorus-31 nuclear magnetic resonance spectral assignments of phosphorus compounds in soil NaOH-EDTA extracts. Soil Sci Soc Am J 67:497–510

Turner BL, Mahieu N, Condron LM, Chen CR (2005) Quantification and bioavailability of scyllo-inositol hexakisphosphate in pasture soils. Soil Biol Biochem 37:2155–2158

Turner BL, Condron LM, Richardson SJ, Pelzer DA, Allison VJ (2007) Soil organic phosphorus transformations during pedogenesis. Ecosystems 10:1166–1181

van Hees PA, Lundstrom US, Giesler R (2000) Low molecular weight organics and their Al complexes in soil solution – composition, distribution, and seasonal variation in three podzolized soils. Geoderma 94:173–200

van Hees PAW, Jones DL, Godbold DL (2002) Biodegradation of low molecular weight organic acids in coniferous podzolic soils. Soil Biol Biochem 34:1261–1272

van Hees PAW, Godbold DL, Jentschke G, Jones DL (2003) Impact of ectomycorrhizas on the concentration and biodegradation of simple organic acids in a forest soil. Eur J Soil Sci 54:697–706

Vance CP, Uhde-Stoen C, Allan DL (2003) Phosphorus acquisition and use: critical adaptations by plants for securing a nonrewable resource. New Phytol 157:423–447

Vose JM, Allen HL (1988) Leaf area, stemwood growth and nutrition relationships in loblolly pine. For Sci 34:547–563

Walbridge M (1991) Phosphorus availability in acid organic soils of the lower North Carolina coastal plain. Ecology 72:2083–2100

Weir D, Gries J (2002) Southern forest resource assessment. General Technical Report SRS-53. USDA Forest Service. Asheville, NC

Wells CG, Jorgensen JR (1975) Nutrient cycling in loblolly pine plantations. In: Bernier WW, Winget CH (eds) Forest soils and forest land management. Les Presses de L'Université, Laval, Quebec, pp 137–158

Wells CG, Craig JR, Kane MB, Allen HL (1986) Foliar and soil tests for the prediction of phosphorus response in loblolly pine. Soil Sci Soc Am J 50:1330–1335

Will G (1985) Nutrient deficiencies and fertilizer use in New Zealand exotic forests. FRI Bulletin 97. Forest Research Institute, Rotorua, New Zealand

Williamson B, Alexander IJ (1975) Acid phosphatases localized in the sheath of beech mycorrhiza. Soil Biol Biochem 7:195–198

Chapter 14
Phosphorus Cycling in Tropical Forests Growing on Highly Weathered Soils

Sasha C. Reed, Alan R. Townsend, Philip G. Taylor, and Cory C. Cleveland

14.1 Introduction

In 1976, Walker and Syers introduced a model describing patterns of soil phosphorus (P) pools and availability during ecosystem development (Fig. 14.1). The model suggested that, early in soil development, the majority of P is in primary mineral forms, mostly as apatite. As apatite is weathered, it releases biologically available forms of P (as PO_4^{3-}). Some P is taken up by plants and microbes and is ultimately returned to inorganic P (P_i) pools in the soil via mineralization, or remains within the soil in organic forms (P_o; Fig. 14.1). However, during each turn of this cycle, some P may also be sorbed by secondary soil minerals, precipitated, or leached in organic or inorganic forms, slowly depleting the total and available P pools (Fig. 14.1).

The Walker and Syers (1976) model predicted that at intermediate levels of soil development, P would be fairly evenly distributed among different pools: primary mineral, secondary mineral, labile P, and soil P_o (Fig. 14.1). However, as soil development reaches more advanced stages, the total amount of P in the system would decline and much of the remaining P would be bound in insoluble or physically protected, nonlabile (i.e., not biologically available) forms. In contrast to carbon (C), nitrogen (N) and sulfur (S), P has no significant gaseous state [phosphane (PH_3) comprises a trivial portion of global P stocks (Toy 1973)], and the primary input of biologically available P is from mineral weathering or from atmospheric dust inputs (Chadwick et al. 1999). In essence, Walker and Syers

S.C. Reed (✉)
U.S. Geological Survey, Southwest Biological Science Center, Moab, UT 84532, USA
e-mail: screed@usgs.gov

A.R. Townsend and P.G. Taylor
Department of Ecology and Evolutionary Biology and the Institute of Arctic and Alpine Research, University of Colorado, Boulder, CO 80303, USA

C.C. Cleveland
Department of Ecosystem and Conservation Sciences, University of Montana, Missoula, MT 59812, USA

E.K. Bünemann et al. (eds.), *Phosphorus in Action*, Soil Biology 26,
DOI 10.1007/978-3-642-15271-9_14, © Springer-Verlag Berlin Heidelberg 2011

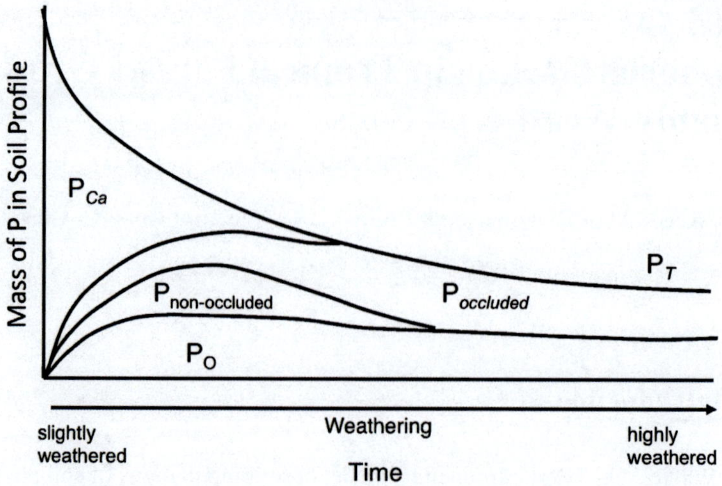

Fig. 14.1 Conceptual model modified from Walker and Syers (1976) showing variations in soil P pools during pedogenesis: P_T total soil P; P_{Ca} calcium phosphates; P_o organic matter P; $P_{occluded}$ sorbed P (relatively unavailable to organisms); $P_{nonoccluded}$ P in the soil solution (relatively available to organisms; often called "labile" P). The model suggests that the highly weathered soils (Ultisols and Oxisols) of many tropical forests have less total P, higher relative proportions of $P_{occluded}$ and P_o, and very little available P ($P_{nonoccluded}$) relative to soils at earlier stages of soil development

(1976) predicted that ecosystems on old, highly weathered soils would eventually reach a "terminal steady state" in which little P would remain available for biological uptake, ultimately leading to P limitation of ecosystem processes.

This model has been further tested and corroborated by multiple studies, including one that assessed nutrient availability and limitation using a fertilization study across a 4 million year substrate age gradient (Vitousek 2004). Results showed that soil P pools behaved as predicted by Walker and Syers (1976) over the ~4 million years of soil development (Crews et al. 1995), and that aboveground primary production was N-limited in the youngest soils (~300 years old), N and P colimited in the intermediate-aged soils (~20,000 years old) and limited by P alone in the oldest soils (~4.1×10^6 years old) (Fig. 14.2) (Vitousek and Farrington 1997; Harrington et al. 2001).

The US Department of Agriculture (USDA) soil classification system (Soil Survey Staff 2006) includes 12 distinct soil orders that are largely defined by the extent of soil weathering. While tropical forests contain all but one of the USDA soil orders (Gelisols; Palm et al. 2007), seven are common in tropical forests: Alfisols, Entisols, Inceptisols, Mollisols, Oxisols, Ultisols, and Vertisols (Table 14.1). The Alfisols and Mollisols represent two of the more fertile soils of the tropics, and contain relatively high quantities of mineral and available P (Table 14.1); these are the soils where much of the most productive tropical agriculture occurs (Sanchez 1976; Vitousek and Sanford 1986). In contrast, the Ultisols and Oxisols occupy the weathered end of the weathering spectrum and are

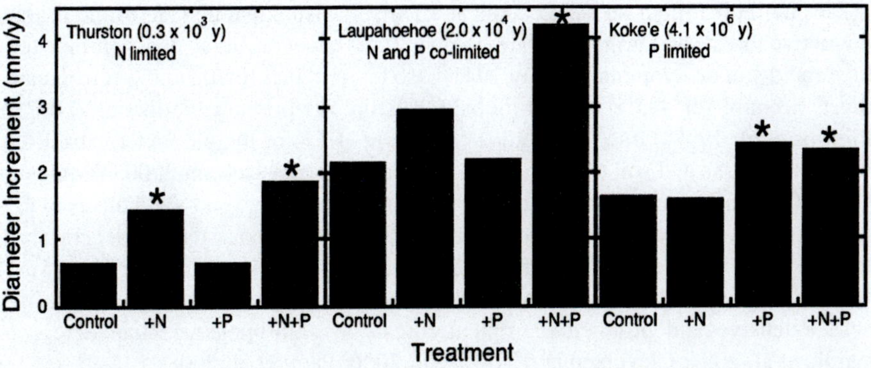

Fig. 14.2 Tree growth rates from a fertilization study along 4.1 million year substrate age gradient (modified from Vitousek and Farrington 1997). *Bars* show the means of tree growth (diameter increment) of *Metrosideros polymorpha* (the dominant tree species) for trees in control plots and in plots fertilized with N, P, and N + P. Within each site, treatment effects that were significantly different from the control are represented by *asterisks*. Data show that tree growth at the youngest site was N-limited, tree growth at the intermediate-aged site was colimited by N and P, and tree growth at the oldest site was P-limited

Table 14.1 Distribution of tropical forest soil orders (from Palm et al. 2007) and associated P pools

Soil order	Area (10^6 ha)	Area (%)	Number (n) of samples tested	Labile P (μg/g)	Total organic P (P_o) (μg/g)	Total P (P_t) (μg/g)	P_o/P_t
Ultisol	654	27.0	7–8[a]	33.3 ± 11.6	80.7 ± 19.6	203.6 ± 38.6	0.40
Oxisol	649	26.8	1–3[b]	29.7 ± 4.5	152.0	321.3 ± 60.9	0.47
Inceptisol	416	17.2	5–9[c]	71.4 ± 19.7	131.9 ± 20.2	753.6 ± 155.8	0.18
Entisol	244	10.1	2	17.5 ± 4.5	36.0 ± 11.0	684.5 ± 80.5	0.05
Alfisol	234	9.7	12	45.6 ± 9.6	83.8 ± 11.0	365.8 ± 101.5	0.23
Vertisol	71	2.9	3	43.3 ± 15.4	57.5 ± 12.5	529.3 ± 130.1	0.11
Mollisol	63	2.6	34–38[d]	47.1 ± 4.4	109.1 ± 8.2	594.2 ± 101.9	0.18
Andisol	37	1.5	–	–	–	–	–
Aridisol	26	1.1	–	–	–	–	–
Histisol	26	1.1	–	–	–	–	–
Spodosol	5	0.2	–	–	–	–	–

Soil P concentrations – assessed using the Hedley-fractionation procedure (Tiessen and Moir 1993) – are provided for soil orders representing >2% of tropical forests and include soil-labile P (resin + bicarbonate-extractable inorganic and organic soil P), total organic P, and total P. The P concentrations represent global averages for each soil order. Soil P values are from Cross and Schlesinger (1995) and Johnson et al. (2003), using only data meeting the soil depth criterion used by Cross and Schlesinger (1995). P concentrations given are means ± s.e.m

[a]Labile P and total organic P values were only available for seven of the eight Ultisols

[b]Total organic P values were only available for one of the three Oxisols

[c]Labile P and total organic P values were only available for five of the nine Inceptisols

[d]Total organic P and total P values were only available for 34 of the 38 Mollisols

characterized by relatively low P availability (Table 14.1). Together these two soil orders are present in more than 50% of tropical forests (Table 14.1) (Vitousek and Sanford 1986; Palm et al. 2007).

Despite their P-poor status, Oxisols and Ultisols also contain some of the most productive forests on Earth. The warm temperatures and wet climate that combine to drive rapid soil development (Jenny 1941) also favor plant growth, and wet tropical forests account for ~35% of total global terrestrial net primary production (NPP) (Phillips et al. 1998; Grace et al. 2001) and store ~25% of the global terrestrial C found in biomass and soil (Schlesinger 1997; Jobbagy and Jackson 2000; Tarnocai et al. 2009). Understanding how P cycling regulates these large pools and fluxes of C is important at the global scale and, although climate and ecosystem models rarely explicitly consider the role of P in regulating C cycling (Parton et al. 2005; Thornton et al. 2009), there is growing evidence that P may strongly constrain the response of these extensive and productive tropical forests to anthropogenic change (e.g., Wardle et al. 2004; Cleveland and Townsend 2006; Paoli et al. 2008).

In the following sections, we describe P cycling in tropical forests, focusing on the most fundamental and unique aspects of the P cycle in forests growing on Ultisols and Oxisols, which are the dominant tropical soil orders (Table 14.1). We make only a few attempts to quantify pools and fluxes, per se, because data are rare and those that do exist suggest that the tropical P cycle is variable in both space and time (Townsend et al. 2008). Instead, we attempt to more generally describe the sources, internal transformations, and losses of P from tropical forests and to highlight the overall importance of P cycling for tropical forest function. In addition, we note that although Ultisols and Oxisols represent the most common tropical forest soil orders and are the focus of this analysis, tropical forests grow on many other soil types and the wide variation in tropical forest soil orders, species assemblages, and biogeochemical cycles results in a heterogeneous biome (Townsend et al. 2008).

14.2 The P Cycle in Tropical Soils

14.2.1 Tropical Soil P: Inputs

14.2.1.1 Parent Material

Of the six elements (C, hydrogen, N, oxygen, P, and S) that comprise 95% of the biosphere, P is somewhat unusual in that its biogeochemical cycle does not include a significant gaseous component (Toy 1973). Thus, as opposed to N that can be biologically "fixed" from the atmosphere, soil P must be supplied almost entirely from the weathering of underlying parent material (Schlesinger 1997) or from dust inputs (see Sect. 14.2.1.2). At early stages of ecosystem and soil development, the majority of P is in primary mineral forms (mostly as calcium apatite minerals) and, through time, weathering processes slowly dissolve the primary mineral P:

$$Ca_5(PO_4)_3OH(\text{apatite}) + 4H_2CO_3 \rightarrow 5Ca^{2+} + 3HPO_4{}^{2-} + 4HCO_3{}^- + H_2O$$

The rate of P weathering is variable and is dependent upon parent material, climate, and other drivers of soil development (Jenny 1941). Moreover, most rocks contain only small amounts of apatite, thus the weathering rate of P-containing minerals can ultimately constrain inputs of available P. In many temperate and high latitude terrestrial ecosystems, evidence suggests that when nutrients limit ecosystem processes, the limitation most commonly comes from N (Vitousek and Howarth 1991; Hooper and Johnson 1999) or from N and P together, not from P alone (see Elser et al. 2007; LeBauer and Treseder 2008). Nitrogen limitation has also been shown in tropical ecosystems where less weathered soils predominate (e.g., montane tropical systems) (Elser et al. 2007; LeBauer and Treseder 2008).

However, tropical forests are unique from many temperate or high latitude forests because they commonly exist on old, highly weathered substrates (Table 14.1). This is due to the high temperatures and rainfall that combine to promote rapid and extensive parent material weathering and because the long periods of intense weathering without large-scale disturbance have depleted much of the primary mineral P (in contrast to ecosystems that undergo episodic large-scale disturbances, e.g., glaciations, which effectively "reset" the soil development clock by exposing unweathered parent material at the surface). In the absence of such disturbances, most P in tropical forests resides in either organic forms (which vary in their availability) or in geochemically protected forms (i.e., P that is "sorbed" onto mineral surfaces or protected within mineral matrices) (occluded P; see Sect. 14.2.2.2). Thus, highly weathered tropical soils receive relatively small P inputs from parent material weathering (Walker and Syers 1976; Chadwick et al. 1999) and only small amounts of P reside in plant-available pools (Table 14.1) (Walker and Syers 1976; Cross and Schlesinger 1995; Johnson et al. 2003).

Although soil-available P is low in many lowland tropical forests, there is one important caveat to this general pattern: at the landscape scale, topographic variation can influence soil P cycling much like large-scale disturbances. For example, even in relatively well-developed soils on geomorphically unstable surfaces – such as slopes – high rates of soil erosion can rapidly expose parent material to weathering and effectively increase the parent material inputs of rock-derived nutrients (like P) (Vitousek et al. 2003; Bern et al. 2005). Thus, over relatively short timescales, P inputs and availability can be subsidized via erosion in forests with significant topographic relief. This does not contradict models of soil development (Jenny 1941; Walker and Syers 1976), but rather explains the significant spatial heterogeneity in soil P related to soil position (e.g., ridge top, slope).

In tropical forests, soils in geomorphically unstable sites can still be P-poor, but P may be high relative to sites occupying more stable surfaces (Silver et al. 1994; Chen et al. 1997; Vitousek et al. 2003). For example, when comparing the foliar element concentrations from soil on a flat, shield volcano in Hawai'i to those of an adjacent slope, Vitousek et al. (2003) used strontium isotopes to show that the parent material inputs of rock-derived elements (like P) on geomorphically unstable slopes could be twice the inputs on stable surfaces (13% vs. 6% of inputs from parent material, respectively).

14.2.1.2 Atmospheric Inputs

In the absence of substantial new inputs of rock-derived P from weathering, what are the dominant inputs of P to tropical forests? Evidence suggests that atmospheric inputs of P may be important in some tropical forests. For example, P can be transferred through the atmosphere over extremely long distances in the form of dust (Swap et al. 1992; Okin et al. 2004; Neff et al. 2006). The importance of dust inputs to tropical forest ecosystems depends on both the existing reservoir of P in the downwind ecosystem as well as on the total flux of dust being transported and deposited (Fig. 14.3). Multiple studies suggest that atmospheric P inputs to tropical forests may be critical for maintaining high NPP (Swap et al. 1992; Artaxo et al. 2002; Okin et al. 2004). For example, Chadwick et al. (1999) showed that, in sites in Hawai'i where rock-derived elements have been depleted, inputs of elements (like P) from dust may be necessary for sustaining NPP (Fig. 14.4). Thus, as soils weather and primary mineral P is depleted, atmospheric inputs may become the primary source of "new" soil P availability, and biological activity in highly weathered soils may increasingly rely upon atmospheric dust inputs (Fig. 14.4). This makes sense: in the absence of such external inputs, tropical ecosystems on

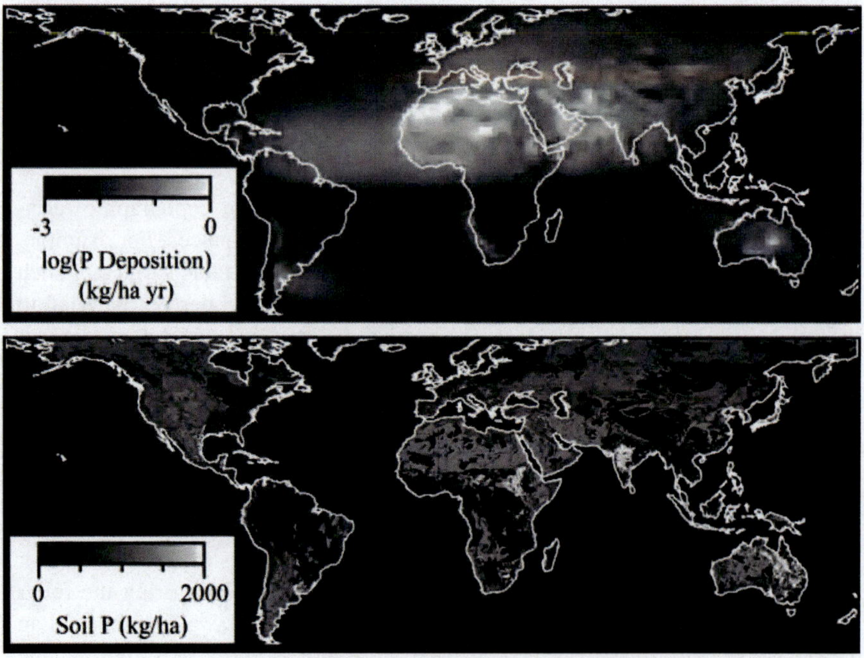

Fig. 14.3 Modeled estimates of global patterns in (**a**) P deposition from dust and (**b**) soil total P concentrations. Figure taken from Okin et al. (2004) and reproduced with the permission of the American Geophysical Union

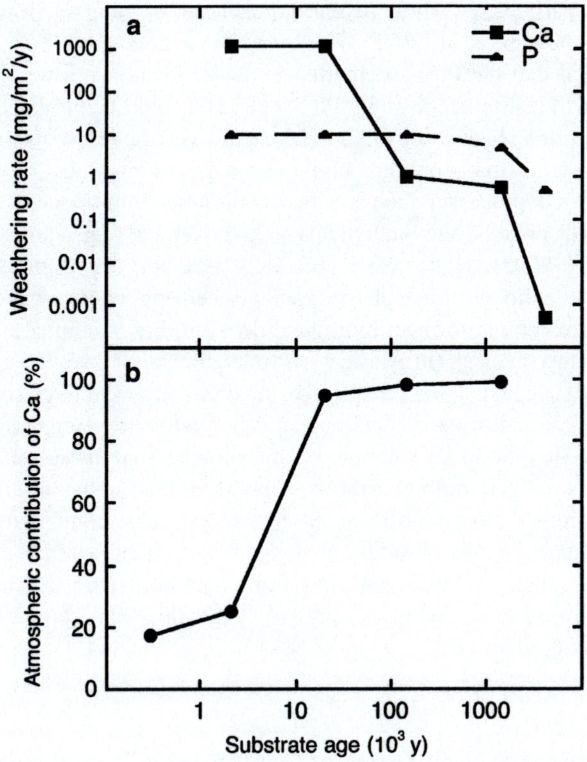

Fig. 14.4 Data from a 4.1 million year chronosequence in Hawai'i (Chadwick et al. 1999) showing (**a**) that rock-derived elements such as Ca and P are less likely to be released from parent material in older soils; and (**b**) an increased contribution of atmospheric inputs in supplying rock-derived elements. Ca and P values in **a** are calculated as the difference in mass of an element lost between two sites and the corresponding difference in age. In **b**, values represent the total percentage of soil Ca that was derived from the atmosphere, calculated by dividing the atmospheric contribution by the sum of atmospheric and weathering contribution. Figure reproduced from Chadwick et al. (1999) with the permission of the Nature Publishing Group

old, highly weathered soils would be predicted to reach a profound and irreversible state of P limitation (as predicted by Walker and Syers 1976).

Research also suggests that P from dust is important in mainland tropical forests (Fig. 14.3). For example, Swap et al. (1992) estimated that in the northeastern Amazon Basin, P inputs via dust range from 1 to 4 kg/ha per year, and they concluded that the high productivity of the Amazon rain forest is fueled, at least in part, by dust inputs originating in the Sahara/Sahel region in Africa. Similarly, Okin et al. (2004) suggested that Amazon Basin rain forests depend on aeolian deposition for the long-term maintenance of NPP.

The large fires now common in the Amazon Basin may also contribute significant amounts of P deposition, as P released from burning biomass may be transferred (as particulates) through the atmosphere to unburned forests (Artaxo et al. 2002; Mahowald et al. 2005). For example, a study from four Amazonian forests suggested that the P requirements estimated for a year's worth of tropical rain forest growth (estimated as 0.6 g P/m^2 per year) could be provided by only 24 years of P deposition (Mahowald et al. 2005). However, data also suggest that these Amazon ecosystems may, on the whole, be experiencing net P losses to the atmosphere through processes such as biomass burning, anthropogenically induced losses of mineral aerosols, and biogenic particle movement (e.g., spores and pollen) (Mahowald et al. 2005).

Nevertheless, although atmospheric P inputs help maintain productivity, they may not be large enough to meet biological demand. For example, a study of six long-term chronosequences (in Alaska, Australia, Hawai'i, Sweden, and two sites in New Zealand) suggested that forests growing on highly weathered soils were in a "decline phase," as indicated by declines in NPP relative to younger sites (Wardle et al. 2004). Soil and foliar chemical data showed that these decline phases correlated with relatively high N:P ratios, suggesting that, in the absence of major disturbance, P constraints on NPP increase as forest ecosystems mature (Wardle et al. 2004). Moreover, this observation suggests that increases in P deposition in tropical forests (such as those resulting from increased land use and biomass burning) (Mahowald et al. 2005; Moulin and Chiapello 2006) could have positive effects on forest productivity.

14.2.2 Tropical Soil P: Internal Cycling and Transformations

14.2.2.1 Mineralization

Plant roots and soil microbes remove P from the soil solution, and use much of that P to build biomass. Subsequently, P is returned to the soil in organic forms as litterfall or dead plant or microbial biomass. Forest fauna are also a source of P_o to tropical soils, and faunal activity can affect P_o cycling in complex ways (Asawalam and Johnson 2007). In tropical forests growing on highly weathered soils, P_o represents a large proportion of total soil P (Ultisols and Oxisols contain ~44% organic P; Table 14.1), particularly when compared with other soil orders in which mineral P is the dominant form (all other soil orders contain ~17% organic P) (Cross and Schlesinger 1995). Large pools of aboveground biomass combined with low inputs of inorganic P from weathering make the return of plant biomass P to the forest floor the dominant input of P into tropical forest soils on an annual basis (Fig. 14.5) (e.g., Tiessen et al. 1984; Newberry et al. 1997). Thus, although P inputs via organic matter do not represent "new" P additions to the ecosystem per se, organic P recycling is a crucial mechanism for maintaining P stocks in tropical rain forests (e.g., Zou et al. 1992; Achat et al. 2009).

Organic P that enters the soil is a complex suite of compounds (e.g., nucleic acids and phospholipids) and variations in the overall chemistry, quantity, and

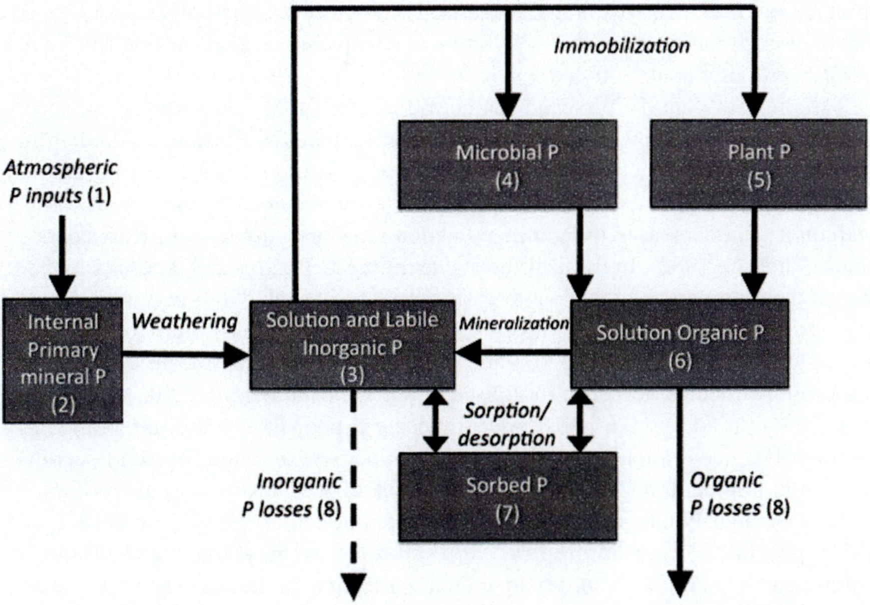

Fig. 14.5 Conceptual model of the terrestrial P cycle. *Boxes* represent the pools of P and *arrows* show the fluxes moving between pools. The size of the *boxes* and *arrows* do not represent pool size or flux rates. We highlight eight notable aspects of P cycling in tropical forests growing on weathered soils related to the pools and fluxes shown (1) Because available soil P pools are low in Ultisols and Oxisols, dust can represent a crucial input of P (see Sect. 14.2.1.2). (2) In contrast to other soil types, relatively low quantities of P are weathered from parent material into the soil matrix in highly weathered Ultisols and Oxisols (see Sect. 14.2.1.1). (3) Due to high rates of sorption (see Sect. 14.2.2.2) and large biological demand, soluble inorganic P pools in tropical forests are often relatively small. (4) Microbial P can comprise a large proportion of total P in tropical forests, and can play a crucial role in both desorbing P and maintaining P in biological cycles via storage within biomass (see Sects. 14.2 and 14.3). (5) Plants in tropical forests store a large amount of P, and tropical trees represent the dominant P source to soil. (6) Organic P represents a relatively large proportion of total soil P in tropical soils compared with many other soil types, and mineralization of P_o is an important biological P source (see Sect. 14.2.2.1). (7) Due to the physical and chemical properties of the clays dominating highly weathered soils, P sorption can effectively remove P from biologically available pools (see Sect. 14.2.2.2). (8) P can be lost from tropical soils in both organic and inorganic forms (see Sect. 14.2.4)

decomposability of organic P influence the chemistry and availability of the P_o pool (Anderson 1967; Condron et al. 1990). For example, organic P varies chemically depending on its source; both foliar and litter P content differ greatly among tropical forest tree species (Reich and Oleksyn 2004; Townsend et al. 2007) and thus the composition of the plant community influences the quantity, quality, and ultimate availability of P_o. Moreover, different plant biomass pools provide variable forms and quantities of P_o to soils. For instance, live leaves lost from trees (e.g., during a wind event) have nearly twice as much P per unit biomass than leaves that drop after senescence (McGroddy et al. 2004). Similarly, wood contains low amounts of P relative to foliage, and wood C:P ratios are typically two to three orders of magnitude higher than C:P ratios of canopy leaves (Likens and Bormann 1995; McGroddy et al. 2004).

Variations in both the type and source of P_o entering the soil strongly influence its fate. For example, low quality litter (i.e., litter with high C:nutrient and/or lignin: nutrient ratios) often decomposes relatively slowly (Swift et al. 1979; Xuluc-Talosa et al. 2003; Wieder et al. 2009). Low quality litter can exacerbate low soil P availability via decreases in P_o mineralization rates and increases in P_o residence times (Vitousek 1982). In tropical forests, leaf litter C:P ratios are generally higher than in their temperate forest counterparts (McGroddy et al. 2004) because tropical trees growing on P-poor soil invest less P per unit biomass into their leaves (Fig. 14.6) (Townsend et al. 2007) and more efficiently retranslocate P from live tissue prior to leaf senescence (Vitousek 1984; Kitayama et al. 2000; McGroddy et al. 2004). In addition, microbes decomposing P-poor litter often immobilize P, and litter P concentrations have been shown to increase over the course of decomposition in tropical forests (Hobbie and Vitousek 2000; Cleveland et al. 2006).

Plants primarily take up P as orthophosphate molecules ($H_2PO_4^-$ and HPO_4^{2-}), which generally occur in tropical soil solutions at very low concentrations (Johnson et al. 2003). In order to increase the rate of transfer from P_o to P_i pools, the plants, their symbionts, and soil microbes may excrete phosphatase enzymes (see Nannipieri et al. 2011). Phosphatases represent a broad class of enzymes (e.g., acid and alkaline phosphomonoesterases) that catalyze the release of P_i by breaking phosphate ester bonds in organic molecules. The frequently observed increase in phosphatase activity in response to low P_i concentration by plants, bryophytes, mycorrhizae, and algae underscores the importance of this type of mineralization to the nutrition of living organisms (Kroehler and Linkins 1991; Duff et al. 1994; Whitton et al. 2005). In particular, data suggest that acid phosphatase activity may account for a significant proportion of plant and microbial P mineralization (and subsequent P uptake) (Kroehler and Linkins 1988; Moorhead et al. 1993). However, because both plants and microbes produce phosphatase enzymes and because roots and elevated microbial biomass often cooccur spatially (Tarafdar and Jungk 1987; Chen et al. 2002), the relative contribution of plants versus microbes to soil phosphatase enzyme production is difficult to discern.

Some data indicate that phosphatase activity is higher in weathered, P-poor soils relative to other soil types (Acosta-Martinez et al. 2007), though this is not always the case (Olander and Vitousek 2000; Treseder and Vitousek 2001). Nonetheless, fertilization studies performed in a variety of ecosystems suggest that phosphatase activity often decreases after P fertilization (e.g., McLachlan 1980; Caradus and Snaydon 1987; Allison and Vitousek 2005). These results make sense in the context of P-poor soils: if inorganic P is scarce but ultimately available in organic forms, then phosphatase production would provide a mechanism for increasing P availability. Alternatively, if P_i is readily available (e.g., after fertilization with P), then organisms may downregulate phosphatase enzyme production (Olander and Vitousek 2000; Houlton et al. 2008).

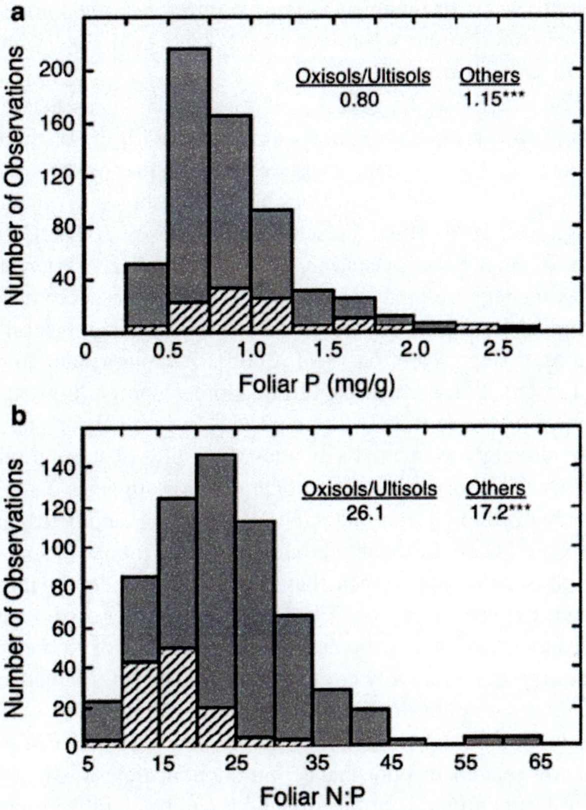

Fig. 14.6 Histograms of (**a**) foliar P and (**b**) N:P ratios (mass basis) from a tropical foliar database. Plants growing on Oxisols and Ultisols are shown in *gray* ($n = 462$) and those from other soil types are *striped* ($n = 110$). Values are means and *asterisks* denotes significant differences at $P < 0.001$. The data show that plants growing on highly weathered soils have lower foliar P concentrations and higher foliar N:P ratios compared with plants growing on other soils types. Figure taken from Townsend et al. (2007) and reproduced with the permission of the Ecological Society of America

More recently, the cloning of genes encoding extracellular phosphatases from plants (Haran et al. 2000; Wasaki et al. 2000; Miller et al. 2001) provides strong evidence for direct secretion and regulation of the expression of these genes in response to P limitation (see Wasaki and Maruyama 2011; George et al. 2011). Thus, P limitation may be "sensed" at the cellular level, driving modifications in gene expression by plants (and perhaps microbes) that result in increased phosphatase production. A recent model of plant P acquisition expands this idea by suggesting that different plant species could respond to competition for P_o by accessing discrete P_o pools (Turner 2008). Different species could produce phosphatase enzymes that target different organic P compounds. If so, plant-specific production of phosphatase enzymes could facilitate the coexistence of plant species in diverse tropical forests growing on P-poor soil. Such a phenomenon has been observed for N in an arctic tundra community (McKane et al. 2002), but has not yet been examined in tropical forests.

A growing body of evidence also suggests strong linkages between N and P cycling in tropical forest soils, and phosphatases appear to be at the heart of this important interaction. As proteins, phosphatase enzyme production requires a significant investment of N, and N additions have been shown to enhance phosphatase activity (e.g., Zou et al. 1995; Treseder and Vitousek 2001). Consistent with that observation, it has been suggested that elevated foliar and litterfall N concentrations in N_2-fixing legumes may promote soil phosphatase production beneath these trees in tropical forests on weathered soil, thus increasing the ability of N-rich species to acquire P (e.g., Houlton et al. 2008). Although data to support this hypothesis are limited, this interaction offers another compelling explanation for the high abundance of symbiotic N_2-fixing species in tropical forests.

Interestingly, although N availability may fuel phosphatase production, one widely accepted model of organic matter mineralization suggests that P mineralization may be largely decoupled from the mineralization of C and N (McGill and Cole 1981). Whereas C and N are intricately bound in organic matter (Asner et al. 1997), P is bound by phosphate ester bonds that can be independently mineralized by phosphatase enzymes (McGill and Cole 1981). Thus, whether organisms mineralize organic substrates to obtain energy (in the form of reduced C) or N, both elements are mineralized, resulting in a relatively complex, coupled set of interactions involving biotic demand and enzyme production. In contrast, linkages between P limitation, phosphatase production, and P mineralization are more direct. Thus, P availability should be inversely related to phosphatase production, unless low availability of soluble organic P limits mineralization (organic P can be stabilized within the soil, thus it is not necessarily available for mineralization or leaching). McGill and Cole (1981) suggested a dichotomous system stabilizing and mobilizing C and N on the one hand, and P on the other. A decoupling of P mineralization from the mineralization of C and N may be particularly prevalent on highly weathered, P-poor soils.

Abiotic P mineralization is also possible via hydrolytic and photolytic reactions (Baldwin et al. 2003). Although photolytic mineralization rates may be relatively low at the forest floor (i.e., where there is little light), photolytic P mineralization reactions may play an important role in some tropical settings (e.g., forest treefall gaps). In addition, abiotic hydrolytic reactions could be important drivers of mineralization because soil water is often abundant in humid tropical forests. However, discriminating between abiotic and biotic hydrolytic reactions is difficult (Baldwin et al. 2003) and the role of abiotic processes in tropical forest mineralization remains poorly understood.

14.2.2.2 Sorption

The chemistry of weathered tropical soils affects nutrient mobility in somewhat unique ways. Mobility of nutrient ions in soil is a major factor controlling nutrient cycling, because it affects the rate at which plant roots can extract nutrients and the rate at which nutrients are lost from the system. "Sorption" broadly describes any process that removes a reactant from a solution, thus reducing its mobility, and includes both adsorption and precipitation (Frossard et al. 1995). More specifically, P sorption refers to the removal of P from the soil solution into less reactive, geochemical sinks. In soils, both organic and inorganic forms of P are susceptible to sorption (e.g., Berg and Joern 2006), but the extent is influenced primarily by the concentration, chemistry, and solubility of soil P. Examples of sorption include P adsorption onto secondary clay minerals (e.g., kaolinite) or phosphate bound with iron ($FePO_4$) or aluminum ($AlPO_4$). Phosphorus sorption to soil particles occurs through different types of chemical bonds; for example, P sorption onto kaolinite occurs via a covalent bond between the oxygen of phosphate and the aluminum of the clay, whereas P precipitation into $FePO_4$ results from a bond between a phosphate oxygen and iron. Different sorption bonds vary in their stability and thus also in how readily the sorbed P can be desorbed.

Phosphorus sorption reactions are particularly important in highly weathered soils because strong soil sorption capacities compete with biological sinks for P, effectively reducing P availability (Uehara and Gillman 1980; Sollins et al. 1988). Intense weathering in Ultisols and Oxisols results in the prevalence of 1:1 clays (e.g., kaolinite) and in Al and Fe oxides and hydroxides that effectively sorb P (Sanchez 1976; Harrison 1987; Pant et al. 1994). As a result, sorbed inorganic P concentrations in many tropical soils often exceed those of soil solution P by several orders of magnitude (Sanchez 1976; Fig. 14.5), and fertilization studies have shown that, in highly weathered soils, P may be rapidly and nearly completely sorbed over short timescales (e.g., Sanchez 1976; Uehara and Gillman 1980; Oberson et al. 1997). Thus, the high P sorption capacity of tropical Ultisols and Oxisols can strongly influence the overall productivity of many forest and agricultural ecosystems in the tropics.

Whereas permanent negative charges are typical in the 2:1 clay-dominated soils common throughout the temperate zone, variable-charge clays are more common in highly weathered tropical soils (Sollins et al. 1988). In variable-charge clays, the electric charge results from protonation and deprotonation of surface hydroxyl groups (Gillman 1984; White and Zelazny 1986) and these hydroxyls occur at the edge of the 1:1 clays and Al and Fe hydroxides that dominate highly weathered soils. Because the charge of these clays varies with soil pH, small-scale changes in soil pH can affect the net soil charge and the mobility of both anions and cations (Sollins et al. 1988). Accordingly, the mobility of P in highly weathered soils will also vary with pH fluctuations, and P_i (PO_4^{3-}) may be less mobile in Ultisols and Oxisols than in soils dominated by clays with more permanent charge because permanent-charge clays typically maintain net cation exchange capacity (Sollins et al. 1988). Decreased P mobility has implications for P limitation (see Sect. 14.3) and also for P losses through leaching (see Sect. 14.2.3).

Most previous research on soil P cycling has suggested that sorbed – or "occluded" forms of P – are relatively unavailable for biological uptake. However, a number of important questions remain about both the biological availability of occluded P and the rate and controlling factors of occluded P return to available pools in tropical soils. Traditional definitions of occluded P suggest that it is unavailable and biologically inactive, at least over short timescales (Walker and Syers 1976; Wada 1985). Yet a growing body of evidence now indicates that occluded P may actually enter available pools on relatively short timescales, and movement from geochemical (i.e., sorbed) to available pools could be influenced by biological demand (Tiessen et al. 1984; Olander and Vitousek 2004; Richter et al. 2006). For example, data from a study of an aggrading forest on an Ultisol (Richter et al. 2006) suggested that traditional definitions of "labile," "passive," or "occluded" P may not be appropriate. Instead, Richter et al. (2006) showed that over 28 years of Piedmont forest regrowth on Ultisols, the increase in biological P came at the expense of the P residing in the Fe-, Al- and occluded-P pools, suggesting that plants may have access to these pools over decadal (or perhaps shorter) timescales.

Thus, the extent to which geochemical and biological sinks compete for soil P remains an ongoing area of research (Fig. 14.7b). For example, following P mineralization, P enters the soil solution before it is taken up by biota (Frossard et al. 1995), and in this way a high P sorption capacity could hinder the chance that high phosphatase activity will result in increased P availability. Yet, soil biota can increase the likelihood that mineralized P does not become geochemically immobilized via the secretion of low-molecular-weight-complexing molecules (such as organic acids) or by modifying the pH around them to promote solubilization and desorption of P. Similarly, some organisms have been shown to secrete organic acids to increase the solubility of organic P (Hayes et al. 2000; George et al. 2004), thus facilitating mineralization by increasing the solubility of the P_o substrate.

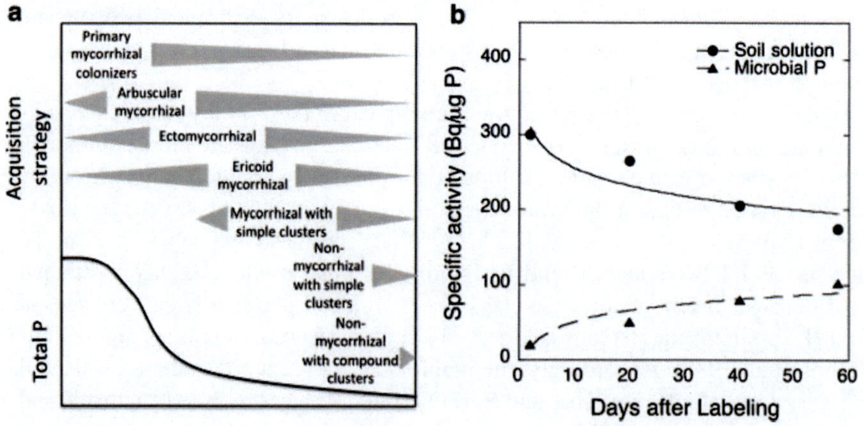

Fig. 14.7 Biological strategies for acquiring P. (**a**) Conceptual model showing changing plant nutrient-acquisition strategies with different levels of soil P. This panel was adapted from Lambers et al. (2008) with the permission of Elsevier. (**b**) The specific activities in the soil solution and in microbial biomass for an Alfisol soil after an initial radioisotope P addition. Data suggest that microbial organisms are able to effectively compete with geochemical sinks and rapidly uptake added radiolabeled P into their biomass. Figure recreated from Oehl et al. (2001) with the permission of Springer

More recent data also suggest that interactions between soil solution P and Fe oxides are more temporally dynamic than previously thought, and are strongly influenced by soil oxidation states (Baldwin and Mitchell 2000; Liptzin and Silver 2009). High biological activity and abundant rainfall in many tropical forests promote episodic anoxia (Silver et al. 1999; Schuur and Matson 2001), and corresponding fluctuations in redox potential can release Fe-bound P (Baldwin and Mitchell 2000; Liptzin and Silver 2009). This P can be subsequently resorbed or taken up by biota, and the sorption–desorption cycle of Fe and P bonding (in part determined by the soil and climate of tropical forests) helps regulate P cycling and availability.

Finally, whereas physical sorption reactions can remove P from the actively cycling pool, aggregate formation in soils dominated by weathered clays can also deplete the available P pool (Oades and Waters 1991; Denef et al. 2002; Six et al. 2002) by stabilizing P_o within aggregates (Merckx et al. 1985; Hassink 1997). Phosphorus may also diffuse from the outer surface of aggregates into the interior (Linquist et al. 1997), where P is physically protected from desorption (and biological uptake). Thus, the mineral soil aggregation found in many highly weathered soils (Oades and Waters 1991; Denef et al. 2002; Six et al. 2002) also contributes to low soil P availability.

14.2.3 Biological Responses to Low P Availability

Some data suggest that P availability limits ecosystem processes in tropical forests growing on highly weathered soil (see Sect. 14.3), yet the rates of many biological processes in this biome are among the highest on Earth (Raich and Schlesinger 1992; Cleveland et al. 1999; Gholz et al. 2000; Grace et al. 2001). Given this apparent contradiction, how do tropical forest organisms maintain high activity in the face of low P availability? A part of the answer lies in the fact that organisms in general – but especially in the tropics – have evolved a number of adaptations that effectively overcome low P availability. These strategies fall into two main categories: those that enhance P conservation and efficiency, and those that enhance P acquisition and uptake (Lajtha and Harrison 1995; Horst et al. 2001; Vance et al. 2003). Key conservation strategies include increased growth per unit P (high P-use efficiency), reallocation of internal P (e.g., foliar P resorption prior to leaf senescence), and modifications in metabolism to those that bypass P-requiring steps [e.g., alternative glycolytic reactions can bypass ATP-requiring steps under P starvation (Theodorou and Plaxton 1996; Schachtman et al. 1998; Raghothama 1999; Uhde-Stone et al. 2004)].

Plants growing on P-poor soils often have high P-use efficiencies, meaning that many tropical plants fix more C per unit P relative to those in other ecosystems (Vitousek 1984; Kitayama et al. 2000, 2004). One mechanism explaining the high P-use efficiency of tropical forest plants is that many show relatively high rates of P resorption from foliage prior to leaf senescence (Kitayama and Aiba 2002; Yuan and Chen 2009). For example, foliar P resorbed from leaves of trees growing on soils with low P availability has been shown to be over 80% (Kitayama and Aiba 2002). Global comparisons suggest that foliar P resorption is highest in the lower latitudes of the tropics, with tropical foliar P resorption values ranging from 35 to 87% of foliar P (averaging 58%) compared with nontropical forests (ranging from 12 to 77% and averaging 47%) (Yuan an Chen 2009). In essence, P resorption provides an effective plant P recycling mechanism, short-circuiting the potential for losses of P during litter decomposition, and reducing plant P demand by conserving P within the plant.

Plants also have mechanisms for maximizing soil P acquisition. Phosphorus is much less mobile in the soil solution than most other major plant nutrients (Barber 1984), and P uptake is often assumed to vary in proportion to the surface area of the plant organs involved in uptake. Accordingly, many plants counter low P availability through symbiotic relationships or a number of morphological adaptations that effectively increase root surface area, including mycorrhizal relationships that increase P transfer to roots (see Jansa et al. 2011) (Fig. 14.7a). Relative to temperate ecosystems, much less is known about the nature and importance of plant symbioses with mycorrhizal fungi, yet the data that do exist suggest that many tropical plants maintain symbiotic relationships with mycorrhizae (Janos 1980; Husband et al. 2002; Aldrich-Wolfe 2007). Similarly, plants often respond to low soil P availability by producing elongated root hairs (Fohse et al. 1991; Ma et al. 2001) or roots with unusual architecture (Al-Ghazi et al. 2003; Lopez-Bucio et al. 2005), both of which have been shown to increase plant access to soil P. For example, some tropical plants make use of lateral roots that scavenge P from the topsoil and litter layers (Herrera et al. 1978; Stark and Jordan 1978; Cuevas and Medina 1986) because these parts of the forest floor contain relatively high P concentrations.

Other plants also produce specialized root structures (e.g., proteoid roots or cluster roots) that allow plants to "mine" insoluble forms of inorganic P from the soil (Fig. 14.7a): cluster roots produce large amounts of carboxylates, which release P from strongly-sorbed forms (Lambers et al. 2008). Cluster-rooted plants have been observed in many tropical forests (Skene 1998), and many of the best-known cluster-rooted species are found in western Australia, an area with some of the most weathered and P-poor soils on Earth. Microbial organisms can also solubilize P from occluded inorganic pools, transforming it into available P. By releasing strong organic acids, certain bacteria can liberate P from inorganic P-bound molecules that are typically thought of as biologically inaccessible, and data suggest that organisms can effectively compete with geochemical sinks for soil P (Fig. 14.7b) (Olander and Vitousek 2004; Bünemann et al. 2004; Richter et al. 2006).

14.2.4 Tropical Soil P: Losses

Understanding controls over P losses from ecosystems is important because losses limit the accumulation of P pools at the ecosystem scale (Hedin et al. 2003). The majority of P losses occur "out the bottom" as dissolved and/or particulate forms that are transported to streams via hydrologic flowpaths. Despite the high capacity for tropical soils to retain P, a combination of factors – hydrologic conditions, parent material age, soil properties, topography, and vegetative dynamics – can drive P loss from tropical forests. Across the tropics, riverine P concentrations tend to be relatively high in places where parent materials are P-rich, where there is rapid geologic uplift and/or erosion, or in relatively unweathered soils (Stallard and Edmond 1983; McDowell et al. 1995; Wilcke et al. 2001).

In tropical forests, water flow generally occurs via vertical infiltration and movement through soil profiles (Wilcke et al. 2001; Biggs et al. 2002). Despite their edaphic and topographic differences, surface water P outputs are dominated by groundwater sources of P in both lowland and montane tropical forests (Lewis et al. 1986; Markewitz et al. 2006). During times of low watershed discharge, P concentrations tend to be too high, which can result from the increased concentration of soluble P and/or increased interaction with the soil matrix (i.e., low rates of macropore flow). Some studies have shown a direct relationship between dissolved silicates and dissolved P (e.g., Hedin et al. 2003), suggesting that P is exported along with other weathered products as water dissolves silicate-based parent material in the soil profile. To date, however, it is unclear what forms of P occupy deep soil P reserves and how they may subsidize plant nutrient demand.

In lowland tropical forests on Oxisols and Ultisols, P concentrations are generally highest in surface soils and decline vertically through the soil profile because water carries mobile P through the hydrologic continuum from leaf litter leachate to stream flow. This probably reflects the strength of geophysical sequestration into refractory or occluded forms vertically through the soil profile, as well as the biological capacity for lowering P loss. Furthermore, compared with montane tropical forests (which are often underlain by less weathered soils and characterized by shorter soil hydrologic residence times), concentrations of dissolved inorganic and organic P in lowland tropical forests on more weathered soils are typically very low.

In some tropical regions, surface soil P pools can serve as export sources when soils permanently or temporarily saturate under rainfall events, and water flow direction shifts to lateral pathways through soil surface horizons. Consequently, the source of P export shifts to upper soil horizons where organically bound P can be horizontally transferred to aquatic systems (Saunders et al. 2006). Also, during periods of high rainfall a brief and rapid flushing of residual P pools can yield a positive relationship between stream discharge and dissolved P traveling via groundwater (Markewitz et al. 2006) and particulate P that probably originates from overland flow (Lewis et al. 1986).

In addition to the physical interactions that control P export, low P losses also reflect efficient biological P recycling in tropical plant–soil systems. Over the course of ecosystem development, Hedin et al. (2003) showed that the amount and stoichiometric composition of P loss changed in concert with ecosystem P status across a 4.1×10^6 year chronosequence. Shifts in ecosystem nutrient status across the Hawai'i chronosequence determine, in part, the concentration of P loss. In forests where N availability limits plant growth, dissolved inorganic P concentrations in soil waters below the active rooting zones and in streams were on average fourfold higher (4–7 μg/L) than P-limited sites. For the older, P-limited forests growing on weathered soils, P concentrations were relatively low, particularly with respect to dissolved N losses.

Yet, among all sites on the chronosequence a positive relationship between silica and inorganic P was uniformly strong, most notably for sites severely limited by P. As mentioned above, this pattern points to a strong connection between hydrologic P losses and weathering among all soil types, even in the face of very low ecosystem P status. Moreover, organic forms of dissolved N and P were found to "leak" from all the sites along the chronosequence, essentially irrespective of ecosystem N or P status (Hedin et al. 2003). Persistent P loss despite high biotic P demand indicates that some P pools are somewhat insensitive to biological retention mechanisms. In pristine temperate systems in Chile, Hedin et al. (1995) found that such losses for the N cycle can constrain ecosystem-scale N accrual, which is a concept that remains largely untested for P in tropical systems.

At the watershed scale, the interactions between biological and hydrological dynamics are not well understood. Generally, hydrologic analyses of P export have focused on how runoff-generating processes (i.e., unsaturated and saturated overland or subsurface flow) control patterns of P loss. However, microbially mediated P transformations certainly play a large role in releasing P stabilized in geophysical and organic forms, yet the biological influence on P mobility at the watershed scale has been largely unexplored. Moreover, tropical land-use change via fire, conversion to pasture, deforestation, agriculture, and urbanization may alter both hydrological and biological controls over P loss from ecosystems.

Few studies have explored each of these disturbance processes, but P cycling and loss patterns are generally consistent with predictions from the successional theory of nutrient cycling and forest regrowth, where P loss and recycling become more conservative in converted landscapes (Vitousek and Reiners 1975; Bormann and Likens 1979). High biological P conservation may also stem from the depletion of available P pools enhanced by hydrologic P loss, or through a combination of high-temperature oxidation or removal of biomass. In old and highly weathered systems where P inputs are low and outputs are small, land conversion reduces P availability through biomass clearing or burning, as well as by driving available P into more recalcitrant pools (Markewitz et al. 2004). For example, in pastures and secondary forests near Paragominas, Brazil, bioavailable (Mehlich-III-extractable) P stocks have been reduced by ~1 kg/ha (despite P fertilization) compared to mature forests, presumably due to irreversible sorption to soil oxides (Markewitz et al. 2004). Also, increases in runoff volume and intensity associated with the removal of trees after land conversion can increase the export of particulate P via overland flow (Williams and Melack 1997; Neill et al. 2001).

14.3 Nutrient Limitation in Tropical Forests

While conceptual models suggesting declining P fertility with soil age have proven useful for understanding pedogenic processes across ecosystems (Fig. 14.1), they also have profound implications for understanding how P may regulate ecosystem processes in tropical forests. Specifically, tropical rain forest systems pose a unique set of problems related to nutrient cycling and limitation. In contrast to temperate systems in which N availability more commonly limits plant growth, decomposition, and organic matter storage, N appears to cycle in relative excess in many tropical ecosystems (e.g., Vitousek 1984; Matson and Vitousek 1990; Neill et al. 1995; Martinelli et al. 1999). The relatively N-rich status of tropical forests may result from the abundance of potentially N_2-fixing leguminous trees in the floras of tropical forests (Cleveland et al. 1999; ter Steege et al. 2006) and perhaps from high rates of free-living N_2 fixation occurring in soils and on leaf litter (Maheswaran and Gunatilleke 1990; Cleveland et al. 1999; Reed et al. 2008). Evidence suggesting relatively rich N economies in tropical forests also includes high rates of N cycling (Keller et al. 1986; Matson et al. 1987), high N trace-gas emissions (Hall and Matson 1999), and enriched foliar and soil ^{15}N isotopic values, owing to increased fractionation rates in N-rich tropical systems and perhaps also high N_2 fixation inputs (Martinelli et al. 1999 and references therein). Thus, in contrast to many temperate forests, current evidence suggests that N may not limit ecosystem processes in many lowland tropical forests on highly weathered soils.

Although direct tests of P limitation in tropical forests have been rare, indirect attempts to assess limitation suggest P constraints to multiple ecosystem processes. For example, Vitousek (1984) investigated patterns of N, P, and calcium (Ca) cycling in litterfall, and found that lowland tropical forests had more N and lower dry litter mass:N ratios than temperate forests. Similarly, Ca return to soil via leaf litter was high in tropical forests. However, Vitousek (1984) found that tropical forests had very low rates of P return and notably high dry litter mass:P ratios relative to temperate forests, suggesting an efficient P cycle characteristic of a P-limited system (Vitousek 1984; Vitousek and Sanford 1986). Data from soil chronosequences (Walker and Syers 1976; Vitousek and Farrington 1997; Wardle et al. 2004) also suggest that P limitation increases with time, as soil P availability declines and foliar N:P ratios and P resorption rates increase over the course of soil development. Thus, soils at the weathered end of the pedogenesis spectrum (like Ultisols and Oxisols) may be sufficiently depleted in P that P limits ecosystem processes.

Although uncommon, direct nutrient manipulations of nutrient fertility in tropical forests on weathered soil also suggest P constraints on ecosystem processes. In Hawai'i, long-term fertilizations with both N and P showed that primary production on old tropical soils was clearly P-limited (Fig. 14.2) (Vitousek and Farrington 1997). Fertilization with P at this older site elicited increases in stand-level wood increment, leaf area index (LAI), and aboveground NPP (Herbert and Fownes 1995; Vitousek and Farrington 1997; Harrington et al. 2001). Similarly, root ingrowth studies following fertilization in a tropical forest site on an Oxisol soil in the northern Amazon Basin suggested that fine root growth was strongly constrained by P (and perhaps Ca) availability (Cuevas and Medina 1988). In contrast, fertilizer additions to a forest on a young Inceptisol soil in the mountains of western Venezuela revealed that litterfall mass responded more positively to N than P additions (Tanner et al. 1992), and similar tree growth responses were observed on young soil types undergoing fertilization with N and P in Hawai'i (Vitousek and Farrington 1997). Thus, although caution must be used when making broad generalizations to describe nutrient limitation in tropical ecosystems as a whole, there is some evidence indicating that low P availability may limit primary production on the highly weathered soils (Ultisols and Oxisols) common throughout the humid tropics.

Data also suggest that microbial processes in tropical rain forests may be limited by soil P availability. For example, a field experiment in a tropical rain forest in Costa Rica showed that soil respiration increased by nearly 40% in soils fertilized with mineral P for 2 years (Fig. 14.8a) (Cleveland and Townsend 2006). These data are noteworthy given that the P-induced increase in soil CO_2 flux was ~60% of the size of the total soil C flux from mid- to high-latitude forests. This increase in soil CO_2 efflux was not related to an increase in litter mass loss rates (Cleveland et al. 2006), but instead data suggested that P fertilization resulted in the heterotrophic mineralization of a larger proportion of the organic C moving from the litter layer into soil (Cleveland and Townsend 2006). Another study at this site suggested that the P content of leaf litter strongly regulated its decomposition rate (Fig. 14.8b) (Wieder et al. 2009), suggesting that both the rate of organic matter release from the litter layer and C efflux to the atmosphere are limited by P availability.

Aspects of the tropical forest N cycle are also affected by P availability. For example, P fertilization has been shown to increase atmospheric N_2 fixation rates in tropical forest epiphytes (Benner et al. 2007), leaf litter, and soil (Fig. 14.8c) (Reed

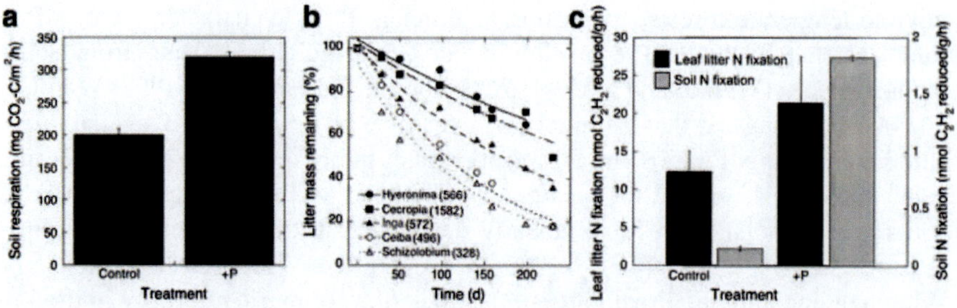

Fig. 14.8 Phosphorus limitation to microbial processes in a Costa Rica tropical rain forest. Data show that (**a**) field soil respiration is increased by P fertilization, (**b**) leaves from species with higher relative P concentrations (e.g., lower C:P ratios) decompose more rapidly on the forest floor, and (**c**) leaf litter and soil free-living N_2 fixation are stimulated by P additions. Leaf C:P ratios in **b** are given in *parentheses* following the plant genus. Data are taken from (**a**) Cleveland and Townsend (2006); (**b**) Wieder et al. (2009); recreated with the permission of Springer; and (**c**) Reed et al. (2007)

et al. 2007). Furthermore, an investigation of symbiotic N_2 fixation across a P gradient in Hawai'i showed that soil P availability and legume nodule N_2 fixation rates were positively related (Pearson and Vitousek 2002). In addition, data suggest that natural, species-specific variation in foliar P concentrations (Townsend et al. 2007) regulate free-living N_2 fixation rates in the canopy as well as on the forest floor, with higher P concentrations correlating with higher free-living N_2 fixation rates (Reed et al. 2008). Phosphorus limitation of free-living N_2 fixation rates is noteworthy given recent research suggesting that tropical forest legumes may not be fixing large amounts of N_2 (e.g., Gehring et al. 2005). Thus, P may directly regulate N inputs into tropical rain forests via limitation to N_2 fixation.

The observation that P limits microbial activity in tropical forests on highly weathered soils (Fig. 14.8) has a number of implications for nutrient cycling, some of which may seem contradictory, and which operate at a variety of timescales. For example, if microbial immobilization of P is high, it could accentuate P constraints on plant growth. However, microbes may ultimately represent a crucial biological valve for P that prevents more permanent losses (see Sect. 14.2.4). In tropical forest ecosystems, the microbial P pool represents an important component of the soil organic P pool (Paul and Clark 1989; Bünemann et al. 2004), and microbial "storage" of P in tropical soils may prevent the movement of soil P into unavailable pools and/or leaching losses. Thus, microbial activity could act to limit more severe P limitation as soils age.

14.4 Global Change and Soil P in the Tropics

Both climate and land-use changes, among others, are affecting tropical forests (e.g., Clark 2004; Asner et al. 2005; Malhi et al. 2009; Friedlingstein et al. 2010), and the P cycle could strongly mediate tropical ecosystem responses to these perturbations (Fig. 14.8) (Vitousek and Farrington 1997; Cleveland and Townsend 2006). For example, multiple studies suggest that intact tropical forests may be acting as significant sinks for atmospheric CO_2 (Stephens et al. 2007; Phillips et al. 2008; Lewis et al. 2009). However, P limitation has the potential to strongly constrain this response both directly (via limitation to NPP) (Vitousek and Farrington 1997) and indirectly (via P regulation over the availability of other nutrients such as N) (Reed et al. 2007). For example, if increased atmospheric CO_2 concentrations stimulate tree growth, ecological stoichiometry suggests that this will drive higher plant nutrient (including P) demand. This could result in increases in organic P in live biomass and decaying organic matter. If such increases come at the expense of lower soil P availability in these already P-poor soils, increasing litter C:P ratios could provide a negative feedback by suppressing decomposition rates (Vitousek 1982). In this way, soil P cycling could strongly regulate forest responses to global change, and these controls will probably be dynamic over multiple timescales. Accordingly, accurate predictions of the future tropical C balance require a more robust incorporation of P cycling, a component of biogeochemical cycling almost wholly absent from ecosystem-scale C cycling models (Parton et al. 2005).

In addition, tropical land use change via fire, forest conversion to pasture, deforestation, agriculture, and urbanization will certainly alter P cycling in tropical ecosystems, and multiple lines of evidence suggest that such perturbations may enhance P losses and exacerbate P limitation. For example, Ultisol soils in Costa Rica showed large losses of P_i and organic matter after conversion from forest to pasture (P_i was reduced from 7.6 ± 0.5 to 4.7 ± 0.6) (Cleveland et al. 2003), and data suggest a postperturbation transition from biologically dominated P cycling to a cycle dominated by geochemical processes (McGrath et al. 2001). What this transition could mean for tropical productivity remains unknown.

Furthermore, many previously deforested tropical landscapes are now recovering from conversion to pasture, and these successional forests occupy an increasingly large proportion of the total tropical forest area. The legacy of land conversion and management has driven profound changes in soil nutrient cycles of these secondary forests, and the trajectory of their recovery will undoubtedly be influenced by P cycling. For example, Davidson et al. (2007) studied two forest chronosequences (using secondary forests that ranged in age from 3 to 70 years and two mature forests) to show that, after agricultural abandonment, secondary forests on highly weathered soils initially showed signs of N limitation. However, over the course of succession, N became relatively more available and P relatively less available, suggesting a transition from N limitation to P limitation during secondary forest succession.

Finally, increased socio-economic demand for agricultural products is driving accelerating rates of deforestation in the tropics and increased expansion of intensive agriculture. For example, soybean production on nutrient-poor soils in Brazil has increased dramatically in the last decade, and requires high P inputs to sustain high crop productivity. Given some of the unique and important roles of P in the biogeochemistry of tropical soils, research on the effects of such activities on P cycling in these emerging ecosystems is crucial (e.g., Sanchez et al. 1982; Oberson et al. 2001) because our understanding of P cycling (and its coupling with other biogeochemical cycles) in active and recovering agricultural Oxisols and Ultisols remains poor. Taken together, data suggest that global change could strongly affect P cycling in tropical landscapes and that these effects could feedback to regulate ecosystem responses to global change.

14.5 Conclusions

Phosphorus availability is crucial to the functioning of all ecosystems, yet could be especially so for tropical forests where P availability appears to limit multiple ecosystem processes. In light of the fact that tropical forests exchange more C with the atmosphere than any other biome and store 25% of global terrestrial C, an enhanced understanding of how P regulates tropical ecosystem function may be disproportionately important for forecasting future climate. Despite significant advances in recent years, we suggest that there is an on-going need for research to be carried out in the following areas to improve our understanding of P cycling in tropical ecosystems:

1. Exploration and quantification of the coupling between the P cycle and the cycles of C and N. In particular, more direct assessments of P limitation to NPP are needed.
2. Understanding of the chemical diversity, turnover, and availability of soil organic P in tropical systems. Organic P represents the largest stocks and losses of P from terrestrial tropical ecosystems, yet we know relatively little about the chemical composition of this important pool.
3. Potential interacting impacts of changing climate, land use, and biogeochemical cycles on P pools and fluxes for tropical rain forests, and the attendant feedbacks to global change.
4. Consideration of how we define and measure "available" or "occluded" P pools. What pools are accessible by what organisms on what timescales?

Acknowledgments Projects supported by the National Science Foundation's Division of Environmental Biology and the Andrew Mellon Foundation contributed to the thoughts and ideas discussed in this chapter. We are grateful to Astrid Oberson and an anonymous reviewer for their insightful suggestions and to Natalie Mahowald and Gregory Okin for providing a high resolution version of Figure 14.3.

References

Achat DL, Bakker MR, Augusto L, Saur E, Sousseron L, Morel C (2009) Evaluation of the phosphorus status of P-deficient podzols in temperate pine stands: combining isotopic dilution and extraction methods. Biogeochemistry 92:183–200

Acosta-Martinez V, Cruz L, Sotomayor-Ramirez D, Perez-Alegria L (2007) Enzyme activities as affected by soil properties and land use in a tropical watershed. Appl Soil Ecol 35:35–45

Aldrich-Wolfe L (2007) Distinct mycorrhizal communities on new and established hosts in a transitional tropical plant community. Ecology 88:559–566

Al-Ghazi Y, Muller B, Pinloche S, Tranbarger T, Nacry P, Rossignol M, Tardieu F, Doumas P (2003) Temporal responses of Arabidopsis root architecture to phosphate starvation: evidence for the involvement of auxin signaling. Plant Cell Environ 26:1053–1066

Allison S, Vitousek P (2005) Responses of extracellular enzymes to simple and complex nutrient inputs. Soil Biol Biochem 37:937–944

Anderson G (1967) Nucleic acids, derivatives and organic phosphates. In: McLaren A, Peterson G (eds) Soil biochemistry, vol 1. Marcel Dekker, New York, NY

Artaxo P, Martins J, Yamasoe M, Procopio A, Pauliquevis T, Andreae M, Guyon P, Gatti L, Leal A (2002) Physical and chemical properties of aerosols in the wet and dry seasons in Rondonia, Amazonia. J Geophys Res 107:D20

Asawalam DO, Johnson S (2007) Physical and chemical characteristics of soils modified by earthworms and termites. Commun Soil Sci Plant Anal 38:513–521

Asner GP, Seastedt TR, Townsend AR (1997) The decoupling of terrestrial carbon and nitrogen cycles. Bioscience 47:226–234

Asner GP, Knapp DE, Broadbent EN, Oliveira PJC, Keller M, Silva JN (2005) Selective logging in the Brazilian Amazon. Science 310:480–482

Baldwin D, Mitchell A (2000) The effects of drying and re-flooding on the sediment and soil nutrient and river-floodplain systems: a synthesis. Regul River 16:457–467

Baldwin D, Howitt J, Beattie J (2003) Abiotic degradation of organic phosphorus compounds in the environment. In: Turner B, Frossard E, Baldwin D (eds) Organic phosphorus in the environment. CABI, Wallingford, UK

Barber S (1984) Soil nutrient bioavailability. Wiley, New York, NY

Benner JW, Conroy S, Lunch CK, Toyoda N, Vitousek PM (2007) Phosphorus fertilization increases the abundance and nitrogenase activity of the cyanolichen *Pseudocyphellaria crocata* in Hawaiian montane forests. Biotropica 39:400–405

Berg AS, Joern BC (2006) Sorption dynamics of organic and inorganic phosphorus compounds in soil. J Environ Qual 35:1855–1862

Bern CR, Townsend AR, Farmer GL (2005) Unexpected dominance of parent-material strontium in a tropical forest on highly weathered soils. Ecology 86:626–632

Biggs TW, Dunne T, Domingues TF, Martinelli LA (2002) Relative influence of natural watershed properties and human disturbance on stream solute concentrations in the southwestern Brazilian Amazon Basin. Water Resour Res 38:1150

Bormann F, Likens G (1979) Catastrophic disturbance and the steady-state in northern hardwood forests. Am Sci 67:660–669

Bünemann E, Smithson PC, Jama B, Frossard E, Oberson A (2004) Maize productivity and nutrient dynamics in maize-fallow rotations in western Kenya. Plant Soil 264:195–208

Caradus J, Snaydon R (1987) Aspects of the phosphorus nutrition of white clover populations. II. Root exocellular acid phosphatase activity. J Plant Nutr 10:287–301

Chadwick OA, Derry LA, Vitousek PM, Huebert BJ, Hedin LO (1999) Changing sources of nutrients during four million years of ecosystem development. Nature 397:491–497

Chen Z, Hsieh C, Jian F, Hsieh T, Sun I (1997) Relations of soil properties to topography and vegetation in a subtropical rain forest in southern Taiwan. Plant Ecol 132:229–241

Chen C, Condon L, Davis M, Sherlock R (2002) Phosphorus dynamics in the rhizosphere of perennial ryegrass (*Lolium perenne* L.) and radiata pine (*Pinus radiata* D. Don.). Soil Biol Biochem 34:487–499

Clark DA (2004) Tropical forests and global warming: slowing it down or speeding it up? Front Ecol Environ 2:73–80

Cleveland CC, Townsend AR (2006) Nutrient additions to a tropical rain forest drive substantial soil carbon dioxide losses to the atmosphere. Proc Natl Acad of Sci USA 103:10316–10321

Cleveland CC, Townsend AR, Schimel DS, Fisher H, Howarth RW, Hedin LO, Perakis SS, Latty EF, Von Fischer JC, Elseroad A, Wasson MF (1999) Global patterns of terrestrial biological nitrogen (N_2) fixation in natural ecosystems. Global Biogeochem Cycles 13: 623–645

Cleveland CC, Townsend AR, Schmidt SK, Constance BC (2003) Soil microbial dynamics and biogeochemistry in tropical forests and pastures, southwestern Costa Rica. Ecol Appl 13:314–326

Cleveland CC, Reed SC, Townsend AR (2006) Nutrient regulation of organic matter decomposition in a tropical rain forest. Ecology 87:492–503

Condron L, Moir J, Tiessen H, Stewart J (1990) Critical evaluation of methods for determining total organic phosphorus in tropical soils. Soil Sci Soc Am J 54:126–1266

Crews TE, Kitayama K, Fownes JH, Riley RH, Herbert DA, Mueller-Dombois D, Vitousek PM (1995) Changes in soil phosphorus fractions an ecosystem dynamics across a long chronosequence in Hawaii. Ecology 76:1407–1424

Cross AF, Schlesinger WH (1995) A literature review and evaluation of the Hedley fractionation: applications to the biogeochemical cycle of soil phosphorus in natural ecosystems. Geoderma 64:197–214

Cuevas E, Medina E (1986) Nutrient dynamics within Amazonian forest ecosystems. I. Nutrient flux in fine litter fall and efficiency of nutrient utilization. Oecologia 68:466–472

Cuevas E, Medina E (1988) Nutrient dynamics with Amazonian forests. Oecologia 76:222–235

Davidson EA, de Carvalho C, Figueira A, Ishida F, Ometto J, Nardoto G, Saba R, Hayashi S, Leal E, Vieira I, Martinelli L (2007) Recuperation of nitrogen cycling in Amazonian forests following agricultural abandonment. Nature 447:995–999

Denef K, Six J, Merckx R, Paustian K (2002) Short-term effects of biological and physical forces on aggregate formation in soils with different clay mineralogy. Plant Soil 246:185–200

Duff S, Sarath G, Plaxton W (1994) The role of acid-phosphatases in plant phosphorus metabolism. Physiol Plant 90:791–800

Elser J, Bracken M, Cleland E, Gruner D, Harpole W, Hillebrand H, Ngai J, Seabloom E, Shurin J, Smith J (2007) Global analysis of nitrogen and phosphorus limitation of primary producers in freshwater, marine and terrestrial ecosystems. Ecol Lett 10:1135–1142

Fohse D, Claassen N, Jungk A (1991) Phosphorus efficiency of plants II. Significance of root radius, root hairs and cation-anion balance for phosphorus influx in seven plant species. Plant Soil 132:261–272

Friedlingstein P, Cadule P, Piao SL, Ciais P, Sitch S (2010) The African contribution to the global climate-carbon cycle feedback of the 21st century. Biogeosciences 7:513–519

Frossard E, Brossard M, Hedley M, Metherell A (1995) Reactions controlling the cycling of P in soils. In: Tiessen H (ed) Phosphorus in the global environment. Wiley, Chichester, UK

Gehring C, Vlek PL, Souza LA, Denich M (2005) Biological nitrogen fixation in secondary regrowth and mature rainforest of central Amazonia. Agric Ecosyst Environ 111:237–252

George T, Richardson A, Hadobas P, Simpson R (2004) Characterization of transgenic *Trifolium subterraneum* L. which expresses phyA and releases extracellular phytase: growth and P nutrition in laboratory media and soil. Plant Cell Environ 27:1351–1361

George TS, Fransson A-M, Hammond JP, White PJ (2011) Phosphorus nutrition: rhizosphere processes, plant response and adaptations. In: Bünemann E, Oberson A, Frossard E (eds) Phosphorus in action: biological processes in soil phosphorus cycling. Soil biology, vol 26. Springer, Heidelberg. doi: 10.1007/978-3-642-15271-9_10

Gholz H, Wedin D, Smitherman S, Harmon M, Parton W (2000) Long-term dynamics of pine and hardwood litter in contrasting environments: toward a global model of decomposition. Glob Change Biol 6:751–765

Gillman G (1984) Using variable charge characteristics to understand the exchangeable cation status of oxic soils. Aust J Agric Res 22:71–80

Grace J, Malhi Y, Higuchi N, Meir P (2001) Productivity of tropical rain forests. In: Roy J, Saugier B, Mooney H (eds) Terrestrial global productivity: past, present and future. Academic, London, pp 401–428

Hall SJ, Matson PA (1999) Nitrogen oxide emissions after nitrogen additions in tropical forests. Nature 400:1997–2000

Haran S, Ogendra S, Seska M, Bratanova M, Raskin I (2000) Characterization of Arabidopsis acid phosphatase promoter and regulation of acid phosphatase expression. Plant Physiol 124:615–626

Harrington RA, Fownes JH, Vitousek PM (2001) Production and resource use efficiencies in N- and P-limited tropical forests: a comparison of responses to long-term fertilzation. Ecosystems 4:646–657

Harrison A (1987) Soil organic phosphorus: a review of world literature. CABI, Wallingford, UK

Hassink J (1997) The capacity of soils to preserve organic C and N by their association with clay and silt particles. Plant Soil 191:77–87

Hayes J, Richardson A, Simpson R (2000) Components of organic phosphorus in soil extracts that are hydrolysed by phytase and acid phosphatase. Biol Fertil Soils 32:279–286

Hedin L, Armesto J, Johnson A (1995) Patterns of nutrient loss from unpolluted, old growth temperate forests-evaluation of biogeochemical theory. Ecology 76:493–509

Hedin LO, Vitousek PM, Matson PA (2003) Nutrient losses over four million years of tropical forest development. Ecology 84:2231–2255

Herbert DA, Fownes JH (1995) Phosphorus limitation of forest leaf area and net primary production on a highly weathered soil. Biogeochemistry 29:223–235

Herrera R, Merida T, Stark N, Jordan C (1978) Direct phosphorus transfer from leaf litter to roots. Naturwissenschaften 65:208–209

Hobbie S, Vitousek PM (2000) Nutrient limitation of decomposition in Hawaiian forests. Ecology 81:1867–1877

Hooper DU, Johnson L (1999) Nitrogen limitation in dryland ecosystems: responses to geographical and temporal variation in precipitation. Biogeochemistry 46:247–293

Horst W, Kamh M, Jibrin J, Chude V (2001) Agronomic measures for increasing P availability to crops. Plant Soil 237:211–223

Houlton BZ, Wang YP, Vitousek PM, Field CB (2008) A unifying framework for dinitrogen fixation in the terrestrial biosphere. Nature 454:327–330

Husband R, Herre E, Turner S, Gallery R, Young J (2002) Molecular diversity of arbuscular mycorrhizal fungi and patterns of host association over time and space in a tropical forest. Mol Ecol 11:2669–2678

Janos DP (1980) Vesicular-arbuscular mycorrhizae affect lowland tropical rain forest plant growth. Ecology 61:151–162

Jansa J, Finlay R, Wallander H, Smith FA, Smith SE (2011) Role of mycorrhizal symbioses in phosphorus cycling. In: Bünemann E, Oberson A, Frossard E (eds) Phosphorus in action: biological processes in soil phosphorus cycling. Soil biology, vol 26. Springer, Heidelberg. doi: 10.1007/978-3-642-15271-9_6

Jenny H (1941) Factors of soil formation. McGraw-Hill, New York, NY

Jobbagy E, Jackson R (2000) The vertical distribution of soil organic carbon and its relation to climate and vegetation. Ecol Appl 10:423–436

Johnson RA, Frizano J, Vann DR (2003) Biogeochemical implications of labile phosphorus in forest soils determined by the Hedley fractionation procedure. Oecologia 135:487–499

Keller M, Kaplan W, Wofsy S (1986) Emissions of N_2O, CH_4, and CO_2 from tropical soils. J Geophys Res 91:11791–11801

Kitayama K, Aiba S (2002) Ecosystem structure and productivity of tropical rain forests along altitudinal gradients with contrasting soil phosphorus pools on Mount Kinabalu, Borneo. J Ecol 90:37–51

Kitayama K, Majalap-Lee N, Aiba S (2000) Soil phosphorus fractionation and phosphorus-use efficiencies of tropical rainforests along altitudinal gradients of Mount Kinabalu, Borneo. Oecologia 123:342–349

Kitayama K, Aiba S, Takyu M, Majalap N, Wagai R (2004) Soil phosphorus fractionation and phosphorus-use efficiency of a Bornean tropical montane rain forest during soil aging with podzolization. Ecosystems 7:259–274

Kroehler C, Linkins A (1988) The root surface phosphatases of *Eriophorum vaginatum* – effects of temperature, pH, substrate concentration and inorganic phosphorus. Plant Soil 105:3–10

Kroehler C, Linkins A (1991) The absorption of inorganic-phosphate from 32P-labeled inositol hexaphosphate by *Eriophorum vaginatum*. Oecologia 85:424–428

Lajtha K, Harrison A (1995) Strategies of phosphorus acquisition and conservation by plant species and communities. In: Tiessen H (ed) Phosphorus in the global environment. Wiley, Chichester, UK

Lambers H, Raven JA, Shaver GR, Smith SE (2008) Plant nutrient-acquisition strategies change with soil age. Trends Ecol Evol 23:95–103

LeBauer DS, Treseder KK (2008) Nitrogen limitation of net primary productivity in terrestrial ecosystems is globally distributed. Ecology 89:371–379

Lewis WM Jr, Saunders JF III, Levine SN, Weibezahn FH (1986) Organic carbon in the Caura River, Venezuela. Limnol Oceanogr 31:653–656

Lewis SL, Lopez-Gonzalez G, Sonke B, Affum-Baffoe K, Baker TR, Ojo LO, Phillips OL, Reitsma JM, White L, Comiskey JA, Djuikouo M-N, Ewango CEN, Feldpausch TR, Hamilton AC, Gloor M, Hart T, Hladik A, Lloyd J, Lovett JC, Makana J-R, Malhi Y, Mbago FM, Ndangalasi HJ, Peacock J, Peh KS-H, Sheil D, Sunderland T, Swaine MD, Taplin J, Taylor D, Thomas SC, Votere R, Wöll H (2009) Increasing carbon storage in intact African tropical forests. Nature 457:1003–1006

Likens G, Bormann F (1995) Biogeochemistry of a forested ecosystem. Springer, New York, NY

Linquist B, Singleton P, Yost R, Cassman K (1997) Aggregate size effects on the sorption and release of phosphorus in an ultisol. Soil Sci Soc Am J 61:160–166

Liptzin D, Silver WL (2009) Effects of carbon additions on iron reduction and phosphorus availability in a humid tropical forest soil. Soil Biol Biochem 41:1696–1702

Lopez-Bucio J, Hernandez-Abreu E, Sanchez-Calderon L, Perez-Torres A, Rampey R, Bartel B, Herrera-Estrella L (2005) An auxin transport independent pathway is involved in phosphate stress-induced root architectural alterations in Arabidopsis. Identification of BIG as a mediator of auxin in pericycle cell activation. Environ Stress Adapt 137:681–691

Ma Z, Bielenberg G, Brown K, Lynch J (2001) Regulation of root hair density by phosphorus availability in *Arabidopsis thaliana*. Plant Cell Environ 24:459–467

Maheswaran J, Gunatilleke I (1990) Nitrogenase activity in soil and litter of a tropical lowland rain forest and an adjacent fernland in Sri Lanka. J Trop Ecol 6:281–289

Mahowald NM, Artaxo P, Baker AR, Jickells TD, Okin GS, Randerson JT, Townsend AR (2005) Impacts of biomass burning emissions and land use change on Amazonian atmospheric phosphorus cycling and deposition. Global Biogeochem Cycles 19:GB4030

Malhi Y, Aragao LEOC, Galbraith D, Huntingford C, Fisher R, Zelazowski P, Sitch S, McSweeney C, Meir P (2009) Exploring the likelihood of a climate-change-induced dieback of the Amazon rainforest. Proc Natl Acad Sci USA 106:20610–20615

Markewitz D, Davidson E, Moutinho P, Nepstad D (2004) Nutrient loss and redistribution after forest clearing on a highly weathered soil in Amazonia. Ecol Appl 14:S177–S199

Markewitz D, Resende JCF, Parron L, Bustamante M, Klink CA, RdO F, Davidson EA (2006) Dissolved rainfall inputs and streamwater outputs in an undisturbed watershed on highly weathered soils in the Brazilian cerrado. Hydrol Process 20:2615–2639

Martinelli L, Piccolo M, Townsend A, Vitousek P, Cuevas E, McDowell W, Robertson G, Santos O, Treseder K (1999) Nitrogen stable isotopic composition of leaves and soil: Tropical versus temperate forests. Biogeochemistry 46:45–65

Matson PA, Vitousek PM (1990) Ecosystem approach to a global nitrous oxide budget. Bioscience 40:667–672

Matson P, Vitousek P, Ewel J, Mazzarino M, Robertson G (1987) Nitrogen transformation following tropical forest felling and burning on a volcanic soil. Ecology 68:491–502

McDowell WH, Lugo AE, James A (1995) Export of nutrients and major ions from Caribbean catchments. J N Am Benthol Soc 14:12–20

McGill WB, Cole CV (1981) Comparative aspects of cycling of organic C, N, S, and P through soil organic matter. Geoderma 26:267–286

McGrath DA, Smith CK, Gholz HL, Oliveira FD (2001) Effects of land-use change on soil nutrient dynamics in Amazonia. Ecosystems 4:625–645

McGroddy M, Daufresne T, Hedin L (2004) Scaling of C:N:P stoichiometry in forests worldwide: implications of terrestrial Redfield-type ratios. Ecology 85:2390–2401

McKane R, Johnson L, Shaver G, Nadelhoffer K, Rastetter E, Fry B, Giblin A, Kielland K, Kwiatkowski B, Laundre J, Murray G, Turner B, Condron L, Richardson S, Peltzer D, Allison V (2002) Resource-based niches provide a basis for plant species diversity and dominance in arctic tundra. Nature 415:68–71

McLachlan K (1980) Acid phosphatase activity of intact roots and phosphorus nutrition in plants.1. Assay conditions and phosphatase activity. Aust J Agric Res 3:429–440

Merckx R, Den Hartog A, van Veen J (1985) Turnover of root-derived material and related microbial biomass formation in soils of different texture. Soil Biol Biochem 17:565–569

Miller S, Liu J, Allan D, Menzhuber C, Fedorova M, Vance C (2001) Molecular control of acid phosphatase secretion into the rhizosphere of proteoid roots from phosphorus-stressed white lupin. Plant Physiol 127:594–606

Moorhead D, Kroehler C, Linkins A, Reynolds J (1993) Extracellular acid phosphatase activities in *Eriophorum vaginatum* tussocks – a modeling synthesis. Arctic Alpine Res 25:50–55

Moulin C, Chiapello I (2006) Impact of human-induced desertification on the intensification of Sahel dust emission and export over the last decades. Geophys Res Lett 33:L18808

Nannipieri P, Giagnoni L, Landi L, Renella G (2011) Role of phosphatase enzymes in soil. In: Bünemann E, Oberson A, Frossard E (eds) Phosphorus in action: biological processes in soil phosphorus cycling. Soil biology, vol 26. Springer, Heidelberg. doi: 10.1007/978-3-642-15271-9_9

Neff JC, Reynolds R, Fernandez D, Lamothe P (2006) Controls of bedrock geochemistry on soil and plant nutrients in southeastern Utah. Ecosystems 9:879–893

Neill C, Piccolo MC, Steudler PA, Cerrf CC (1995) Nitrogen dynamics in soils of forests and active pastures in the western Brazilian Amazon basin. Soil Biol Biochem 27:1167–1175

Neill C, Deegan LA, Thomas SM, Cerri CC (2001) Deforestation for pasture alters nitrogen and phosphorus in small Amazonian streams. Ecol Appl 11:1817–1828

Newberry D, Alexander L, Rother J (1997) Phosphorus dynamics in a lowland African rain forest: the influence of ectomycorrhizal trees. Ecol Monogr 67:367–409

Oades J, Waters A (1991) Aggregate hierarchy in soils. Aust J Agric Res 29:815–828

Oberson A, Friesen DK, Morel C, Tiessen H (1997) Determination of phosphorus released by chloroform fumigation from microbial biomass in high P sorbing tropical soils. Soil Biol Biochem 29:1579–1583

Oberson A, Friesen DK, Rao IM, Bühler S, Frossard E (2001) Phosphorus transformations in an Oxisol under contrasting land-use systems: the role of soil microbial biomass. Plant Soil 237:197–210

Oehl F, Oberson A, Sinaj S, Frossard E (2001) Organic phosphorus mineralization studies using isotopic dilution techniques. Soil Sci Soc Am J 65:780–787

Okin GS, Mahowald N, Chadwick OA, Artaxo P (2004) Impact of desert dust on the biogeochemistry of phosphorus in terrestrial ecosystems. Glob Biogeochem Cycles 18:GB2005

Olander L, Vitousek P (2000) Regulation of soil phosphatase and chitinase activity by N and P availability. Biogeochemistry 49:175–190

Olander L, Vitousek PM (2004) Biological and geochemical sinks for phosphorus in soil from a wet tropical forest. Ecosystems 7:404–419

Palm C, Sanchez P, Ahamed S, Awiti A (2007) Soils: a contemporary perspective. Annu Rev Environ Resour 32:99–129

Pant H, Edwards A, Vaughan D (1994) Extraction, molecular fractionation and enzyme degradation of organically associated phosphorus in soil solutions. Biol Fertil Soils 17:196–200

Paoli GD, Curran LM, Slik JWF (2008) Soil nutrients affect spatial patterns of aboveground biomass an emergent tree density in southwestern Borneo. Oecologia 155:287–299

Parton W, Neff J, Vitousek P (2005) Modeling phosphorus, carbon, and nitrogen dynamics in terrestrial ecosystems. In: Turner B, Frossard E, Baldwin D (eds) Organic phosphorus in the environment. CABI, Wallingford, UK

Paul E, Clark F (1989) Soil microbiology and biochemistry. Academic, San Diego, CA

Pearson HL, Vitousek PM (2002) Soil phosphorus fractions and symbiotic nitrogen fixation across a substrate-age gradient in Hawaii. Ecosystems 5:587–596

Phillips OL, Malhi Y, Higuchi N, Laurance WF, Nu PV, Va RM, Laurance SG, Ferreira LV, Stern M, Brown S, Grace J (1998) Changes in the carbon balance of tropical forests: evidence from long-term plots. Science 282:439–442

Phillips OL, Lewis SL, Baker TR, Chao K-J, Higuchi N (2008) The changing Amazon forest. Philos Trans R Soc B 363:1819–1827

Raghothama K (1999) Phosphate acquisition. Annu Rev Plant Physiol Plant Mol Biol 50:665–693

Raich J, Schlesinger W (1992) The global carbon dioxide flux in soil respiration and its relationship to vegetation and climate. Tellus 44B:81–99

Reed SC, Cleveland CC, Townsend AR (2007) Controls over leaf litter and soil nitrogen fixation in two lowland tropical rain forests. Biotropica 39:585–592

Reed SC, Cleveland CC, Townsend AR (2008) Tree species control rates of free-living nitrogen fixation in a tropical rain forest. Ecology 89:2924–2934

Reich PB, Oleksyn J (2004) Global patterns of plant leaf N and P in relation to temperature and latitude. Proc Natl Acad Sci USA 101:11001–11006

Richter DD, Allen L, Li J, Markewitz D, Raikes J (2006) Bioavailability of slowly cycling soil phosphorus: major restructuring of soil P fractions over four decades in an aggrading forest. Oecologia 150:259–271

Sanchez P (1976) Properties and management of soils in the tropics. Wiley, New York, NY

Sanchez PA, Bandy DE, Villachica JH, Nicholaides JJ (1982) Amazon basin soils: management for continuous crop production. Science 216:821–827

Saunders T, McClain M, Llerena C (2006) The biogeochemistry of dissolved nitrogen, phosphorus, and organic carbon along terrestrial-aquatic flowpaths of a montane headwater catchment in the Peruvian Amazon. Hydrol Process 20:2549–2562

Schachtman D, Reid R, Ayling S (1998) Phosphorus uptake by plants: from soil to cell. Plant Physiol 116:447–453

Schlesinger W (1997) Biogeochemistry: an analysis of global change. Academic, San Diego, CA

Schuur EAG, Matson PA (2001) Net primary productivity an nutrient cycling across a mesic to wet precipitation gradient in Hawaiian montane forest. Oecologia 128:431–442

Silver W, Scatena F, Johnson A, Siccama T, Sanchez M (1994) Nutrient availability in a montane wet tropical forest: Spatial patterns and methodological considerations. Plant Soil 164:129–145

Silver WL, Lugo AE, Keller M (1999) Soil oxygen availability and biogeochemistry along rainfall and topographic gradients in upland wet tropical forest soils. Biogeochemistry 44:301–328

Six J, Conant RT, Paul EA, Paustian K (2002) Stabilization mechanisms of soil organic matter: Implications for C-saturation of soils. Plant Soil 241:155–176

Skene KR (1998) Cluster roots: some ecological considerations. J Ecol 86:1060–1064

Soil Survey Staff (2006) Keys to soil taxonomy. Natural Resources Conservation Service, US Department of Agriculture, Washington, DC

Sollins P, Robertson GP, Uehara G (1988) Nutrient mobility in variable- and permanent-charge soils. Biogeochemistry 6:181–199

Stallard RF, Edmond JM (1983) Geochemistry of the Amazon. 2. The influence of geology and weathering environment on the dissolved load. J Geophys Res 88(C14):9671–9688

Stark NM, Jordan CF (1978) Nutrient retention by the root mat of an Amazonian rain forest. Ecology 59:434–437

Stephens BB, Gurney KR, Tans PP, Sweeney C, Peters W, Bruhwiler L, Ciais P, Ramonet M, Bousquet P, Nakazawa T, Aoki S, Michida T, Inoue G, Vinnichenko N, Lloyd J, Jordan A, Heimann M, Shibistova O, Langenfelds R, Steele LP, Francey RJ, Denning AS (2007) Weak northern and strong tropical land carbon uptake from vertical profiles of atmospheric CO_2. Science 316:1732–1735

Swap R, Garstang M, Greco S, Talbot R, Kallberg P (1992) Saharan dust in the Amazon basin. Tellus B Chem Phys Meteorol 44:133–1149

Swift M, Heal O, Anderson J (1979) Decomposition in terrestrial ecosystems. Studies in ecology, vol 5. University of California Press, Berkeley, CA

Tanner E, Kapos V, Franco W (1992) Nitrogen and phosphorus fertilization effects on Venezuelan montane forest trunk growth and litterfall. Ecology 73:78–86

Tarafdar J, Jungk A (1987) Phosphatase activity in the rhizosphere and its relation to the depletion of soil organic phosphorus. Biol Fertil Soils 3:199–204

Tarnocai C, Canadell J, Schuur E, Kuhry P, Mazhitova G, Zimov S (2009) Soil organic carbon pools in the northern circumpolar permafrost region. Glob Biogeochem Cycles 23:GB2023 ter Steege H, Pitman NC, Phillips OL, Chave J, Sabatier D, Duque A, Spichiger R, Va R (2006) Continental-scale patterns of canopy tree composition and function across Amazonia. Nature 443:444–447

Theodorou ME, Plaxton WC (1996) Metabolic adaptations of plant respiration to nutritional phosphate deprivation. Plant Physiol 101:339–344

Thornton P, Doney S, Lindsay K, Moore J, Mahowald N, Randerson J, Fung I, Lamarque J, Feddema J, Lee Y (2009) Carbon-nitrogen interactions regulate climate-carbon feedbacks: results from an atmosphere-ocean general circulation model. Biogeosciences 6:2099–2120

Tiessen H, Moir JO (1993) Characterization of available P by sequential extraction. In: Carter M (ed) Soil sampling and methods of analysis. Lewis, Ann Arbor, MI

Tiessen H, Stewart J, Cole C (1984) Pathways of phosphorus transformations in soils of differing pedogenesis. Soil Sci Soc Am J 48:853–858

Townsend AR, Cleveland CC, Asner GP, Bustamante MM (2007) Controls over foliar N:P ratios in tropical rain forests. Ecology 88:107–118

Townsend AR, Asner GP, Cleveland CC (2008) The biogeochemical heterogeneity of tropical forests. Trends Ecol Evol 23:424–431

Toy ADF (1973) The chemistry of phosphorus. Pergamon, Oxford

Treseder K, Vitousek P (2001) Effects of soil nutrient availability on investment in acquisition of N and P in Hawaiian rain forests. Ecology 82:946–954

Turner BL (2008) Resource partitioning for soil phosphorus: a hypothesis. J Ecol 96:698–702

Uehara G, Gillman G (1980) Charge characteristics of soils with variable and permanent charge minerals. 1. Theory. Soil Sci Soc Am J 44:250–252

Uhde-Stone C, Gilbert G, Johnson J, Litjens R, Zinn K, Temple S, Vance C, Allan D (2004) Acclimation of white lupin to phosphorus deficiency involves enhanced expression of genes related to organic acid metabolism. Plant Soil 248:99–116

Vance C, Uhde-Stone C, Allan D (2003) Phosphorus acquisition and use: critical adaptations by plants for securing a nonrenewable resource. New Phytol 157:423–447

Vitousek PM (1982) Nutrient cycling and nutrient use efficiency. Am Nat 119:553–572

Vitousek PM (1984) Litterfall, nutrient cycling and nutrient limitation in tropical forests. Ecology 65:285–298

Vitousek PM (2004) Nutrient cycling and limitation: Hawai'i as a model system. Princeton University Press, Princeton, NJ

Vitousek PM, Farrington H (1997) Nutrient limitation and soil development: Experimental test of a biogeochemical theory. Biogeochemistry 37:63–75

Vitousek PM, Howarth RW (1991) Nitrogen limitation on land and in the sea: How can it occur? Biogeochemistry 13:87–115

Vitousek PM, Reiners W (1975) Ecosystem succession and nutrient retention hypothesis. Bioscience 25:376–381

Vitousek PM, Sanford RL (1986) Nutrient cycling in moist tropical forest. Annu Rev Ecol Syst 17:137–167

Vitousek P, Chadwick O, Matson P, Allison S, Derry L, Kettley L, Luers A, Mecking E, Monastra V, Porder S (2003) Erosion and the rejuvenation of weathering-derived nutrient supply in an old tropical landscape. Ecosystems 6:762–772

Wada K (1985) The distinctive properties of andisols. Adv Soil Sci 2:174–229

Walker T, Syers J (1976) The fate of phosphorus during pedogenesis. Geoderma 15:1–19

Wardle D, Walker L, Bardgett R (2004) Ecosystem properties and forest decline in contrasting long-term chronosequences. Science 305:509–513

Wasaki J, Maruyama H (2011) Molecular approaches to the study of biological phosphorus cycling. In: Bünemann E, Oberson A, Frossard E (eds) Phosphorus in action: biological processes in soil phosphorus cycling. Soil biology, vol 26. Springer, Heidelberg. doi: 10.1007/978-3-642-15271-9_4

Wasaki J, Omura M, Ando M, Dateki H, Shinano T, Osaka M, Ito H, Matsui H, Tadano T (2000) Molecular cloning and root specific expression of secretory acid phosphatases form phosphate-deficient lupin (*Lipinus albus* L.). Soil Sci Plant Nutr 46:427–437

White G, Zelazny L (1986) Charge properties of soil colloids. In: Sparks D (ed) Soil physical chemistry. CRC, Boca Raton, FL, pp 39–81

Whitton B, Al-Shehri A, Ellwood N, Turner B (2005) Ecological aspects of phosphatase activity in cyanobacteria, eukaryotic algae and bryophytes. In: Turner BL, Frossard E, Baldwin D (eds) Organic phosphorus in the environment. CABI, Wallingford, UK, pp 205–241

Wieder W, Cleveland CC, Townsend AR (2009) Controls over leaf litter decomposition in wet tropical forests. Ecology 90:3333–3341

Wilcke W, Yasin S, Valarezo C, Zech W (2001) Change in water quality during the passage through a tropical montane rain forest in Ecuador. Biogeochemistry 55:45–72

Williams MR, Melack JM (1997) Solute export from forested and partially deforested catchments in the central Amazon. Biogeochemistry 38:67–102

Xuluc-Talosa F, Vester H, Ramirez-Marcial N, Castellanos-Albores J, Lawrence D (2003) Leaf litter decomposition of tree species in three successional phases of tropical dry secondary forest in Campeche, Mexico. For Ecol Manag 174:401–412

Yuan ZY, Chen HY (2009) Global-scale patterns of nutrient resorption associated with latitude, temperature and precipitation. Glob Ecol Biogeogr 18:11–18

Zou X, Binkley D, Doxtader K (1992) A new method for estimating gross phosphorus mineralization and immobilization rates in soils. Plant Soil 147:243–250

Zou X, Binkley D, Caldwell B (1995) Effects of dinitrogen-fixing trees on phosphorus biogeochemical cycling in contrasting forests. Soil Sci Soc Am J 59:1452–1458

Chapter 15
Biological Phosphorus Cycling in Dryland Regions

Jayne Belnap

15.1 Introduction

Drylands consist of arid, semiarid, and subhumid grasslands and shrublands (Fig. 15.1). These landscapes comprise over 40% of terrestrial lands, with over 1 billion people depending on them for their livelihoods. Thus, understanding and maintaining the fertility of these ecosystems is essential to human well-being. Water and nitrogen (N) have long been thought to be the major factors limiting primary productivity in dryland ecosystems (Hooper and Johnson 1999). However, studies are now showing that phosphorus (P) can be at least equally, or even more, limiting than N in many dryland ecosystems, especially in high-pH and calcareous soils (e.g., Ma et al. 2007; James et al. 2005; Lajtha and Schlesinger 1986). The biological cycling of P in drylands has many differences from cycling in more mesic ecosystems. In this chapter, I will discuss the major sources of P in drylands, how this P is redistributed, the abiotic and biotic controls on P availability (measured as resin or bicarbonate-extractable P), how available P affects native plant distribution and exotic plant invasion, and how climate change is expected to influence P availability in dryland soils.

15.2 Inputs and Losses of P in Drylands

In dryland regions, P inputs into soils are primarily the result of the deposition of atmospheric dust, and secondarily are a result of the new weathering of parent materials (Fig. 15.2). Dust deposition into these regions is highly variable in space and time. Dust inputs to the Colorado Plateau and Mojave Deserts, USA, for the past 25 years show annual inputs of 20–40 g m^2 per year. Inputs in other dryland

J. Belnap
US Geological Survey, Canyonlands Research Station, 2290 S. West Resource Boulevard, Moab, UT 84532, USA
e-mail: jayne_belnap@usgs.gov

E.K. Bünemann et al. (eds.), *Phosphorus in Action*, Soil Biology 26,
DOI 10.1007/978-3-642-15271-9_15, © Springer-Verlag Berlin Heidelberg 2011

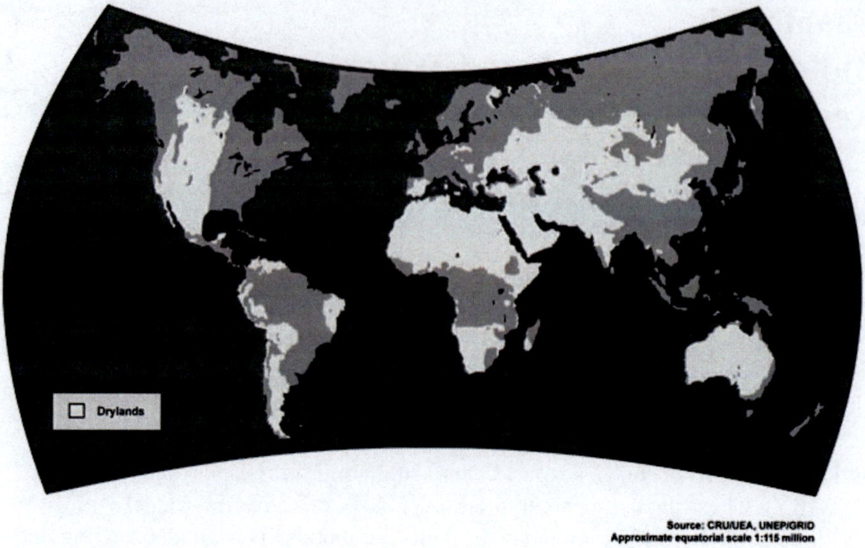

Fig. 15.1 Drylands of the world. The *light-colored* regions indicate drylands, which include the hyperarid, arid, semiarid and subhumid grasslands and shrublands of the world

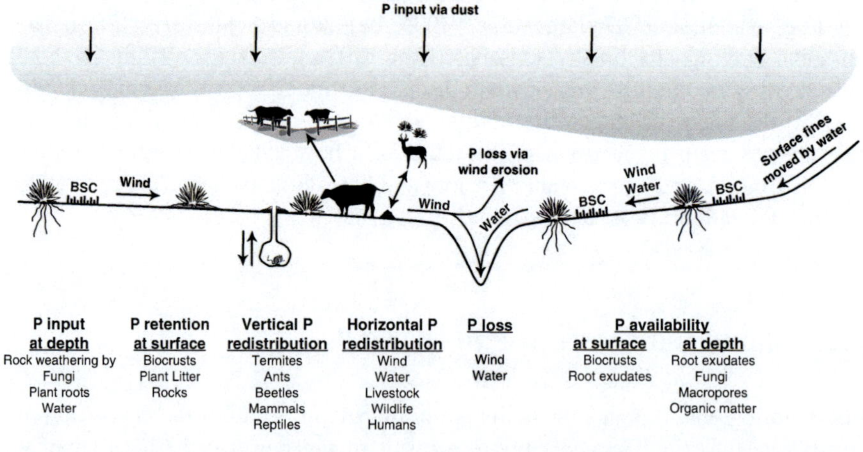

Fig. 15.2 Important processes that control the distribution and availability of P in dryland regions. These include the input and losses of P, vertical and horizontal redistribution of P-containing materials, and the factors that control P bioavailability (BSC = biological soil crust)

regions range from <1 to 276 g m^2 per year, with most values falling between 6 and 20 g m^2 per year (for a review, see Reheis and Kihl 1995). As most of this dust consists of fine-textured materials, it can contain a significant amount of P, ranging from 0.01% to 5% in the western USA (Reynolds et al. 2006a, b, c; Reheis et al. 1999).

Dust also collects on plant leaves, and when washed off it can significantly increase P in soils under plant canopies, with one study showing 24% of soil P coming from canopy dust capture (Callaway 2007). On a global scale, dust deposition is far higher in the semiarid steppes of Africa and Eurasia than North America (Okin et al. 2004; Field et al. 2009).

Deposited dust is retained at sites where soil surfaces are protected by high plant cover or where soils are covered by rocks, physical crusts (formed by silt, clay, and/or salts), or biological soil crusts. High cover of small gravel (2.5 cm) maximizes dust capture (Li and Liu 2003). Physical crusts that are generally hard and flat tend to not capture and retain dust unless wet. Biocrusts are also important in capturing dust. Biocrusts are communities of cyanobacteria, mosses, and lichens that cover the top 1 cm of most dryland soil surfaces. In SE Utah, USA, biocrusted soil surfaces had 2.6 times greater total P than soils 2–5 cm below them (271 vs. 104 ppm P, respectively) (Reynolds et al. 2001). Biocrusts have been shown to effectively capture dust in a variety of desert environments (Verrecchia et al. 1995; also reviewed in Belnap 2003a) because the organisms are sticky. Depending on the environment, they can create soil pinnacles up to 15 cm high (Fig. 15.3) and often have moss and lichen tissue projecting upwards from the soil surface (Fig. 15.3). In California, USA, lichens and mosses were observed to trap 0.15 kg sediment ha^{-1} per year (Nash 1996).

Rock weathering occurs at the surface because differential insolation cracks. In cooler deserts, rocks at the surface experience freeze–thaw action as well. Wind abrasion is also high in this environment. The erosion rate of rock by water is low in drylands because overland water flows are infrequent and highly localized. Fungi, cyanobacteria, and lichens contribute to weathering of surface rock, and fungi to weathering at depth. These organisms tunnel along crystal planes, cleavages, cracks, and grain boundaries in rocks, especially in sandstone, dolomite, calcite, and some granites. They crack rock with osmotically generated turgor pressure (Gadd et al. 2007). They also form carbonic acid during respiratory activity and excrete metal-complexing metabolites and organic acids (e.g., oxalate, citrate,

Fig. 15.3 *Left*: well-developed biological soil crusts. Note the many crevices that trap P-containing materials, including organic matter and dust. *Right*: a thin section of moss, showing dust trapped within the moss shoots. (Photo courtesy of D. Eldridge)

malate, succinate, gluconate), all of which dissolve rock material while also increasing bioavailable P (Jones and Oburger 2011).

15.3 Distribution and Redistribution of P in Drylands

Because most P inputs occur at the soil surface via dust, weathering, and the deposition of plant litter, P is concentrated in surface soils in most environments (Jobbágy and Jackson 2001). This is especially pronounced in drylands because there is insufficient rainfall for subsequent downward leaching of P into soils (Charley and Cowling 1968; Jobbágy and Jackson 2001; Charley 1977). For instance, P concentration was 900–1,100 ppm at 3 cm, decreasing to 500 ppm by 29 cm depth in northern Utah, USA (Jurinak and Griffin 1972). However, one study in the arid regions of southwestern USA found equally high levels of P at 2–3 m depth as found at the surface at one of their three sites (McCulley et al. 2004).

Despite the input of P via dust occurring fairly evenly over a given area, redistribution of soil P by wind, water, and animals is common in dryland regions. The large spaces between plants allow for soil movement by wind and water. The consumption of forage in one place & movement to free water can result in moving P over large distances. This redistribution of P can determine the composition, biomass, and distribution of a given plant community, which in turn influences animal distributions. In an amplifying feedback cycle, dryland plants heavily influence the spatial distribution and the availability of P through capture of surface materials and the exudation of compounds that solubilize otherwise biounavailable P.

15.3.1 Horizontal Redistribution of P and Materials Affecting P Availability

15.3.1.1 Wind

Dryland regions are dominated by a low cover of short plants, and are generally characterized by constant and often high winds. Left undisturbed, most desert soils are well-stabilized by the presence of rocks, physical crusts, and/or biocrusts, which confer substantial resistance to movement of surficial materials (Belnap et al. 2006). However, when plant cover is low (whether naturally or postdisturbance) and these protective covers are disturbed by grazing, plowing, off-road vehicles, or other activities that disturb the soil surface, the soils are easily mobilized and high amounts of P can be lost or redistributed (Fig. 15.4) (Neff et al. 2005; Hiernaux et al. 1999).

Disturbed dryland soil surfaces can produce up to 15 times or more sediment compared to undisturbed soils (Field et al. 2009). Larger particles are moved to the

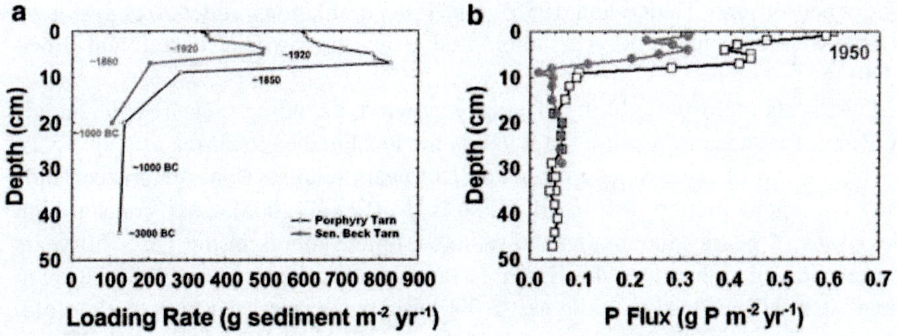

Fig. 15.4 (**a**) Total sediment loading increased in high elevation lakes after about 1850, coinciding with the intensification of agricultural activities (e.g., plowing, livestock grazing) in drylands of the western USA. The Taylor Grazing Act, enacted in 1920, controlled livestock numbers and resulted in a decline in dust deposition. These data are from two lake cores, Porphyry Tarn and Senator Beck Tarn. (**b**) Levels of P showed a similar sudden surge in sediment loading with the arrival of intense agriculture, showing how land disturbance has accelerated the vertical transport of P. However, unlike sediment, P loading has continued to increase over time (after Neff et al. 2008)

nearest obstruction (e.g., rocks, plants, biocrust mounds, base of hills, or depressions, or are lost from the system if plant materials accumulate under plants). This can lead to the often observed "islands of fertility" under shrubs (Schlesinger et al. 1990). However, in a review of 49 studies comparing soil chemistry of 127 plant/interspace pairs, soil P (some studies measured total P, some measured available P) was more often elevated in the interspace soil than under the plant canopy, whereas carbon and N were elevated under the plant canopy in 82 and 68% of the comparisons, respectively (Belnap unpublished results). In the studies where the increase in resources did occur under plants, many sites had highly disturbed interspaces. Under these circumstances, the increase in available P under the plant was probably correlated not only with the collection of P-containing plant and soil materials under the plant, but also with the increase in organic material because organic matter can increase P availability through four main mechanisms (1) increased soil C increases the abundance and activity of soil biota, which then decompose plant litter more rapidly; (2) greater organic matter increases water infiltration and soil moisture retention, thus allowing for longer times of microbial and plant activity (Santos et al. 1978; Callaway 2007); (3) competition of organic anions for adsorption sites on calcite increases available P (Holford et al. 1990); and (4) organic matter can complex with Al and Fe, thus increasing bioavailable P (von Wandruszka 2006; Ma et al. 2009).

As organic matter content is often <0.3% in bulk dryland soils, its importance in keeping P bioavailable may be low (Lajtha and Schlesinger 1988), except in the localized patches under shrubs (Ma et al. 2009). When P does accumulate under plants, it can persist long after the plant is removed. Soils under shrubs in the southwestern USA that had been dead for 13 years still showed elevated P

(Klemmedson and Tiedemann 1986), and P accumulations under *Acacia papyrocarpa* were unchanged 50 years after shrub death in Australia (Facelli and Brock 2000).

Dust is also transported long distances by wind, including to nearby mountains or even across oceans, resulting in P loss at the local to the global scales (Fig. 15.2). A comparison of landscapes grazed for 150 years relative to never grazed lands shows a large depletion of P in grazed soils (Neff et al. 2005). Lake cores at high elevations of nearby mountains show that sediment inputs in the past 150 years (corresponding to the time when intense grazing and agricultural activities began in the western USA) relative to the past 3,000 years have increased five- to eightfold, with a concomitant four to five times increase in P levels (Fig. 15.4) (Neff et al. 2008), thus partially accounting for P lost from lower elevation soils.

15.3.1.2 Water

Over long time periods, water redistributes smaller soil surficial materials from the tops of slopes to slope bases, resulting in higher concentrations of silt and clay, organic matter, and nutrients (including P) at lower landscape positions (Fig. 15.2) (e.g., Reynolds et al. 2006a; Bestlemeyer et al. 2006; Venter et al. 2003). This facilitates the growth of nutrient-demanding plants, especially annuals, and thus can determine plant community structure. Overland flow can also redistribute large amounts of surficial material in a short time period with intense rain events. Over days to millennia this materal is lost from the local ecosystems. Much of this material eventually ends up in large rivers, which represent the main source of P for the oceans (Baturin 2003).

15.3.1.3 Wildlife, Livestock, and Human Settlements

The activities of wildlife, livestock, and human settlements can also result in a substantial horizontal redistribution of P (Fig. 15.2). Wildlife and livestock often consume vegetation in one location whereas excretory processes may occur kilometers distant. Feces contain P, increase soil organic matter and water infiltration/retention, thus increasing total soil and bio available P. As P is highly conserved in soils and vegetation (Charley 1977; Woodmansee 1978), especially in places where animals congregate, including under trees, in feedlots, within wildlife territorial boundaries, along fence lines, and in preferred plant patches at water or mineral resources (Naiman et al. 2003; Bestlemeyer et al. 2006), P stays elevated in these patches for long time periods. Protective nighttime enclosures in Africa have up to 14 times the P concentration found in surrounding soils and, after abandonment, soil and foliar concentrations of P can remain elevated for 40 years or more (Augustine 2003; Charley 1977; Hilder and Mottershead 1963). In contrast, soils in the surrounding landscape experience a slow depletion in P over time, as it is translocated to the animal enclosures (Augustine 2003). After abandonment, these areas of elevated P (and N) are preferentially used by livestock and wildlife, including birds, reptiles, and large and small mammals, whose fecal material helps maintain high P levels over time, reinforcing the landscape-scale heterogeneity of nutrient distributions (Bestlemeyer et al. 2006; Naiman et al. 2003; Palmer et al. 1999; Augustine 2003; Scholes 1990).

A wide variety of small mammals, large mammals, and reptiles also create many shallow surface pits or burrows that trap wind or waterborne litter, feces, seeds, and fine soil particles. These materials increase soil P. They also indirectly increase available P by increasing organic matter (Eldridge and Rath 2002; Whitford 2002).

Ants can move large numbers of seeds, leaves, and other materials from the surrounding landscape to their nests. For instance, it is estimated that the genus *Pheidole* moved ~1.6×10^9 seeds in one season at a site in New Mexico, USA (Whitford and Bestlemeyer 2006). In Wyoming, ants were estimated to have denuded 36,450 ha of dryland vegetation, while greatly increasing nutrients, including P, and productivity at nest edges (MacMahon et al. 2000). These ants generally leave the outer seed coat on the surface and take the rest of the seed underground. This seed material at the surface directly increases soil P, as well as enhancing available P through increased organic matter. Dung beetles, found in most dryland regions, can move large amounts of fecal material substantial distances from where the material was originally deposited (Naiman et al. 2003) (For more information on the role of macrofauna in soil P cycling see Chapuis-Lardy et al. 2011).

15.3.2 Vertical Redistribution of P

The soil depth at which P occurs can determine its availability to biota because various organisms have differing abilities to reach various soil depths. For example, plant roots can explore soils much more deeply than soil surface mosses and lichens. Thus, vertical mixing of soils can have substantial ecosystem-level impacts. Although most vertical redistribution of P is due to animal activity, plants can also influence where P is located in the soil profile.

15.3.2.1 Animal Activity

Digging activities of animals can carry soil, plant litter, seeds, feces, and cadavers down to >1 m depth or more. On the other hand, these same materials are often brought to the surface. Burrowing also creates large macropores that enhance water infiltration and, during overland flow events, allow surface plant litter and feces to enter the soil. However, dryland ecosystems vary widely in the number of animals that mix soils to depth. For instance, the Colorado Plateau region, USA, has very low numbers of rodents, termites, or ants and therefore very little vertical mixing compared to many other dryland ecosystems such as the Sonoran or Chihuahuan deserts (Belnap and Phillips 2001; Belnap et al. 2005; Whitford 2002). In addition, the occurrence of these animals can be spread widely or concentrated in specific areas. Therefore, the relative amount of P found at depth due to animal mixing varies widely at the local to regional scales.

15.3.2.2 Ants and Termites

Ants and termites occur worldwide in drylands. They are among the most abundant animals in terrestrial habitats and are the dominant insects in many ecosystems (MacKay 1991). Various ant and termite species move different amounts and types of surficial materials and, as species vary among regions, the relative impact on vertical P distributions varies as well (see Chapuis-Lardy et al. 2011; Whitford and Bestlemeyer 2006). Globally, termites and ants move prodigious amounts of soils in dryland regions and are probably the most important group in the vertical redistribution of P in dryland soils. In Argentina, ants can move 1,100 kg soil ha^{-1} per year from 150 cm depth to the surface (Lee 1977). In the Chihuahuan desert, ants carry soil from 200 cm deep to the surface and, combined with vertebrate burrowing, move sufficient subsurface soils to cover 20% of the soil surface (Whitford 2002). Large quantities of seeds and plant material are carried underground by ants, and their excretory products also accumulate at depth. Almost all studies show that long-lasting ant mounds have higher N, P, and organic matter than nearby nonmound soils, regardless of the continent or desert region, landscape position, soil texture, or other environmental factors (e.g., Wagner 1997; Lobry de Bruyn and Conacher 1990; Whitford 2002; Palmer et al. 1999). The creation of soil macropores results in higher infiltration rates than in surrounding soils (Whitford 1999, 2002). Ant nests have also been shown to support a higher number of soil organisms (e.g., bacteria, fungi, protozoa, mites, and collembola), which results in faster decomposition, thus resulting in higher soil available P (Wagner 1997; Boulton et al. 2003).

Termites can move up to 16,000 kg soil ha^{-1} per year and affect up to 30% of the soil surface (MacKay 1991; Lobry de Bruyn and Conacher 1990; Wood and Sands 1978; McClaran and Van Devender 1995). They consume or move up to 90% of plant material and 100% of dung found on soil surfaces to depth, as well as depositing their own feces and carcasses at depth (McClaran and Van Devender 1995; Whitford 2002). As with ants, termite mounds have higher N, P, and organic matter than nearby nonmound soils, regardless of location or other environmental factors. Unlike ant mounds, some termite mounds will decrease water infiltration whereas others increase it (Whitford 2002). Many studies have shown that large herbivores preferentially utilize plants growing on the N- and P-rich soils of recently abandoned termite and ant mounds (e.g., Augustine 2003; Scholes 1990). As mentioned above, this results in further P enhancement due to dung and urine from these animals.

15.3.2.3 Dung Beetles

Dung beetles remove voluminous amounts of dung from the soil surface and bury it to different depths, depending on the species (Naiman et al. 2003). In Kruger National Park, South Africa, there are over 120 species of dung beetles. Despite the vast quantity of dung deposited daily by wildlife, these insects remove the bulk of newly deposited dung, burying it underground as food for their larvae. This burrowing also creates soil macropores that facilitate water infiltration.

15.3.2.4 Mammals and Reptiles

Mammals and reptiles (e.g., kangaroo rats, springhares, molerats, gerbils, aardvarks, badgers, foxes, porcupines, lizards, aardwolves, wombats, coyotes, woylies) create burrows for shelter and to find or cache food, thus vertically mixing soils and other materials to 1 m or more (Scholes et al. 2003; Palmer et al. 1999). Burrows can be so numerous in small areas that they are visible on LANDSAT images (Whitford 2002). Large amounts of subsurface soils are moved to the surface during burrowing. Australian woylies (also known as brush-tailed bettongs) can dig 5,000–16,000 new holes ha^{-1} per year, moving over 13 tonnes of soil (Garkaklis et al. 2004; soil turnover estimations for other mammals are also listed in this publication). Heteromyid rodents in the Chihuahuan desert can create over 100,000 holes ha^{-1} per year (Whitford 2002). Porcupines alone have been estimated to have impacted up to 4% of soil surfaces in the Negev Desert in Israel (Alkon 1999). Dust-bathing and mud-wallowing by large mammals such as elephants and hippos can also vertically mix soils down to 3 m or more (Naiman et al. 2003).

15.3.2.5 Plants

Plants can also vertically redistribute P because roots collect P from soils at depth and use it to create tissue. However, P appears to be tightly conserved in desert plants, with resorption of up to 90% of the P before leaves or even woody stems are dropped. This occurs in many diverse plant genera and life forms (Lajtha 1987; Killingbeck and Whitford 2001; Charley 1977; Charley and Cowling 1968). Greater resorption appears to occur during drought stress and is greater in obligate riparian species than in facultative riparian plants and in soils with lower P availability. Resorption efficiency varies greatly among sites and years. Therefore, contribution of P to surface soils via litterfall leaching and litter decomposition is probably minimal. In addition, very low rainfall results in low leaching of P from leaves into the soil (Jobbágy and Jackson 2004). However, substantial redistribution of P can occur when roots decompose because the P found in the roots may have come from many meters away, and fine root turnover rates are high in drylands (Whitford 2002).

15.4 Controls on P Bioavailability in Drylands

In dryland settings, soil P ranges from ~200 to 1,200 mg P kg^{-1} (Turner et al. 2003a; Tiessen et al. 1984). Generally, >50% of P in dryland soils is inorganic P, in contrast to more mesic soils, which often contain a higher proportion of organic P. Dryland soils have high levels of calcium (Ca), carbonates, aluminum (Al), and iron (Fe), all of which complex with what P is present to make much of it bio-unavailable. For instance, in northern Utah, USA, 75–90% of the soil P was inorganic, mostly as Ca phosphates (80%), followed by Al phosphates (8–10%), and Fe phosphates (1–2%) (Jurinak and Griffin 1972).

Carbonate accumulation zones in dryland soils can range from 1 to 93% $CaCO_3$, with up to 220 kg C m^{-2}, a value similar to organic carbon in peat bogs (Monger 2006). Roots, cyanobacteria, bacteria, and other structures all contribute to pedogenic carbonate formation, along with warm, dry soil conditions (Breecker et al. 2009). Plant roots are often sheathed in amorphous or microcrystalline secondary compounds (e.g., Ca, silica) that bind P and/or form insoluble P minerals. Calcrete root coatings have up to ten times the level of P as surrounding soils (Verboom and Pate 2006). As a result, the solubility of P is mostly controlled by the sorption of P on calcite or by formation of di-calcium phosphate (Ma et al. 2009; Nadeau et al. 2007). Microbes immobilize P as well, although the levels can be quite low in dryland soils compared to mesic soils. In a Chihuahuan desert study, Lajtha and Schlesinger (1988) never found microbial biomass to be more than 3% of the total P content, even in the wet season, in contrast to peat soils where microbial biomass can be 36–55% of the total P (Walbridge 1991). Dry conditions also lead to the inactivity of microbes and enzymes and low diffusivity of soil P, and thus little degradation of organic P can occur except during pulses of soil moisture availability (Turner et al. 2003a, b).

Many studies, including those from the Great Basin, Mojave, Sonoran, and Chihuahuan deserts, USA (e.g., Schlesinger et al. 1989; Parker 1995), show P is often a limiting nutrient to dryland plants. In addition, P is needed for maximum N mineralization and nitrification (West et al. 1984). Nitrogen additions can also stimulate phosphatase activity, which is purported to increase available P (Phuyal et al. 2008; Collins et al. 2008). Soil concentrations of P, and the differential ability of plants to uptake soil P, can have a large influence on plant community composition (Midgley and van der Heyden 1999).

The positive feedback between water and nutrient uptake is especially important in dryland plants (Caldwell and Richards 1986; Radin and Eidenbock 1984). Because water is the most limiting resource in drylands (Smith and Nowak 1990), plants that are more effective at water uptake or use are likely to have a competitive advantage over other plants, including their access to nutrients. Better nutrition, including P, then facilitates the plants' capacity to acquire more nutrients and water (Caldwell and Richards 1986; Radin and Eidenbock 1984).

Wright and Mooney (1965) showed that P influences plant distributions in high-pH dolomite soils, and Billings (1950) found that low P or Ca limited where sagebrush occurred in their study region. Klemmedson and Tiedemann (1998) showed a high positive correlation between available P and cover of the grass *Stipa lettermanii* ($r = 0.88$), whereas *Cymopterus lemmonii* was negatively correlated ($r = -0.15$ to -0.37) with available P. Lei and Walker (1997) showed that density of the shrub *Coleogyne ramosissima* was highly correlated with stem and foliar P ($r = 0.99$ and 0.80, respectively) and soil P at 7–15 cm ($r = 0.68$). Exotic plant species richness has also been correlated with soil available P ($r = 0.84$; Bashkin et al. 2003; for further discussion, see Sects. 15.4.6 and 15.5.1). Accordingly, DeLucia et al. (1989) found that *Artemisia tridentata* and *Bromus tectorum* were excluded from low-P soils. As annual plants generally have a higher nutrient demand than perennials, they are more limited by low available P than are perennial plants (Epstein 1961; Marschner 1995). Harner and Harper (1973) showed that forb cover of plant communities increases with increasing foliar P. In addition to the vascular plants described above, the species composition of lichen communities is also influenced by P levels (Bowker et al. 2006).

15.4.1 Abiotic and Microbial Controls on P Bioavailability

Ultimately, the amount of biologically cycled P in dryland regions is controlled by the timing, intensity, and amount of precipitation. Soil moisture directly affects the availability of P by influencing rates of geochemical reactions, ion diffusion, and biotic activity. Because precipitation is highly variable in drylands, pulses of P-releasing activities vary on both spatial and temporal scales.

The interaction of precipitation and temperature can affect the release of bound P. The rate of H_2CO_3 formation is controlled by soil water content, CO_2 concentrations, and temperature (Krauskopf and Bird 1995). The solubility of carbonates and CO_2 in water increases with decreasing temperature (as long as temperatures are above freezing). Thus, assuming sufficient soil moisture, H_2CO_3 production should occur along spatial and temporal gradients of soil temperature. Cool, wet conditions would result in the greatest concentrations of H_2CO_3, thus decreasing soil pH, dissolving carbonates, and increasing the transition of solid-phase P to solution-phase P (Jungk and Claassen 1997). This scenario is supported by several studies. In-situ resin bags at Colorado Plateau sites showed increased soil available P during the cold, moist conditions found in winter (Miller et al. 2006a, b). Lajtha and Schlesinger (1988) also found that resin bag P peaked during cool winter conditions in the Chihuahuan desert. Magid and Nielsen (1992) showed that laboratory extractions of soils at 4°C recovered significantly more P from the soils than those done at 25°C.

The timing of rainfall relative to biotic processes is also crucial to concentrations of soil P. Precipitation that occurs when temperatures are warm results in rapid increases in microbial populations and the rate of microbial processes that affect P cycling. If precipitation occurs at the appropriate time to stimulate annual plant activity for a few weeks to months, the subsequent C inputs stimulate soil biotic activity, including processes that liberate bound P (Whitford 1999).

Most precipitation events in drylands are less than 5 mm, and thus soil surface organisms and the processes and environments they influence are able to respond to precipitation events far more often than vascular plants, because vascular plants generally require events of a larger size (depending on rooting depths) or specific seasonal timing (depending on species; Schwinning and Sala 2004). The result is a temporal decoupling between microbial and plant response to precipitation, nutrients, and carbon (Stursova et al. 2006; Belnap et al. 2004) (Fig. 15.5) and an accumulation of nutrients, including hydrolyzable organic P, and carbon at the soil surface between large rain events (Whitford 2002; White et al. 2004). Experimental addition of phosphatases showed that up to 87% of organic P in dryland soils is hydrolyzable if soils are wetted (Nadeau et al. 2007; Turner et al. 2003a).

The length of time between precipitation events and rapid soil wetting and drying influences available P in multiple ways:

1. During dry periods, P accumulates on the soil surface due to high microbial mortality from desiccation and radiation damage, UV degradation of organic matter (Castenholz and Garcia-Pichel 2000), and dust accumulation (Verrecchia et al. 1995). This contributes to a flush of available P in subsurface soils when precipitation events occur.
2. In drylands, rewetting of dry soils and rapid drying due to high air temperatures can kill up to 58% of the soil microbial biomass (Kieft et al. 1987). The contents

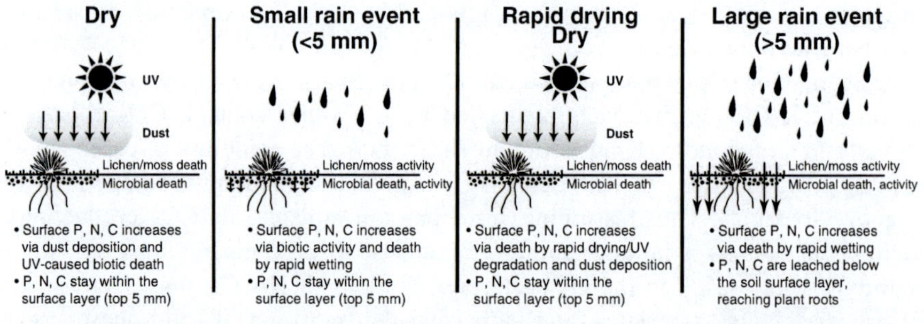

Fig. 15.5 Precipitation events are infrequent in drylands, and thus soil surfaces are most often dry. During dry times, nutrients and carbon build up on the soil surface due to dust deposition and the degradation and/or death of organisms from UV exposure. When rain occurs, most events give less than 5 mm precipitation. Whereas these small events elicit responses from surface organisms and the processes and environments they influence (including P, N, and C cycling; the death of microbes with rapid wetting/drying cycles), such low precipitation levels result in very shallow penetration of water into the soil. Thus, nutrients and carbon stay at the soil surface. Vascular plants generally require much larger events (depending on rooting depths) or specific seasonal timing (depending on species) to respond (Schwinning and Sala 2004). In addition, larger events are required to move materials such as P, N, and C to depth in the soil. Because there are many small events in between large events, substantial buildup of nutrients and carbon at the soil surface can occur. The end result is a temporal decoupling between microbial and plant response to precipitation, nutrients, and carbon

of these lysed cells, which have most of the P in the form of nucleic acids and phospholipids, can represent up to 95% of water-soluble P in Australian soils (Turner et al. 2003c) and can increase water-soluble soil P up to 1,900% (Turner and Haygarth 2001).

3. Independent of microbial activity, wetting/drying cycles can increase organic P solubility by disrupting organic matter coatings and detaching and mobilizing soil colloids, leading to increased P in soil solution (Blackwell et al. 2009). However, pulses of dissolved organic P do not always occur (Butterly et al. 2009).

15.4.2 Biocrust Controls on P Availability

Biocrust communities, which cover a substantial portion of dryland soil surfaces (Belnap et al. 2003c), are an especially important determinant of available P in dryland soils. As discussed above, the presence of well-developed crusts increases the capture of P-containing dust and prevents loss or redistribution of P-containing soils by wind or water. The silt and nutrient-rich clay particles captured by the crusts increase the fertility and water-holding capacity of the soil, keeping crust organisms metabolically active for a longer period of time. Biocrusts also affect available P in many other ways, as discussed below. As a result, plant uptake of P increases when plants are growing in well-developed soil crusts relative to bare sand (reviewed in Belnap et al. 2003a). For example, two sites on the same soil type in SE Utah had 82 and 81 mg P kg^{-1} in the undisturbed, biocrusted soil versus 51 and 31 mg P kg^{-1} in adjacent disturbed, noncrusted soils. Foliar concentrations of the shallow rooted *Festuca octoflora* at these same sites were 2.5 mg P g^{-1} soil in crusted soils versus 1.4 mg P g^{-1} soil in uncrusted soils. However, the two perennial species tested at this site showed no difference in P foliar concentrations. Before conclusions are drawn, many plant species remain to be studied.

15.4.2.1 Secretion of Extracellular Polysaccharides, Organic Acids, and Chelators

Most biocrust components fix C, which enters the soil 1) during the formation of polysaccharide sheaths, 2) when the organisms die (Fogg 1966; Lewin 1956) or 3) C is secreted within minutes to a few days of C acquisition. It can represent up to 50% of total fixed C and increase soil carbon by up to 300% (Fogg 1966; Lewin 1956; Rogers and Burns 1994). The increase in organic matter increases the availability of P (see Sect. 15.3.1.1). Exopolymers secreted by the crustal organisms also modulate metal-ion concentrations at the microbial cell surface. Lipid, protein, and carbohydrate components combine to create a mosaic of many polyfunctional metal-binding sites that differ in affinity and specificity. Both cations and anions, including P, can be bound (Greene and Darnall 1990). Most adsorbed metals stay on or within the extracellular sheath and are not absorbed by the cell, thus remaining available to plants while reducing leaching losses to subsurface soils (Geesey and Jang 1990; Belnap 2003a; Verrecchia et al. 1995).

Soil crust organisms also increase available P via the excretion of H^+ during respiration, which decreases soil pH and frees carbonate-bound P. However, the activity of biocrust organisms can also significantly increase soil and rock pH. In SE Utah, biocrusts raised soil pH from ~8 to ~10.5 (Garcia-Pichel and Belnap 1996), and endolithic cyanobacteria in Venezuela and South Africa raised the pH from ~8 to ~10 (Büdel 2000).

Many biocrust organisms secrete organic acids such as citrate, malate, acetate, pyruvate, lactate, formate, and fumerate (Belnap 2003b; Whitton et al. 2005). Specific lichen-secreted organic acids include physodate, lobarate, salzinate, stictate, evernate, lecanorate, roccelate, atranorate, norstictate, oxalate, and usnate acids. All these compounds can solubilize bound P (see Jones and Oburger 2011; Jones and Wilson 1985). What compounds are released is often specific to a particular species or genera. For instance, mat-forming ectomycorrhizae in dryland riparian areas secrete oxalate, but the common upland mycorrhizal species *Glomus* and *Acaulospora* do not (Allen et al. 1996). Whereas the fungus *Aspergillus niger* and the bacterium *Penicillium simplicissiumum* produce citrate, many other species of fungi and bacteria cannot solubilize bound P (Barroso and Nahas 2005). Fungi can also secrete acids within rocks, penetrating to depths of up to 4 m (Bornyasz et al. 2005), solubilizing nutrients and transferring these nutrients, including P, directly to plant roots (Van Breemen et al. 2000).

Biocrust organisms also secrete powerful metal chelators, such as siderochromes, that increase available P by maintaining metals in solution (Lange 1974; McLean and Beveridge 1990; Schelske et al. 1962; Belnap 2003b). Cyanobacteria also secrete peptide N and riboflavin. Together with other chelators, these substances form complexes with tricalcium phosphate, copper, zinc, nickel, and ferric iron, keeping the nutrients plant-available. Because chelators are water-soluble, these nutrients are also available to associated nonchelating plants or microbes (Lange 1974; Geesey and Jang 1990; Gadd et al. 2007). Soil cyanobacteria also secrete glycollate, which stimulates phosphate uptake (Fogg 1966).

15.4.2.2 Secretion of Phosphatases

Most soil and hypolithic cyanobacteria, green algae, lichens, and mosses have phosphatases in their cell walls and mucilaginous sheaths. Most of the cyanobacteria and fungi tested release extracellular phosphatases into the surrounding soil. Phosphatases hydrolyze organic phosphates, liberating P (Turner et al. 2003b; Nannipieri et al. 2011). Once released, these compounds can then be immobilized by microbes, transferred to plant host roots, or stabilized by humic substances (Sinsabaugh 1994; Lindahl et al. 2005). Phosphatase activity was found to be higher under well-developed biocrusts than under nearby bare soil (Bolton et al. 1993). However, as phosphatase is highly correlated with soil organic matter and organic matter is relatively low in dryland soils, phosphatase activity is expected to be relatively low in drylands relative to more mesic ecosystems (Sinsabaugh et al. 2008).

The expression of phosphatase is somewhat inducible. As Whitton et al. (2005) discuss, whereas P in the medium induces phosphatase activity in small unicellular organisms (e.g., bacteria), larger multicellular cyanobacteria appear instead to respond to internal P concentrations. Phosphatase activity can also be inhibited by P-binding metal ions (e.g., zinc, iron, manganese). Temperature optima appear to vary with species and phosphatase form. For instance, *Nostoc commune* UTEX 584 shows optima of 32°C for phosphomonoesterase and 42°C for phosphodiesterase. There also is light sensitivity in this species, with activity consistently highest in the dark, followed by activity in low and then in high light. Phosphatase activity in cyanobacteria increased with increasing Ca and decreased with magnesium, whereas potassium and sodium had little effect. Dried *Nostoc* samples show phosphatase activity even after many months or after organisms die or are removed. Phosphatase activity can be increased in soil by up to 27% by adding extracellular polysaccharides from cyanobacteria without the living organisms being present (de Caire et al. 2000). Free-living and lichenized cyanobacteria also fix N. When soil N is elevated, phosphatase amount and activity also increases, as discussed in Sect. 15.4 (Collins et al. 2008; Phuyal et al. 2008), increasing labile (resin- and bicarbonate-extractable) soil P as well (Zou et al. 1995).

In environments with low or highly variable levels of P, such as dryland soils, phosphatase-coated, multicellular hairs are found in cyanobacteria and sometimes green algae (Whitton et al. 2005). These hairs have long tapered ends that are highly vacuolated, providing a high surface area for P absorption. In colony-forming cyanobacteria, these hairs can extend out beyond the colony. Smaller hairs are also formed in response to Fe and N limitation, but do not show phosphatase activity. It is not known if the hairs can be used to access inorganic P.

15.4.3 Fungi as Connectors

Biocrust fungi and subsurface fungi are clearly crucial in P cycling in drylands. Most desert plants are mycorrhizal (Whitford 2002), a relationship that is well-known to increase P uptake and tissue concentration in vascular plants (see Jansa et al. 2011). Strong positive correlations have been shown between mycorrhizal infections of desert seed plants and biocrusted surfaces, as plants in biocrusted soils from SE Utah have up to three times higher mycorrhizal infection rates in crusts compared to plants in bare soil (reviewed in Belnap et al. 2003a). However, mycorrhizal abundance decreases with aridity as they are replaced with proportionally more dark septate ascomycetes in dryland soils (Green et al. 2008). These dark septate fungi form endophytic and ectophytic associations with plants, nondestructively colonizing the root cortex and surface, shoot and stem surfaces (including the inside of hairs), and the vascular cylinder (Barrow and Osuna 2002). The same species found on roots can be found free-living in the soils. These organisms can solubilize rock phosphate and tricalcium phosphate and have been shown to increase P in roots and shoots in the sedge *Carex*, the grass *Vulpia ciliate,* and the shrub *Atriplex canescens* (Mandyam and Jumpponen 2005). Fungal networks, dominated by dark septate fungi, transfer C and N (and probably P as well, although it was not measured in this experiment) up to 100 cm per day between plants and cyanobacterial biocrusts (Green et al. 2008).

15.4.4 Soil Fauna Controls on P Availability

In drylands, soil microfauna have greater abundance, activity, and richness under well-developed biocrusts than under less well-crusted soils (Belnap 2003b; Darby et al. 2007). The selective feeding of soil microfauna on microbes affects factors that influence P cycling, including microbial abundance and community composition, microbial enzyme production, and release of P when the microbes are eaten (Bardgett 2005). Multiple studies show that the nutrients released by the death of microorganisms are utilized by plants (as discussed in Bardgett 2005). Soil microfauna and macrofauna also fragment litter, increasing decomposition rates that indirectly increase available P. Because populations vary seasonally, the release of microbial P will vary seasonally as well. However, as discussed in Sect. 15.3.2, the importance of soil fauna in P cycling varies widely among different dryland regions.

15.4.5 Plant Controls on Soil P

Plants utilize multiple mechanisms to increase their access to soil P in all environments, as discussed by George et al. (2011). In dryland regions, some of these mechanisms are employed more often than others, as discussed below.

First, alteration of key root characteristics can increase plant P uptake. Plants with higher root cation exchange capacities (CEC) can more readily access recalcitrant P (Drake and Steckel 1995), and as dicotyledonous plants generally have higher root CECs than monocotyledonous plants, they should be favored in P-limited soils (Marschner 1995). This may partially explain why dicotyledonous plants heavily dominate dryland regions. Roots generally contact only 0.5% of dryland soil volume at any one time (Lynch and Deikman 1998) and thus an increase in this contact is probably advantageous for P uptake. Most dryland plants utilize fungi to enhance their ability to exploit the soil volume (Lajtha and Harrison 1995; Gadd et al. 2007; Quiquampoix and Mousain 2005; also see Sect. 15.4). Although root uptake capacities could theoretically be increased, P uptake is probably more limited by low soil diffusion rates than root uptake capacity (Lajtha and Harrison 1995; Clarkson 1985). In addition, some plants such as *Larrea* actively maintain space around their roots via root exudates (root exclusion), whereas others such as *Ambrosia* utilize a root-recognition response to maintain space (Schenk et al. 1999). Root segregation in dryland soils allows plants to dominate belowground space, giving increased access to nutrients, especially immobile nutrients such as P, and thus promoting the plant's competitive ability (Casper and Jackson 1997). The root-recognition response appears more effective at accessing soil P because root self-incompatibility may limit flexibility in exploring the soil volume (Caldwell and Richards 1986).

A second way by which plants affect soil available P is through hydraulic redistribution, which is exudation of water by roots into dry soil pockets (Caldwell and Richards 1986). This is a common phenomenon in dryland plants. The exuded water enhances soil available P and both the water and nutrients are available for re-uptake. Third, dryland plants also have P-conserving strategies. Woody shrubs and evergreen life forms with long-lived tissue dominate these regions. As discussed in Sect. 15.3.2.5, large amounts of P are also resorbed in these plants before tissue is dropped. Both these mechanisms reduce the need for additional P uptake to construct new tissue (Jonasson and Chapin 1991).

Fourth, the most common way for plants to access bound P in dryland soils is to acidify the rhizosphere via release of H or OH/HCO_3 to counterbalance net excess of cations or anions, and all dryland shrubs tested showed a decrease in rhizosphere pH relative to bulk soil (Ma et al. 2009). Fifth, plant roots can also exude a wide range of complex mixtures of organic acids (e.g., oxalate, malate, succinate), phytosiderophores, sugars, vitamins, amino acids, purines, nucleosides, inorganic ions, gaseous molecules, a wide array of chelators, enzymes such as phosphatases and phytases, root border cells, and phospholipid root surfactants, all of which can act to make bound P bioavailable (Callaway 2007; Jurinak et al. 1986; Quiquampoix and Mousain 2005). The secretion of many of these compounds can be enhanced or suppressed, depending on soil concentrations of P. In addition, as with cyanobacteria, plant phosphatase and phytase activity can be inhibited by polyvalent anions such as phosphate and some forms of P-binding metal ions (zinc, iron, manganese). Plant phosphatases are also secreted in response to water stress and salinity, with drier sites showing much higher soil and shrub root surface phosphatase activity than more mesic sites (Sardans et al. 2006; Estiarte et al. 2008; Li and Sarah 2003). Various plant species, and litter from these plants, have differential effects on stimulation of soil phosphatase and thus on P availability (Dornbush 2007).

A sixth way by which plants can influence available P is via leaching of leaves or litter. In drylands and seasonally dry areas of the world, plant leaves of the *Casuarinaceae* and *Proteaceae* contain carboxylicates and hydroxylic acids that free bound P (Verboom and Pate 2006). Over 200 families of angiosperms and gymnosperms, representing 74% of the angiosperm families, contain oxalate in their tissues. Oxalate also occurs in the wood of more than 1,000 species and some cacti contain up to 80% dry weight calcium oxalate (Horner and Wagner 1995). Many exotic plant invaders secrete compounds to access recalcitrant P (for further discussion, see Sect. 15.4.6). In the Karoo Desert of South Africa, many plant leaves contain malate, citrate, and oxalate which, upon leaching, increase soil available P (Whitford 1999). Because most of these compounds are water-soluble, exudates from one plant can benefit neighboring plants and microbes. This has been shown in agricultural systems: the P-releasing activity of *Vicia faba* facilitates P uptake in *Zea mays*, and the P-releasing activities of *Cicer arietinum* facilitate P uptake by *Triticum aestivum* (Zhang and Li 2003). The presence of the invasive plants *Bromus tectorum*, *Halegeton glomeratus*, *Salsola* spp., and *Centaurea maculosum* in the western USA have all been shown to facilitate P availability for neighboring native plants (Allen and Allen 1988; Fox and Comerford 1990; Herron et al. 2001; Thorpe et al. 2006; Belnap and Sherrod 2009).

Lastly, N-fixing plants have higher tissue levels of N than nonfixing plants and thus their tissue often decomposes quickly, releasing P (Naiman et al. 2003). As discussed above in this section, high soil N can also elevate phosphatase activity and thus soil available P. This can be an amplifying feedback because fertilization with P often increases N as well (Reed et al. 2007).

There are also amplifying feedbacks between soils and plants in dryland ecosystems that affect soil available P. Large woody plants of dryland ecosystems and the microorganisms associated with them often alter soils to optimize the capture and utilization of limiting water and nutrients, including P, which in turn alters plant community composition (Verboom and Pate 2006). However, the strategies to increase the availability of P in soils can be quite costly in terms of plant carbon (Lynch and Ho 2005; DeLucia et al. 1997; James et al. 2005). Because dryland plants are already water and nutrient limited and thus carbon is scarce, the cost of making P bioavailable may be much more challenging than for plants growing in more mesic regions (James et al. 2005). For example, root exudates have been estimated to represent up to 50% of all belowground carbon allocation, up to 25% of photosynthetic production, and up to 20% of total plant dry weight (James et al. 2005). Because available P is very low in dryland soils, the proportion of carbon allocated to P uptake is likely to be even higher in these plants. Mycorrhizal associations also cost the infected plant carbon, which can range from 4 to 20% of daily net photosynthesis (James et al. 2005). In addition, the need for soil moisture may place temporal restrictions on when such mechanisms can be employed, as they may be possible only during rainy seasons or in years with above-average precipitation.

15.4.6 Interactions Between Invasive Plants and P Availability

Scientists have long attempted to understand what plant properties allow particular species to become highly invasive (e.g., Lonsdale 1999). In the low available P soils of the drylands of western USA, the most pervasive and threatening invasive species include *Centaurea diffusa*, *C. maculosa*, *Halogeton glomeratus*, *Salsola* spp., *Lepidium latifolium*, *Bromus tectorum*, *B. madretensis*, *Schismus barbatus*, and *Taeniatherum caput-medusae*. The success of these plants is at least partially due to their ability to (1) access nonbioavailable P more efficiently than native plants, (2) outcompete native plants for soil P, and/or (3) be more efficient in P uptake or utilization than native plants. The strategies used are those listed in Sect. 15.4.5, including root exudates and leachates of aboveground tissue that contain organic acids, more efficient P uptake and/or utilization, and being overall better competitors when P is not limiting to the invasive plants (Table 15.1). The distribution of some of these species is also significantly correlated with soil available P. Many invasive plants in dryland ecosystems are early successional annual species, colonizing areas with few other plants or fungi to assist in P uptake. Thus, the ability of these species to develop means of accessing relatively unavailable P in such situations may have, at least partially, led to their superior competiveness.

15.5 Case Study: Interaction Between Exotic Annual Grasses and Soil P Availability in the Western United States

15.5.1 Controls of P on Exotic Annual Plant Distribution

How soil physical and chemical characteristics influence the cover of *B. tectorum, B. madretensis,* and *S. barbatus* has been investigated by the Belnap laboratory (unpublished results). We used maps of soils, geology, topography, and geologic surfaces (delineating the age and composition of the soil surface) to identify sites representing a broad range of soil texture, chemistry, and elevation within four of the six US deserts: Chihuahuan, Mojave, Colorado Plateau, and the Great Basin (we did not sample the Sonoran or Columbia Plateau). Climate maps were then used to sample these representative conditions across a gradient of temperature and rainfall timing and amount (hot Chihuahuan Desert with relatively high amounts of summer-dominant rainfall; hot Mojave Desert with low amounts of winter-dominant rainfall; cool Colorado Plateau Desert with higher amounts of summer/winter mixed rainfall; cool Great Basin Desert with high amounts of winter-dominant rainfall). Within each selected unit, we sampled 0–10 cm soils in adjacent invaded and uninvaded patches (Fig. 15.6). In the Chihuahuan Desert, where no annual grass patches were found, we sampled the approximate center of the desired soil type as outlined on the soil survey.

Our results suggest that as the ratio of cool season to total precipitation increases, as well as the total amount of cold season precipitation, so do conditions that allow the conversion of bound P into bioavailable P (Fig. 15.7). In the hot Chihuahuan Desert, where the percentage and total amount of winter rainfall is very low, we found no annual grasses at our sites. Our hypothesis is that soils are seldom cool and wet in this desert, and thus soil P remains bio-unavailable. In the hot Mojave Desert, where total winter rainfall is higher than in the Chihuahuan but still low compared

Table 15.1 The more problematic exotic invasive plants in US dryland regions and the ways in which they affect soil P availability

Plant	Root exudates	Tissue leachates	Better competitor	Uptake efficiency	Utilization efficiency	Distribution control	References
Centaurea diffusa	×		×	×			Callaway (2007), LeJeune and Seastedt (2001), LeJeune et al. (2006), Suding et al. (2004), Callaway and Aschehoug (2000)
Centaurea maculosa				×	×		Harvey and Nowierski (1989), Herron et al. (2001), Thorpe et al. (2006)
Halogeton glomeratus		×					Allen and Allen (1988), Cook and Gates (1960), Duda et al. (2003)
Salsola spp.		×					Allen and Allen (1988), Cannon et al. (1995), Fox and Comerford (1990)
Lepidium latifolium	×						Blank and Young (2002)
Bromus tectorum	×					×	Belnap unpublished results
Bromus madretensis						×	Belnap unpublished results, Yoder and Nowak (2000)
Schismus barbatus						×	Belnap unpublished results
Taeniatherum caput-medusae	×					×	Blank and Sforza (2007)

× indicates a positive effect

Fig. 15.6 An illustration of the patchy nature of *Bromus tectorum* invasion in many parts of the western USA (*arrows* mark the *light-colored Bromus* patches)

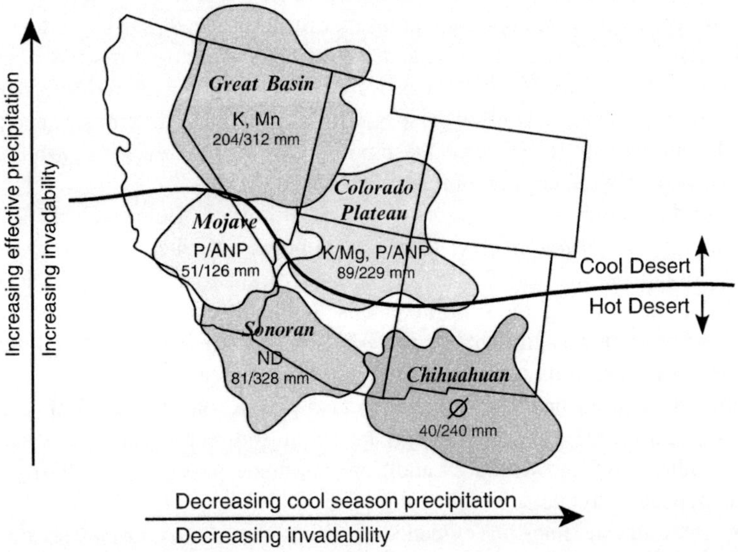

Fig. 15.7 Conceptual model of the relationship between climate and soil factors (nutrients and water) controlling annual grass distribution in the different US deserts. Limiting nutrients are listed in order of importance within regional boundaries. The *numbers* within the regional boundaries indicate the average annual amount of precipitation (mm) when air temperatures are below 10°C relative to total precipitation. We suggest that as this ratio and the total amount of cool season precipitation increase, so do conditions that allow the conversion of bound P into bioavailable P. As P becomes less limiting, other cations become more important to annual grass distribution. Although we do not have data (*ND*) on the limiting soil factors in the Sonoran desert, soils are expected to show strong P limitation. We found no (*slashed O*) annual grasses in the Chihuahuan desert

to other deserts, available P has strong control over annual plant distribution ($r^2 = 0.83$). On the Colorado Plateau, available P becomes less limiting when soils are cool and wet, especially at lower elevation (1,400–1,585 m elevation, $r^2 = 0.29$; >1,585 m elevation, $r^2 = 0.63$), and other cations (potassium and magnesium) and water become more important to annual grass establishment. In the Great Basin, with an even greater time of cool, wet soils, P does not appear to affect annual grass distribution. Although we do not have data on the limiting soil factors in the Sonoran Desert, our hypothesis would suggest that soils will be strongly P-limited, and annual grasses thus controlled by available P.

15.5.2 *The Interaction of Soil P and* Bromus tectorum

Bromus tectorum is an annual grass that currently dominates vast regions of the western USA, and many studies have been done on how soil nutrients affect this species. In SE Utah, USA, field measurements showed that resin-extractable P best predicted relative growth rates of *B. tectorum* in fall ($r = 0.40$ in watered and 0.59 in unwatered plots) and winter ($r = 0.53$ in watered and $r = 0.65$ in unwatered plots) (Miller et al. 2006a, b). Other soil factors best predicted germination, and spring and overall growth rates, indicating that resource needs shifted with season and soil moisture, similar to the regional scale resource limitations discussed above. In these soils, although available P levels were above those considered "critical" for plant growth, they were still below those considered adequate for unrestrained plant growth. In another study, *B. tectorum* again showed high levels of winter root growth when soils were cold and moist (Harris 1967).

Other studies show that *B. tectorum* can be P-limited. Belnap et al. (2003b) showed that *B. tectorum* germination was suppressed when available P was reduced. DeLucia et al. (1989) found that *B. tectorum* biomass was reduced by over 90% when growing on P-limited soils. Gundale et al. (2008) also found *B. tectorum* growth was limited by low soil P. Bashkin et al. (2003) found a landscape-level correlation between *B. tectorum* biomass and available P.

Nitrogen deposition has been thought to also play a role in annual grass invasions (Fenn et al. 2003), although it may be via an indirect effect on available P. Multiple studies have shown that N additions stimulate phosphatase activity, thus probably increasing available P (Phuyal et al. 2008; Collins et al. 2008).

There are multiple lines of evidence showing that *B. tectorum* can convert recalcitrant P to labile forms. A patchy invasion of *B. tectorum* into a never-grazed (by livestock) grassland showed that labile P was much higher in invaded plots than in adjacent uninvaded plots at the same site (45.6 vs. 14.6 $\mu g\ P\ g^{-1}$ soil, respectively) (Hansen 1999) whereas there were no significant differences in soil available P before the invasion (Kleiner and Harper 1977). In addition, the higher the *B. tectorum* cover, the higher the increase in labile P; as *B. tectorum* cover increased from 0 to 10 to >40% cover, labile P increased from 14.6 to 19.5 to 28.2 $\mu g\ P\ g^{-1}$ soil.

Miller et al. (2006a, b) also found greater available P in plots with *B. tectorum* compared to those without *B. tectorum*.

Similar results have been obtained from plots in SE Utah at various locations (Belnap and Robert Sanford Jr, University of Denver unpublished results). In a 30-month study where plots were sampled monthly, native plant plots (dominated either by the perennial C_4 grass *Hilaria jamesii* or the perennial C_3 grasses *Stipa hymenoides* and *S. comata*) that had been invaded by *B. tectorum* had higher resin-extractable and labile P than adjacent native-only plots. Similarly, a 12-month study (Fig. 15.8) at another site again showed resin-extractable and labile P higher in *Stipa* plots invaded by *B. tectorum* than in native-only plots. This difference was most pronounced in June when all plants were dormant, and in October when native grasses were dormant and *B. tectorum* was in the seedling stage.

In the same never-grazed grassland as the Hansen (1999) and the 30-month labile P studies, Belnap and colleagues (unpublished results) also sampled available P annually in the spring from 1995 to 2006 (Fig. 15.9). As seen at the other study sites, *Stipa–B. tectorum* plots showed elevated available P at over half of the sample

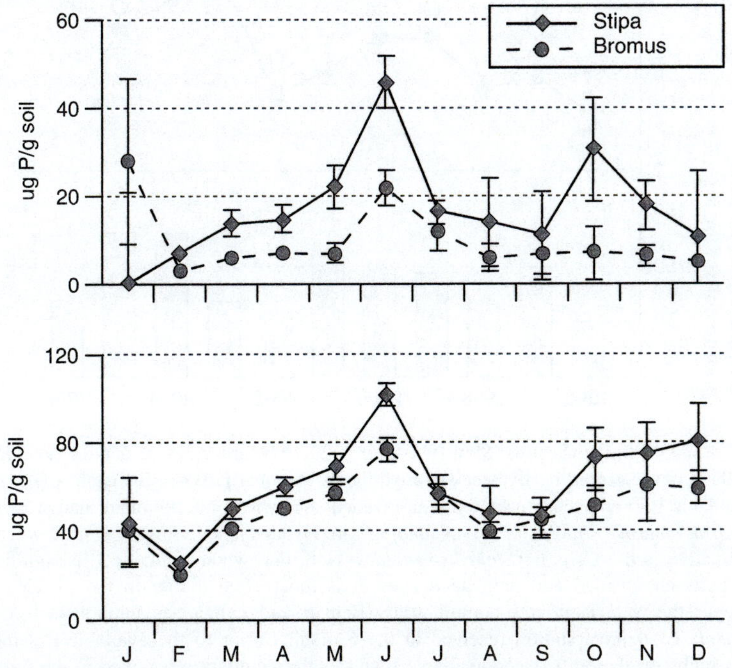

Fig. 15.8 Monthly measures (Jan – Dec 2002) of bioavailable P in a grassland grazed for over 100 years outside Canyonlands National Park in SE Utah, USA. Samples were taken from adjacent invaded and uninvaded sites where the dominant native is the C_3 grass *Stipa hymenoides*. *Upper panel*: resin extractable P. *Lower panel*: labile P (resin- and bicarbonate-extractable total P). Note that resin-extractable and labile P concentrations are almost always elevated in the invaded sites relative to the uninvaded sites, especially in June when all plants are dormant, and in October, when natives are dormant (Belnap and Sanford unpublished results)

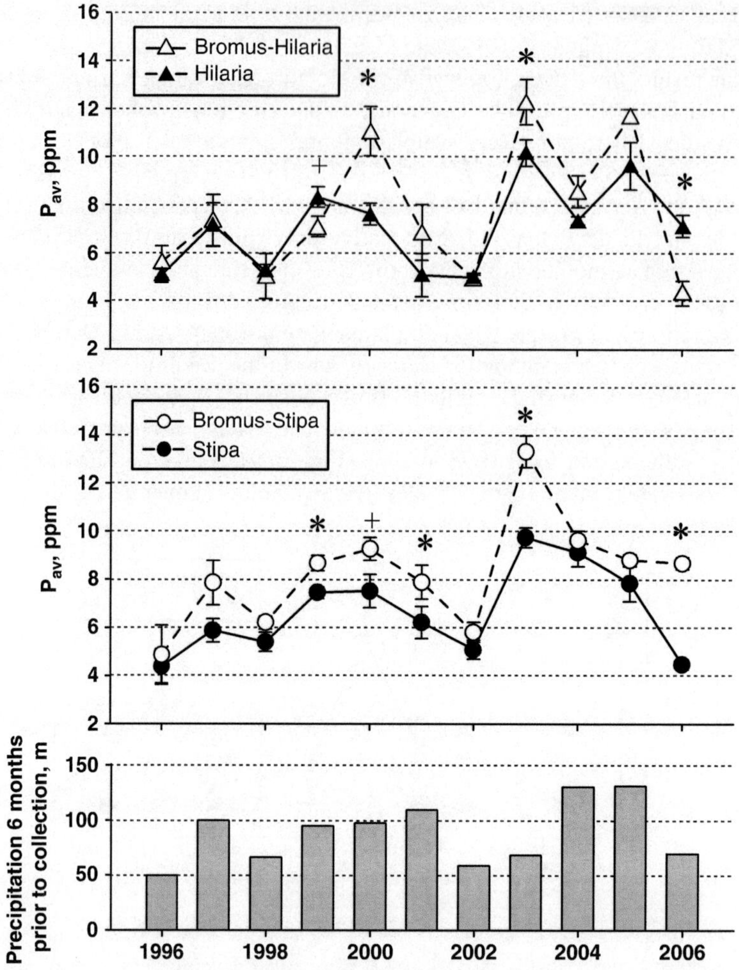

Fig. 15.9 Annual spring measures (May 1996–Apr 2006) of bicarbonate-extractable (available P) in a grassland never grazed by livestock (Canyonlands National Park) in SE Utah, USA. *Upper panel*: available P in adjacent invaded and uninvaded sites where the dominant native is the C_4 grass *Hilaria jamesii*. *Middle panel*: available P in invaded and uninvaded sites where the dominant native is the C_3 grass *Stipa hymenoides*. Note that when available P is significantly different between invaded and uninvaded sites, it is most often higher in the invaded sites, especially in the *S. hymenoides* communities (Belnap and colleagues unpublished results). *Lower panel*: Total precipitation received in the 6 months prior to the available P sampling. There is a high correlation between available P and total precipitation received in the 6 months prior to sampling (Belnap and colleagues unpublished results)

times relative to *Stipa*-only plots, whereas the results in the *H. jamesii–B. tectorum* versus *H. jamesii*-only plots were more mixed. Levels of available P were highly correlated with total precipitation that occurred in the 6 months prior to the sampling time (*H. jamesii* dominated plots, $r = 0.49$; *H. jamesii* with *B. tectorum* plots, $r = 0.55$; *Stipa* plots, $r = 0.62$; *Stipa* with *B. tectorum* plots, $r = 0.56$). This is similar to the observations of Miller et al. (2006a, b) that greater precipitation results in greater available P in soils dominated by *B. tectorum*.

In the greenhouse, where pots were watered and thus soil moisture was relatively high, P fractions in four different soil types (high and low sand, high and low $CaCO_3$) were analyzed before and after planting with *B. tectorum* (Fig. 15.10) (Sanford and Belnap unpublished results). Soils before planting and those in control pots (watered, no *B. tectorum*) showed very high percentages of HCl-extractable P, whereas soils with *B. tectorum* present for 100 days had almost no HCl-extractable P remaining, indicating that *B. tectorum* had facilitated the conversion of almost all the recalcitrant P to bioavailable P.

All these lines of evidence indicate that *B. tectorum* is able to access P that is otherwise unavailable to native plants. This is most likely via root exudates, as the greenhouse studies did not allow for plant litter or leaf leachates to influence soil P (there was no litter, and plants were not watered from above). This hypothesis is also supported by the correlation between the increase in available P and precipitation (either in the field or the greenhouse), because increased soil moisture would be present at the time when *B. tectorum* would have more carbon available to produce root exudates.

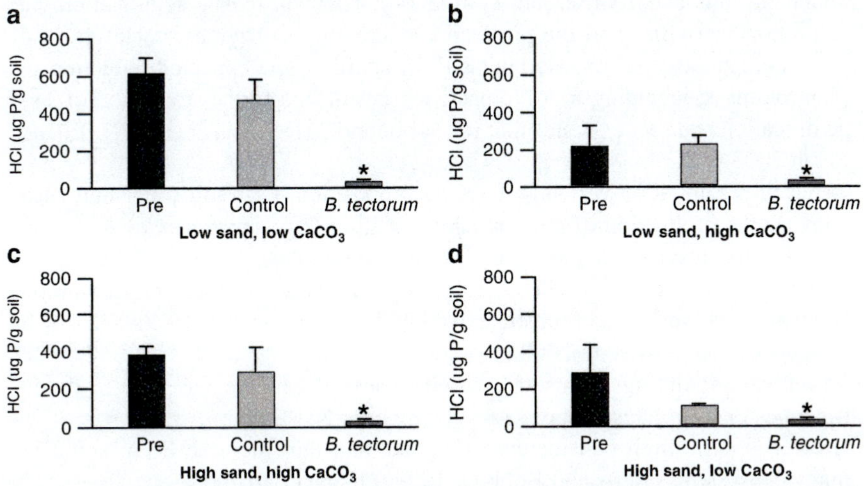

Fig. 15.10 Changes in HCl-extractable (recalcitrant) P for each of four soils (**a**–**d**), representing high and low sand and $CaCO_3$ values. *Pre* indicates measurements from soils before they were planted with *Bromus tectorum*. *Control* indicates measurements from soils that had water added but were not planted with *B. tectorum*. *B. tectorum* measurements are from soils where *B. tectorum* was planted, watered, and grown for 100 days. *Bars* indicate one standard deviation. Note the almost total conversion of the highly recalcitrant P in *B. tectorum* pots to more labile forms of P. Pre and Control samples were not significantly different from each other, whereas the *B. tectorum* treatment was different from both of them in all soil types (*$P < 0.05$) (Belnap and Sanford unpublished results)

15.6 Climate Change Effects on Biological P Cycling in Drylands

Because climate change is expected to affect both temperature and precipitation in dryland regions and both of these factors have a strong influence on the availability of P through both direct and indirect mechanisms, changes in climate will have a substantial impact on P cycling in drylands. As discussed above, the interaction of these two factors can directly affect concentrations of available P in soils. Plants whose distributions are limited by available P will probably see habitats shift in space. For instance, as temperatures increase and soil P becomes less available (due to loss of wet, cool conditions), we might expect the invasive annual grasses found in the western USA to expand their range into more northern habitats, such as the Colorado Plateau, while disappearing from parts of the southern Mojave desert. In regions where temperatures are predicted to rise but precipitation to stay the same or decrease, such as the western USA, a decline in soil moisture will result. This, in turn, will lead to a reduction in biological cycling of P in dryland soils due to multiple factors:

1. A decline in soil moisture will slow all abiotic processes that release bio-unavailable P because these processes depend on soils being moist.
2. Whereas increased temperature could increase microbial activity, reduced soil moisture will limit activity time and probably lead to a decrease in microbial abundance and activity (Sardans et al. 2006). This will reduce acidification via respiratory activities and the production/excretion of enzymes, chelators, and other compounds that release bound P. In addition, the expected reduction in plant biomass, including root biomass, will result in a further reduction of soil acidification and root exudates that release bound P (DeLucia et al. 1997; Li and Sarah 2003).
3. Reduced precipitation will slow litter decomposition. Less soil water may also increase the production of more recalcitrant litter (e.g., more wax on leaves to reduce water loss), further slowing decomposition rates.
4. The activity of many compounds, such as phosphatases, is more dependent on soil water availability than substrate availability and thus their effectiveness will be reduced (Sardans et al. 2008).
5. Fewer wetting/drying events will result in less P being released from dying microbes or from physical processes that release P.
6. Greater N deposition will increase P uptake and thus increase P limitation in many ecosystems (Aerts and Bobbink 1999; Phuyal et al. 2008).
7. Drought and warming will increase plant resorption of P, resulting in less P in litter (Killingbeck and Whitford 1996; Sardans et al. 2006).
8. If soil surfaces are disturbed, the expected decrease in plant biomass will lead to higher sediment movement, and thus an accelerated loss of soil P via wind and water erosion.

However, there will be some factors offsetting these expected declines in soil available P concentrations in dryland regions. Multiple factors will also reduce P uptake, such as reduced microbial and plant biomass, soil moisture, diffusion of P to roots, and root P uptake (Bradford and Hsiao 1982). Increased N deposition is also likely to stimulate the release of bound P.

15.7 Conclusion

There are many factors that distinguish biological P cycling in drylands from P cycling in more mesic regions (Table 15.2). In drylands, P inputs and losses are due more to the deposition and loss of dust than to factors such as weathering and downward leaching. In these sparsely vegetated landscapes, the redistribution of P is a major driving force in determining soil P levels and thus plant and animal community structure. In addition, concentration of P at the soil surface leaves these ecosystems highly vulnerable to loss of P via wind and water erosion. In these landscapes, the high pH and the abundance of $CaCO_3$ and other compounds in soils that complex with P result in low bioavailability. Although microbes and plants have many varied mechanisms to free bound P, low rainfall in these regions limits both the biomass of these organisms and the efficacy of these strategies, while also conferring a proportionally high carbon cost. Many of the more successful invasive plants in dryland regions are those able to access recalcitrant P more effectively than native plants. Because P availability is ultimately dependent on soil moisture and temperature, climate change is expected to have large impacts on P cycling in dryland regions.

Table 15.2 Comparison of different aspects of P cycling between drylands and more mesic areas

Aspect of P cycling	Drylands	Mesic areas
P redistribution	High	Low
Hours of dry soils	High	Low
Size and frequency of plant gaps	High	Low
Number of intense rain events	High	Low
Distance to free surface water	High	Low
Horizontal distribution	High	Low
Animal vertical mixing	Low	High
Plant vertical mixing	Low	High
P availability	Low	High
Inorganic P	High	Low
P complexed with $CaCO_3$	High	Low
Organic matter	Low	High
Microbial immobilization	Low	High
Soil diffusivity	Low	High
Decomposition rates	Low	High
Times of soil moisture	Low	High
Phosphatase release/activity	Low/high	High/low
Root exudates	Low	High
Carbon for mycorrhizae	Low	High

Acknowledgments Many thanks to Sue Phillips and many technicians for data collection and analysis; Robert Sanford for providing data; Mark Miller for helpful comments on the manuscript; and Kymm Herring, Christy Parry, and Jeff Martin for editing and graphics.

References

Aerts R, Bobbink R (1999) The impact of atmospheric nitrogen deposition on vegetation in terrestrial non-forest ecosystems. In: Langan S (ed) The impacts of nitrogen deposition on natural and semi-natural ecosystems. Kluwer, Dordrecht, pp 85–122

Alkon PU (1999) Microhabitat to landscape impacts: crested porcupine digs in the Negev Desert highlands. J Arid Environ 41:183–202

Allen EB, Allen MF (1988) Facilitation of succession by the nonmycotrophic colonizer *Salsola kali* (Chenopodiaceae) on a harsh site: effects of mycorrhizal fungi. Am J Bot 75:257–266

Allen MF, Figueroa C, Weinbaum BS, Barlow SB, Allen EB (1996) Differential production of oxalate by mycorrhizal fungi in arid ecosystems. Biol Fertil Soils 22:287–292

Augustine DJ (2003) Long-term, livestock-mediated redistribution of nitrogen and phosphorus in an East African savanna. J Appl Ecol 40:137–149

Bardgett RD (2005) The biology of soil: a community and ecosystem approach. Oxford University Press, New York

Barroso CB, Nahas E (2005) The status of soil phosphate fractions and the ability of fungi to dissolve hardly soluble phosphates. Appl Soil Ecol 29:73–83

Barrow JR, Osuna P (2002) Phosphorus solubilization and uptake by dark septate fungi in fourwing saltbush, *Atriplex canescens* (Pursh) Nutt. J Arid Environ 51:449–459

Bashkin M, Stohlgren TJ, Otsuki Y, Lee M, Evangelista P, Belnap J (2003) Soil characteristics and plant exotic species invasions in the Grand Staircase-Escalante National Monument, Utah, USA. Appl Soil Ecol 22:67–77

Baturin GN (2003) Phosphorus cycle in the ocean. Lithol Miner Resour 38:101–119

Belnap J (2003a) Comparative structure of physical and biological soil crusts. In: Belnap J, Lange OL (eds) Biological soil crusts: structure, function, and management. Springer, Berlin, pp 177–191

Belnap J (2003b) Microbes and microfauna associated with biological soil crusts. In: Belnap J, Lange OL (eds) Biological soil crusts: structure, function, and management. Springer, Berlin, pp 167–174

Belnap J, Phillips SL (2001) Soil biota in an ungrazed grassland: response to annual grass (*Bromus tectorum)* invasion. Ecol Appl 11:1261–1275

Belnap J, Sherrod SK (2009) Soil amendment effects on the exotic annual grass *Bromus tectorum* L. and facilitation of its growth by the native perennial grass *Hilaria jamesii* (Torr.) Benth. Plant Ecol 201(2):709–721

Belnap J, Prasse R, Harper KT (2003a) Influence of biological soil crusts on soil environments and vascular plants. In: Belnap J, Lange OL (eds) Biological soil crusts: structure, function, and management. Springer, Berlin, pp 281–300

Belnap J, Sherrod SK, Miller ME (2003b) Effects of soil amendments on germination and emergence of downy brome (*Bromus tectorum*) and *Hilaria jamesii*. Weed Sci 51:371–378

Belnap J, Büdel B, Lange OL (2003c) Biological soil crusts: characteristics and distribution. In: Belnap J, Lange OL (eds) Biological soil crusts: structure, function, and management. Springer, Berlin, pp 3–30

Belnap J, Phillips SL, Miller ME (2004) Response of desert biological soil crusts to alterations in precipitation frequency. Oecologia 141:306–316

Belnap J, Phillips SL, Sherrod S, Moldenke A (2005) Soil biota can change after exotic plant invasion: does this affect ecosystem processes? Ecology 86:3007–3017

Belnap J, Phillips SL, Troxler T (2006) Soil lichen and moss cover and species richness can be highly dynamic: the effects of invasion by the annual exotic grass *Bromus tectorum* and the effects of climate on biological soil crusts. Appl Soil Ecol 32:63–76

Bestlemeyer BT, Brown JR, Havstad KM, Fredrickson EL (2006) A holistic view of an arid ecosystem: a synthesis of research an dits applications. In: Havstad KM, Huenneke LF, Schlesinger WH (eds) Structure and function of a Chihuahuan desert ecosystem: the Jornada Basin long-term ecological research site. Oxford University Press, Oxford, pp 354–368

Billings WD (1950) Vegetation and plant growth as affected by chemically altered rocks in the western Great Basin. Ecology 31:62–74

Blackwell MSA, Brookes PC, de la Fuente-Martinez N, Murray PJ, Snars KE, Williams JK, Haygarth PM (2009) Effects of soil drying and rate of re-wetting on concentrations and forms of phosphorus in leachate. Biol Fertil Soils 45:635–643

Blank R, Sforza R (2007) Plant-soil relationships of the invasive annual grass *Taeniatherum caput-medusae*: a reciprocal transplant experiment. Plant Soil 298:7–19

Blank R, Young JA (2002) Influence of the exotic invasive crucifer, *Lepidium latifolium*, on soil properties and elemental cycling. Soil Sci 167:821–829

Bolton H Jr, Smith JL, Link SO (1993) Soil microbial biomass and activity of a disturbed and undisturbed shrub-steppe ecosystem. Soil Biol Biochem 25:545–552

Bornyasz MA, Graham RC, Allen MF (2005) Ectomycorrhizae in a soil-weathered granitic bedrock regolith: linking matrix resources to plants. Geoderma 126:141–160

Boulton AM, Jaffee BA, Scow KM (2003) Effects of a common harvester ant (*Messor andrei*) on richness and abundance of soil biota. Appl Soil Ecol 23:257–265

Bowker MA, Belnap J, Davidson DW, Goldstein H (2006) Correlates of biological soil crust abundance across a continuum of spatial scales: support for a hierarchical conceptual model. J Appl Ecol 43:152–163

Bradford KJ, Hsiao TC (1982) Physiological responses to moderate water stress. In: Lange OL, Nobel PS, Osmond CB, Ziegler H (eds) Physiological plant ecology II. Springer, Berlin, pp 263–324

Breecker DO, Sharp ZD, McFadden LD (2009) Seasonal bias in the formation and stable isotopic composition of pedogenic carbonate in modern soils from central New Mexico, USA. Geol Soc Am Bull 121:630–640

Büdel B (2000) Symbioses (living one inside the other). In: Seckbach J (ed) Journey to diverse microbial worlds: adaptation to exotic environments. Kluwer, Dordrecht, The Netherlands, pp 257–266

Butterly CR, Bünemann EK, McNeill AM, Baldock JA, Marschner P (2009) Carbon pulses but not phosphorus pulses are related to decreases in microbial biomass during repeated drying and rewetting of soils. Soil Biol Biochem 41:1406–1416

Caldwell MM, Richards JH (1986) Competing root systems: morphology and models of absorption. In: Givnish TJ (ed) On the economy of plant form and function. Cambridge University Press, Cambridge, pp 251–273

Callaway RM (2007) Positive interactions and interdependence in plant communities. Springer, Dordrecht, The Netherlands

Callaway RM, Aschehoug ET (2000) Invasive plants versus their new and old neighbors: a mechanism for exotic invasion. Science 290:521–523

Cannon JP, Allen EB, Allen MF, Dudley LM, Jurinak JJ (1995) The effects of oxalates produced by *Salsola tragus* on the phosphorus nutrition of *Stipa pulchra*. Oecologia 102:265–272

Casper BB, Jackson RB (1997) Plant competition underground. Annu Rev Ecol Syst 28:545–570

Castenholz RW, Garcia-Pichel F (2000) Cyanobacterial responses to UV-radiation. In: Whitton BA, Potts M (eds) The ecology of cyanobacteria. Kluwer, Dordrecht, The Netherlands, pp 591–611

Chapuis-Lardy L, Le Bayon R-C, Brossard M, López-Hernández D, Blanchart E (2011) Role of soil macrofauna in phosphorus cycling. In: Bünemann E, Oberson A, Frossard E (eds)

Phosphorus in action: biological processes in soil phosphorus cycling. Soil biology, vol 26. Springer, Heidelberg. doi: 10.1007/978-3-642-15271-9_8

Charley JL (1977) Mineral cycling in rangeland ecosystems. In: Sosebee RE (ed) Rangeland plant physiology. Society of Range Management, Denver, CO, pp 215–256

Charley JL, Cowling SW (1968) Changes in soil nutrient status resulting from overgrazing and their consequences in plant communities of semi-arid areas. Proc Ecol Soc Aust 3:28–38

Clarkson DT (1985) Factors affecting mineral nutrient acquisition by plants. Annu Rev Plant Physiol 26:77–115

Collins SL, Sinsabaugh RL, Crenshaw C, Green L, Porras-Alfaro A, Stursova M, Zeglin LH (2008) Pulse dynamics and microbial processes in aridland ecosystems. J Ecol 96:413–420

Cook CW, Gates DH (1960) Effects of site and season on oxalate content of halogeton. J Range Manage 13:97–101

Darby BJ, Neher DA, Belnap J (2007) Soil nematode communities are ecologically more mature beneath late- than early-successional stage biological soil crusts. Appl Soil Ecol 35:203–212

de Caire GZ, de Cano MS, Palma RM, de Mulé CZ (2000) Changes in soil enzyme activities following additions of cyanobacterial biomass and exopolysaccharide. Soil Biol Biochem 32:1985–1987

DeLucia EH, Schlesinger WH, Billings WD (1989) Edaphic limitations to growth and photosynthesis in Sierran and Great Basin vegetation. Oecologia 78:184–190

DeLucia EH, Callaway RM, Thomas EM, Schlesinger WH (1997) Mechanisms of phosphorus acquisition for ponderosa pine seedlings under high CO2 and temperature. Ann Bot 79:111–120

Dornbush ME (2007) Grasses, litter, and their interaction affect microbial biomass and soil enzyme activity. Soil Biol Biochem 39:2241–2249

Drake M, Steckel JE (1995) Solubilization of soil and rock phosphate as related to root cation exchange capacity. Proc Soil Sci Soc Am 19:449–450

Duda JJ, Freeman DC, Emlen JM, Belnap J, Kitchen SG, Zak JC, Sobek E, Tracy M, Montante J (2003) Differences in native soil ecology associated with invasion of the exotic annual chenopod, Halogeton glomeratus. Biol Fertil Soils 38:72–77

Eldridge DJ, Rath D (2002) Hip holes: Kangaroo (*Macropus* spp.) resting sites modify the physical and chemical environment of woodland soils. Aust Ecol 27:527–536

Epstein E (1961) The essential role of calcium in selective cation transport by plant cells. Plant Physiol 36:437–444

Estiarte M, Penuelas J, Sardans J, Emmett BA, Sowerby A, Beier C, Schmidt IK, Tietema A, Van Meeteren MJM, Kovacs Lang E, Mathe P, De Angelis P, De Dato G (2008) Root-surface phosphatase activity in shrublands across a European gradient: effects of warming. J Environ Biol 29:25–29

Facelli JM, Brock DJ (2000) Patch dynamics in arid lands: localized effects of *Acacia papyrocarpa* on soils and vegetation of open woodlands of south Australia. Ecography 23:479–491

Fenn ME, Baron JS, Allen EB, Rueth HM, Nydick KR, Geiser L, Bowman WD, Sickman JO, Meixner T, Johnson DW, Neitlich P (2003) Ecological effects of nitrogen deposition in the Western United States. Bioscience 53:404–420

Field JP, Belnap J, Breshears DD, Neff JC, Okin GS, Whicker JJ, Painter TH, Ravi S, Reheis MC, Reynolds RL (2009) The ecology of dust. Front Ecol Environ. doi:10.1890/090050

Fogg GE (1966) The extracellular products of algae. Oceanogr Mar Biol 4:195–212

Fox TR, Comerford NB (1990) Low-molecular-weight organic acids in selected forest soils of the southeastern USA. Soil Sci Soc Am J 54:1139–1144

Gadd GM, Watkinson SC, Dyer PS (2007) Fungi in the environment. Cambridge University Press, New York

Garcia-Pichel F, Belnap J (1996) Microenvironments and microscale productivity of cyanobacterial desert crusts. J Phycol 32:774–782

Garkaklis MJ, Bradley JS, Wooller RD (2004) Digging and soil turnover by a mycophagous marsupial. J Arid Environ 56:569–578

Geesey G, Jang L (1990) Extracellular polymers for metal binding. In: Ehrlich HL, Brierley CL (eds) Microbial mineral recovery. McGraw-Hill, New York, pp 223–247

George TS, Fransson A-M, Hammond JP, White PJ (2011) Phosphorus nutrition: rhizosphere processes, plant response and adaptations. In: Bünemann E, Oberson A, Frossard E (eds) Phosphorus in action: biological processes in soil phosphorus cycling. Soil biology, vol 26. Springer, Heidelberg. doi: 10.1007/978-3-642-15271-9_10

Green LE, Porras-Alfaro A, Sinsabaugh RL (2008) Translocation of nitrogen and carbon integrates biotic crust and grass production in desert grassland. J Ecol 96:1076–1085

Greene B, Darnall DW (1990) Microbial oxygenic photoautotrophs (cyanobacteria and algae) for metal-ion binding. In: Ehrlich HL, Brierley CL (eds) Microbial mineral recovery. McGraw-Hill, New York, pp 277–302

Gundale MJ, Sutherland S, DeLuca TH (2008) Fire, native species, and soil resource interactions influence the spatio-temporal invasion pattern of *Bromus tectorum*. Ecography 31:201–210

Hansen ES (1999) Epilithic lichens on iron- and copper-containing crusts at Qeqertarsuaq, Central West Greenland. Graphis Scripta 10:7–12

Harner RF, Harper KT (1973) Mineral composition of grassland species of the eastern great basin in relation to stand productivity. Can J Bot 51:2037–2046

Harris GA (1967) Some competitive relationships between *Agropyron spicatum* and *Bromus tectorum*. Ecol Monogr 37:89–111

Harvey SJ, Nowierski RM (1989) Spotted knapweed: allelopathy or nutrient depletion. In: Fay PK, Lacey JR (eds) Knapweed symposium. Plant and Soil Science Department and Extension Service, Montana State University, Bozeman, p 118

Herron GJ, Sheley RL, Maxwell BD, Jacobsen JS (2001) Influence of nutrient availability on the interaction between spotted knapweed and bluebunch wheatgrass. Restor Ecol 9:326–331

Hiernaux P, Bielders CL, Valentin C, Bationo A, Fernandez-Rivera S (1999) Effects of livestock grazing on physical and chemical properties of sandy soils in Sahelian rangelands. J Arid Environ 41:231–245

Hilder EJ, Mottershead BE (1963) The redistribution of plant nutrients through free-grazing sheep. Aust J Sci 26:88–89

Holford ICR, Chater M, Mattingly GEG (1990) Effects of decalcification on the phosphate sorption characteristics of eight calcareous soils. Aust J Soil Res 28:919–928

Hooper DU, Johnson LC (1999) Nitrogen limitation in dryland ecosystems: responses to geographical and temporal variation in precipitation. Biogeochemistry 46:247–293

Horner HT, Wagner BL (1995) Calcium oxalate formation in higher plants. In: Khan SR (ed) Calcium oxalate in biological systems. CRC, Boca Raton, FL, pp 53–72

James JJ, Tiller RL, Richards JH (2005) Multiple resources limit plant growth and function in a saline-alkaline desert community. J Ecol 93:113–126

Jansa J, Finlay R, Wallander H, Smith FA, Smith SE (2011) Role of mycorrhizal symbioses in phosphorus cycling. In: Bünemann E, Oberson A, Frossard E (eds) Phosphorus in action: biological processes in soil phosphorus cycling. Soil biology, vol 26. Springer, Heidelberg. doi: 10.1007/978-3-642-15271-9_6

Jobbágy EG, Jackson RB (2001) The distribution of soil nutrients with depth: global patterns and the imprint of plants. Biogeochemistry 53:51–77

Jobbágy EG, Jackson RB (2004) The uplift of soil nutrients by plants: biogeochemical consequences across scales. Ecology 85:2369–2379

Jonasson S, Chapin FS III (1991) Seasonal uptake and allocation of phosphorus in *Eriophorum vaginatum L.* measured by labelling with ^{32}P. New Phytol 118:349–357

Jones DL, Oburger E (2011) Solubilization of phosphorus by soil microorganisms. In: Bünemann E, Oberson A, Frossard E (eds) Phosphorus in action: biological processes in soil phosphorus cycling. Soil biology, vol 26. Springer, Heidelberg. doi: 10.1007/978-3-642-15271-9_7

Jones D, Wilson MJ (1985) Chemical activity of lichens on mineral surfaces – a review. Int Biodeter 21:99–104

Jungk A, Claassen N (1997) Ion diffusion in the soil-root system. Adv Agron 61:53–110

Jurinak JJ, Griffin RA (1972) Factors affecting the movement and distribution of anions in desert soils. US/IBP desert biome research memorandum 72-38. Utah State University, Logan, UT

Jurinak JJ, Dudley LM, Allen MF, Knight WG (1986) The role of calcium oxalate in the availability of phosphorus in soils of semiarid regions: a thermodynamic study. Soil Sci 142:255–261

Kieft TL, Soroker E, Firestone MK (1987) Microbial biomass response to a rapid increase in water potential when dry soil is wetted. Soil Biol Biochem 19:119–126

Killingbeck KT, Whitford WG (1996) High foliar nitrogen in desert shrubs: an important ecosystem trait or defective desert doctrine? Ecology 77:1728–1737

Killingbeck K, Whitford W (2001) Nutrient resorption in shrubs growing by design, and by default in Chihuahuan Desert arroyos. Oecologia 128:351–359

Kleiner EF, Harper KT (1977) Occurrence of four major perennial grasses in relation to edaphic factors in a pristine community. J Range Manage 30:286–289

Klemmedson JO, Tiedemann AR (1986) Long-term effects of mesquite removal on soil characteristics: II. Nutrient availability. Soil Sci Soc Am J 50:476–480

Klemmedson JO, Tiedemann AR (1998). Lithosequence of soils and associated vegetation on subalpine range of the Wasatch Plateau, Utah. USDA Forest Service, Pacific Northwest Research Station, Portland, Oregon, pp 1–16

Krauskopf KB, Bird DK (1995) Introduction to geochemistry. McGraw Hill, New York

Lajtha K (1987) Nutrient reabsorption efficiency and the response to phosphorus fertilization in the desert shrub *Larrea tridentata* (DC.) Cov. Biogeochemistry 4:265–276

Lajtha K, Harrison AF (1995) Strategies of phosphorus acquisition and conservation by plant species and communities. In: Tiessen H (ed) Phosphorus in the global environment. Wiley, Chichester, UK, pp 140–147

Lajtha K, Schlesinger WH (1986) Plant response to variations in nitrogen availability in a desert shrubland community. Biogeochemistry 2:29–37

Lajtha K, Schlesinger WH (1988) The biogeochemistry of phosphorus cycling and phosphorus availability along a desert soil chronosequence. Ecology 69:24–39

Lange W (1974) Chelating agents and blue-green algae. Can J Microbiol 20:1311–1321

Lee KE (1977) Physical effects of herbivores on arid and semi-arid rangeland ecosystems. The impact of herbivores on arid and semi-arid rangelands. In: Proceedings 2nd US/Aus Rangeland Panel, Adelaide, 1972. Australian Rangeland Society, Perth, West Australia

Lei SA, Walker LR (1997) Biotic and abiotic factors influencing the distribution of *Coleogyne* communities in southern Nevada. Great Basin Nat 57:163–171

LeJeune KD, Seastedt TR (2001) *Centaurea* species: the forb that won the West. Conserv Biol 15:1568–1574

LeJeune KD, Suding KN, Seastedt TR (2006) Nutrient availability does not explain invasion and dominance of a mixed grass prairie by the exotic forb *Centaurea diffusa* Lam. Appl Soil Ecol 32:98–110

Lewin RA (1956) Extracellular polysaccharides of green algae. Can J Microbiol 2:665–672

Li XY, Liu LY (2003) Effect of gravel mulch on aeolian dust accumulation in the semiarid region of northwest China. Soil Tillage Res 70:73–81

Li X, Sarah P (2003) Enzyme activities along a climatic transect in the Judean Desert. CATENA 53:349–363

Lindahl BD, Finlay RD, Cairney JWG (2005) Enzymatic activities of mycelia in mycorrhizal fungal communities. In: Dighton J, White JF, Oudemans P (eds) The fungal community: its organization and role in the ecosystem, vol 23. Taylor & Francis, Boca Raton, FL, pp 331–348

Lobry de Bruyn LA, Conacher AJ (1990) The role of termites and ants in soil modification: a review. Aust J Soil Res 28:55–93

Lonsdale WM (1999) Global patterns of plant invasions and the concept of invasibility. Ecology 80:1522–1536

Lynch JP, Deikman J (1998) Phosphorus in plant biology: regulatory roles in molecular, cellular, organismic and ecosystem processes. American Society of Plant Physiologists, Rockville, MD

Lynch JP, Ho MD (2005) Rhizoeconomics: carbon costs of phosphorus acquisition. Plant Soil 269:45–56

Ma B, Zhou ZY, Zhang LL, Gao WX (2007) The principal component analysis of soil and population growth status of *Artemisia sphaerocephala* in arid region of Alex Desert. Xibei Zhiwu Xuebao 27:995–999

Ma B, Zhou ZY, Zhang CP, Zhang G, Hu YJ (2009) Inorganic phosphorus fractions in the rhizosphere of xerophytic shrubs in the Alxa Desert. J Arid Environ 73:55–61

MacKay WP (1991) The role of ants and termites in desert communities. In: Polis G (ed) The ecology of desert communities. University of Arizona Press, Tuscon, pp 113–150

MacMahon JA, Mull JF, Crist TO (2000) Harvester ants (*Pogonomyrmex* SPP.): their community and ecosystem influences. Annu Rev Ecol Syst 31:265–291

Magid J, Nielsen NE (1992) Seasonal variation in organic and inorganic phosphorus fractions of temperate-climate sandy soils. Plant Soil 144:155–165

Mandyam K, Jumpponen A (2005) Seeking the elusive function of the root-colonising dark septate endophytic fungi. Stud Mycol 53:173–189

Marschner H (1995) Ion uptake mechanisms of individual cells and roots: short-distance transport. In: Marschner H (ed) Mineral nutrition of higher plants. Academic, San Diego, pp 6–78

McClaran MP, Van Devender TR (1995) The desert grassland. University of Arizona Press, Tucson

McCulley RL, Jobbagy EG, Pockman WT, Jackson RB (2004) Nutrient uptake as a contributing explanation for deep rooting in arid and semi-arid ecosystems. Oecologia 141:620–628

McLean RJC, Beveridge TJ (1990) Metal-binding capacity of bacterial surfaces and their ability to form mineralized aggregates. In: Ehrlich HL, Brierley CL (eds) Microbial mineral recovery. McGraw-Hill, New York, pp 185–222

Midgley GF, van der Heyden F (1999) Form and function in perennial plants. In: Dean WRJ, Milton SJ (eds) The Karoo: ecological patterns and processes. Cambridge University Press, Cambridge, pp 91–106

Miller ME, Belnap J, Beatty SW, Webb BL (2006a) Effects of water additions, chemical amendments, and plants on in situ measures of nutrient bioavailability in calcareous soils of southeastern Utah, USA. Plant Soil 288:19–29

Miller ME, Belnap J, Beatty S, Reynolds RL (2006b) Performance of *Bromus tectorum* L. in relation to soil properties, water additions, and chemical amendments in calcareous soils of southeastern Utah, USA. Plant Soil 288:1–18

Monger HC (2006) Soil development in the Jornada Basin. In: Havstad KM, Huenneke LF, Schlesinger WH (eds) Structure and function of a Chihuahuan Desert ecosystem: the Jornada Basin long-term ecological research site. Oxford University Press, Oxford, pp 81–106

Nadeau J, Qualls R, Nowak R, Blank R (2007) The potential bioavailability of organic C, N, and P through enzyme hydrolysis in soils of the Mojave Desert. Biogeochemistry 82:305–320

Naiman RJ, Braack L, Grant R, Kemp AC, Du Toit JT, Venter FJ (2003) Interactions between species and ecosystem characteristics. In: Du Toit JT, Rogers KH, Biggs HC (eds) The Kruger experience: ecology and management of savanna heterogeneity. Island, Washington, pp 221–241

Nannipieri P, Giagnoni L, Landi L, Renella G (2011) Role of phosphatase enzymes in soil. In: Bünemann E, Oberson A, Frossard E (eds) Phosphorus in action: biological processes in soil phosphorus cycling. Soil biology, vol 26. Springer, Heidelberg, Berlin. doi: 10.1007/978-3-642-15271-9_9

Nash TH III (1996) Lichen biology. Cambridge University Press, Cambridge

Neff JC, Reynolds R, Belnap J, Lamothe P (2005) Multi-decadal impacts of grazing on soil physical and biogeochemical properties in southeast Utah. Ecol Appl 15:87–95

Neff JC, Ballantyne AP, Farmer GL, Mahowald NM, Conroy JL, Landry CC, Overpeck JT, Painter TH, Lawrence CR, Reynolds RL (2008) Increasing eolian dust deposition in the western United States linked to human activity. Nat Geosci 1:189–195

Okin GS, Mahowald N, Chadwick OA, Artaxo P (2004) Impact of desert dust on the biogeochemistry of phosphorus in terrestrial ecosystems. Glob Biogeochem Cycles 18:GB2005

Palmer AR, Novellie PA, Lloyd JW (1999) Community patterns and dynamics. In: Dean WRJ, Milton SJ (eds) The Karoo: ecological patterns and processes. Cambridge University Press, Cambridge, pp 208–222

Parker KC (1995) Effects of complex geomorphic history on soil and vegetation patterns on arid alluvial fans. J Arid Environ 30:19–39

Phuyal M, Artz R, Sheppard L, Leith I, Johnson D (2008) Long-term nitrogen deposition increases phosphorus limitation of bryophytes in an ombrotrophic bog. Plant Ecol 196:111–121

Quiquampoix H, Mousain D (2005) Enzymatic hydrolysis of organic phosphorus. In: Turner BL, Frossard E, Baldwin DS (eds) Organic phosphorus in the environment. CABI, Cambridge, MA, pp 89–112

Radin JW, Eidenbock MP (1984) Hydraulic conductance as a factor limiting leaf expansion of phosphorus-deficient cotton plants. Plant Physiol 75:372–377

Reed SC, Seastedt TR, Mann CM, Suding KN, Townsend AR, Cherwin KL (2007) Phosphorus fertilization stimulates nitrogen fixation and increases inorganic nitrogen concentrations in a restored prairie. Appl Soil Ecol 36:238–242

Reheis MC, Kihl R (1995) Dust deposition in southern Nevada and California, 1984-1989 – Relations to climate, source area, and source lithology. J Geophys Res 100:8893–8918

Reheis MC, Budahn JR, Lamothe PJ (1999) Elemental analyses of modern dust in southern Nevada and California. USGS OFR 99-0531

Reynolds R, Belnap J, Reheis M, Lamothe P, Luiszer F (2001) Aeolian dust in Colorado Plateau soils: nutrient inputs and recent change in source. Proc Natl Acad Sci USA 98:7123–7127

Reynolds R, Neff JC, Reheis M, Lamothe P (2006a) Atmospheric dust in modern soil on aeolian sandstone, Colorado Plateau (USA): Variation with landscape position and contribution to potential plant nutrients. Geoderma 130:108–123

Reynolds RL, Reheis M, Yount J, Lamothe P (2006b) Composition of aeolian dust in natural traps on isolated surfaces of the central Mojave Desert – insights to mixing, sources, and nutrient inputs. J Arid Environ 66:42–61

Reynolds RL, Reheis MC, Neff JC, Goldstein H, Yount J (2006c) Late Quaternary eolian dust in surficial deposits of a Colorado Plateau grassland: controls on distribution and ecologic effects. CATENA 66:251–266

Rogers SL, Burns RG (1994) Changes in aggregate stability, nutrient status, indigenous microbial populations, and seedling emergence, following inoculation of soil with *Nostoc muscorum*. Biol Fertil Soils 18:209–215

Santos PF, DePree E, Whitford WG (1978) Spatial distribution of litter and microarthropods in a Chihuahuan desert ecosystem. J Arid Environ 1:41–48

Sardans J, Peñuelas J, Estiarte M (2006) Warming and drought alter soil phosphatase activity and soil P availability in a Mediterranean shrubland. Plant Soil 289:227–238

Sardans J, Peñuelas J, Ogaya R (2008) Experimental drought reduced acid and alkaline phosphatase activity and increased organic extractable P in soil in a *Quercus ilex* Mediterranean forest. Eur J Soil Biol 44:509–520

Schelske CL, Hooper FF, Haertl EJ (1962) Responses of a marl lake to chelated iron and fertilizer. Ecology 43:646–653

Schenk HJ, Callaway RM, Mahall BE (1999) Spatial root segregation: are roots territorial? Adv Ecol Res 28:145–180

Schlesinger WH, DeLucia EH, Billings WD (1989) Nutrient-use efficiency of woody plants on contrasting soils in the western Great Basin, Nevada. Ecology 70:105–113

Schlesinger WH, Reynolds JF, Cunningham GL, Huenneke LF, Jarrell WM, Virginia RA, Whitford WG (1990) Biological feedbacks in global desertification. Science 247:1043–1048

Scholes RJ (1990) The influence of soil fertility on the ecology of Southern African dry savannas. J Biogeogr 17:415–419

Scholes RJ, Bond WJ, Eckhardt HC (2003) Vegetation dynamics in the Kruger ecosystem. In: Du Toit JT, Rogers KH, Biggs HC (eds) The Kruger experience: ecology and management of savanna heterogeneity. Island, Washington, pp 242–262

Schwinning S, Sala OE (2004) Hierarchy of responses to resource pulses in arid and semi-arid ecosystems. Oecologia 141:211–220

Sinsabaugh RS (1994) Enzymic analysis of microbial pattern and process. Biol Fertil Soils 17:69–74

Sinsabaugh RL, Lauber CL, Weintraub MN, Ahmed B, Allison SD, Crenshaw C, Contosta AR, Cusack D, Frey S, Gallo ME, Gartner TB, Hobbie SE, Holland K, Keeler BL, Powers JS, Stursova M, Takacs-Vesbach C, Waldrop MP, Wallenstein MD, Zak DR, Zeglin LH (2008) Stoichiometry of soil enzyme activity at global scale. Ecol Lett 11:1252–1264

Smith SD, Nowak RS (1990) Ecophysiology of plants in the intermountain lowlands. In: Osmond CB, Pitelka LF, Hidy GM (eds) Plant biology of the basin and range. Springer, New York, pp 179–241

Stursova M, Crenshaw CL, Sinsabaugh RL (2006) Microbial responses to long-term N deposition in a semiarid grassland. Microb Ecol 51:90–98

Suding KN, LeJeune KD, Seastedt TR (2004) Competitive impacts and responses of an invasive weed: dependencies on nitrogen and phosphorus availability. Oecologia 141:526–535

Thorpe AS, Archer V, DeLuca TH (2006) The invasive forb, *Centaurea maculosa*, increases phosphorus availability in Montana grasslands. Appl Soil Ecol 32:118–122

Tiessen H, Stewart JWB, Cole CV (1984) Pathways of phosphorus transformations in soils of differing pedogenesis. Soil Sci Soc Am J 48:853–858

Turner BL, Haygarth PM (2001) Biogeochemistry: phosphorus solubilization in rewetted soils. Nature 411:258–258

Turner BL, Cade-Menun BJ, Westermann DT (2003a) Organic phosphorus composition and potential bioavailability in semi-arid arable soils of the western United States. Soil Sci Soc Am J 67:1168–1179

Turner BL, Baxter R, Ellwood NTW, Whitton BA (2003b) Seasonal phosphatase activities of mosses from Upper Teesdale northern England. J Bryol 25:189–200

Turner BL, Driessen JP, Haygarth PM, McKelvie ID (2003c) Potential contribution of lysed bacterial cells to phosphorus solubilisation in two rewetted Australian pasture soils. Soil Biol Biochem 35:187–189

Van Breemen N, Finlay RD, Lundström US, Jongmans AG, Giesler R, Melderud P-A (2000) Mycorrhizal weathering: a true case of mineral plant nutrition? Biogeochemistry 49:53–67

Venter FJ, Scholes RJ, Eckhardt HC (2003) The abiotic template and its associated vegetation pattern. In: Du Toit JT, Rogers KH, Biggs HC (eds) The Kruger experience: ecology and management of savanna heterogeneity. Island, Washington, pp 83–129

Verboom WH, Pate JS (2006) Bioengineering of soil profiles in semiarid ecosystems: the 'phytotarium' concept. Plant Soil 289:71–102

Verrecchia E, Yair A, Kidron GJ, Verrecchia K (1995) Physical properties of the psammophile cryptogamic crust and their consequences to the water regime of sandy soils, north-western Negev Desert, Israel. J Arid Environ 29:427–437

von Wandruszka R (2006) Phosphorus retention in calcareous soils and the effect of organic matter on its mobility. Geochem Trans 7:6

Wagner D (1997) Harvester ant nests, soil biota and soil chemistry. Oecologia 112:232–236

Walbridge MR (1991) Phosphorus availability in acid organic soils of the lower North Carolina coastal plain. Ecology 72:2083–2100

West N, Griffin R, Jurinak J (1984) Comparison of phosphorus distribution and cycling between adjacent native semidesert shrub and cultivated grass-dominated ecosystems. Plant Soil 81:151–164

White CS, Moore DI, Craig JA (2004) Regional-scale drought increases potential soil fertility in semiarid grasslands. Biol Fertil Soils 40:73–78

Whitford WG (1999) Comparison of ecosystem processes in the Nama-karoo and other deserts. In: Deand WRJ, Milton SJ (eds) The Karoo: ecological patterns and processes. Cambridge University Press, Cambridge, pp 291–313

Whitford WG (2002) Ecology of desert systems. Academic, San Diego

Whitford WG, Bestlemeyer BT (2006) Chihuahuan desert fauna: effects on ecosystem properties and processes. In: Havstad KM, Huenneke LF, Schlesinger WH (eds) Structure and function of a Chihuahuan Desert ecosystem: the Jornada Basin long-term ecological research site. Oxford University Press, Oxford, pp 247–265

Whitton BA, Al-Shehri AM, Ellwood NTW, Turner BL (2005) Ecological aspects of phosphatase activity in cyanobacteria, eukaryotic algae and bryophytes. In: Turner BL, Frossard E, Baldwin DS (eds) Organic phosphorus in the environment. CABI, Cambridge, MA, pp 205–241

Wood TG, Sands WA (1978) The role of termites in ecosystems. In: Brian MV (ed) Production ecology of ants and termites. Cambridge University Press, Cambridge, pp 245–292

Woodmansee RG (1978) Additions and losses of nitrogen in grassland ecosystems. Bioscience 28:448–453

Wright RD, Mooney HA (1965) Substrate-oriented distribution of bristlecone pine in the White Mountains of California. Am Midl Nat 73:257–284

Yoder CK, Nowak RS (2000) Phosphorus acquisition by *Bromus madritensis* ssp. *rubens* from soil interspaces shared with Mojave Desert shrubs. Funct Ecol 14:685–692

Zhang F, Li L (2003) Using competitive and facilitative interactions in intercropping systems enhances crop productivity and nutrient-use efficiency. Plant Soil 248:305–312

Zou X, Binkley D, Caldwell BA (1995) Effects of dinitrogen-fixing trees on phosphorus biogeochemical cycling in two experimental forests. Soil Sci Soc Am J 59:1452–1458

Chapter 16
Effects of Manure Management on Phosphorus Biotransformations and Losses During Animal Production

Thanh H. Dao and Robert C. Schwartz

16.1 Introduction

Phosphorus (P) continues to be a significant pollutant of surface waters and estuaries in the USA and around the world despite major research and education efforts, the guidelines and rules of comprehensive nutrient management given by federal and state agricultural and regulatory institutions, and the scrutiny of public interest groups (Tamminga 1992; Zhu et al. 2006; Kurz et al. 2006; Withers et al. 2007; USEPA 2008a; Trollea et al. 2008). Livestock production is a major contributor to P loading of surface- and groundwater sources in livestock production-intensive watersheds. Nutrients are imported in large quantities in animal feed, while nutrients excreted in manure accumulate locally. Environmental discharges of P include runoff from animal pens and open feedlot surfaces, seepage of effluents from manure storage areas or overflow and spills, as well as diffuse discharges from manure-treated landscapes and inadequately managed riparian zones and stream banks. An improved understanding of the composition or accurate distribution of the inorganic and organic P forms in animal manure, their respective transformations and modes of dispersal, and the receiving landscapes' vulnerability to losses of P inputs to natural terrestrial and aquatic biospheres is crucial to the success of restoring impacted aquatic ecosystems and protecting them from further impairment. In a recent analysis of the interactions between reactive surfaces of soils and manure soluble and particulate organic P forms, we have explored the multiple strategies that plants and microorganisms use to detect and obtain needed P from animal manure and the soil to meet their nutritional needs. Underlying these strategies, frequent interweaving of biophysical and biochemical processes are observed in the turnover of organic P forms (Dao 2010). In this review, particular

T.H. Dao (✉)
US Department of Agriculture, Agricultural Research Service, Beltsville, MD, USA
e-mail: thanh.dao@ars.usda.gov

R.C. Schwartz
US Department of Agriculture, Agricultural Research Service, Bushland, TX, USA
e-mail: robert.schwartz@ars.usda.gov

E.K. Bünemann et al. (eds.), *Phosphorus in Action*, Soil Biology 26,
DOI 10.1007/978-3-642-15271-9_16, © Springer-Verlag Berlin Heidelberg 2011

attention is given to the linkages between feed inputs and the excreted P characteristics and transformations, so as to detect the relationships that exist between manure management practices and the characteristics of the waste products. This knowledge is essential for an improved understanding of management-induced biotransformations of P forms found in animal manures in order to beneficially reuse this non-renewable nutrient in plant agriculture. Shortages of the finite resource are predicted worldwide with dire consequences for global agricultural productivity (Déry and Anderson 2007; Gilbert 2009). The knowledge is also vital to the development of mitigation strategies and nutrient conservation practices to attenuate the likely losses associated with systems of manure handling, storage structures, and manure-amended agricultural fields.

16.2 Major Types of Animal Production Systems and Associated Manure Management Systems

Understanding the fate and transformations of dietary P inputs, and the choice of approach aimed at mitigating the environmental impacts of manure P lost from animal production, depend largely on the type of animal production system and particularly on the settings in which animals are raised or finished for market.

16.2.1 Extensive Production Systems on Pastures, Rangelands, and Forested Lands

In many parts of the world, small-size livestock that include sheep, goats, free-range poultry and swine can be found foraging on pastures and rangelands. Large grazing ruminants are also raised roaming freely on improved pastures, grass and woodlands in parts of North America, the pampas of South America, and the plateau of China (Heitschmidt et al. 2004; Rotolo et al. 2007). Manure management is incidental to the management of available forage and grazing intensity, and oftentimes consists of simply adjusting animal density and managing the location of drinking water sources and shade. A mixture of animal and plant production can be found in small, more intensively managed operations or subsistence farming systems where a farm has a herd of livestock, and arable lands for crop and animal feed production. Balance of P in these systems is nearly closed. The manure and bedding materials are recycled into the land used for production of crops and forages in order to improve the fertility and tilth of the soil. This type of production system is extensively used in dairy cattle production in New Zealand or Southern Australia, where production is based on a year-round pasture-based system by virtue of a temperate climate (Verkerk 2003). Similarly, a common practice in the southern Great Plains of the USA is the grazing of crop residues and the forage of winter cereals, in particular winter wheat, prior to the onset of winter dormancy or the initiation of reproductive stages of wheat. The combined economic return from winter wheat includes both that of the grain crop and that of animal products and may exceed that of the grain crop without grazing (Baumhardt et al. 2009). During the summer months, available forages include many warm-season grasses. Optimizing system performance is focused on grazing management and supplementation with dietary minerals and harvested feed during periods of low forage availability during a typical annual plant growth cycle.

16.2.2 Confined Livestock and Poultry Production Systems

Animal production has shown an increasing trend for consolidation, where annual animal concentrations of 25,000–50,000 heads of livestock or 60,000–120,000 birds per operation are becoming the norm in the southwestern USA or in the Mid-Atlantic states. Key efficiency considerations in these confined animal feeding operations are daily weight gains and feed conversion efficiency (NRC 1994, 1998, 2001). Common to all large confined feeding operations, feed is introduced from outside the farm. Decisions concerning feed uses are separated from those of production, and particularly of manure utilization on fields to produce the feed and/or cash crops. Therefore, nutrient flow in these systems is open-ended.

16.2.2.1 Ruminant Production and Dietary P Balance

In a typical system involved in the finishing of beef cattle in the USA, young calves are raised on pastures and rangelands for about 12–18 months. They are then gathered and transported to centrally located feedlots to be given a high-energy grain diet and attain a uniform marketable weight. Cattle usually stay for 4–6 months, during which time they can gain 1–1.5 kg day^{-1}. On the other hand, dairy cattle are raised in loafing barns that can accommodate 1,500–2,000 cows. The production of milk requires that the cow be in lactation, following the birth of a calf. The cycle of insemination, pregnancy, parturition, and lactation spans a period of 12–16 months. Dairy operations, therefore, include both the production of milk and the production of calves.

The P requirements for domesticated animals are well established, including those for beef cattle (NRC 2000) and dairy cattle (NRC 2001). The effects of dietary P sources and intake on P partitioning in ruminants have been evaluated in numerous studies over the last 2 decades (Wu et al. 2000; CAST 2002; Bravo et al. 2003), with more emphasis put explicitly on reducing excretion of P and on water quality impairments associated with the excess P used in animal production in the last 10 years (Morse et al. 1992; Wu et al. 2000; Erickson et al. 2002; Knowlton et al. 2001; Estermann et al. 2002; Jewell et al. 2007; Brokman et al. 2008). Dietary P requirements have been refined and lowered to values of 3.4–3.8 g kg^{-1} of dry matter for lactating dairy cows or a total P intake of 75–95 g day^{-1} (NRC 2001). Because of the high P concentration in milk (i.e., 0.95 g kg^{-1}), a continuous supply of feed P is needed to sustain a high level of milk production and animal health. However, P intake in excess of this level results in excessive blood, urine, and fecal P concentrations. For example, Wu et al. (2001) have shown that high-producing dairy cows (>55 kg milk day^{-1}) consuming 77, 97, and 115 g P day^{-1} or rations containing 3.1, 3.9, and 4.7 g kg^{-1} of dietary P, excreted 43, 66, and 88 g P day^{-1} in their feces, respectively. Fecal P concentrations increased linearly with those of dietary P intake, as shown in Fig. 16.1.

A similar relationship between P intake and excretion is found in grazing cattle (Erickson et al. 2002; Knowlton and Herbein 2002; Brokman et al. 2008). Betteridge et al. (1986) showed that steers grazing high-quality pastures excreted 10–23 g P day^{-1} or 44–74% of P intake from the grazed forage dry matter. Holstein steers grazing cool-season grass and legume mixtures excreted 39–55% of the ingested P (Brokman et al. 2008). As a matter of fact, Erickson et al. (2002) observed no change in performance and bone density of 45 feedlot steers fed rations containing 14.2–35.5 g P day^{-1} (i.e., 1.6–4.0 g kg^{-1} of dry matter). However, blood P levels increased with increased total P intake in excess of the basal daily intake of 14.2 g day^{-1} during a 204-day period. In summary, these results all showed that cattle inefficiently utilize dietary P and excrete between 40% and well over 70% of ingested P in high-P diets.

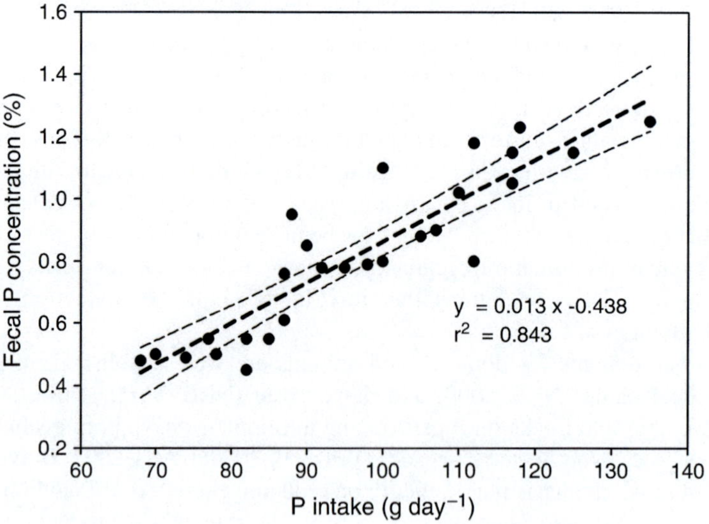

Fig. 16.1 Relationship between dietary P intake and P concentration in feces of dairy cattle fed varying levels of dietary P. *Heavy dashed line* represents the best fit and *dashed lines* on either side represent the 95% confidence intervals for predicted values (adapted from Wu et al. 2001)

16.2.2.2 Production of Monogastric Livestock and Dietary P Balance

Industrial poultry operations can involve 25,000–30,000 birds in the confined quarters of a chicken housing, and an average farm has from three to five units. Today, poultry farms produce over 900 million birds for meat and 72 billion eggs per year in the USA (NASS 2009). Feeding to requirements, growth stage or phase-feeding programs, and more recently, the use of dietary phosphohydrolases (i.e., phytases) are some of the new avenues towards reducing dietary P excretion in poultry and swine. In the latter approach, the goal is to increase feed organic P availability via the hydrolytic activity of added microbial phosphohydrolases, coupled with a decrease in inorganic P (Pi) supplementation (Kornegay et al. 1996; Bedford 2000; Boling et al. 2000; CAST 2002).

Lacking the enzymes to hydrolyze phytic acid (*myo*-inositol 1,2,3,5/4,6-hexa*kis* dihydrogenphosphate; *m*IHP), which is found in abundance in feed grains, the rations of layers and broilers are formulated on the basis of Pi content by assuming that most of the feed P is in an organically bound, indigestible *m*IHP form. For broiler chicks (<3 weeks old), Waldroup et al. (2000) observed that fecal P concentration was 12.1 g kg^{-1} when the chicks were fed at the 1994 NRC recommended dietary level of 4.5 g Pi kg^{-1} of feed dry matter. Dietary P concentrations between 3.2 and 3.4 g kg^{-1} appeared adequate for maximum bone strength (i.e., tibia) and body weight, and optimal feed conversion efficiency in broiler and egg-laying hens (Boling et al. 2000). Fecal P concentrations averaged 10.9 g kg^{-1}, and were reduced by 25% upon supplementation with phosphohydrolases (EC 3.1.3.8) at 800 units kg^{-1} of feed dry matter. Fecal P concentrations increased with increasing Pi intake, a trend similar to that observed in ruminant livestock (Fig. 16.1) (Waldroup et al. 2000). Total fecal production on a dry weight basis was about 1.3 kg per finished bird over a 48-day period (ASAE 2006). Miles et al. (2003) also found reductions in total fecal P excreted upon substitution of 1 g kg^{-1} Pi with the addition of phosphohydrolases at 600 units kg^{-1} (i.e., 9.1 g kg^{-1}, compared to 12 g kg^{-1} with a corn-based ration containing 3.5 g kg^{-1} Pi). However, water-soluble P concentration in the feces increased with enzyme supplementation and was highest (2.85 g kg^{-1}) with the supplementation treatment, compared to the NRC recommended diet of 3.5 g kg^{-1} Pi (i.e., 2.17 g kg^{-1}). Therefore, poultry were inefficient at utilizing feed P, excreting 0.48 g day^{-1} and well over 70% of feed *m*IHP even with dietary phosphohydrolases.

Swine do not produce phosphohydrolase enzymes to break down *m*IHP in the ingested feed and would excrete most of this source of P, if not for the supplementation of microbial phosphohydrolases (Cromwell et al. 1993; Powers et al. 2006). The use of exogenous phosphohydrolases can lower P excretion by about 30% (Jongbloed et al. 1992; Cromwell et al. 1993; Kemme et al. 1997; Adeola et al. 2004), limiting P excretion to between 20 and 55 g P day^{-1} (ASAE 2006). However, inclusion of dietary phosphohydrolases increased the proportion of fecal total P that was water-soluble, for example, from 55 to 59% of fecal total P (Powers et al. 2006). Thus, the practice may increase the assimilation efficiency of feed P by livestock and poultry, although it may inadvertently increase the risk of loss of water-soluble P from the site of manure application or storage stockpiles via runoff following rainfall or irrigation.

16.3 Manure P Forms and Management-Induced Biotransformations

Two major groups of manure management systems can be distinguished (1) systems to collect and store liquid manure suspensions and wastewaters and (2) systems to manage dry solid manures. Each system with their unique features adds to the complexity of the transformations of manure-borne nutrients. Similarities and differences in P transformations based on manure physical forms are discussed together to improve our understanding of the linkages between manure P speciation and potential loss pathways from major confined animal production systems.

16.3.1 Transformations of P in Semisolid and Liquid Manure Management Systems

To characterize the biotransformation processes and fate of P forms in manure, their relationship to P forms that were present in the feed must be re-examined, as well as the environment in which the manure was excreted, collected, and stored. In a small group of commercial dairy farms, Toor et al. (2005) reported that fecal samples contained on average 62% of the total P as Pi, 27% as phosphomonoesters, and 9% as phosphodiesters, where the rations contained 55% of total P as Pi, 41% as phosphomonoesters, and 5% as phosphodiesters. These chemical forms were excreted in feces, along with phosphonates that can be traced back to rumen microbes. It appears that the three farms that used high levels of dietary P (4.8–5.3 g P kg^{-1}) had cattle excreting higher fecal concentrations of *m*IHP (1.6–2.3 g P kg^{-1}) than the other three farms feeding their cattle rations containing 3.6–4.0 g P kg^{-1} (0.8–1.1 g P kg^{-1}). In excreta samples collected from distinctively different animal production systems (i.e., a confined production system, a grazing system, or a hybrid system with partial confinement and grazing for dairy cattle production), a large but variable proportion of fecal total P was present as Pi (25–68%) when Pi also made up a higher percentage of the total P in the feed (46–70%) (McDowell et al. 2008). The proportion of fecal *m*IHP declined, but decreased less in production systems feeding mixed rations with a high Pi content.

These studies and others (e.g., Knowlton et al. 2001; Dou et al. 2002) have reinforced the notion that P forms in animal manures are mostly inorganic forms, primarily because Pi is added liberally to the ration. Uncertainties in the exact composition of inorganic and organic P forms of feed ingredients, and gross generalizations used in estimating organic P availability, often caused the producer to use extra inexpensive Pi supplement in the ration as insurance against losses in animal productivity (CAST 1996, 2002; Karn 2001; Pew Commission IFAPS 2008). The supplement is highly soluble, and the excess is expelled in the animal's excreta. Thus, for all practical purposes, manure is often managed and land applied as if it contains only Pi. Widespread efforts have been expended on assessing the water-extractable Pi content of manures (Wolf et al. 2005) and in developing practices for lowering soluble P concentration in manures with P-immobilizing additives (Dao 1999; Miles et al. 2003). Land application of P-enriched manure has often been linked to temporary increases in water-extractable P (WEP) in runoff from treated fields, thus threatening to impair the water quality of nearby aquatic ecosystems (Pierson et al. 2001; Toor et al. 2003; Green et al. 2007; Dao et al. 2008). Patterns of manure P release have been quantitatively described by a log-normal distribution function, i.e., the released P concentration profile was shaped as an asymmetric peak, skewed to the left (Dao 1999; Dao and Cavigelli, 2003; Dao et al. 2008).

Manure P, however, includes many forms other than the familiar H_xPO_4, where x is 0, 1, or 2. Forms include *m*IHP and its degradation intermediates (i.e., inositol mono-, di-, tri-, tetra-, and pentaphosphates), and phosphodiesters (i.e., nucleic acids, and phospholipids) (Koopmans et al. 2003; Turner 2004; Jayasundera et al. 2005; McDowell et al. 2008; Fuentes et al. 2009). To assess conditions on the farm, Dao et al. (2006) collected 107 samples of dairy manure suspensions from storage structures on working dairy farms across five states of the northeastern USA. The manures showed a wide range of total P (i.e., 3–20 g kg^{-1}), and there were outliers that exceeded 33 g kg^{-1} in total P (Fig. 16.2). The study differed from the published literature in that water-extractable P (WEP) did not make up such a large proportion of manure suspensions, in fact, ranging only between 11 and 20% of the manure, and in that most of the manure P was associated with the colloidal (>0.1 μm) and larger particulate fraction (Dao et al. 2008). More importantly, the results showed that dairy manure contained a wide range of concentrations of enzyme-labile organic P. These forms were either derived from the feed ingredients that were not assimilated by the cattle (Dao 2003, 2004a, 2007), or were biologically synthesized during storage (Dao and Schwartz 2010).

Total P and WEP concentrations varied considerably with sampling location and reflected the high variability of management practices, manure storage conditions, and manure age between these dairy farms, as illustrated in Fig. 16.2 and Table 16.1. These two P forms were weakly correlated ($r^2 = 0.29$) (Fig. 16.3). These results suggested that the solubility of phosphates and some organic P in the manure solution phase was not exceeded in a complex manure suspension, which exhibited a wide range of electrochemical properties and suspended solid content. Solubility and speciation of P in such a complex solution phase are a function of a number of factors that include valency of the counterions for the phosphate species, chelate stability, ligand exchange, pH of medium, and dissociation equilibria, as detailed in Dao (2007). Although manure WEP content can be affected by the nature and amount of dietary Pi (as previously discussed), the solution-phase chemistry of the

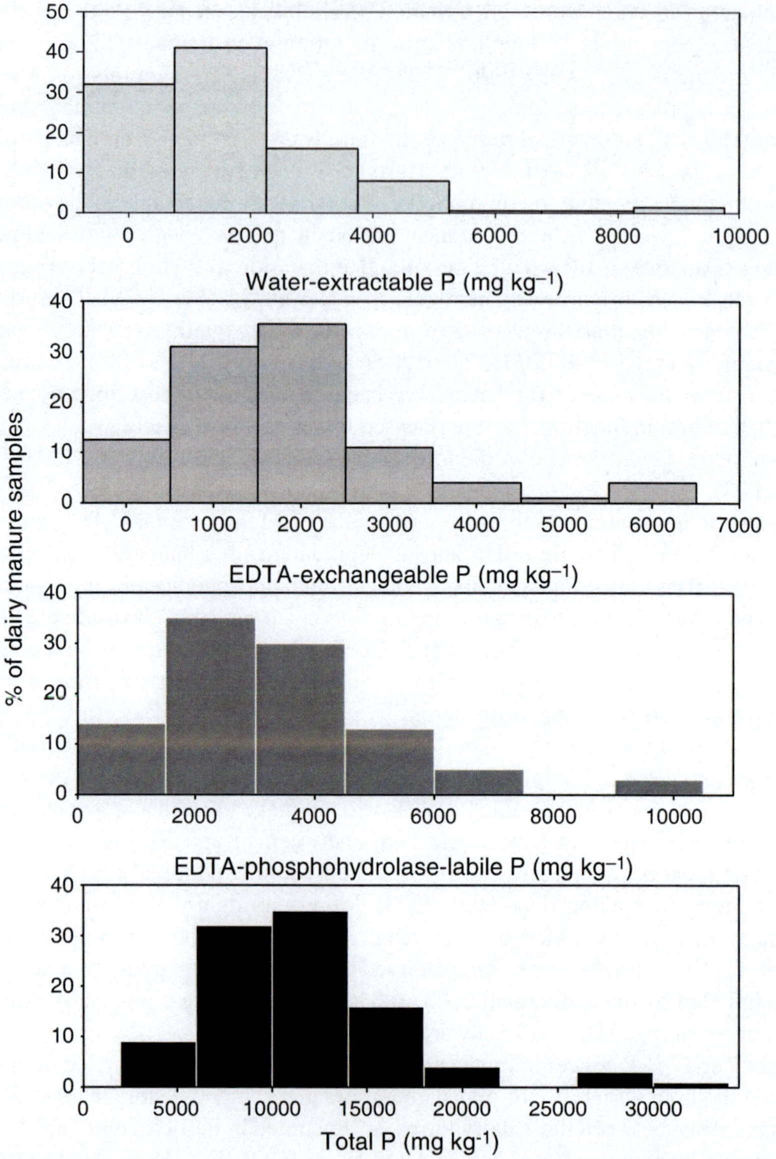

Fig. 16.2 Frequency distributions of bioactive P and total P concentrations in 107 manure suspensions collected from dairy farms across five states of the northeastern USA (adapted from Dao et al. 2006)

external environment where manure was deposited or stored (e.g., pH) appeared to be an equally important factor (Dao et al. 2006; Dao 2007, 2010).

An additional 15 (+/−8%) of manure total P was desorbable as Pi by a ligand-exchange process (shown as EDTA-exchangeable P, EEPi, in Fig. 16.2 and

Table 16.1 Mean extractability of three inorganic and enzyme-labile P fractions within groups of dairy manures varying in total P contents collected from dairy farms across five states of the northeastern USA

Manure total P range (g kg^{-1})	Number of observations	Manure P forms		
		WEP	EEPi	EDTA-PHP
		(Fraction of manure total P, g g^{-1})		
3–8	20	0.200	0.141	0.294
8–12	44	0.172	0.151	0.282
12–16	26	0.131	0.136	0.272
16–20	11	0.137	0.162	0.289
20–32	6	0.108	0.121	0.300
$LSD_{0.05}$	–	0.125	0.062	0.073

WEP water-extractable P, *EEPi* inorganic EDTA-exchangeable P, *EDTA-PHP* EDTA-exchangeable phosphohydrolase-labile P, $LSD_{0.05}$ least significant difference at $P=0.05$
Adapted from Dao et al. 2006

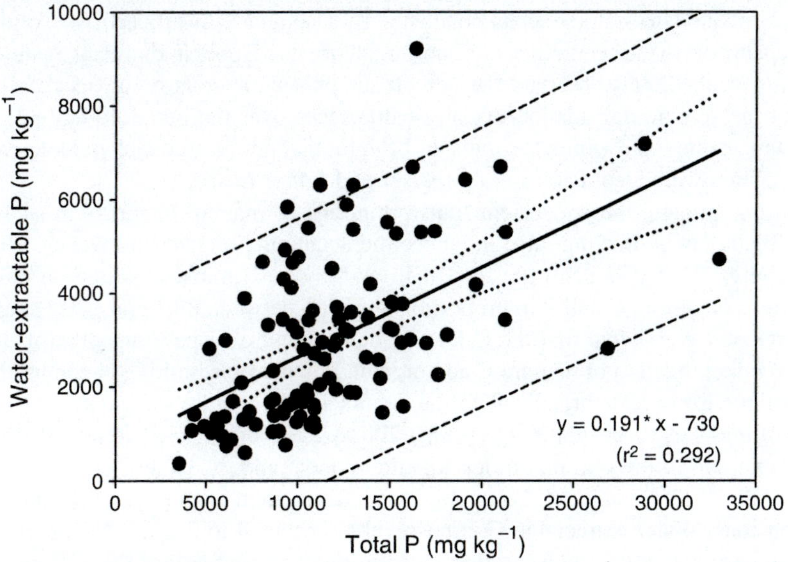

Fig. 16.3 Relationship between WEP and total P concentrations in manure suspensions collected from 107 dairy farms across five states of the northeastern USA. *Dotted lines* represent 95% confidence intervals for predicted values of WEP and *dashed lines* represent the prediction 95% confidence intervals. *Solid line* is best fit (Dao et al. 2006)

Table 16.1). Adding a small amount of the ligand EDTA to the water extractant enhanced the mobilization of this pool of complexed Pi in the manures. Total extractable Ca^{2+} in the EDTA-extracting solution was correlated to EEPi concentration. A slightly stronger relationship existed between these EEPi forms and the sum of Ca^{2+} and Mg^{2+}. The linear relationship suggested that primary forms of the EEPi were derived from Ca and Mg phosphates that are bound to the suspended manure solids.

The variability in feed P composition is exacerbated by the variability in P intake, retention, and assimilation by the animal. The net results are highly variable manure organic P contents. The net ligand-exchangeable phosphohydrolase-labile P (EDTA-PHP) content of all samples of the manure set averaged 32 (+/−16%) of manure total P of this heterogeneous collection of dairy manures. These results confirmed the presence of organic P substrates that include *m*IHP. Complexation with polyvalent cations was previously shown capable of inhibiting *m*IHP dephosphorylation and would explain the excretion and presence of intact *m*IHP in manure of ruminant livestock (Dao 2004a; Toor et al. 2005; Jayasundera et al. 2005). Table 16.1 summarizes the extractability and proportion of dairy manure P forms as a function of total P content of this collection of manure samples. In spite of the heterogeneity of the manure set, these results suggested a marked consistent extraction efficiency of bioactive P forms in all five manure groups, except for WEP as previously discussed. Manure organic P forms are hydrolyzed to yield Pi by endogenous phosphohydrolases in the stored manure slurries, just as in the long-term incubation-fractionation study of fresh dairy cattle excreta conducted by Dao and Schwartz (2010). To give further support to the occurrence of a substantial organic P pool in ruminant manures, Fuentes et al. (2009) observed the activity of phosphohydrolases on fecal *m*IHP during the aerobic degradation of dairy cattle feces over the initial 40-day period following excretion. Dominant groups of bacteria that produced the phytate-degrading enzymes were identified as *Enterobacter* and *Rahnella* species.

During storage, the concurrent transformations of manure C and N in such a microbially active medium also influence the accumulation and turnover of WEP in manures (Dao and Schwartz 2010). In a study of P transformations in dairy manures containing C and P in proportions ranging between 83:1 and 130:1, these researchers have shown that the C load of dairy manure suspensions controls the rate of mineralization of manure C and organic P under non-limiting N conditions. The mineralization of organic P forms occurs in all manure C:P treatments, in spite of initial Pi levels that range from 1,200 to 2,250 mg kg^{-1}. The rate of WEP accumulation follows a first-order kinetic model, and is largest at wide C:P conditions (130:1) and can be up to 156% faster than in narrow C:P manure suspensions. Water-extractable C is positively correlated to manure N:P, particularly at narrow C:P (i.e., non-limiting P conditions). These findings have important implications for dairy slurry management in confined animal production operations. Manure is collected and stored in a central location, undergoing biological transformations before disposal. The manures contain bacterial and fungal populations in sufficient numbers to secrete and release extracellular phosphohydrolases and oxidative enzymes to degrade organic C and N substrates. Wide C:P enhances organic P dephosphorylation and N immobilization, given the biological need to assimilate C substrates and obtain metabolic energy for cell growth and development of manure-borne microorganisms. Manure handling and storage conditions can be managed to minimize gaseous C losses (which reduce enrichment of nutrients in manures) and to moderate the transformation of insoluble complexed Pi and organic P forms to soluble bioactive ones (Dao 2007, 2010; Dao and Schwartz 2010).

It would appear that the mineralization of organic P during storage and the production of Pi enhances the availability of manure P and its value as a biofertilizer in plant production systems; however, a well-cured manure may not be a desirable P source from an environmental perspective. High-available-P manure would be more prone to loss via convective transport, should a plant not be present to absorb and assimilate the soluble P. Maintaining a stable pool of organic P forms in stored manure would provide a slow-release effect of delaying the availability of the organic P fraction to meet the continual need of the plant during the growing season. The potential drawback of this approach, as in managing manure organic N, resides in our inability to accurately predict the rate of dephosphorylation under various soils and climatic conditions. In any case, the phosphorylation–dephosphorylation (i.e., immobilization–mineralization) is a dynamic equilibrium phenomenon, following a biochemical first-order kinetic law. Therefore, P forms in manure will continually be a mixture of Pi and organic P, as observed in a year-long incubation-fractionation study of dairy manure (Dao and Schwartz 2010). In addition, the conservation of the manure C and nutrients under aerobic fermentation conditions commonly found in on-farm liquid manure storage structures would minimize atmospheric emissions of harmful greenhouse gases, and mitigate the confined animal feeding operations' neighbor health concerns associated with current methods of manure management (Schiffman et al. 1995; Pew Commission IFAPS 2008; USEPA 2008b).

16.3.1.1 Loss Pathways of P in Liquid Manure Management Systems

Losses of P occur primarily via its transport in rainwater or snowmelt running off from animal housing, open surfaces of animal pens, manure storage areas, and/or leaching from slurry-amended fields and grasslands (Turner and Haygarth 2000; Toor et al. 2003; Koopmans et al. 2003; Green et al. 2007; Dao et al. 2008). The animal wastewaters and runoff are channeled to detention ponds or large lagoons for settling and periodic cleanup of the sediments upon evaporation. Seepage losses and overflows from manure storage structures and lagoons can contribute to P lost to the environment. Manure slurries stored in underground pits undergo constant mineralization, while Pi forms are being re-immobilized in microbial tissues. These processes continue when the effluent is flushed to an outside detention pond or lagoon. Although notorious for emitting ammonia and other greenhouse gases and malodorous compounds, earthen lagoons are also sources of losses to the subsurface via seepage and failure of the clay liner (Ham and DeSutter 1999; DeSutter et al. 2005). Coarse manure solids, particularly spilled feeds, and bedding materials are mechanically separated before the wastewater enters the lagoon. However, typical mechanical separation efficiencies have ranged from 5 to 30% of particulate removal, which resulted in rapid loss of capacity in these manure storage structures. Chemical coagulants used in the drinking water treatment industry have been shown to improve the separation and removal of particulate P and organic matter from the wastewaters, thus resulting in the transport and reuse of a relatively smaller and lighter volume of nutrient-enriched solids as a soil amendment or as feedstock for the production of bioenergy and chars (Zhang and Lei 1998; Dao and Daniel 2002; Sørensen and Thomsen 2005). The heating of plant materials such as manure solids or crop residues to moderate temperatures to produce a charcoal-like material (called char) that can be incorporated into soil to more permanently sequester the C is well beyond the scope of this review. The reader is referred to selected recent works on the role of chars in soil C sequestration and bioenergy production (Laird 2008; Ro et al. 2009). Much remains to be learned about the potential availability of P forms contained within these materials, depending upon how they were generated (Dao, personal communication).

Anaerobic digestion, a fermentation process in which microorganisms break down organic matter in the absence of oxygen, is an alternative method for treating dairy or swine manure slurries. Biological nutrient reduction technology is used to concentrate nutrients from dilute wastewaters, and generates biogas as a source of renewable energy (Hill 1984; Møller et al. 2004). Bacterial decomposers and methanogens convert the organic matter into methane, carbon dioxide, and low molecular weight organic acids under anaerobic conditions. The residual solids typically contain mineral and complex organic compounds and are rich in P and N, particularly ammonia. Lignin and other residual macromolecules are available for degradation by aerobic microorganisms and for recycling in agricultural soils as previously mentioned. The digester's effluent contains a high concentration of Pi and an elevated level of chemical and biochemical oxygen demand, and must therefore be aerated to reduce the reactivity of the wastewater prior to land application.

16.3.2 Transformations of P in Dry Solid Manure Management

16.3.2.1 Beef Cattle and Dry Manure Management

On open feedlots, 10.5×10^6 Mg of manure solids were generated by the 11.6 million head of beef cattle on feed across the USA in 2007 (NASS 2009). In semi-arid climates, such as the Panhandle of Oklahoma,Texas, or southern Alberta, Canada, the manure in feeding pens is scraped every 4–6 months, and stored in huge uncovered piles. This has resulted in intensive land applications in the immediate vicinity of the feedlots because of the high cost of transporting such material for distances greater than 30 km. Otherwise, the manure is placed in stockpiles that dot the region to contain nutrient escapes to the surrounding area, because transport and recycling of manure in plant production becomes cost-prohibitive, in comparison to nutrient-dense fertilizers. Mineralization of labile C fractions and P-containing organic matter results in an enrichment of P in the stockpiled manure (Hao et al. 2001; Dao and Cavigelli 2003).

Composition and rate of P released from stockpiled and composted manure solids have been quantified in numerous studies (Barnett 1994; Dao 1999; Larney and Hao 2007). Dao and Cavigelli (2003) observed initial large flushes of CO_2–C, exceeding 100 mg kg^{-1} day^{-1}, in stockpiled and composted cattle manure and manure-amended soils. Net mineralizable C, N, and WEP flux densities were log-normally distributed during the incubation period. Significant nonlinear relationships exist between cumulative CO_2–C and inorganic N and between cumulative CO_2–C and WEP and suggest the presence of multiple organic substrate pools of variable stability, as shown in Fig. 16.4.

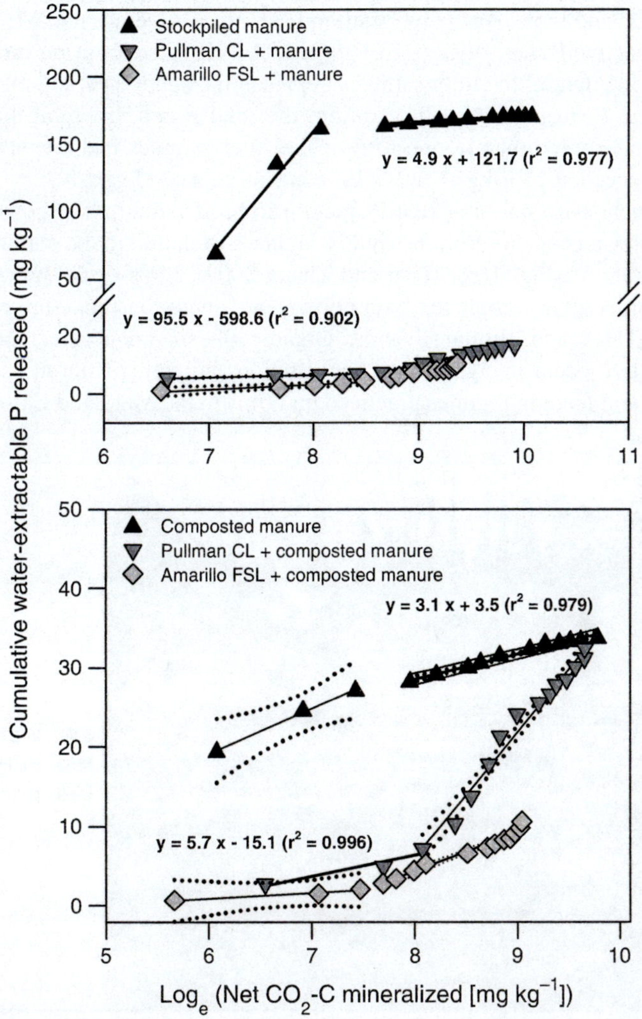

Fig. 16.4 Relationships between WEP released and cumulative CO_2–C evolved from stockpiled manure (*top*) and composted manure (*bottom*) and manure-amended Amarillo and Pullman soils during a 322-day incubation at 35°C. *Dashed lines* represent the 95% confidence limits and *solid lines* the best fit (adapted from Dao and Cavigelli 2003)

16.3.2.2 Poultry Manure and Litter Management

In a study of 87 samples of poultry litter collected across two major regions of industrial poultry production in the USA, an even larger fraction of poultry manure total P in comparison to that of cattle manure, was made up of organic P species, which were susceptible to enzymatic hydrolysis to release Pi, just as previously discussed in Sect. 16.3.1 (Dao and Zhang 2007; Dao and Hoang 2008). From a typical dietary concentration of 3–4 g P kg^{-1}, the variability in litter management practices across farms and producing regions resulted in highly enriched and variable WEP and bioactive P composition (Fig. 16.5). Overall, the WEP fraction and total bioactive P (i.e., WEP + EEP_i + EDTA-PHP) fraction were proportional to the total P content of the litter samples, representing about 24% and over 60% of the litter total P, respectively. In addition, the total P concentration distribution suggested that approximately one-fifth of the litter samples had elevated total P levels that exceeded 21 g kg^{-1} and WEP concentrations $\geq$5 g kg^{-1}.

The all-inclusive total bioactive P fraction in broiler litter also showed a high correlation to Ca concentration or slightly higher correlation to the sum of Ca and Mg concentrations (Fig. 16.6) (Dao and Zhang 2007). Most of the inorganic and organic P forms were complexed with polyvalent counterions but were exchangeable with EDTA and similar ligands. Organic phosphomonoesters or organic enzyme-labile P occur in significant concentrations in poultry litter at current diet formulation and feeding regimes practiced on farms in the Maryland Eastern Shore

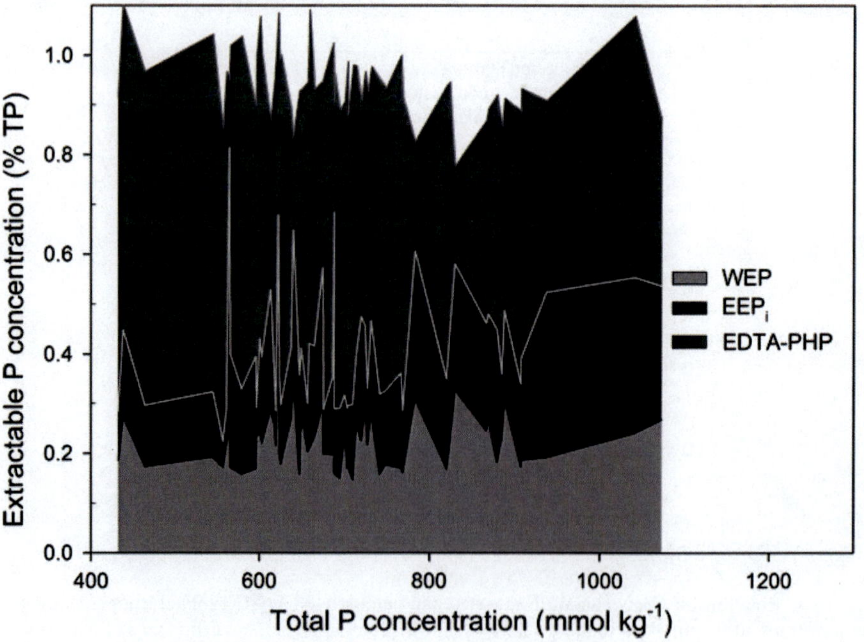

Fig. 16.5 Extractable bioactive P fractions in 87 poultry litter samples (Dao and Hoang 2008)

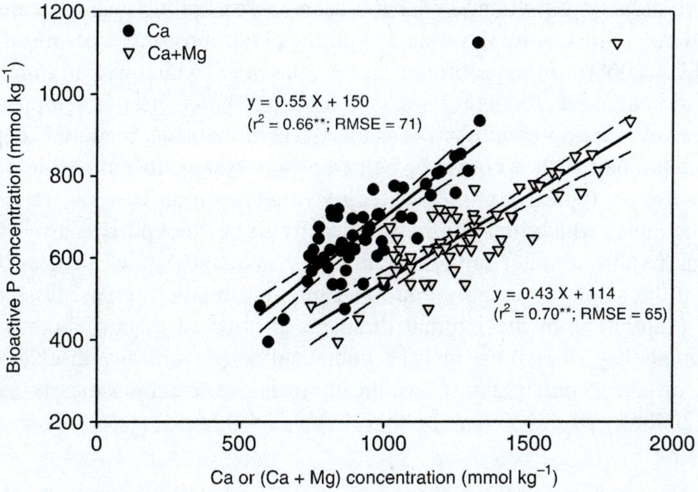

Fig. 16.6 Extractable calcium–magnesium–P relationships in 87 poultry litter samples. *Dashed lines* represent the 95% confidence limits and *solid lines* the best fit (Dao and Hoang 2008)

and the Arkansas–Oklahoma region of the USA Because of the inorganic–organic P composition of poultry litter, there are serious consequences of current feeding management practices and land-based litter options to the quality of the environment. The reduction in dietary P in feed to essential levels must continue to decrease P excretion. Recent research has demonstrated that organic species are highly dynamic (Dao et al. 2005; Dao and Schwartz 2010) and hence potentially biologically active in the environment, particularly in a carbon-rich environment. Previous research has shown internal rearrangement and interpool transfers over time that affected the mineralization of the enzyme-labile organic P pool in manure and manure-amended field soils (Dao 1999, 2004b; Dao et al. 2001, 2005). Such transformations have been observed during the composting of livestock manure and poultry litter. These management-induced alterations of P forms in manure include chemical shifts from the WEP and ligand-exchangeable EEPi fractions to the formation of more stable calcium- and iron-associated phosphates (Traoré et al. 1999; Dao et al. 2001). The changes in P exchangeability occur during the early stage of the composting process, i.e., during the initial intense period of organic matter mineralization (Traoré et al. 1999; Dao 1999; Dao and Schwartz 2010).

16.3.3 Loss Pathways of P in Solid Manure Management Systems

16.3.3.1 Open Feedlots and Dry Cattle and Poultry Manure Stockpiles

On large cattle feedlots, manure solids in animal pens are removed and stacked in the open to dry and, in essence, undergo a slow aerobic fermentation. There is a progressive reduction in volume over the years of stockpiling and containment of manure-borne pollutants in accordance with the conditions of the permit of operation (USEPA 2008b). In large broiler houses, the litter, which is a mixture of bird excreta and bedding materials (i.e., wood chips, rice hulls, etc.), is scrapped every two to three cycles of production a year. In egg-layer housing, manure is deposited on floors equipped with a conveyer-belt transport system to remove it from the poultry house. A typical commercial egg-laying operation can generate 70 Mg day^{-1} of manure, which must be removed daily to be stockpiled nearby. Current regulations require a water-impermeable cover of manure piles to keep rainfall away from the stockpile in sub-humid and humid climatic regions. Loss of concentrated "liquor" from the internal drainage of covered piles of high-moisture manure or poultry litter must also be contained to prevent any discharge from escaping off-site, contaminating wellhead areas, or reaching nearby streams (USEPA 2008b).

16.3.3.2 Losses of P from Grazed Lands

In a grazingland-based system, the P extracted from the forage is returned to the soil from which the forage was grown, except for the 2–5% that leaves the cycle in exported meat products and/or milk. Approximately 75–90% of the nutrients consumed by grazing animals are cycled back to the soil in urine and primarily in feces, which accounts for over 70% of the total excreted P (Betteridge et al. 1986). However, offsite loss of P is known to occur via erosional processes affecting the soil subject to grazing animal traction (Nash et al. 2000; Haynes and Williams 1993; Blackshaw and Blackshaw 1994). By selective grazing, livestock produce a highly heterogeneous distribution of plant species and, consequently, a spatially variable distribution of excreta. Manure nutrients are also concentrated in certain small areas of the pastures where the animals gather to drink, rest, and get shade or shelter from the wind during cold weather. Livestock also cause substantial increases in soil bulk density, which restricts water, air, and plant root penetration, and thereby restricts vegetation growth (Dao et al. 1994; Drewry 2006; Bilotta et al. 2007). Reduced water infiltration leads to increased overland flow, which varies with time of the year and with precipitation frequency and intensity (Nash et al. 2000; McDowell et al. 2006). In alpine meadows, losses of P and other nutrients occur primarily through catastrophic events because these areas are prone to disturbance through soil erosion, landslides, and avalanches (Jewell et al. 2007). Grazing livestock with access to streams and waterways can drop manure P and other contaminants directly into the water; they can damage stream banks and accelerate the loss of vegetation in buffer zones along the streams. However, grazingland-based systems contribute relatively little to the P-induced impairment of surface waters, provided that best grazing management practices, soil erosion control measures, and riparian buffer zones around waterways are used.

16.3.3.3 Loss of P from Land-Applied Manure in Mixed Crop–Livestock Systems

Where plant and animal production coexist on a farm, manure and bedding materials are recycled back to the land to improve the fertility and tilth of the soil for plant production. For all practical purposes, animal manure is managed and land-applied as if manure contains only Pi. It is used to reduce reliance on expensive inorganic fertilizers. However, manure inorganic and organic P forms build up in the soil when applications are made to meet the N requirements of crops and forages, and also when P rates exceed P removal by plants. In a study conducted on a permanent orchardgrass–red clover stand, long-term additions of dairy manure increased total soil organic C, storing the manure C in the near-surface zone (Dao 2004b). Increases in total P, comprising enzyme-labile P as well as Pi (i.e., WEP and Mehlich 3-extractable P, i.e., plant-available P) fractions paralleled the increase in organic C. Added organic P also became increasingly resistant to enzymatic hydrolysis, explaining its preferential accumulation in these soils (Dao 2004b). Sorption mechanisms of *m*IHP and Pi have been postulated to include a binuclear surface complex and monodentate complexes, as summarized in a recent review (Dao 2010). Although Fe and Al hydroxides have a high affinity for Pi and phosphomonoesters, an increase in the desorption of these P forms is observed when polydentate ligands or organic complexing agents are present in the soil solution (Pavinato et al. 2010). Soluble Pi and easily exchangeable P forms that are embodied in WEP assays play an important role in plant nutrition, transfer and loss to runoff, and potential dispersal in the environment (Turner and Haygarth 2000; Toor et al. 2003; Dao 2004b; Dao et al. 2008; Green et al. 2007; Rao and Dao 2008). Green et al. (2007) showed that the concentration and mass distributions of P forms in runoff over time were log-normally distributed. The more inclusive total bioactive P fraction was found in greater concentration and mass than WEP. Peak concentrations and mass loads were greater from soil amended with 30 kg ha^{-1} of manure P than from untreated soil and from soil under orchardgrass–clover than from soil under soybean–wheat rotation, where the application rate was well within the annual P removal rate by these crops. Moreover, the correlations observed between soil bioactive P and P distribution in runoff suggested that runoff P forms were directly associated with soil available P fractions that were partly derived from enzyme-mediated processes in the soil.

Once the quantities and forms of P subject to transport off a particular field location are established, fate and transport models can be used to route P losses to water bodies to permit the assessment of the larger impacts on the watershed and surface water quality. Models can be either a simple approach, such as using the NRCS curve numbers to estimate runoff, or more complex ones such as surface water and sediment routing algorithms contained in fate and transport models (e.g., GLEAMS, AGNAP, or SWAT). Thus, tracking the runoff and soil leachate will reveal the mechanisms of the dispersal and convective transport of the various forms of P described in the previous paragraph.

16.4 Conclusions

Livestock and poultry production play a vital role in agricultural and rural communities of many developing and developed economies and goes well beyond the direct production of eggs, meat, and meat by-products. Whereas mixed plant–animal production systems strive to maintain a closed loop for nutrients used and nutrient outputs, the on-going intensification of industrial animal production has led to a regional imbalance in nutrient distribution. Their local buildup in livestock manures is the result of the geographical separation of the feed production and feed utilization sectors. An understanding of the fate and biological transformations of feed P inputs in an animal and the fraction excreted in manure largely depends upon the settings in which livestock are raised and fed. Recent studies have shown that, whether it is a monogastric or a ruminant species, livestock utilize feed P very inefficiently because they retain 30% or less of the ingested total P. A strong relationship exists between feed P intake and fecal forms and concentrations. As nutritionists and producers attempt to optimize diet formulations, unavoidable loss of dietary P remains large. Though Pi additions are reduced, feed organic P is not fully utilized, short of novel transgenic sources of low-*m*IHP grains. Use of dietary phosphohydrolases can improve feed digestibility and recovery of these organic forms, but the practice also increases the proportion of water-soluble P excreted in feces. Management-induced transformations of the excreted P further increase the risks of soluble P losses from either grassland-based systems or, especially, in the large confined animal feeding operations. Although manure P is influenced by the composition and amount of feed P intake, the chemistry of the solution phase and the solid matrix, as well as the microbiology of the external environment where manure P is deposited, become equally important in P turnover and environmental dispersal. In a nutrient-rich and microbially active environment, manure C and N also influence the fate of P in manure and manure-amended soils, and therefore influence the eventual protracted release of agricultural P in impaired watersheds.

Once manure is disposed on land, the bioactivity of manure P forms is highly dynamic and depends on interactions with the reactive surfaces of soil particles and on biologically mediated transformations. Thus, the contribution of biological processes to the release, transformations, and transport of organic P species must be well understood to develop sustainable management practices for recycling bionutrients in plant production systems. Biological tools that combine ligand exchange and enzyme-mediated mineralization of organic P can mimic plants and microorganisms in their ways of acquiring P from their environment; they can be valuable tools in the exploration and study of the environmental behavior of these organic P sources. They can also provide insights into the loss of P from animal production systems that can occur via its transport in rainwater or snowmelt running off animal housings, manure storage areas, and/or leaching from manure-amended fields, pastures, and rangelands. Off-site P transport is dominated by the runoff–erosion processes, where it is conveyed in dissolved forms or in association with suspended colloidal and larger particulate matter. These transport processes interact in a complex way with the composition of manure, the timing of manure applications, the hydrological characteristics of the application site, and the conservation practices in place. Together, these factors will ultimately determine the extent of P export.

References

Adeola O, Sands JS, Simmins PH, Schulze H (2004) The efficacy of an *Escherichia coli*-derived phytase preparation. J Animal Sci 82:2657–2666

ASAE (2006) Manure production and characteristics. American Society of Agricultural Engineers Standard ASAE D384.2. March 2005. American Society of Agricultural and Biological Engineers, St. Joseph, MI, pp 709–727

Barnett GM (1994) Phosphorus forms in animal manure. Bioresour Technol 49:139–147

Baumhardt RL, Schwartz RC, Greene LW, McDonald JC (2009) Cattle gain and crop yield for a dryland wheat–sorghum–fallow rotation. Agron J 101:150–158

Bedford MR (2000) Exogenous enzymes in monogastric nutrition – their current value and future benefits. Anim Feed Sci Technol 86:1–13

Betteridge K, Andrewes WGK, Sedcole JR (1986) Intake and excretion of nitrogen, potassium and phosphorus by grazing steers. J Agric Sci Camb 106:393–404

Bilotta GS, Brazier RE, Haygarth PM (2007) The impacts of grazing animals on the quality soils, vegetation, and surface waters in intensively managed grasslands. Adv Agron 94:237–280

Blackshaw JK, Blackshaw AW (1994) Heat stress in cattle and the effect of shade on production and behavior: a review. Aust J Exp Agric 34:285–295

Boling SD, Douglas MW, Shirley RB, Parsons CM, Koelkebeck KW (2000) The effect of various levels of phytase and available phosphorus on performance of laying hens. Poult Sci 79:535–538

Bravo D, Sauvant D, Bogaert C, Meschy F (2003) I. A bibliographic database for quantitative analysis of phosphorus flow in ruminants. Reprod Nutr Dev 43:251–269

Brokman AM, Lehmkuhler JW, Undersander DJ (2008) Reducing phosphorus inputs for grazing Holstein steers. J Animal Sci 86:712–719

CAST (1996) Integrated animal waste management. Council for Agricultural Science and Technology, Task Force Report 128. CAST, Ames, IA

CAST (2002) Animal diet modification to decrease the potential for nitrogen and phosphorus pollution. Council for Agricultural Science and Technology, Issue Paper 21. CAST, Ames, IA

Cromwell GL, Stahly TS, Coffey RD, Moneque HJ, Randolph JH (1993) Efficacy of phytase in improving the bioavailability of P in soybean meal and corn-soybean meal diet for pigs. J Animal Sci 71:1831–1840

Dao TH (1999) Co-amendments to modify phosphorus extractability and N:P ratio in feedlot manure and composted manure. J Environ Qual 28:1114–1121

Dao TH (2003) Polyvalent cation effects on *myo*-inositol hexa*kis* dihydrogen- phosphate enzymatic dephosphorylation in dairy wastewater. J Environ Qual 32:694–701

Dao TH (2004a) Organic ligand effects on the enzymatic dephosphorylation of *myo*-inositol hexakis dihydrogenphosphate in dairy wastewater. J Environ Qual 33:349–358

Dao TH (2004b) Ligands and phytase hydrolysis of organic phosphorus in soils amended with dairy manure. Agron J 96:1188–1195

Dao TH (2007) Ligand effects on inositol phosphate solubility and bioavailability in animal manures. In: Turner BL, Richardson AE, Mullaney EJ (eds) Inositol phosphates: linking agriculture and environment. CABI, Wallingford, pp 169–185

Dao TH (2010) Extracellular enzymes in sensing environmental nutrients and ecosystem changes: ligand mediation in organic phosphorus cycling. In: Shukla GC, Varma A (eds) Soil enzymology. Soil biology, vol 22. Springer, New York, Chap. 5. doi: 10.1007/978-3-642-14224-6_5 (in press)

Dao TH, Cavigelli MA (2003) Mineralizable carbon, nitrogen, and water-extractable phosphorus release from stockpiled and composted manure, and manure-amended soils. Agron J 95:405–413

Dao TH, Daniel TC (2002) Particulate and dissolved phosphorus chemical separation and phosphorus release from treated dairy manure. J Environ Qual 31:1388–1398

Dao TH, Hoang KQ (2008) Dephosphorylation and quantification of organic phosphorus in poultry litter by purified phytic-acid high affinity *Aspergillus* phosphohydrolases. Chemosphere 72:1782–1787

Dao TH, Schwartz RC (2010) Mineralizable phosphorus, nitrogen, and carbon relationships in dairy manure at various carbon-to-phosphorus ratios. Bioresour Technol 101:3567–3574

Dao TH, Zhang H (2007) Rapid composition and source screening of heterogeneous poultry litter by energy-dispersive x-ray fluorescence spectrometry. Ann Environ Sci 1:69–79

Dao TH, Morrison J, Unger PW (1994) Soil compaction and bearing strength. In: Stewart BA, Moldenhauer WC (eds) Crop residue management to reduce erosion and improve soil quality. USDA, Agricultural Research Service Conservation Research Report 37. Government Publication Office, Washington DC, pp 40–44

Dao TH, Sikora L, Hamasaki A, Chaney R (2001) Manure P extractability as affected by aluminum and iron by-products and aerobic composting. J Environ Qual 30:1693–1698

Dao TH, Codling EE, Schwartz RC (2005) Time-dependent phosphorus extractability in calcium- and iron-treated high-phosphorus soils. Soil Sci 170:810–821

Dao TH, Lugo-Ospina A, Reeves JB, Zhang H (2006) Wastewater chemistry and fractionation of bioactive phosphorus in dairy manure. Commun Soil Sci Plant Anal 37:907–924

Dao TH, Guber AK, Sadeghi AM, Karns JS, van Kessel JS, Shelton DR, Pachepsky YA, McCarty G (2008) Loss of bioactive phosphorus and enteric bacteria in runoff from dairy manure applied to sod. Soil Sci 173:511–521

Déry P, Anderson B (2007) Peak phosphorus. In: Energy Bulletin Aug 13 2007. Post Carbon Institute. Available at http://www.energybulletin.net/node/33164. Last accessed 10 Apr 2010

DeSutter TM, Pierzynski GM, Ham JM (2005) Movement of lagoon-liquor constituents below four animal waste lagoons. J Environ Qual 34:1234–1242

Dou Z, Knowlton KF, Kohn RA, Wu Z, Satter LD, Zhang G, Toth JD, Ferguson JD (2002) Phosphorus characteristics of dairy feces affected by diets. J Environ Qual 31:2058–2065

Drewry JJ (2006) Natural recovery of soil physical properties from treading damage of pastoral soil New Zealand and Australia: a review. Agric Ecosyst Environ 114:159–169

Erickson GE, Klopfenstein TJ, Milton CT, Brink D, Orth MW, Whittet KM (2002) Phosphorus requirement of finishing feedlot calves. J Animal Sci 80:1690–1695

Estermann BL, Sutter F, Schlegel PO, Erdin D, Wettstein HR, Kreuzer M (2002) Effect of calf age and dam breed on intake, energy expenditure, and excretion of nitrogen, phosphorus, and methane of beef cows with calves. J Animal Sci 80:1124–1134

Fuentes B, Milko Jorquera M, de la Luz MM (2009) Dynamics of phosphorus and phytate-utilizing bacteria during aerobic degradation of dairy cattle dung. Chemosphere 74:325–331

Gilbert N (2009) The disappearing nutrient. Nature 46:716–718

Green VS, Dao TH, Stone G, Cavigelli MA, Baumhardt RL, Devine TE (2007) Bioactive phosphorus loss in simulated runoff from a P-enriched soil under two forage management systems. Soil Sci 172:721–732

Ham JM, DeSutter TM (1999) Seepage losses and N export from swine waste lagoons: a water balance study. J Environ Qual 28:1090–1099

Hao X, Chang C, Larney FJ, Travis GR (2001) Greenhouse gas emissions during cattle feedlot manure composting. J Environ Qual 30:376–386

Haynes RJ, Williams PH (1993) Nutrient cycling and soil fertility in the grazed pasture ecosystem. Adv Agron 49:119–197

Heitschmidt RK, Vermeire LT, Grings EE (2004) Is rangeland agriculture sustainable? J Animal Sci 82(Suppl):138–146

Hill DT (1984) Methane productivity of the major animal waste types. Trans ASAE 27:530–534
Jayasundera S, Schmidt W, Reeves JB III, Dao TH (2005) Direct P-31 NMR spectroscopic measurement of phosphorous forms in dairy manures. J Food Agric Environ 3:328–333
Jewell PL, Käuferle D, Güsewell S, Berry NR, Kreuzer M, Edwards PJ (2007) Redistribution of phosphorus by cattle on a traditional mountain pasture in the Alps. Agric Ecosyst Environ 122:377–386
Jongbloed AW, Mroz Z, Kemme PA (1992) The effect of supplementary *Aspergillus niger* phytase in diets for pigs on concentration and apparent digestibility of dry matter, total phosphorus, and phytic acid in different sections of the alimentary tract. J Animal Sci 70:1159–1168
Karn JF (2001) Phosphorus nutrition of grazing cattle: a review. Anim Feed Sci Technol 89:133–153
Kemme PA, Radcliffe JS, Jongbloed A, Mroz Z (1997) Factors affecting phosphorus and calcium digestibility in diets for growing-finishing pigs. J Animal Sci 75:2139–2146
Knowlton KF, Herbein JH (2002) Phosphorus partitioning during early lactation in dairy cows fed diets varying in phosphorus content. J Dairy Sci 85:1227–1236
Knowlton KF, Herbein JH, Meister-Weisbarth MA, Wark WA (2001) Nitrogen and phosphorus partitioning in lactating Holstein cows fed different sources of dietary protein and phosphorus. J Dairy Sci 84:1210–1217
Koopmans GF, Chardon WJ, Dolfing J, Oenema O, van der Meer P, van Riemsdijk WH (2003) Wet chemical and phosphorus-31 nuclear magnetic resonance analysis of phosphorus speciation in a sandy soil receiving long-term fertilizer or animal manure applications. J Environ Qual 32:287–295
Kornegay ET, Denbow DM, Yi Z, Ravindran V (1996) Response of broilers to graded levels of microbial phytase added to maize-soyabean meal-based diets containing three levels of non-phytate phosphorus. Br J Nutr 75:839–852
Kurz I, O'Reilly CD, Tunney H (2006) Impact of cattle on soil physical properties and nutrient concentrations in overland flow from pasture in Ireland. Agric Ecosyst Environ 113:378–390
Laird DA (2008) The charcoal vision: a win–win–win scenario for simultaneously producing bioenergy, permanently sequestering carbon, while improving soil and water quality. Agron J 100:178–181
Larney FJ, Hao X (2007) A review of composting as a management alternative for beef cattle feedlot manure in southern Alberta, Canada. Bioresour Technol 98:3221–3227
McDowell RW, Muirhead RW, Monaghan RM (2006) Nutrient, sediment, and bacterial losses in overland flow from pasture and cropping soils following cattle dung deposition. Comm Soil Sci Plant Anal 37:93–108
McDowell RW, Dou Z, Toth J, Cade-Menun B, Kleinman P, Soder K, Saporito L (2008) A comparison of phosphorus speciation and potential bioavailability in feed and feces of different dairy herds using 31P nuclear magnetic resonance spectroscopy. J Environ Qual 37:741–752
Miles DM, Moore PA Jr, Smith DR, Rice DW, Stilborn HL, Rowe DR, Lott BD, Branton SL, Simmons JD (2003) Total and water-soluble phosphorus in broiler litter over three flocks with alum litter treatment and dietary inclusion of high available phosphorus corn and phytase supplementation. Poult Sci 82:1544–1549
Møller HB, Sommer SG, Ahring BK (2004) Methane productivity of manure, straw and solid fractions of manure. Biomass Bioenergy 26:485–495
Morse D, Head HH, Wilcox CJ, van Horn HH, Hissem CD, Harris B Jr (1992) Effects of concentration of dietary phosphorus on amount and route of excretion. J Dairy Sci 75:3039–3049
Nash D, Hannah M, Halliwell D, Murdoch C (2000) Factors affecting phosphorus export from a pasture-based grazing system. J Environ Qual 29:1160–1166
NASS (2009) The 2007 Census of agriculture. Department of Agriculture, National Agricultural Statistics Service. Available at http://www.agcensus.usda.gov/. Last accessed 10 Apr 2010
NRC (1994) Nutrient requirements of poultry, 9th edn. National Acadamies Press, National Research Council, Washington, DC

NRC (1998) Nutrient requirements of swine, 10th edn. National Research Council, National Acadamies Press, Washington, DC
NRC (2000) Nutrient requirements of beef cattle, 7th revised edn. Update 2000. National Research Council, National Acadamies Press, Washington, DC
NRC (2001) Nutrient requirements of dairy cattle, 7th revised edn. National Research Council, National Acadamies Press, Washington, DC
Pavinato PS, Dao TH, Rosolem CA (2010) Tillage and phosphorus management effects on enzyme-labile bioactive phosphorus availability in Brazilian Cerrado Oxisols. Geoderma 156:207–215
Pew Commission IFAPS (2008) Pew Commission on Industrial Farm Production. Putting meat on the table: industrial farm animal production in America. Available at http://www. ncifap.org/bin/e/j/ PCIFAPFin.pdf. Last accessed 10 Apr 2010
Pierson ST, Cabrera ML, Evanylo GK, Kuykendall HA, Hoveland CS, McCann MA, West LT (2001) Phosphorus and ammonium concentrations in surface runoff from grasslands fertilized with broiler litter. J Environ Qual 30:1784–1789
Powers WJ, Fritz ER, Fehr W, Angel R (2006) Total and water-soluble phosphorus excretion from swine fed low-phytate soybeans. J Animal Sci 84:1907–1915
Rao SC, Dao TH (2008) Relationships between phosphorus uptake in two grain legumes and soil bioactive P pools in fertilized and manure-amended soil. Agron J 100:1535–1540
Ro KS, Cantrell KB, Hunt PG, Ducey TF, Vanotti MB, Szogi AA (2009) Thermochemical conversion of livestock wastes: carbonization of swine solids. Bioresour Technol 100:5466–5471
Rotolo GC, Rydberg T, Lieblein G, Francis C (2007) Emergy evaluation of grazing cattle in Argentina's Pampas. Agric Ecosyst Environ 119:383–395
Schiffman S, Miller ES, Suggs MS, Graham BG (1995) The effect of environmental odors emanating from commercial swine operations on the mood of nearby residents. Brain Res Bull 37:369–375
Sørensen P, Thomsen IK (2005) Separation of pig slurry and plant utilization and loss of nitrogen-15-labeled slurry nitrogen. Soil Sci Soc Am J 69:1644–1651
Tamminga S (1992) Nutrition management of dairy cows as a contribution to pollution control. J Dairy Sci 75:345–357
Toor GS, Condron LM, Di HJ, Cameron KC, Cade-Menun BJ (2003) Characterization of organic phosphorus in leachate from a grassland soil. Soil Biol Biochem 35:1317–1323
Toor GS, Cade-Menun BJ, Sims JT (2005) Establishing a linkage between phosphorus forms in dairy diets, feces, and manure. J Environ Qual 34:1380–1391
Traoré O, Sinag S, Frossard E, Van De Kerkhove JM (1999) Effects of composting time on phosphate exchangeability. Nutr Cycl Agroecosyst 55:123–131
Trollea D, Skovgaardc H, Jeppesen E (2008) The Water Framework Directive: setting the phosphorus loading target for a deep lake in Denmark using the 1D lake ecosystem model DYRESM-CAEDYM. Ecol Modell 219:138–152
Turner BL (2004) Optimizing phosphorus characterization in animal manures by solution phosphorus-31 nuclear magnetic resonance spectroscopy. J Environ Qual 33:757–766
Turner BL, Haygarth PM (2000) Phosphorus forms and concentrations in leachate under four grassland soil types. Soil Sci Soc Am J 64:1090–1097
USEPA (2008a) Chesapeake Bay Program. The Chesapeake Action Plan. US Environmental Protection Agency, Washington DC. Available at http://cap.chesapeakebay.net/goal3.htm. Last accessed 10 Apr 2010
USEPA (2008b) 40 CFR Parts 9,122 and 412 revised national pollutant discharge elimination system permit regulation and effluent limitations guidelines for concentrated animal feeding operations in response to the Waterkeeper decision. US Environmental Protection Agency, Washington DC. Federal Register 73(225):70418-70486
Verkerk G (2003) Pasture-based dairying: challenges and rewards for New Zealand producers. Theriogenology 59:553–561

Waldroup PW, Kersey JH, Saleh EA, Fritts CA, Yan F, Stillborn HL, Crum RC Jr, Raboy V (2000) Non-phytate phosphorus requirement and phosphorus excretion of broiler chicks fed diets composed of normal or high available phosphate corn with and without microbial phytase. Poult Sci 79:1451–1459

Withers PJA, Hodgkinson RH, Adamson H, Green G (2007) The impact of pasture improvement on phosphorus concentrations in soils and streams in an upland catchment in Northern England. Agric Ecosyst Environ 122:220–232

Wolf AM, Kleinman PJA, Sharpley AN, Beegle DB (2005) Development of a water-extractable phosphorus test for manure: an interlaboratory study. Soil Sci Soc Am J 69:695–700

Wu Z, Satter LD, Sojo R (2000) Milk production, reproductive performance, and fecal excretion of phosphorus by dairy cows fed three amounts of phosphorus. J Dairy Sci 83:1028–1041

Wu Z, Satter LD, Blohowiak AJ, Stauffacher RH, Wilson JH (2001) Milk production, estimated phosphorus excretion, and bone characteristics of dairy cows fed different amounts of phosphorus for two or three years. J Dairy Sci 84:1738–1748

Zhang RH, Lei F (1998) Chemical treatment of animal manure for solid-liquid separation. Trans ASAE 41:1103–1108

Zhu GW, Qin BQ, Zhang L, Luo LC (2006) Geochemical forms of phosphorus in sediments of three large, shallow lakes of China. Pedosphere 16:726–734

Chapter 17
Management Impacts on Biological Phosphorus Cycling in Cropped Soils

Astrid Oberson, Pieter Pypers, Else K. Bünemann, and Emmanuel Frossard

17.1 Introduction

Global population is forecast to increase to 7.6 billion by the year 2020 and to about 9 billion in 2050 (UNPP 2008). The FAO (2009) estimates that food production will have to increase by 70% by 2050 to meet the needs of the increasing world population. About 80% of the increase in agricultural production will have to come from increased productivity (FAO 2009). This will be related to a strongly growing demand for fertilizers (see Tiessen et al. 2011). In view of the projection that phosphorus (P) deposits, which can be mined at a relatively low cost to be processed for fertilizers, will be exhausted in about one century (Cordell et al. 2009), P management in cropping systems will have an important impact on future food security.

Harvested products are removed from cropped fields for human or animal consumption. With these products, P is removed and has to be restituted through fertilizer P inputs to maintain soil fertility, except in soils with a high P fertility where soil P can serve as the sole P source for a limited time. Another specific trait of cropped soils is the cyclic presence of crops, with related changes in soil cover and rooting, and the continuous disturbance through interventions by farmers. These interventions include soil tillage, sowing or planting, water management such as irrigation or drainage, fertilizer application, plant protection measures, and harvesting. The farmers select the crops and varieties, thus managing crop diversity because soils can be repeatedly cropped by the same crop (monocropping), by a sequence of crops (crop rotation), or by mixed stands (e.g., intercropping). Soil tillage affects soil P dynamics through modification of the spatial distribution of P between soil layers and soil aggregates, and through the incorporation of crop

A. Oberson (✉), E.K. Bünemann, and E. Frossard
Institute of Plant, Animal and Agroecosystem Sciences, Group of Plant Nutrition, ETH Zurich, Research Station Eschikon, 8315 Lindau, Switzerland
e-mail: astrid.oberson@ipw.agrl.ethz.ch

P. Pypers
Tropical Soil Biology and Fertility Institute of the International Centre for Tropical Agriculture (TSBF-CIAT), Nairobi, Kenya

E.K. Bünemann et al. (eds.), *Phosphorus in Action*, Soil Biology 26,
DOI 10.1007/978-3-642-15271-9_17, © Springer-Verlag Berlin Heidelberg 2011

residues and fertilizers. It also affects soil microbial activity through changes in environmental conditions such as infiltration, aeration, and access to organic substrates. Organic P sources (animal manure, plant residues) will undergo microbiological transformations to a higher degree than mineral P fertilizers and may affect organic P contents in the soil. Crops affect biological P cycling in several ways, ranging from root-induced processes to the impact of above- and belowground residues on soil microbial activity and P availability.

This chapter summarizes current knowledge on the impact of management and crop-specific factors on biological P cycling in cropped soils of temperate and tropical regions. Cropped soils can belong to a mixed crop–livestock farm or an arable farm; the term "cropping system" as used in this chapter does not preclude linkage to livestock production. Emphasis is put on microbial functions in soil P dynamics, which are presented in Sect. 17.2. We then address the effect of soil tillage on microbial functions and on the forms and distribution of P in soils. In Sect. 17.4 we analyze the impact of long term application of organic versus mineral fertilizer inputs on microbial functions, and the role of biological processes in the use of P by crops. Finally, in Sect. 17.5 we analyze the importance of crop rotations, with emphasis on the impact of legumes on processes of P dynamics when introduced into tropical cropping systems with low-P soils. Overall, we scrutinize whether increased soil microbial activity results in a more efficient use of soil and fertilizer P.

17.2 Microbial Functions in Cropped Soils

Microorganisms mediate several key processes in the P cycle (Fig. 17.1), which affect P availability to plants (Oberson and Joner 2005). Mycorrhizal fungi colonize the roots of most crops, among them major food crops such as wheat, maize, rice, potatoes, and cassava (Jansa et al. 2006). Their role in P absorption and transport to plant roots is described in Jansa et al. (2011). Microorganisms can solubilize or desorb poorly available inorganic P by releasing organic acids and inducing changes in pH (see Jones and Oburger 2011). By their ability to synthesize extracellular phosphatases to hydrolyze organic P, microorganisms take a key role in the mineralization of organic P (Jones and Oburger 2011; Nannipieri et al. 2011). If these processes occur in the rhizosphere or mycorrhizosphere of crops, then plants may benefit from increased P availability (George et al. 2011; Jansa et al. 2011). The constantly ongoing processes of soil organic matter decomposition and turnover of microbial biomass represent a less specific but highly important process in soil P dynamics (Oberson and Joner 2005).

Organic inputs such as manures and plant residues are a major source of organic carbon (C) to the soil microorganisms. The availability of organic substrates, either specific substrates like glucose or cellulose (Bünemann et al. 2004a), or complex sources such as animal manure (Wichern et al. 2004) and plant residues (Bünemann et al. 2004a, c), induces microbial growth. Plants also

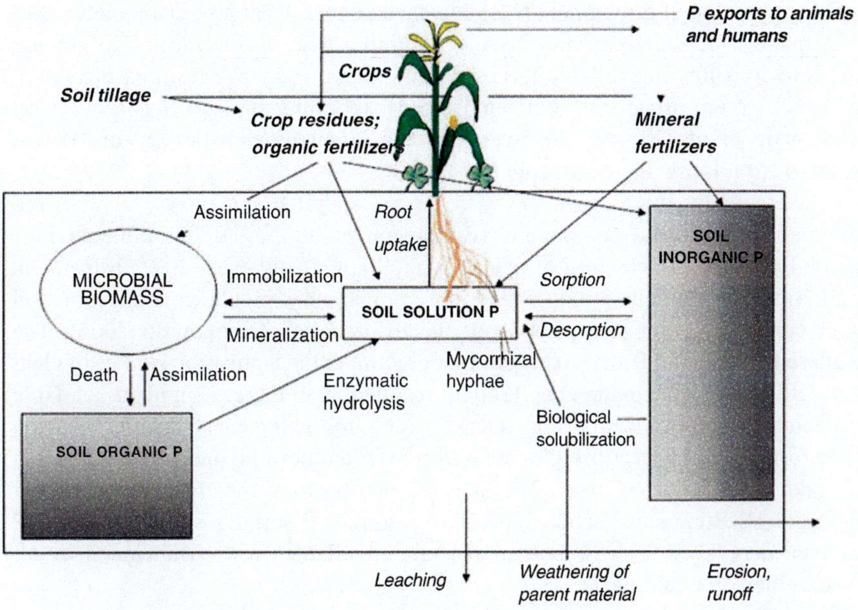

Fig. 17.1 Microbial functions in soil P dynamics of cropped soils. The microbial biomass includes microorganisms living in the rhizosphere of crops. *Italics* nonmicrobial processes, *bold italics* farmers' interventions

promote microbial populations and subsequent turnover by exuding organic C and nitrogen (N) from the roots (Wichern et al. 2007). Even under low-P conditions the microbial biomass can grow rapidly and take up substantial P amounts when readily degradable C is available (Bünemann et al. 2004a). Such an increase in microbial P is usually connected with a rapid decrease of soil solution P concentration (Oehl et al. 2001), suggesting that microorganisms may compete with plants for available P (Olander and Vitousek 2004). If P-containing organic inputs such as plant residues are added to the soil, then microorganisms immobilize P from soil and residues, and the respective contribution of the two sources can be derived using P radioisotope techniques (McLaughlin et al. 1988). For instance, residue P and soil P contributed similar parts to the increase in microbial P following the addition of the legume residues in Bünemann et al. (2004c). The recovery in the microbial biomass of P added with the legume residues was about 15% according to Bünemann et al. (2004c) and 22–28% according to McLaughlin et al. (1988). In comparison, only about 5% of mineral fertilizer P was incorporated into the soil microbial biomass (Bünemann et al. 2004c; McLaughlin et al. 1988). This is because even in soils with low plant-available P content, soil microorganisms are limited by C and N rather than by P availability (Bünemann et al. 2004b; Ehlers et al. 2010). Most P (64–68%) added with the legume residues and the mineral P was recovered in the NaOH-extractable inorganic P fraction of the soil 10 days after amendment (Bünemann et al. 2004c).

During microbial growth, microorganisms take up P from less-available P pools (Bünemann et al. 2004a, c). For instance, within 2 days of the addition of glucose plus N to soils low in available P (i.e., 2–7 mg anion exchange resin-extractable P kg^{-1} soil), microorganisms took up from 18 to more than 40 mg P kg^{-1} soil (Bünemann et al. 2004a). However, it remains unknown whether this P was acquired from inorganic or organic soil P pools.

P taken up by microorganisms will be incorporated into newly synthesized inorganic and organic compounds (Bünemann et al. 2011). Immobilized P is released because of depletion of readily available C (Oehl et al. 2001; Bünemann et al. 2004a) when cells are disrupted, e.g., in response to sudden changes in soil water content (Turner et al. 2003) or due to predation (Bonkowski 2004). The decrease in microbial P after exhaustion of glucose in the incubation studies of Oehl et al. (2001) and Bünemann et al. (2004a) resulted in an increase in plant-available P. Microbial immobilization and release cycles following plant residue addition differ from those induced by glucose addition in temporal dynamics and differ in the amount of P turned over, but likewise demonstrate the dynamic nature of microbial P (Bünemann et al. 2004a). In strongly P-sorbing soils, microbial P turnover may keep P in a potentially plant-available form (Bünemann et al. 2004b; Oberson et al. 2001).

The functions of microorganisms in the biogeochemical cycling of P are not specific to cropped soils, but they are affected to a high degree by farmers' interventions, and the microbial P pool size is usually lower in cropped soils than in grassland or forest soils, both in relative and absolute terms. A compilation by Oberson and Joner (2005) shows that topsoil layers of cropped soils contain 2–21 mg microbial P kg^{-1} (median 5.5 mg P kg^{-1}), which corresponds to 0.4–2.5% of total soil P. In comparison, grassland soils contain 4–77 mg microbial P kg^{-1} (median 18 mg P kg^{-1}), which is 0.5–7.5% of total P. From their compilation, Oberson and Joner (2005) deduced that soil type, texture, and soil organic matter content are the major factors producing this broad range in microbial P content, and that lower soil organic matter content is the main reason for the usually lower microbial P content in cropped soils than in grassland soils. The close relationship between soil organic matter content and microbial biomass has also been demonstrated using microbial C as a measure of microbial biomass (Houot and Chaussod 1995; Moore et al. 2000). Although soil type and texture cannot be modified by the farmer, management can affect soil organic matter content, its location, and its dynamics in cropped soils.

17.3 Tillage Impacts on Biological P Cycling

To reduce the risk of soil erosion and to decrease energy consumption and labor related to ploughing, systems of no-tillage and other conservation tillage systems have been established under a wide range of climates and farm types, and are being further developed (Huggins and Reganold 2008; Séguy et al. 2006). Soil tillage affects the distribution of P along the soil profile and changes the environment of soil microorganisms. Clearest demonstrations of such effects arise from the comparison of soils under conventional tillage using a plough versus no-tillage systems, which are characterized by minimal disturbance to the soil at seeding (direct seeding) and usually by the management of crop residues at or near the soil surface.

17.3.1 Tillage Effects on P Distribution and P Dynamics

Under no-tillage, total and available P concentrations are higher in the topsoil than in the lower soil horizons (Jansa et al. 2003) because fertilizer P as well as P released from plant residues remain at the soil surface. Organic C content is also higher in the topsoil layer of no-tillage soils than in ploughed soils (Daroub et al. 2000; Jansa et al. 2003; Zibilske and Bradford 2003). This is paralleled by increased organic P in the topsoil layer of no-tillage soils (Bünemann et al. 2006a). The greater offer of organic matter for microbial growth results in higher soil microbial biomass and activity (Rabary et al. 2008). In turn, microbial P is higher in the topsoil of no-tillage soils than in ploughed soils (Balota et al. 2003; Daroub et al. 2000). In both studies, plant-available inorganic P also tended to increase, although not significantly. This agrees with the accumulation of total P in the topsoil, but also suggests that greater microbial P immobilization did not reduce available P content, presumably because of microbial P turnover. Zibilske and Bradford (2003) found higher plant-available P, increased phosphatase activity, higher soluble organic C, and greater soil respiration in soils under no-tillage than in ploughed soils, and concluded that greater organic P mineralization occurs under no-tillage. Organic P mineralization may explain why labile organic P fractions were not significantly increased under no-tillage (Daroub et al. 2000).

17.3.2 Interactions Between Tillage, Specific Organisms, and P Dynamics

Soil tillage affects specific organism groups such as earthworms (Ernst and Emmerling 2009) and mycorrhizal fungi (Jansa et al. 2003). Effects can be indirect through a modification of the environment, e.g., the amount and location of organic residues presenting the nutritional basis for some earthworm species. Direct effects are mechanical damage to earthworms or cutting of mycorrhizal hyphae at ploughing. Density, biomass, and community composition of earthworm populations are affected by the tillage system, with decreased tillage intensity having a positive effect on biodiversity (Ernst and Emmerling 2009) but not necessarily on earthworm density and biomass (Berner et al. 2008). Earthworms create structures with characteristics different to those of the surrounding soil (Chapuis-Lardy et al. 2011). Plant-available P concentrations are usually higher in earthworm casts (Chapuis-Lardy et al. 2009; Jiménez et al. 2003) and in burrow walls (Tiunov et al. 2001). Location and composition of these structures varies depending on the species involved and can be affected by tillage. Likewise, Jansa et al. (2003) showed an effect of soil tillage on the community structure of mycorrhizal fungi in maize roots, and there is functional complementarity among species within the arbuscular mycorrhizal fungi community colonizing a single root system (Jansa et al. 2008). A series of studies carried out at Guelph (Canada) revealed larger early-season P uptake by maize due to a more effective arbuscular mycorrhizal symbiosis when the soil was not disturbed. The larger P uptake was mostly a result of the undisrupted mycelium present in an undisturbed soil, rather than of the increased colonization (Miller 2000). Thus, mycorrhizal symbioses might compensate for the poorer root growth during the early growth stages of maize grown under no-tillage than under conventional tillage (Chassot et al. 2001).

17.3.3 Tillage or No-Tillage: Trade-Offs in the Use of Crop Residues

Direct seeding combined with a permanent soil cover is currently being promoted for small-scale farmers in the tropics (Husson et al. 2006; Séguy et al. 2006). In such systems, the soil is never ploughed and the staple crops (maize, rice, soybean, or cassava) are grown either in a mulch of plant residues (e.g., cereal straw or legume residues) or intercropped with another plant (e.g., *Brachiaria ruziziensis* or *Stylosanthes guianensis*). The mulch or the cover crop play a key role in soil protection, organic matter input, nutrient recycling, and weed control. A successful implementation of direct seeding systems requires that plant residues are largely retained in the field. Without sufficient surface cover, soils will compact, infiltration will reduce, maximum temperatures will increase in the surface layer, and crop growth will be negatively affected (Giller et al. 2009). Under such conditions, crops often show signs of P deficiency because of impaired root growth, even though soils may contain sufficient P. Small-scale farmers in sub-Saharan Africa often use crop residues as livestock feed, as firewood for cooking, or traditionally burn them for weed and pest management. When applied to the soil, they are often broken down within a few weeks by termites (Giller et al. 2009). This hampers adoption of no-tillage systems with direct seeding systems into living or dead mulch. Before direct seeding systems can be recommended to resource-poor farmers in the tropics, trade-offs between different uses of crop residues need to be evaluated.

17.4 Fertilizer Inputs: The Form of Nutrient Sources Matters

The repeated input of organic amendments such as animal manure, composts, or plant residues usually results in higher soil organic matter content and in larger soil microbial biomass and activity compared to soils receiving mineral fertilizers

(Bünemann et al. 2006b; Fliessbach et al. 2007; Liebig and Doran 1999; Wells et al. 2000). The resulting feedback on P cycling has been studied under field situations in different climate zones and on different soil types. Here we present a case study using animal manure.

17.4.1 *Animal Manure Promotes Microbial P Cycling: A Case Study on Organic Versus Conventional Farming in Switzerland*

The recycling of animal manure is of increasing importance globally (Tiessen et al. 2011). Organic farming largely depends on organic P sources (Oberson and Frossard 2005). Manure P is composed of inorganic and organic forms (Toor et al. 2006). After application to the soil, these compounds undergo abiotic and biotic reactions, which determine the availability of manure P to crops.

In 1978, a long-term experiment was started in Switzerland to study the effects of organic and conventional farming systems on crop performance and soil fertility (Mäder et al. 2002). The trial includes two organic systems (biodynamic and bio-organic), which are both fertilized exclusively with animal manure and receive no synthetic pesticides, in contrast to the conventional systems. The two organic systems differ in the treatment of manure during storage on the respective farms, with composting of farmyard manure and aeration of slurry in the biodynamic system, whereas slightly decomposed manure and slurry are applied in the bio-organic system. Specific to the biodynamic system are the biodynamic preparations described in Mäder et al. (2002), which are used as composting additives or for plant protection. In the two conventional systems, either mixed mineral and organic inputs or exclusively water-soluble mineral fertilizers are applied. The systems differ also in applied P amounts. In both organic systems, amounts are based on manure production by 1.4 livestock units per hectare, whereas in the conventional systems amounts correspond to the Swiss fertilization guidelines, resulting in the average P inputs indicated in Table 17.1. The field experiment includes a control without nutrient inputs. All systems follow the same crop rotation sequence, which lasts for 7 years. Likewise, all systems are under conventional tillage. Because of different plant protection and nutrient input strategies, the farming systems differ in productivity (Mäder et al. 2002). Potato yields in the organic systems were around 60% of those in the conventional plots whereas cereal yields were around 80% and grass-clover yields nearly the same as under conventional management.

Systems that regularly receive animal manure, i.e., both organic systems and one of the conventional systems, maintained a higher soil organic C level than the nonfertilized control or the conventional system with exclusively mineral fertilizers (Leifeld et al. 2009). It was shown that soil microbial biomass and activity is highest in soils of the biodynamic system, followed by the bio-organic and the conventional system receiving manure inputs (Fliessbach et al. 2007; Mäder et al. 2002). On the

Table 17.1 P status and indicators of biological P cycling in soils of a field experiment in Switzerland under organic and conventional cropping

	Cropping system				
	Control	Biodynamic	Bio-organic	Conventional mixed	Conventional mineral
Type of fertilizer	None	Organic	Organic	Organic + mineral	Mineral
Fertilizer P input (kg^{-1} ha^{-1} per year)[a]	0	24	27	43	28[b] (41)
P balance (kg^{-1} ha^{-1} per year)[a]	−21	−8	−6	+4	−5[b] (+6)
Water-soluble P (mg^{-1} kg^{-1})[c]	0.05[*]	0.4[†]	0.6[‡]	1.9[¶]	1.0[§]
Microbial P (mg kg^{-1})[c]	4.7[*]	11.8[†]	13.0[†]	12.1[†]	6.6[*]
Organic P (mg kg^{-1})[c]	339 ns	379 ns	364 ns	352 ns	349 ns
Total P (mg kg^{-1})[c]	563[*]	640[†,‡]	629[†]	683[§]	658[‡,§]
Acid phosphatase activity (mg paranitrophenol kg^{-1} h^{-1})[d]	nd	182[*]	172[†]	nd	148[‡]
P mineralization (mg kg^{-1} per day)[e]	nd	2.5[†]	1.7[*]	nd	1.5[*]
Exchangeable P (mg kg^{-1} per day)[f]	nd	27.6[*]	24.8[*]	nd	31.8[†]
Mineralizable P/exchangable P (%)[g]	nd	9.1[†]	7.0[*,†]	nd	4.9[*]
$^{33}PO_4$ incorporation (%)[h]	nd	6.1[†]	3.8[*]	nd	2.5[*]
Manure P recovery in ryegrass[i]	35[‡]	28[*,†]	30[†,‡]	24[*]	29[*,†,‡]
Mineral P recovery in ryegrass[i]	39 ns	39 ns	43 ns	37 ns	40 ns
Residual P recovery in ryegrass[j]	na	9[*]	12[†]	15[§]	13[‡]

Values within a line followed by different symbols ([*], [†], [‡], [§], [¶]) are significantly different (Duncan's multiple range test)

ns not significant, *nd* not determined, *na* not applicable

[a]Average annual fertilizer P input (mineral and/or organic) and P balance (difference between P inputs by fertilizers and outputs by harvested products) for 21 years of field experimentation (Oberson et al. 2010)

[b]During the first crop rotation period lasting 7 years, the conventional mineral system was used as an unfertilized control. The value in brackets therefore shows the average for the second and third crop rotation periods when it was fertilized as conventional mineral system

[c]Data from Oberson et al. (2010)

[d]Data from Oehl et al. (2004)

[e]Basal organic P mineralization rate per day assessed using isotopic dilution techniques; data from Oehl et al. (2004)

[f]Quantity of inorganic P exchangeable within 1 day determined by isotopic exchange kinetics; data from Oehl et al. (2004)

[g]Ratio between quantities of daily mineralized organic P and isotopically exchangeable P

[h]Percentage of applied $^{33}PO_4$ taken up by microorganisms 5 days after soil labeling; data from Oehl et al. (2001)

[i]Percentage of manure P and mineral P, respectively, taken up by four harvests of ryegrass (Oberson et al. 2010)

[j]Percentage of residual P taken up by four harvests of ryegrass (Oberson et al. 2010)

other hand, during the course of the field experiment, the total and available P contents decreased more in both organic systems than in the conventional systems because of their P budget deficit (Oehl et al. 2002) (Table 17.1). Thus, in the organic systems, yields were partly attained at the expense of soil P reserves. In the control, soil P reserves were the only source for crop P uptake.

Several biological processes were shown to be involved in the use of soil P reserves. Monitoring of changes in total P stocks in different soil layers gave evidence that with increasing depletion of available P in the topsoil, crops increasingly took up P from deeper soil layers (Oehl et al. 2002). This process was manifested in all systems having a P deficit. In contrast, microbial functions that promote access to P and availability of soil P were more important under organic than conventional cropping. The percentage of root length colonized by mycorrhizal fungi was 30–60% higher in crops growing in soils from the organic rather than the conventional systems (Mäder et al. 2000). The activity of acid phosphatase, which can be of plant or microbial origin, was also increased under organic cropping (Oehl et al. 2004). A radioisotope P dilution experiment revealed a higher basal mineralization of soil organic P in soils from both organic systems than in the soil under the conventional system without manure (Oehl et al. 2004). Also, soils of the organic systems had a larger microbial P pool with a faster turnover (Oehl et al. 2001) (Table 17.1).

Greater microbial activity in the soils of the organic systems, however, did not significantly affect the P uptake by ryegrass from fresh fertilizers (Oberson et al. 2010). In a pot experiment using radioactive P labeling techniques, we studied the uptake of P applied with animal manure or water-soluble mineral P. In each soil, manure addition increased microbial P content, with similar amounts of P immobilized. Recovery of manure P in ryegrass was lower than that of water-soluble mineral P. It ranged from 24 to 35% for manure, and from 37 to 43% for mineral P (Table 17.1). Differences in microbial activity among soils had little importance in the use of these fertilizers. However, recovery of manure P was affected by soil available P contents, with lower recovery of manure P at higher soil P availability. Because of their lower available P content, organically cropped soils therefore have the potential for higher efficiency of manure use.

In the same experiment, labeling of available P allowed the assessment of the uptake from residual P (composed of plant-available soil P depleted in the control soil but not in the fertilized soils, and of residual fertilizer P remaining in the fertilized soils). Residual P uptake was lowest for the biodynamic soils (Table 17.1), probably because their lower residual P contents were composed of stable P forms (Keller et al. 2009; Oberson et al. 2010). The treatment of manure during storage affects the availability of P (Dao and Schwartz 2011). Presumably, composting of manure prior to application in the biodynamic plots resulted in more stable inorganic and organic P forms.

In conclusion, manure addition stimulates microbial activity, which translates into greater basal soil P mineralization and higher microbial P turnover. Decreases in available P in the topsoil may increase the use of soil P by P uptake from deeper soil layers, particularly in situations where past fertilizer P inputs surplus to crop uptake also increased available P in deeper soil layers. Plant roots as well as mycorrhizae could be involved in this process. On the other hand, we have no evidence that higher microbial activity rendered the poorly soluble soil P fractions available for crop uptake (Oberson et al. 1993; Keller et al. 2009). Also, Dann et al.

(1996) reported that uptake from rock phosphate was equally low in soils under organic and conventional cropping.

17.4.2 Availability and Quality of Animal Manure and Recycling Fertilizers

P limitation in organic farming systems may increasingly limit crop production, particularly on large-scale arable farms with little organic P resources and with neutral to alkaline soils where rock phosphate (the only nonorganic P source permitted in organic farming) is virtually plant-unavailable (Cornish 2009). The use of animal manure is far from specific to organic farming systems but is common in conventional mixed crop–livestock farms. Also, many stockless, conventional arable farms use organic inputs such as organic waste composts or sewage sludge. A compilation on comparative effects of inorganic and organic fertilizers (farmyard manure, sewage sludge, composts) shows that organic fertilizers usually increase the microbial biomass and enzyme activities, whereas the effects of mineral fertilizers are more variable (Bünemann et al. 2006b). They induce no change or enhance biological activity via increases in system productivity and crop residue return or, in some cases, have negative impacts on soil organisms, e.g., through soil acidification (Bünemann et al. 2006b). Thus, the use of organic inputs increases biological P cycling. In industrialized countries, new trends of manure management in large livestock production farms arise, such as gasification or incineration of animal manure (Kuligowski and Poulsen 2009). These treatments remove organic matter and leave a P-rich ash. The products permit the recycling of P, but direct beneficial effects on soil organic matter content and microbial activity are lost.

Small-scale farmers in sub-Saharan Africa, like most subsistence farmers everywhere, use manure, but only in limited quantities. Manure is a rather scarce resource and is of very variable quality (Rufino et al. 2006), whereas mineral fertilizer (particularly P fertilizer) is often expensive. Integrated soil fertility management options for smallholders should consider manure quality and availability as well as access to mineral fertilizer and competing uses for crop residues (Tittonell et al. 2008). Purely organic farming systems are often not viable under conditions where soil P stocks have been depleted and manure is scarce. Therefore, combining mineral and organic resources at practicable and economic rates is recommended (see Sect. 17.6) (Bassala et al. 2008; Vanlauwe et al. 2010).

Interactions between manure and mineral fertilizer application are not clear-cut. Effects of both resources can be additive (Akponikpe et al. 2008) or may interact positively (Abunyewa et al. 2007; Onduru et al. 2008; Opala et al. 2007). However, at the same P dose, water-soluble P fertilizer application often results in a superior response relative to manure addition (Materechera and Morutse 2009). Much depends on the quality of the manure as well as on the inherent soil fertility and the crop. Manure application has positive impacts on the soil organic C content and microbial activity and is essential for the rehabilitation of degraded soils (Zingore et al. 2008), but if the manure is of inferior quality, soil restoration can only be achieved by combining manure and mineral fertilizer application (Tittonell et al. 2008). Given the limited P resources and the differential resource endowments of farmers, appropriate strategies are required for improving manure storage, quantity, and quality, and for efficient use of both organic and inorganic resources (Rufino et al. 2007).

17.5 Crop Rotation: Higher P-Use Efficiency Through the Integration of Legumes?

Dinitrogen fixation by the symbiotic associations between legumes and rhizobia is a major source of N input into agricultural soils (Herridge et al. 2008). Important arable crops like soybean can largely rely on symbiotically fixed N (Alves et al. 2003). Legumes are an essential part of crop rotations in organic farming systems, but also in many conventional farming systems such as in southern and western Australia or in Switzerland. Nonfixing crops planted after legumes can benefit from the improved N availability (Peoples et al. 2009). Rotational effects of legumes on P availability have also been observed (Bünemann et al. 2004b; Pypers et al. 2007; Muchane et al. 2010).

Here, we focus on the integration of legumes into the crop rotation of smallholder farms in developing countries in the tropics for the following reasons: Population growth in the poor rural areas of developing countries has resulted in land use intensification. Fallow phases formerly used to restore soil fertility have increasingly been reduced and continuous cropping of the major crop (such as maize) has become widespread (Bünemann et al. 2004b; Douxchamps et al. 2010). Often farmers can afford only small amounts of mineral fertilizer or none at all. Soil nutrients are being depleted, and soil erosion due to poor soil cover further increases nutrient losses and soil degradation (Tan et al. 2005). N input through symbiotic fixation is recognized as a key component in the development of sustainable cropping systems (Boddey et al. 2006; Ojiem et al. 2007). P deficiency, however, is seen as the major factor limiting legume growth and symbiotic fixation (Hogh-Jensen et al. 2002; Mafongoya et al. 2004) and is widespread in highly weathered tropical soils. Under these conditions, legumes need specific strategies to satisfy their P requirements.

17.5.1 Legume Strategies for Acquiring P

Plant strategies for acquiring P in P-deficient soils can be grouped as follows:

1. Morphological root traits to improve spatial access to soil P
2. Increased spatial access to soil P through association with mycorrhizal fungi (Jansa et al. 2011)

3. Solubilization of recalcitrant inorganic P and mineralization of organic P in the rhizosphere (George et al. 2011)
4. Efficient mechanisms for uptake of solubilized P into the root (George et al. 2011)

Such strategies are not specific to legumes and have also been reported, e.g., for members of the families of Brassicaceae or Asteraceae (Hedley et al. 1982; Smestad et al. 2002).

Deep-rooting legumes such as *Sesbania sesban* (L) Merr can access nutrients in deep soil layers, e.g., N from a depth of 1 m (Gathumbi et al. 2003). This mechanism may also be significant for P uptake, particularly in highly weathered tropical soils where the topsoil is often P depleted and where gradients in total and available P along the soil profile are less pronounced than in P-enriched soils in industrialized countries (Bünemann et al. 2004b; Friesen et al. 1997). Arbuscular mycorrhizal fungi play an important role in the P uptake and growth of many legumes (Smith and Read 2008) although under low soil P conditions, the yields of crops depending solely on the arbuscular mycorrhizal fungi without any P inputs are lower than upon application of moderate P fertilizer level (Muchane et al. 2010). Arbuscular mycorrhizal fungi appear to access mainly orthophosphate in the soil solution, unlike the fungi forming other types of mycorrhizal symbioses (Jansa et al. 2011). Thus, there is no evidence that they can access significant amounts of recalcitrant inorganic or organic P. Under low-P conditions, P fertilizer application may improve colonization of roots by mycorrhizal fungi (Muchane et al. 2010; Jansa et al. 2011, and references therein), suggesting that arbuscular mycorrhizal fungi can increase fertilizer P-use efficiency by foraging for P within a greater soil volume than the roots, and by building up a strong gradient in P concentration between soil and the mycorrhizosphere. By doing this, plants can utilize P that could otherwise not be taken up by the roots or root hairs. As a particularity of legumes, the tripartite symbiosis between legumes, rhizobia, and mycorrhizal fungi reacts to the P status (Vanlauwe et al. 2000a), thus feeding back on N and P cycling.

Some legume species, including *Lupinus albus*, are nonmycorrhizal but form cluster or proteoid roots in P-deficient soils (Shane and Lambers 2005). Cluster roots are very densely branched roots that excrete large amounts of organic anions (mostly citrate and malate), protons, and acid phosphatases to mobilize phosphate in the rhizosphere (Neumann and Martinoia 2002). Organic anions are, however, not only exuded from cluster roots, but are detected in the rhizosphere of many legumes growing under low P availability. For instance, faba bean (*Vicia faba* L.) acidifies the rhizosphere via release of organic acids and protons (Li et al. 2007). The roots of some P-efficient genotypes of cowpea (*Vigna unguiculata* Walp) and soybean (*Glycine max*. L.) react towards P stress with the exudation of organic acid anions, whereas another P-efficient soybean genotype responds with a higher activity of root surface phosphatase (Jemo et al. 2006). Not all genotypes express these strategies, showing that there is scope for selecting and breeding P-efficient genotypes (Lynch 2007).

It is noteworthy that P-solubilizing microorganisms living in the rhizosphere can also mobilize organic P through the liberation of extracellular enzymes and/or inorganic P through the release of complexing or mineral-dissolving compounds (Jones and Oburger 2011). Jones and Oburger (2011) point out the great potential of coinoculating legumes with dinitrogen-fixing *Rhizobium* sp. and P-solubilizing microorganisms, but also stress the prevailing lack in mechanistic understanding, which currently renders success of inoculation in the field quite erratic.

Because of P acquisition strategies, legumes have access to nonlabile soil P, as demonstrated for cowpea compared to maize (Pypers et al. 2006). They also result in a better use of poorly available P inputs, such as rock phosphate (Vanlauwe et al. 2000b) or phytate (Li et al. 2003). Presumably, neighboring nonlegumes can also benefit from the legume's P acquisition strategies. Results from pot experiments using separated root systems suggest that legumes facilitate P uptake of a neighboring cereal crop from recalcitrant sources. Wheat took up more P from phytate when its roots were intermingled with cowpea roots than when it was growing alone (Li et al. 2003). However, wheat also took up more calcium, magnesium, and microelements (Li et al. 2004). This suggests that enhanced P uptake may result from a better supply of other elements. Thus, radioisotope P techniques should be used to clearly identify the P source used by the crop (Frossard et al. 2011).

17.5.2 Crop Rotations: Can Succeeding Crops Benefit from Legume P Acquisition Strategies?

Many field studies carried out on soils low in available P have shown higher yields for cereals growing after a legume crop than when monocropped (Bagayoko et al. 2000; Bünemann et al. 2004b; Horst et al. 2001; Jemo et al. 2006; Kamh et al. 2002; Pypers et al. 2007). Crops might have attained higher yields because of improved P nutrition through P contained in legume residues or because of better soil P availability because of changed soil properties. Higher crop yields are also attained through enhanced supply of other elements (particularly N), or through weed, disease, or pest suppression. Under field conditions, it is difficult to separate these effects.

17.5.2.1 Legume Residues as a P Source

In most cropping systems, legume residues are produced in situ. They are rarely a net P input because cut and carry systems have never been widely practiced. After decomposition of the legume residues, P taken up by the legume can become available to the succeeding crop. Residue decomposition is mediated by the activity of soil fauna and flora. The release of P contained in organic materials depends on their quality, particularly on total P, total N, lignin, and soluble polyphenol concentrations (Kwabiah et al. 2003; Vanlauwe et al. 2008). P concentrations in legume residues can vary widely, as a function of species, plant age, and plant part. The organic resource database created by the Tropical Soil Biology and Fertility Program (TSBF) contains information on the quality of plant residues (macronutrients, lignin, and polyphenol contents) (Palm et al. 2001). The median P concentration in 550 legume leaves was 1.7 g kg^{-1}, with 50% of analyzed samples in the range 0.4–3.25 g kg^{-1}. In spite of legumes often having special P acquisition strategies, P concentrations in their leaves was not higher than those of other plant families (Palm et al. 2001).

Plant residue addition induces microbial immobilization and mineralization, which occur simultaneously and involve turnover of residue and soil P (see Sect. 17.2). If mineralization exceeds immobilization, then a net P release results. In several laboratory studies, total P concentration was the best predictor for residue P release, followed by the C:P and N:P ratios of the plant material (Kwabiah et al. 2003; Mukuralinda et al. 2009). Materials with total P concentrations higher than 2.0–2.7 g kg^{-1}, C:P below 156:1 to 252:1, and N:P higher than 7:1 to 14:1 were found to result in a net P release, i.e., an increase in available P in the soil at different times during the 56 days of incubation (Kwabiah et al. 2003). Other materials result in transient net P immobilization. Along the soil food web, C:P ratios decrease because CO_2 is produced, which will finally result in P release. Bünemann et al. (2004a) and Ehlers et al. (2010) have shown that organic substrate decomposition is only retarded by low P availability, but not prevented. Thus, microorganisms cover their need by the uptake of soil P, which involves microbial uptake from less-available soil P pools (Bünemann et al. 2004a, c). Still, experimental evidence that crops benefit from microbial P release is lacking. The higher microbial P pool in cropping systems with more organic inputs is probably rather like a turning wheel that keeps a higher proportion of soil P in a potentially available form (Stewart and Sharpley 1987). This function is important, particularly in highly P sorbing soils (Oberson et al. 2001, 2006).

The application of legume residues increased P uptake by maize (Nziguheba et al. 2000), and the P uptake by cereals growing after a legume was higher than when monocropped (Bünemann et al. 2004b; Horst et al. 2001; Jemo et al. 2006; Pypers et al. 2007), as stated above. Bünemann et al. (2004b) reported that P uptake by maize grown in the maize–legume rotation was about 150% of the P uptake when grown in monoculture, at the same fertilizer P application rate. However, under field conditions, without labeling of P contained in the legume residues or of available soil P, the two sources cannot be separated. In a pot experiment, radioisotope P-labeled residues from faba bean (*Vicia faba*) and field peas (*Pisum sativum*) contributed up to 10 and 5% of the total P uptake by corn, respectively, while water-soluble mineral P contributed 20–50% (Nachimuthu et al. 2009). However, in this experiment total plant uptake was low for all treatments, resulting in a fertilizer P recovery of only 3% for mineral P and less than 0.5% for legume residues.

The seminal work of McLaughlin and Alston (1986) on wheat grown in pots showed a significantly lower contribution of P applied with ^{33}P-labeled legume (*Medicago truncülata* cv. Paraggio) residues than with ^{32}P-labeled monocalcium phosphate. About 6–9% of P taken up by wheat was derived from the legume residues, while up to 40% was derived from the mineral P source. The addition of residues depressed wheat biomass production, which McLaughlin and Alston (1986) assigned to microbial P immobilization. Because many legumes contain secondary metabolites with allelopathic activity (Caamal-Maldonado et al. 2001), legume residues may depress plant growth, particularly in pot experiments where the soil volume is limited. Although the pot experiments using radioisotope-labeled inputs indicate that the recovery of residue P in the crop is about six times lower than that of mineral P, recoveries from legume N and mineral N differ somewhat less. Field studies using the application of ^{15}N-labeled legume residues to maize under tropical conditions indicate that 9–15% of residue N is recovered in the crop (Douxchamps 2010; Vanlauwe et al. 1998), while mineral N recovery in the crop is around 35% (Douxchamps 2010) (see also compilation in Crews and Peoples 2005). Combined N and P tracer studies will improve our understanding of the fate of legume residue-derived N and P in the soil–plant system, but remain limited to studies of a few months duration because of the short half-life of P radioisotopes (Frossard et al. 2011). There is need for long-term field studies on the rotational effects of legumes on N- and P-use efficiency in cropping systems, including the effect on nutrient stocks and forms in the soil.

17.5.2.2 Improved Soil P Availability Because of Modified Soil Conditions

The integration of legumes in the rotation affects the size, activity, and community composition of the soil microbial biomass. Levels of microbial C and P were higher under maize–crotalaria rotation than under maize monocropping (Table 17.2) (Bünemann et al. 2004b). In spite of higher P immobilization in the microbial biomass, maize following crotalaria took up more P than monocropped maize, although it is unclear whether this was driven by increased P supply or by increased demand for P from more vigorous maize because of, e.g., increased N supply. Higher microbial biomass was connected with higher total amounts of phospholipid fatty acids and an increase in the relative abundance of indicators for fungi and Gram-negative bacteria (Bünemann et al. 2004a). Although the species diversity of arbuscular mycorrhizal fungal spores in the soil was not affected by the maize–crotalaria rotation in this field trial, the composition (relative species abundance) of spore communities was significantly altered by crop rotation, with two of the arbuscular mycorrhizal fungi species (*Acaulospora scrobiculata* and *Scutellospora verrucosa*) spores being more abundant in the rotated as compared to continuous maize plots (Mathimaran et al. 2007). The integration of legumes in a rotation increased mycorrhizal infection of maize (Horst et al. 2001) and other cereals such as pearl millet and sorghum (Bagayoko et al. 2000). Crop rotations versus monocropping also caused a significant shift in the rhizosphere bacterial community

Table 17.2 P status and indicators of microbial P cycling in soils of a field experiment in western Kenya under different crop rotations (maize monocropping versus maize–fallow rotations) without (−P) and with (+P) mineral P fertilizer input

	Crop rotation					
	Continuous maize		Maize–crotalaria fallow		Maize-natural fallow	
	−P	+P	−P	+P	−P	+P
Fertilizer P input (kg P ha^{-1} per year)[a]	0	50	0	50	0	50
P output (kg ha^{-1} per year)[b]	4.0–4.5	8.4–13.8	5.3–9.7	15.6–18.5	3.7–5.3	7.6–9.5
Annual P balance (kg ha^{-1} per year)[c]	−4.2 ± 0.3	38.9 ± 3.8	−7.5 ± 3.1	32.9 ± 2.1	−4.5 ± 1.2	41.4 ± 1.4
Available P (mg kg^{-1})[d]	1.9 ± 0.1	7.9 ± 1.6	1.6 ± 0.1	8.3 ± 2.0	1.6 ± 0.4	11.0 ± 3.3
Microbial P (mg kg^{-1})[e]	2.8 ± 1.2	2.4 ± 0.9	6.7 ± 0.7	6.8 ± 0.8	5.2 ± 1.3	4.9 ± 1.0
Organic P (mg kg^{-1})[e]	310 ± 38	302 ± 29	328 ± 20	357 ± 17	327 ± 35	335 ± 28
Total P (mg kg^{-1})[e]	720 ± 26	838 ± 51	721 ± 38	829 ± 19	703 ± 26	837 ± 54
Residue P recovery in microbial P (%)[f]	11.0 ± 1.4	nd	17.5 ± 2.1	nd	nd	nd
Mineral P recovery in microbial P (%)[f]	5.1 ± 3.9	nd	3.3 ± 1.8	nd	nd	nd

Values are average and standard deviations
nd not determined
[a]Applied as triple superphosphate
[b]P export in maize grains, maize stover, and fallow wood; average of 2 years; modified from Bünemann et al. (2004b)
[c]Average of 2 years [two cycles of short rainy (fallow or maize) and long rainy (maize) seasons]; modified from Bünemann et al. (2004b)
[d]Determined using anion exchange membranes; data from Bünemann et al. (2004b)
[e]Data from Bünemann et al. (2004b)
[f]Recovery of 33-P labeled amendments (residues from *Crotalaria grahamiana* Wight & Arn.) and water-soluble mineral (P, respectively) in microbial P 10 days after amendment application; modified from Bünemann et al. (2004c)

(Alvey et al. 2003), and the increased phosphatase activity in the rhizosphere of cereals grown after legumes was related to microbial activity (Alvey et al. 2001). These results show that several microbial functions modifying P availability to crops are affected by the integration of legumes.

Soils in legume rotations have higher soil organic C content in the topsoil when legume residues are returned to the soil (Bünemann et al. 2004b). Many legumes exude protons into their rhizosphere when actively fixing N. Some tropical legumes, however, do not acidify their rhizosphere as much as temperate legumes do because their N assimilation products are ureides (Bolan et al. 1991). The pH of bulk soil and in the rhizosphere of cereals was higher when cereals (millet or sorghum) were grown after a legume than when monocropped (Alvey et al. 2001). Changes in soil organic C content and pH can influence microbial activity and directly affect the availability of P and other nutrients. Probably also because of indirect effects on P availability, the recovery of mineral P fertilizer by maize was higher when applied in combination with organic residues than when applied alone (Nziguheba et al. 2002).

Finally, adapted germplasm of crops affects P losses. For instance, genotypic differences in root architecture of the common bean (*Phaseolus vulgaris* L.) led to a 20–50% variation in groundcover by shoots, which was associated with a 50–80% reduction in soil loss by erosion (Henry et al. 2010).

17.5.2.3 Effects Not Related to P

Physical conditions such as soil moisture and temperature are modified when legume residues are used as mulch and/or when soil cover is increased. Soil chemical conditions can also be changed. For example, Alvey et al. (2001) reported that the higher soil pH fed back to cause lower Al availability and enhanced Mg and Ca availability. Most importantly, the integration of legumes in the rotation improves the N supply to the succeeding crop (Bagayoko et al. 2000; Yusuf et al. 2009). Of high impact, however, are phytosanitary improvements such as weed control and suppression of parasitic nematodes (Cherr et al. 2006). These effects can be more important than changes in nutrient supply. For instance, in maize fields in western Kenya, a white lupin cover crop significantly inhibited infestation by *Striga hermonthica* (Weisskopf et al. 2009). Likewise, the infestation of cereals by plant parasitic nematodes has been reduced in rotations (Alvey et al. 2001; Bagayoko et al. 2000). Microbiological changes in soils under legume–cereal cropping systems may have positive effects on crop performance, but it is often difficult to distinguish causes from effects (Marschner et al. 2004). Still, if repeatedly grown, the legumes themselves can become affected by pests and diseases, and their biomass production and resulting benefits can decrease dramatically (Bünemann et al. 2004b).

17.5.3 Limitations of the Approach to Integrate Legumes in Crop Rotations

To benefit from the symbiotic N fixation potential of legumes, an external P fertilizer input is required on low-P soils (Okogun et al. 2005). On P-deficient sites, legume–cereal rotations (Bünemann et al. 2004b) or legume–maize intercropping systems (Mucheru-Muna et al. 2010) receiving P fertilizers are more productive than without fertilizer P input (Table 17.2). On strongly P sorbing soils, the repeated addition of low doses of P fertilizers placed in the vicinity of the crop is recommended over single, high input doses (van der Eijk et al. 2006).

Legumes can accelerate soil P depletion if greater P export in the legume–cereal rotation is not balanced by fertilizer inputs (Table 17.2) (Bünemann et al. 2004b). The extent depends on the use of legumes and the productivity of the system. In the case of green manure cover crops, soil P taken up by the legume is recycled on the plot, but increased P output by the cereal (Table 17.2) needs to be compensated. In the case of grain or forage legumes, both the legume and the cereal remove P. Thus, P removal by the legume and the main crop (often a cereal) is usually higher in rotations than under monocropping and has to be compensated unless total soil P stocks are high and a limited reduction is justifiable. When legumes are used as animal forage, management of animal excreta becomes a key issue because, e.g., growing beef cattle and lactating cows convert approximately 15 and 25% of dietary P into carcass and milk P, respectively, and the rest goes into manure (Satter et al. 2005). Thus, under grazing a large proportion of legume P is recycled on the plots although spatial distribution of animal excreta is usually not homogenous. If animals are kept in paddocks or in a cattle shed, excreta can be collected, stored, and used as fertilizer. However, the infrastructure to do so is often poor in smallholder systems and measures to improve manure management should receive more attention (Rufino et al. 2006).

17.6 Integrated P Management for Sustainable P Use

Because of the P resource scarcity and the environmental and economic implications of P use (Tiessen et al. 2011), we need to optimize P-use efficiency in soil–plant systems, i.e., attain highest yields for a given P application rate. This requires that fertilizer P is either taken up by the crop and efficiently converted into biomass, or that it is kept in the soil in a potentially plant-available form for a subsequent crop. Based on the definitions of Vanlauwe et al. (2010) for Integrated Soil Fertility Management, and of Frossard et al. (2009) for Integrated Nutrient Management, we define Integrated P Management in cropping systems as follows: Integrated P Management aims at maximizing P-use efficiency in the cropping system while minimizing P losses to the environment and optimizing economic benefits. It considers all the biophysical components involved in P cycling as well as the relevant socioeconomic factors such as the production preferences of farmers, the food preferences of consumers, the markets, and trade policy. It includes the use of mineral and organic P sources, of improved germplasm, and of measures to control P losses to the environment, combined with the knowledge of how to adapt these practices to local conditions (Fig. 17.2).

The availability of soil and fertilizer P manifests the global divide in the distribution of this vital resource. The national P budgets of agriculture in industrialized countries often have a P surplus (Sharpley et al. 2005) whereas the P budgets of developing countries often have a deficit (Lesschen et al. 2007). In turn,

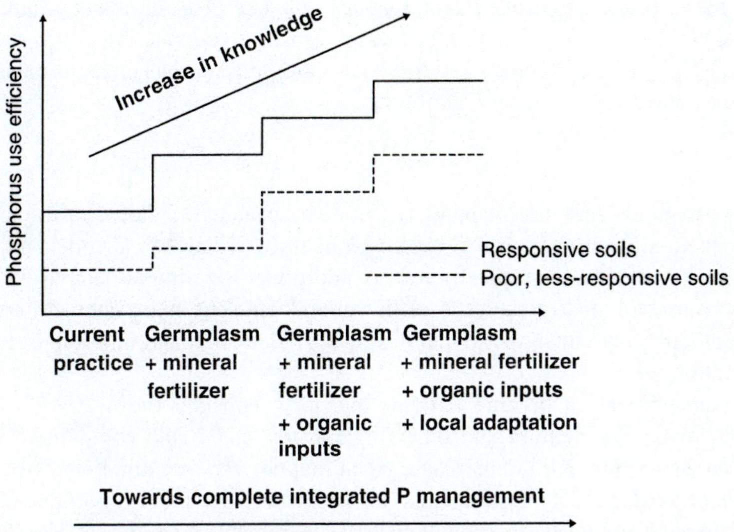

Fig. 17.2 Conceptual relationship between the P-use efficiency in smallholder cropping systems, where P limits crop production, and the implementation of various components of integrated P management. Soils that are responsive to P fertilizer (*solid line*) and those that are poor and less-responsive (*dotted line*) are distinguished. At constant fertilizer application rates, yield is linearly related to P-use efficiency (adapted from Vanlauwe et al. 2010)

soils of industrialized countries, particularly in areas with high livestock densities, are over-fertilized with P while soils in developing countries in the tropics are often P depleted. There is variation within countries, regions, and even within farms (Cobo et al. 2010). Different fields of a single farming household respond differently to fertilizer, which results in differential P-use efficiency (Vanlauwe et al. 2006). This variability needs to be accounted for by considering the local conditions, as suggested in the definition of integrated P management.

For cropping systems where soil nutrients are mined using the current practice, we can deduce the following recommendations:

1. Mineral fertilizer use needs to increase to improve crop production.
2. P use by crops needs to be maximized through the appropriate timing and placement of the fertilizer in the vicinity of the crop roots, and by the use of crop germplasm with a high P acquisition and/or a high P conversion efficiency (Fig. 17.2).
3. The introduction of legume fallows or green manures (which are system internal organic resources) can enhance P cycling and P use in the cropping system.
4. The proper reuse of animal manure and urban solid and liquid wastes is an important part of sustainable P use.
5. P losses by erosion and run-off have to be minimized. Minimal tillage and retention of groundcover, including legume cover crops, increases soil cover and reduces the risk of P losses.

In general, there are positive relationships between P-use efficiency, environmental friendliness, and profitability, but trade-offs may exist, e.g., as shown in Sect. 17.3.3 for the use of plant residues. The Integrated P management measures need to be adapted to the local conditions:

1. Input rates need to be tuned to the fertility level of the soil in each field and for each crop.
2. Degraded fields may not respond to P inputs because they are constrained by other factors, such as high soil acidity, moisture availability, or other nutrient deficiencies. Such soils typically require additional investments such as liming, erosion control measures, or organic amendments at high rates. Microbial inoculations may enhance P dynamics in degraded soils and thus support their restoration.
3. The management of organic P inputs needs to consider the availability and quality of animal manure and other organic resources, and competitive uses, e.g., the use of legume biomass as a green manure versus animal feed.
4. If reduced tillage is practiced, then sufficient biomass must be produced and retained to ensure adequate groundcover. The applicability of these technologies will therefore vary across agro-ecological zones, farming systems, and niches within the farm.

On soils where plant-available P concentration is sufficiently high for yields to be at the economic optimum, no additional P inputs are needed. In fields with very high concentrations of soil P, reflecting many years of over-fertilization or animal manure disposal, measures to reduce P losses are crucial. In farms and regions with surplus manure P, animal manure P should be applied to fields where P inputs are required and where the risk of P loss to the environment is low. However, because of the disintegration of agriculture into specialized farms, specialized regions, and even specialized countries, local nutrient cycles have been disrupted (Schröder 2005; Dao and Schwartz 2011). The P surpluses in industrialized countries are caused by feed and fertilizers imports, often from continents with predominantly low-P soils (Tiessen et al. 2011). Also, the global trend of urbanization results in sinks for P exported from arable fields, unless cities have an infrastructure for wastewater treatment that enables the proper reuse of sewage P. Sustainable P use requires recycling of P excreted by animals and humans back to the agricultural land in need of P.

17.7 Final Remarks and Research Needs

Farmer intervention affects biological processes in P cycling. Enhanced microbial P cycling is closely linked to the presence of organic matter and the use of organic inputs. Substantial amounts of P can be immobilized and released by microorganisms, with temporal dynamics depending on the characteristics of the inputs and on environmental conditions. We do not yet have sufficient knowledge about the source of microbial P uptake. In particular, we do not know how to enhance access to recalcitrant inorganic and organic P. Also, there is a need to improve the understanding of the pathways that render microbial P available to crops. This knowledge is required to manage microbial immobilization and release of P so that there can be greater synchrony and synlocation with plant demand.

Tillage clearly affects P cycling, but the impacts of direct seeding systems on P dynamics and on the use efficiency of soil and fertilizer P need to be studied in more detail under different agro-ecological and socio-economic conditions, and with different scenarios of crop residue use.

We do not yet sufficiently understand how germplasm, which is efficient in P acquisition, can contribute to overall use efficiency of P inputs in rotation or intercropping systems. The N benefits of legume varieties with low N harvest indices and high N fixation capacity have been demonstrated and verified, but the contribution of P-efficient legume varieties with specific root architecture and P mobilization capacity to the recovery of P inputs is much less understood, especially if they influence microbial diversity and P pools.

Recently, great progress has been made in analyzing the biodiversity of micro-organisms, but the significance of the microbial biodiversity in P dynamics is not understood. Often, large proportions of soil P are kept in recalcitrant forms, and also residual fertilizer and rock phosphate have limited availability for crops. There is a need to establish a better linkage between the microbial community composition and microbial P functions in cropping systems, and to understand under which conditions the combination of organic inputs with inoculants can improve P use.

Acknowledgments We gratefully acknowledge Jan Jansa (ETH Zurich) and two reviewers for their helpful comments on our chapter.

References

Abunyewa AA, Osei D, Asiedu EK, Safo EY (2007) Integrated manure and fertilizer use, maize production and sustainable soil fertility in sub humid zone of West Africa. J Agron 6:302–309

Akponikpe PBI, Michels K, Bielders CL (2008) Integrated nutrient management of pearl millet in the Sahel combining cattle manure, crop residue and mineral fertilizer. Exp Agric 44:453–472

Alves BJR, Boddey RM, Urquiaga S (2003) The success of BNF in soybean in Brazil. Plant Soil 252:1–9

Alvey S, Bagayoko M, Neumann G, Buerkert A (2001) Cereal/legume rotations affect chemical properties and biological activities in two West African soils. Plant Soil 231:45–54

Alvey S, Yang CH, Buerkert A, Crowley DE (2003) Cereal/legume rotation effects on rhizosphere bacterial community structure in West African soils. Biol Fertil Soils 37:73–82

Bagayoko M, Buerkert A, Lung G, Bationo A, Romheld V (2000) Cereal/legume rotation effects on cereal growth in Sudano-Sahelian West Africa: soil mineral nitrogen, mycorrhizae and nematodes. Plant Soil 218:103–116

Balota EL, Colozzi A, Andrade DS, Dick RP (2003) Microbial biomass in soils under different tillage and crop rotation systems. Biol Fertil Soils 38:15–20

Bassala JPO, M'Biandoun M, Ekorong JA, Asfom P (2008) Changes in soil fertility under the cotton and cereal farming system in North Cameroon: diagnostic and perspectives. Tropicultura 26:240–245

Berner A, Hildermann I, Fliessbach A, Pfiffner L, Niggli U, Mader P (2008) Crop yield and soil fertility response to reduced tillage under organic management. Soil Till Res 101:89–96

Boddey RM, Alves BJR, Urquiaga S (2006) Leguminous biological nitrogen fixation in sustainable tropical agroecosystems. In: Uphoff N, Ball AS, Fernandes E, Herren H, Husson O, Laing M, Palm C, Pretty J, Sanchez P, Sanginga N, Thies J (eds) Biological approaches to sustainable soil systems. CRC, Boca Raton, FL, pp 401–408

Bolan N, Hedley M, White R (1991) Processes of soil acidification during nitrogen cycling with emphasis on legume based pastures. Plant Soil 134:53–63

Bonkowski M (2004) Protozoa and plant growth: the microbial loop in soil revisited. New Phytol 162:617–631

Bünemann E, Bossio DA, Smithson PC, Frossard E, Oberson A (2004a) Microbial community composition and substrate use in a highly weathered soil as affected by crop rotation and P fertilization. Soil Biol Biochem 36:889–901

Bünemann E, Smithson PC, Jama B, Frossard E, Oberson A (2004b) Maize productivity and nutrient dynamics in maize-fallow rotations in western Kenya. Plant Soil 264:195–208

Bünemann EK, Steinebrunner F, Smithson PC, Frossard E, Oberson A (2004c) Phosphorus dynamics in a highly weathered soil as revealed by isotopic labeling techniques. Soil Sci Soc Am J 68:1645–1655

Bünemann EK, Heenan DP, Marschner P, McNeill AM (2006a) Long-term effects of crop rotation, stubble management and tillage on soil phosphorus dynamics. Aust J Soil Res 44:611–618

Bünemann EK, Schwenke GD, Van Zwieten L (2006b) Impact of agricultural inputs on soil organisms – a review. Aust J Soil Res 44:379–406

Bünemann EK, Prusisz B, Ehlers K (2011) Characterization of phosphorus forms in soil microorganisms. In: Bünemann E, Oberson A, Frossard E (eds) Phosphorus in action: biological processes in soil phosphorus cycling. Soil biology, vol 26. Springer, Heidelberg. doi: 10.1007/978-3-642-15271-9_2

Caamal-Maldonado JA, Jiménez-Osornio JJ, Torres-Barragán A, Anaya AL (2001) The use of allelopathic legume cover and mulch species for weed control in cropping systems. Agron J 93:27–36

Chapuis-Lardy L, Ramiandrisoa RS, Randriamanantsoa L, Morel C, Rabeharisoa L, Blanchart E (2009) Modification of P availability by endogeic earthworms (*Glossoscolecidae*) in Ferralsols of the Malagasy Highlands. Biol Fertil Soils 45:415–422. doi:10.1007/s00374-00008-00350-y

Chapuis-Lardy L, Le Bayon R-C, Brossard M, López-Hernández D, Blanchart E (2011) Role of soil macrofauna in phosphorus cycling. In: Bünemann E, Oberson A, Frossard E (eds) Phosphorus in action: biological processes in soil phosphorus cycling. Soil biology, vol 26. Springer, Heidelberg. doi: 10.1007/978-3-642-15271-9_8

Chassot A, Stamp P, Richner W (2001) Root distribution and morphology of maize seedlings as affected by tillage and fertilizer placement. Plant Soil 231:123–135

Cherr CM, Scholberg JMS, McSorley R (2006) Green manure approaches to crop production: a synthesis. Agron J 98:302–319

Cobo JG, Dercon G, Cadisch G (2010) Nutrient balances in African land use systems across different spatial scales: a review of approaches, challenges and progress. Agric Ecosyst Environ 136:1–15

Cordell D, Drangert JO, White S (2009) The story of phosphorus: global food security and food for thought. Glob Environ Change 19:292–305

Cornish PS (2009) Phosphorus management on extensive organic and low-input farms. Crop Pasture Sci 60:105–115

Crews TE, Peoples MB (2005) Can the synchrony of nitrogen supply and crop demand be improved in legume and fertilizer-based agroecosystems? A review. Nutr Cycl Agroecosyst 72:101–120

Dann PR, Derrick JW, Dumaresq DC, Ryan MH (1996) The response of organic and conventionally grown wheat to superphosphate and reactive phosphate rock. Aust J Exp Agric 36:71–78

Dao TH, Schwartz RC (2011) Effects of manure management on phosphorus biotransformations and losses during animal production. In: Bünemann E, Oberson A, Frossard E (eds) Phosphorus in action: biological processes in soil phosphorus cycling. Soil biology, vol 26. Springer, Heidelberg. doi: 10.1007/978-3-642-15271-9_16

Daroub SH, Pierce FJ, Ellis BG (2000) Phosphorus fractions and fate of phosphorus-33 in soils under plowing and no-tillage. Soil Sci Soc Am J 64:170–176

Douxchamps S (2010) Integration of Canavalia brasiliensis into the crop-livestock system of the Nicaraguan hillsides: environmental adaption and nitrogen dynamics. Dissertation, ETH Zurich

Douxchamps S, Humbert FL, Van der Hoek R, Mena M, Bernasconi S, Schmidt A, Rao IM, Frossard E, Oberson A (2010) Nitrogen balances in farmers fields under alternative uses of a cover crop legume – a case study from Nicaragua. Nutr Cycl Agroecosyst. doi:10.1007/s10705-010-9368-2

Ehlers K, Bakken LR, Frostegard A, Frossard E, Bünemann E (2010) Phosphorus limitation in a Ferralsol: impact on microbial activity and cell internal P-pools. Soil Biol Biochem 42:558–566

Ernst G, Emmerling C (2009) Impact of five different tillage systems on soil organic carbon content and the density, biomass, and community composition of earthworms after a ten year period. Eur J Soil Biol 45:247–251

FAO (2009) How to feed the world in 2050. Food and Agriculture Organization of the United Nations, Rome. http://www.fao.org/fileadmin/templates/wsfs/docs/expert_paper/How_to_Feed_the_World_in_2050.pdf. Accessed 13 Aug 2010

Fliessbach A, Oberholzer HR, Gunst L, Mader P (2007) Soil organic matter and biological soil quality indicators after 21 years of organic and conventional farming. Agric Ecosyst Environ 118:273–284

Friesen DK, Rao IM, Thomas RJ, Oberson A, Sanz JI (1997) Phosphorus acquisition and cycling in crop and pasture systems in low fertility tropical soils. Plant Soil 196:289–294

Frossard E, Bünemann E, Jansa J, Oberson A, Feller C (2009) Concepts and practices of nutrient management in agro-ecosystems: can we draw lessons from history to design future sustainable agricultural production systems? Bodenkultur 60:5–22

Frossard E, Achat DL, Bernasconi SM, Bünemann EK, Fardeau J-C, Jansa J, Morel C, Rabeharisoa L, Randriamanantsoa L, Sinaj S, Tamburini F, Oberson A (2011) The use of tracers to investigate phosphate cycling in soil–plant systems. In: Bünemann E, Oberson A, Frossard E (eds) Phosphorus in action: biological processes in soil phosphorus cycling. Soil biology, vol 26. Springer, Heidelberg. doi: 10.1007/978-3-642-15271-9_3

Gathumbi SM, Cadisch G, Buresh RJ, Giller KE (2003) Subsoil nitrogen capture in mixed legume stands as assessed by deep nitrogen-15 placement. Soil Sci Soc Am J 67:573–582

George TS, Fransson A-M, Hammond JP, White PJ (2011) Phosphorus nutrition: rhizosphere processes, plant response and adaptations. In: Bünemann E, Oberson A, Frossard E (eds) Phosphorus in action: biological processes in soil phosphorus cycling. Soil biology, vol 26. Springer, Heidelberg. doi: 10.1007/978-3-642-15271-9_10

Giller KE, Witter E, Corbeels M, Tittonell P (2009) Conservation agriculture and smallholder farming in Africa: the heretics' view. Field Crop Res 114:23–34

Hedley MJ, White RE, Nye PH (1982) Plant-induced changes in the rhizosphere of rape (*Brassica napus* var. Emerald) seedlings III: changes in L value, soil phosphate fractions and phosphatase activity. New Phytol 91:45–56

Henry A, Chaves NF, Kleinman PJA, Lynch JP (2010) Will nutrient-efficient genotypes mine the soil? Effects of genetic differences in root architecture in common bean (*Phaseolus vulgaris* L.) on soil phosphorus depletion in a low-input agro-ecosystem in Central America. Field Crop Res 115:67–78

Herridge D, Peoples M, Boddey R (2008) Global inputs of biological nitrogen fixation in agricultural systems. Plant Soil 311:1–18

Hogh-Jensen H, Schjoerring JK, Soussana JF (2002) The influence of phosphorus deficiency on growth and nitrogen fixation of white clover plants. Ann Bot 90:745–753

Horst WJ, Kamh M, Jibrin JM, Chude VO (2001) Agronomic measures for increasing P availability to crops. Plant Soil 237:211–223

Houot S, Chaussod R (1995) Impact of agricultural practices on the size and activity of the microbial biomass in a long-term field experiment. Biol Fertil Soils 19:309–316

Huggins DR, Reganold JP (2008) No-till: the quiet revolution. Sci Am 299:70–77

Husson O, Séguy L, Michellon R, Boulakia S (2006) Restoration of acid soil systems through agroecological management. In: Uphoff N, Ball AS, Fernandes E, Herren H, Husson O, Laing M, Palm C, Pretty J, Sanchez P, Sanginga N, Thies J (eds) Biological approaches to sustainable soil systems. CRC, Boca Raton, FL, pp 343–356

Jansa J, Mozafar A, Kuhn G, Anken T, Ruh R, Sanders IR, Frossard E (2003) Soil tillage affects the community structure of mycorrhizal fungi in maize roots. Ecol Appl 13:1164–1176

Jansa J, Wiemken A, Frossard E (2006) The effects of agricultural practices on arbuscular mycorrhizal fungi. In: Frossard E, Blum WEH, Warkentin BP (eds) Function of soils for human societies and the environment. Geological Society Special Publications, London, pp 89–115

Jansa J, Smith FA, Smith SE (2008) Are there benefits of simultaneous root colonization by different arbuscular mycorrhizal fungi? New Phytol 177:779–789

Jansa J, Finlay R, Wallander H, Smith FA, Smith SE (2011) Role of mycorrhizal symbioses in phosphorus cycling. In: Bünemann E, Oberson A, Frossard E (eds) Phosphorus in action: biological processes in soil phosphorus cycling. Soil biology, vol 26. Springer, Heidelberg. doi: 10.1007/978-3-642-15271-9_6

Jemo M, Abaidoo R, Nolte C, Tchienkoua M, Sanginga N, Horst W (2006) Phosphorus benefits from grain-legume crops to subsequent maize grown on acid soils of southern Cameroon. Plant Soil 284:385–397

Jiménez JJ, Cepeda A, Decaëns T, Oberson A, Friesen DK (2003) Phosphorus fractions and dynamics in surface earthworm casts under native and improved grasslands in a Colombian savanna oxisol. Soil Biol Biochem 35:715–727

Jones DL, Oburger E (2011) Solubilization of phosphorus by soil microorganisms. In: Bünemann E, Oberson A, Frossard E (eds) Phosphorus in action: biological processes in soil phosphorus cycling. Soil biology, vol 26. Springer, Heidelberg. doi: 10.1007/978-3-642-15271-9_7

Kamh M, Abdou M, Chude V, Wiesler F, Horst WJ (2002) Mobilization of phosphorus contributes to positive rotational effects of leguminous cover crops on maize grown on soils from northern Nigeria. J Plant Nutr Soil Sci 165:566–572

Keller M, Oberson A, Frossard E, Mäder P, Mayer J and Bünemann EK (2009) Einfluss unterschiedlicher Bewirtschaftungsverfahren auf P-Formen und P-Dynamik im Boden. In: Mayer J, Alföldi T, Leiber F, Dubois D, Fried P, Heckendorn F, Hillmann E, Klocke P, Lüscher A, Riedel S, Stolze M, Strasser F, van der Heijden M, Willer H (eds) 10. Wissenschaftstagung Ökologischer Landbau. ETH Zürich, pp 73–74

Kuligowski K, Poulsen TG (2009) Phosphorus leaching from soils amended with thermally gasified piggery waste ash. Waste Manage 29:2500–2508

Kwabiah AB, Stoskopf NC, Palm CA, Voroney RP (2003) Soil P availability as affected by the chemical composition of plant materials: implications for P-limiting agriculture in tropical Africa. Agric Ecosyst Environ 100:53–61

Leifeld J, Reiser R, Oberholzer HR (2009) Consequences of conventional versus organic farming on soil carbon: results from a 27-year field experiment. Agron J 101:1204–1218

Lesschen JP, Stoorvogel JJ, Smaling EMA, Heuvelink GBM, Veldkamp A (2007) A spatially explicit methodology to quantify soil nutrient balances and their uncertainties at the national level. Nutr Cycl Agroecosyst 78:111–131

Li L, Tang C, Rengel Z, Zhang F (2003) Chickpea facilitates phosphorus uptake by intercropped wheat from an organic phosphorus source. Plant Soil 248:297–303

Li L, Tang C, Rengel Z, Zhang FS (2004) Calcium, magnesium and microelement uptake as affected by phosphorus sources and interspecific root interactions between wheat and chickpea. Plant Soil 261:29–37

Li L, Li SM, Sun JH, Zhou LL, Bao XG, Zhang HG, Zhang FS (2007) Diversity enhances agricultural productivity via rhizosphere phosphorus facilitation on phosphorus-deficient soils. Proc Natl Acad Sci USA 104:11192–11196

Liebig MA, Doran JW (1999) Impact of organic production practices on soil quality indicators. J Environ Qual 28:1601–1609

Lynch JP (2007) Roots of the second green revolution. Aust J Bot 55:493–512

Mäder P, Edenhofer S, Boller T, Wiemken A, Niggli U (2000) Arbuscular mycorrhizae in a long-term field trial comparing low-input (organic, biological) and high-input (conventional) farming systems in a crop rotation. Biol Fertil Soils 31:150–156

Mäder P, Fließbach A, Dubois D, Gunst L, Fried P, Niggli U (2002) Soil fertility and biodiversity in organic farming. Science 296:1694–1697

Mafongoya PL, Giller KE, Odee D, Gathumbi S, Ndufa SK, Sitompul SM (2004) Benefiting from N2-fixation and managing rhizobia. In: Van Noordwijk M, Cadisch G, Ong CK (eds) Below-ground interactions in tropical agroecosystems. CABI, Wallingford, UK, pp 227–242

Marschner P, Joergensen RG, Piepho HP, Buerkert A (2004) Legume rotation effects on early growth and rhizosphere microbiology of sorghum in West African soils. Plant Soil 264:325–334

Materechera SA, Morutse HM (2009) Response of maize to phosphorus from fertilizer and chicken manure in a semi-arid environment of South Africa. Exp Agric 45:261–273

Mathimaran N, Ruh R, Jama B, Verchot L, Frossard E, Jansa J (2007) Impact of agricultural management on arbuscular mycorrhizal fungal communities in Kenyan ferralsol. Agric Ecosyst Environ 119:22–32

McLaughlin MJ, Alston AM (1986) The relative contribution of plant residues and fertilizer to the P nutrition of wheat in a pasture/cereal system. Aust J Soil Res 24:517–526

McLaughlin MJ, Alston AM, Martin JK (1988) Phosphorus cycling in wheat-pasture rotations II. The role of the microbial biomass in phosphorus cycling. Aust J Soil Res 26:333–342

Miller MH (2000) Arbuscular mycorrhizae and the phosphorus nutrition of maize: a review of Guelph studies. Can J Plant Sci 80:47–52

Moore JM, Klose S, Tabatabai MA (2000) Soil microbial biomass carbon and nitrogen as affected by cropping systems. Biol Fertil Soils 31:200–210

Muchane MN, Jama B, Othieno C, Okalebo R, Odee D, Machuna J, Jansa J (2010) Influence of improved fallow systems and phosphorus application on arbuscular mycorrhizal fungi symbiosis in maize grown in western Kenya. Agrofor Syst 78:139–150

Mucheru-Muna M, Pypers P, Mugendi D, Kung'u J, Mugwe J, Merckx R, Vanlauwe B (2010) A staggered maize-legume intercrop arrangement robustly increases crop yields and economic returns in the highlands of Central Kenya. Field Crop Res 115:132–139

Mukuralinda A, Tenywa JS, Verchot L, Obua J, Namirembe S (2009) Decomposition and phosphorus release of agroforestry shrub residues and the effect on maize yield in acidic soils of Rubona, southern Rwanda. Nutr Cycl Agroecosyst 84:155–166

Nachimuthu G, Guppy C, Kristiansen P, Lockwood P (2009) Isotopic tracing of phosphorus uptake in corn from P-33 labelled legume residues and P-32 labelled fertilisers applied to a sandy loam soil. Plant Soil 314:303–310

Nannipieri P, Giagnoni L, Landi L, Renella G (2011) Role of phosphatase enzymes in soil. In: Bünemann E, Oberson A, Frossard E (eds) Phosphorus in action: biological processes in soil phosphorus cycling. Soil biology, vol 26. Springer, Heidelberg. doi: 10.1007/978-3-642-15271-9_9

Neumann G, Martinoia E (2002) Cluster roots – an underground adaptation for survival in extreme environments. Trends Plant Sci 7:162–167

Nziguheba G, Merckx R, Palm CA, Rao MR (2000) Organic residues affect phosphorus availability and maize yields in a Nitisol of western Kenya. Biol Fertil Soils 32:328–339

Nziguheba G, Merckx R, Palm CA, Mutuo P (2002) Combining Tithonia diversifolia and fertilizers for maize production in a phosphorus deficient soil in Kenya. Agrofor Syst 55:165–174

Oberson A, Frossard E (2005) Phosphorus management for organic agriculture. In: Sims T, Sharpley AN (eds) Phosphorus: agriculture and the environment. Agronomy monograph 46. ASA/CSSA/SSSA, Madison, WI, pp 761-779

Oberson A, Joner EJ (2005) Microbial turnover of phosphorus in soil. In: Turner BL, Frossard E, Baldwin DS (eds) Organic phosphorus in the environment. CABI, Wallingford, UK, pp 133–164

Oberson A, Fardeau JC, Besson JM, Sticher H (1993) Soil-phosphorus dynamics in cropping systems managed according to conventional and biological agricultural methods. Biol Fertil Soils 16:111–117

Oberson A, Friesen DK, Rao IM, Bühler S, Frossard E (2001) Phosphorus transformations in an Oxisol under contrasting land- use systems: the role of the soil microbial biomass. Plant Soil 237:197–210

Oberson A, Bünemann EK, Friesen DK, Rao IM, Smithson PC, Turner BL, Frossard E (2006) Improving phosphorus fertility in tropical soils through biological interventions. In: Uphoff N, Ball AS, Fernandes E, Herren H, Husson O, Laing M, Palm C, Pretty J, Sanchez P, Sanginga N, Thies J (eds) Biological approaches to sustainable soil systems. CRC, Boca Raton, FL, pp 531–546

Oberson A, Tagmann H, Langmeier M, Dubois D, Mäder P, Frossard E (2010) Fresh and residual phosphorus uptake by ryegrass from soils with different fertilization histories. Plant Soil 334:391–407

Oehl F, Oberson A, Probst M, Fliessbach A, Roth HR, Frossard E (2001) Kinetics of microbial phosphorus uptake in cultivated soils. Biol Fertil Soils 34:31–41

Oehl F, Oberson A, Tagmann HU, Besson JM, Dubois D, Mäder P, Roth HR, Frossard E (2002) Phosphorus budget and phosphorus availability in soils under organic and conventional farming. Nutr Cycl Agroecosyst 62:25–35

Oehl F, Frossard E, Fliessbach A, Dubois D, Oberson A (2004) Basal organic phosphorus mineralization in soils under different farming systems. Soil Biol Biochem 36:667–675

Ojiem JO, Vanlauwe B, De Ridder N, Giller KE (2007) Niche-based assessment of contributions of legumes to the nitrogen economy of Western Kenya smallholder farms. Plant Soil 292:119–135

Okogun JA, Sanginga N, Abaidoo R, Dashiell KE, Diels J (2005) On-farm evaluation of biological nitrogen fixation potential and grain yield of Lablab and two soybean varieties in the northern Guinea savanna of Nigeria. Nutr Cycl Agroecosyst 73:267–275

Olander LP, Vitousek PM (2004) Biological and geochemical sinks for phosphorus in soil from a wet tropical forest. Ecosystems 7:404–419

Onduru DD, du Preez CC, Muchena FN, Gachimbi LN, de Jager A, Gachini GN (2008) Exploring options for integrated nutrient management in semi-arid tropics using farmer field schools: a case study in Mbeere District, eastern Kenya. Int J Agric Sustain 6:208–228

Opala PA, Jama BA, Othieno CO, Okalebo JR (2007) Effect of phosphate fertilizer application methods and nitrogen sources on maize in western Kenya: an agronomic and economic evaluation. Exp Agric 43:477–487

Palm CA, Gachengo CN, Delve RJ, Cadisch G, Giller KE (2001) Organic inputs for soil fertility management in tropical agroecosystems: application of an organic resource database. Agric Ecosyst Environ 83:27–42

Peoples MB, Brockwell J, Herridge DF, Rochester IJ, Alves BJR, Urquiaga S, Boddey RM, Dakora FD, Bhattarai S, Maskey SL, Sampet C, Rerkasem B, Khan DF, Hauggaard-Nielsen H, Jensen ES (2009) The contributions of nitrogen-fixing crop legumes to the productivity of agricultural systems. Symbiosis 48:1–17

Pypers P, Van Loon L, Diels J, Abaidoo R, Smolders E, Merckx R (2006) Plant-available P for maize and cowpea in P-deficient soils from the Nigerian Northern Guinea Savanna – comparison of E- and L-values. Plant Soil 283:251–264

Pypers P, Huybrighs M, Diels J, Abaidoo R, Smolders E, Merckx R (2007) Does the enhanced P acquisition by maize following legumes in a rotation result from improved soil P availability? Soil Biol Biochem 39:2555–2566

Rabary B, Sall S, Letourmy P, Husson O, Ralambofetra E, Moussa N, Chotte J-L (2008) Effects of living mulches or residue amendments on soil microbial properties in direct seeded cropping systems of Madagascar. Appl Soil Ecol 39:236–243

Rufino MC, Rowe EC, Delve RJ, Giller KE (2006) Nitrogen cycling efficiencies through resource-poor African crop-livestock systems. Agric Ecosyst Environ 112:261–282

Rufino MC, Tittonell P, van Wijk MT, Castellanos-Navarrete A, Delve RJ, de Ridder N, Giller KE (2007) Manure as a key resource within smallholder farming systems: analysing farm-scale nutrient cycling efficiencies with the NUANCES framework. Livest Sci 112:273–287

Satter LD, Klopfenstein TJ, Erickson GE, Powell JM (2005) Phosphorus and dairy/beef nutrition. In: Sims T, Sharpley AN (eds) Phosphorus: agriculture and the environment. Agronomy monograph 46. ASA/CSSA/SSSA, Madison, WI, pp 587–606

Schröder J (2005) Revisiting the agronomic benefits of manure: a correct assessment and exploitation of its fertilizer value spares the environment. Bioresour Technol 96:253–261

Séguy L, Bouzinac S, Husson O (2006) Direct-seeded tropical soil systems with permanent soil cover: learning from the Brazilian experience. In: Uphoff N, Ball AS, Fernandes E, Herren H, Husson O, Laing M, Palm C, Pretty J, Sanchez P, Sanginga N, Thies J (eds) Biological approaches to sustainable soil systems. CRC, Boca Raton, FL, pp 323–342

Shane MW, Lambers H (2005) Cluster roots: a curiosity in context. Plant Soil 274:101–125

Sharpley AN, Withers PJA, Abdalla CW, Dodd AR (2005) Strategies for the sustainable management of phosphorus. In: Sims T, Sharpley AN (eds) Phosphorus: agriculture and the environment. Agronomy monograph 46. ASA/CSSA/SSSA, Madison, WI, pp 1069–1101

Smestad TB, Tiessen H, Buresh RJ (2002) Short fallows of *Tithonia diversifolia* and *Crotalaria grahamiana* for soil fertility improvement in western Kenya. Agrofor Syst 55:181–194

Smith SE, Read D (2008) Mycorrhizal symbiosis. Academic, New York

Stewart JWB, Sharpley AN (1987) Controls on dynamics of soil and fertilizer phosphorus and sulfur. In: Follett RF, Stewart JWB, Cole CV (eds) Soil fertility and organic matter as critical components of production systems. SSSA Special Publication 19. American Society of Agronomy, Madison, WI, pp 101–121

Tan ZX, Lal R, Wiebe KD (2005) Global soil nutrient depletion and yield reduction. J Sustain Agric 26:123–146

Tiessen H, Ballester MV, Salcedo I (2011) Phosphorus and global change. In: Bünemann E, Oberson A, Frossard E (eds) Phosphorus in action: biological processes in soil phosphorus cycling. Soil biology, vol 26. Springer, Heidelberg. doi: 10.1007/978-3-642-15271-9_18

Tittonell P, Corbeels M, Van Wijk MT, Vanlauwe B, Giller KE (2008) Combining organic and mineral fertilizers for integrated soil fertility management in smallholder farming systems of Kenya: explorations using the crop-soil model FIELD. Agron J 100:1511–1526

Tiunov AV, Bonkowski M, Alphei J, Scheu S (2001) Microflora, protozoa and nematoda in *Lumbricus terrestris* burrow walls: a laboratory experiment. Pedobiologia 45:46–60

Toor GS, Hunger S, Peak JD, Sims JT, Sparks DL, Donald LS (2006) Advances in the characterization of phosphorus in organic wastes: environmental and agronomic applications. Adv Agron 89:1–72

Turner BL, Driessen JP, Haygarth PM, McKelvie ID (2003) Potential contribution of lysed bacterial cells to phosphorus solubilisation in two rewetted Australian pasture soils. Soil Biol Biochem 35:187–189

UNPP (2008) World population prospects: the 2008 revision. United Nations Population Division, New York

van der Eijk D, Janssen BH, Oenema O (2006) Initial and residual effects of fertilizer phosphorus on soil phosphorus and maize yields on phosphorus fixing soils – a case study in south-west Kenya. Agric Ecosyst Environ 116:104–120

Vanlauwe B, Sanginga N, Merckx R (1998) Recovery of leucaena and dactyladenia residue nitrogen-15 in alley cropping systems. Soil Sci Soc Am J 62:454–460

Vanlauwe B, Diels J, Sanginga N, Carsky RJ, Deckers J, Merckx R (2000a) Utilization of rock phosphate by crops on a representative toposequence in the Northern Guinea savanna zone of Nigeria: response by maize to previous herbaceous legume cropping and rock phosphate treatments. Soil Biol Biochem 32:2079–2090

Vanlauwe B, Nwoke OC, Diels J, Sanginga N, Carsky RJ, Deckers J, Merckx R (2000b) Utilization of rock phosphate by crops on a representative toposequence in the Northern Guinea savanna zone of Nigeria: response by Mucuna pruriens, Lablab purpureus and maize. Soil Biol Biochem 32:2063–2077

Vanlauwe B, Tittonell P, Mukalama J (2006) Within-farm soil fertility gradients affect response of maize to fertiliser application in western Kenya. Nutr Cycl Agroecosyst 76:171–182

Vanlauwe B, Idrissa A, Diels J, Sanginga N, Merckx R (2008) Plant age and rock phosphate effects on the organic resource quality of herbaceous legume residues and their N and P release dynamics. Agron Sustain Dev 28:429–437

Vanlauwe B, Bationo A, Chianu J, Giller KE, Mercks R, Mokwunye U, Ohiokpehai O, Pypers P, Tabo R, Shepherd K, Smaling E, Woomer PL, Sanginga N (2010) Integrated soil fertility management: operational definition and consequences for implementation and dissemination. Outlook Agric 39:17–24

Weisskopf L, Akello P, Milleret R, Khan Z, Schulthess F, Gobat J-M, Le Bayon R-C (2009) White lupin leads to increased maize yield through a soil fertility-independent mechanism: a new candidate for fighting Striga hermonthica infestation? Plant Soil 319:101–114

Wells AT, Chan KY, Cornish PS (2000) Comparison of conventional and alternative vegetable farming systems on the properties of a yellow earth in New South Wales. Agric Ecosyst Environ 80:47–60

Wichern F, Muller T, Joergensen RG, Buerkert A (2004) Effects of manure quality and application forms on soil C and N turnover of a subtropical oasis soil under laboratory conditions. Biol Fertil Soils 39:165–171

Wichern F, Mayer J, Joergensen RG, Muller T (2007) Release of C and N from roots of peas and oats and their availability to soil microorganisms. Soil Biol Biochem 39:2829–2839

Yusuf AA, Iwuafor ENO, Abaidoo RC, Olufajo OO, Sanginga N (2009) Grain legume rotation benefits to maize in the northern Guinea savanna of Nigeria: fixed-nitrogen versus other rotation effects. Nutr Cycl Agroecosyst 84:129–139

Zibilske LM, Bradford JM (2003) Tillage effects on phosphorus mineralization and microbial activity. Soil Sci 168:677–685

Zingore S, Delve RJ, Nyamangara J, Giller KE (2008) Multiple benefits of manure: the key to maintenance of soil fertility and restoration of depleted sandy soils on African smallholder farms. Nutr Cycl Agroecosyst 80:267–282

Chapter 18
Phosphorus and Global Change

Holm Tiessen, Maria Victoria Ballester, and Ignacio Salcedo

18.1 Human Appropriation of P Deposits

Phosphorus (P) is related to the issues of global change in two ways: the approximately 1,000 million tonnes (t) of P that have been mined and added to the environment over the past 150 years are part of environmental change at a global scale; and global climate and land use changes affect how P is being moved and used in the environment.

Prior to the middle of the nineteenth century, phosphate was largely derived by recycling bones, and its use was necessarily limited. In the 1840s, John Bennett Lawes patented the production of superphosphate, treating a phosphate ore found in Britain with sulfuric acid. By the 1870–1880s British superphosphate production reached some 15,000 t of P and the large phosphate ore deposits of South Carolina and Florida were being discovered, which are now responsible for about 20% of world production. The history of industrial phosphate use is therefore very similar to that of petroleum, which had been used for thousands of years in limited applications (such as asphalt) but was first transformed into liquid fuels at industrial scale in the middle of the nineteenth century. Like oil use, P use has increased exponentially and played a major role in human development. Phosphate deposits are unevenly distributed around the globe: Western Sahara, Morocco and China

H. Tiessen (✉)
Inter-American Institute for Global Change Research, Av. dos Astronautas 1758, 12227-010 São José dos Campos, SP, Brasil
e-mail: htiessen@dir.iai.int

M.V. Ballester
Centro de Energia Nuclear na Agricultura, Universidade de São Paulo, Av. Centenário 303, 13416-000 Piracicaba, SP, Brasil
e-mail: vicky@cena.usp.br.

I. Salcedo
Departamento de Energia Nuclear, Ignacio Salcedo. Universidade Federal de Pernambuco, Centro de Tecnologia, Av. Prof. Luiz Freire 1000, Cidade Universitária, 50740-540 Recife, PE, Brasil
e-mail: salcedo@ufpe.br

E.K. Bünemann et al. (eds.), *Phosphorus in Action*, Soil Biology 26,
DOI 10.1007/978-3-642-15271-9_18, © Springer-Verlag Berlin Heidelberg 2011

have the largest economic minerable reserves (5,700 and 6,600 million t, respectively, followed by South Africa, USA, and Jordan with 1,500, 1,200, and 900 million t, respectively. Brazil has reserves of 260 million t of mostly sedimentary deposits, but India does not have sizeable reserves (USGS 2007). For some countries, the reserve base can be two to four times greater than the actual reserves (USGS 2007) and it includes those resources that are currently economic (reserves), marginally economic (marginal reserves), and some of those that are currently subeconomic (subeconomic resources).

The Government of China has already imposed limits on exports of phosphate rock to maintain supply for domestic consumption (Jasinski 2004), and other phosphate-producing countries may follow suit in the near future. Hence, phosphate has been classed as a strategic resource. The "International Strategic Minerals Inventory Summary Report" on phosphate published by the US Geological Service (USGS Circular 930-C) in 1984 was prepared by a group of earth science and mineral resource agencies from Australia, Canada, Germany, South Africa, and the USA (Krauss et al. 1984). This report still remains the source for most assessments on the future availability of phosphate, and for arguments on "peak phosphorus," although numbers were reviewed in 2007 and the International Fertilizer Association is now in the process of compiling new data. It is an indication of phosphate's strategic importance that the USGS report was prepared as part of the first series of strategic mineral assessments that included P, Cr, Mn, and Ni – the latter all essential to metallurgy and therefore to military and industrial applications. The potential for geochemical constraints of these and additional elements has also been analyzed by Pickard (2008).

Phosphate is a finite resource and irreplaceable as an essential nutrient for all living organisms. The 1984 USGS strategic inventory report provides a summary of production and deposit information that may guide understanding of "how finite" phosphate ore deposits are and therefore how urgent conservative and renewable technologies are. The earth's crust contains on average 0.12% P. Mined deposits contain 2–20% P, i.e. 15–150 times the crustal average. In 1945, world P production was 1.8 million t P increasing to 21 million t P by 1981 (11 and 142 million t phosphate concentrate respectively, Krauss et al. 1984). Based on these numbers, and assessments by Cordell et al. (2009) and Tenkorang and Lowenburg-DeBoer (2009), Craswell et al. (2010) quote peak production estimates of 23–29 million t around the year 2030. The 1984 report estimates supply in known deposits and their extensions to last for about 100 years, assuming 5% consumption growth per year. These estimates depend on a large number of economic and regulatory assumptions, transport costs, elasticity of the agricultural input markets, and mandated limits on contaminants such as cadmium or fluoride. Significant reductions in P use occurred after the collapse of the Soviet Union with its centrally mandated fertilizer applications, and overall consumption has leveled off since the 1980s, particularly in richer countries where soil P reserves have been built up during 30 years of applications in excess of replacement levels (MEA 2005). A 7% drop in sales in 2008 is an indication of economic pressure on P use, which is likely to cause a re-evaluation of P needs over several years. Fixen (2009) estimates reserves to last for potentially another 300 years, although that includes currently uneconomic ores. Regardless of when peak production is expected and when mined P will become too expensive to serve as an input to crop production at present costs of food, P needs to be treated as a finite resource that is not renewable, only recyclable.

18.2 P Cycling in Natural and Agro-Ecosystems

In natural terrestrial ecosystems, P is relatively closely cycled between soils and biota (Smil 2000). The absence of a gaseous phase, the low solubility of calcium-bound P, the relatively strong retention of the orthophosphate anion by Fe and Al oxy-hydroxides, and the limited soil erosion under good vegetation cover, are key determinants in this close cycling. In such natural systems, mineralization and immobilization transformations between inorganic and organic forms are relatively more important than transport processes. In agro-ecosystems, human intervention opens the P cycle and transport processes become relatively more important. Fertilizer, livestock feeds, manure and/or bedding, agricultural products, human wastes, and P leaching and runoff have become important global P flows (Smil 2000; Liu et al. 2008). In addition to such transfers, P transformation processes also occur in managed lands, particularly in soils that receive continuous manure applications, which modify the proportions of organic and inorganic soil P fractions (Galvão and Salcedo 2009; Hao et al. 2008).

P pollution of surface waters by farming activities is largely due to soil erosion and surface runoff (overland flow) from croplands (Sharpley et al. 1995; Liu et al 2008). Grazing of pastures can also be a source of significant P pollution, particularly when stocking rates are high (Soupir et al. 2006; Chardon et al. 2007; Bilotta et al. 2008). Subsurface transport or leaching is a lesser but still important route for water pollution (McGechan et al. 2005; Kleinman et al. 2006). P overloads from manures and fertilizers, which enrich the surface layers of soil, contribute to pollution of surface waters. On the basis of data collected between 1998 and 2000 in Southern and Eastern Asia, Gerber et al. (2005) estimated that almost 40% of the cropped area had P overloads of 4.4–16 kg ha^{-1}. These two regions of Asia have the world's highest P consumption while Africa's consumption is the lowest (Table 18.1). Rates in Table 18.1 were calculated using total agricultural areas (arable + pasture lands), which certainly underestimate fertilizer use in croplands because in many regions pastures are less fertilized, if at all. Asia is the least favored region when the ratio of population to arable land was considered (Table 18.1) and therefore has the most fertilizer-intensive land management.

In addition to enhanced transport processes at the agro-ecosystem scale, P enters global transfers involving agricultural and industrial commodities (grains, meat, fruits, vegetables, seeds, phosphate rock, etc), processed food, fertilizers, and biofuels. Part of this global circulation is beneficial to people (Beaton et al. 1995; Villalba et al. 2008). However, excessive or uncontrolled transfers of P that occur as a consequence of intensive animal husbandry based on imported feeds result in contamination of lakes, rivers (Sharpley et al. 1995) and, ultimately, estuaries (Salcedo and Medeiros 1995) and coastal seas (Howarth et al. 1995), resulting in the environmental dispersion of P at a

Table 18.1 World mineral fertilizer-P consumption, population, arable land, meadows and pastures areas subdivided by continents and main regions (data for 2007)

Continents and regions	Consumed P $(t \times 10^3)$	Population (Inhabitants $\times 10^6$)	Arable land (ha $\times 10^6$)	Meadows and pastures (ha $\times 10^6$)	P consumed in agricultural land[a] (kg ha^{-1})	Population per arable land (Inhabitants ha^{-1})
Africa	348	943	219	911	0.3	4.3
Northern	182	300	56	239	0.3	5.3
America	5,396	900	364	803	4.5	2.5
North	2,677	336	215	253	5.6	1.6
South	2,461	379	113	454	4.2	3.4
Asia	9,871	3,983	504	1,089	5.9	7.9
Eastern	5,464	1,530	150	515	7.9	10.2
Southern	3,116	1,612	217	78	10.1	7.5
Europe	1,963	731	277	180	4.1	2.6
Eastern	751	296	194	116	2.4	1.5
Western	452	187	34	198.2	5.5	5.5
Oceania	671	34	45	393	1.5	0.7
World	18,250	6,593	1,411	3,378	3.7	4.7

Source: FAOSTAT (2009)

[a]Arable land plus meadows and pastures

massive scale. This highlights one fallacy of the literature on P. The often repeated statement that P is "not mobile" and the inference that excess P additions to agricultural land will eventually all contribute to residual fertilization effects is wrong. P is indeed relatively immobile compared to other elements such as N, but it does move in and through soils (Letkeman et al. 1996). Phosphate does dissolve, is found in ground and surface waters, and gets delivered to the oceans. When global reserves and global budgets are being assessed, that mobility is an important issue.

Transfers of P into aquatic environments can be classified by the nature of the P source as point (industrial, urban, feedlots), diffuse (agriculture and atmospheric deposition) and background (natural land) sources (EEA 2005). About 73% of the total P load (60 kt year^{-1}) in waters of Great Britain was attributed to households, 20% to agriculture, 3% to industry and 4% to background sources (White and Hammond 2009). The agricultural share in most source apportionments in Europe varies between 25 and 75% of the total load (0.3–2.5 kg ha^{-1} year^{-1}). The reason for this variability is the uneven point source share, which involves differences in population density, industrial activities, and wastewater treatment among the various catchments (EEA 2005). Comparison of area-specific indices to compute P loads into the aquatic system places Belgium and Netherlands at the top, with approximately 2.5 kg P ha^{-1} (EEA 2005). Over all land uses, China's area-specific P load was estimated at 1.4 kg ha^{-1} (Chen et al. 2008) but when only arable land was considered, it attained 2.8 kg P ha^{-1}. On the high end of area-specific P loads, analysis of the total suspended solids (TSS) load in 126 rivers pointed to 15 basins with particulate phosphorus (PP) loads greater than 20 kg P ha^{-1} year^{-1}, eight of which with loads above 30 kg P ha^{-1}: the Purari and Fly (New Guinea), Ord (Australia), Ganges and Tapti (India), Mahakam (Indonesia), and Irrawady (Myanmar) (Beusen et al. (2005)). Most of this will be from agricultural sources, although some of these basins also release PP by erosion of non-agricultural areas.

Estimates of world P transfers into the aquatic system from all sources vary between 12 and 21 million t year^{-1} (Liu et al. 2008; Beusen et al. 2005; Smil 2000; Howarth et al. 1995; Meybeck and Helmer 1989). The almost twofold variability in the estimates is due to the complexity of large scale data integration (Krueger et al. 2007). The relevant point is that estimated P losses are of the same order of magnitude as the annual world consumption of fertilizer-P (18.3 million t year^{-1}, Table 18.1). Two highly negative aspects of P flows into aquatic ecosystems are (1) the enhancement of biological activity in fresh water bodies (eutrophication), impairing water quality (USEPA 1996), and (2) the loss by burial of P in sediments of coastal ocean waters, except for a small fraction that is harvested in fish catch (Howarth et al. 1995).

18.3 Drivers of Changes in Reservoirs and Fluxes

One major global change process is the increasing urbanization of human populations. Changes to P flows in moving from rural to urban societies are illustrated by the city of Linköping, Sweden, whose population increased from 7,300 in 1870 to

130,000 in 2000 (Neset et al. 2008): increasing amounts of P reaching consumers and hence the waste handling system; increasing flow of products of animal origin, which are the main sources of P; and, most notably, increasing inputs of fertilizer-P in peri-urban areas.

Projections of world population are 9,100 million by 2050. Almost all of the increase will occur in developing countries (UN 2004) with much of the growth concentrated in urban areas (Steinfeld and Wassenaar 2007). The consequence of this urban population growth will be the expansion not only of food crops, but also of feed crops to satisfy demand for livestock products because increases in urbanization are correlated with increases in per capita consumption of animal products (Delgado et al. 1999). Although the projected increase in population for the next 40 years is approximately 50%, meat and milk productions are expected to double, reaching 465 million t of meat and 1,043 million t of milk. These projections are based on the expansion observed between 1980 and 2002, when the annual per capita consumption of meat in developing countries tripled and milk production more than doubled (Steinfeld et al. 2006).

Traditionally, the distribution of livestock production, particularly ruminants, was determined by the availability of natural pastures and crop residues. Pigs and poultry were typical of smallholders, and still are in some regions, because they re-process wastes. Market concentration has largely disengaged livestock production from grazing systems and replaced them with "landless" meat production systems that depend on outside supplies of feed (Steinfeld and Wassenaar 2007; Tamminga 2003). Poultry and pig production, ruminant feedlots and large-scale dairy production are examples of such landless systems (Seré and Steinfeld 1995). Nowadays, livestock production is more associated with access to output and input markets than the availability of grazing lands (Steinfeld et al. 2006). Almost 44% of the P fertilizer consumption in the USA was used for maize production and more than half of this crop production went into livestock feed (Steinfeld and Wassenaar 2007). The large soybean production of Brazil and Argentina is based on imported P fertilizer (42% of total fertilizer P consumption in the case of Brazil, Table 18.2) and most of the grain and by-products are exported (2nd and 3rd world exporters, respectively, FAO, 2009) since the nomenclature refers to the classification given in the FAO publication. These exports represent P flows mostly directed to animal production; they are examples of how globalization drives land-use changes and P fertilizer use at the national level.

The livestock sector is expanding at a faster rate than the rest of agriculture in almost all countries (Steinfeld and Wassenaar 2007). The shift of diets towards more meat and other animal products increases land and fertilizer needs beyond the compensation for growing population numbers. Case studies of the effects of increasing world consumption of beef for Queensland, Colombia and Brazil led McAlpine et al (2009) to point to the global beef market as a driver of regional and global change and to the need for reducing beef consumption, a view also shared by other authors (McMichael et al. 2007). The proximity of landless animal production to urban centers, particularly in Asia (Gerber et al. 2005), determine that large amounts of P contained in the feed drawn from agricultural land end up in the sewer

Table 18.2 Amount of mineral fertilizer-P (%) used for different crops in the world, selected countries, and the European Union, and the percentage mineral fertilizer-P consumption of these regions in relation to the world consumption (data for 2007)

Crop	World	China	India	USA	Brazil	EU-27	Egypt
Wheat	16.2	16.0	20.0	14.7	2.1	21.8	14.5
Rice	12.3	15.0	25.0	1.1	3.9	0.5	8.5
Maize	12.4	7.0	1.5	43.9	20.0	13.0	10.5
Soybean	7.5	3.0	2.5	10.5	42.4	0.3	0.2
Sugar crops	3.9	2.3	4.5	1.1	8.6	3.1	3.0
Fruits and vegetables	17.9	34.0	11.0	6.0	5.0	12.4	45.0
Other	29.8	22.7	36.5	22.7	18.0	48.9	18.3
% of world consumption[a]	100	30.5	14.6	10.4	9.3	8.8	0.6

Source: IFA (2009)
[a]These percentages allow calculation of the amount of fertilizer-P (t × 10^3) used by each country, based on the world consumption shown in Table 18.1

systems of cities (Weikard and Seyhan 2009). Citing Schröder (2005), "the ongoing disintegration of agriculture into specialized farms, specialized regions and even specialized countries has disrupted local nutrient cycles."

18.4 P Cycle and Biofuels

In addition to the expansion of food and feed production for growing populations and livestock production, there is also substantial growth in biofuel production in some countries, further increasing P demand. Interest in biofuels is increasing as a result of growing energy demand, geographical concentration of known petroleum reserves, increasing costs of finding and producing additional oil, and due to concerns about fossil fuel contributions to atmospheric CO_2 and climate change. Bioethanol and biodiesel production to substitute fossil carbon with biologically fixed carbon is being promoted to limit CO_2-driven climate change, because it returns to the atmosphere recently sequestered carbon dioxide (Ruth 2008; Field et al. 2007). A globally escalating demand for biofuels raises concerns about the potential of these biofuels to be sustainable, abundant, and environmentally beneficial energy sources (Tilman et al. 2006) because large areas of forest and grass lands, especially in the Americas and Southeast Asia, are being converted to produce biofuel (Klink and Machado 2005). Concurrently, lands typically used for food production are also being diverted to biofuel production (Fargione et al. 2008).

Most current biofuels are based on food crops: ethanol is produced by fermenting starch or sugar, mainly from corn or sugarcane, and biodiesel is made from oil seeds such as rape, from soybean, or from palm nuts (Ruth 2008). The expansion of biofuel production will result in increased applications of N and P fertilizers when natural or grazing lands are converted to sugarcane, corn, or soybean production. For instance, in 2007 the two main sources for biofuel production in Brazil, sugarcane with 8 million and soybean with 22 million ha (IBGE 2007), were responsible for the consumption of almost 9% (146,000 t) and 42% (712,000 t) of fertilizer-P, respectively (Table 18.2). Of the areas cultivated to these crops 54.6% (IBGE 2010) and 8%, respectively, were for fuel production. Approximately 44% of P consumption in USA is for corn (Table 18.2), some 15% of which is directed to biofuel production. Residue recycling will not greatly reduce P fertilizer demand. For instance, vinasse, a often re-cycled by-product of sugarcane alcohol production has a high organic matter and potassium content, but is poor in P and other nutrients (Ferreira and Monteiro 1987).

Global production of fuel ethanol tripled between 2000 and 2007, with the USA and Brazil accounting for most of the growth (OECD-FAO 2008). In Brazil, 25–30% of road transportation uses sugarcane-derived ethanol (Somerville 2006). In the last 5 years, sugarcane and soybean production have been growing at an annual rate of 7–8 % (IBGE 2007) and the land-use changes to accommodate the expansion of biofuel and feed production have affected large extensions of savannahs and subtropical forests and increased P flows. The increase in P-fertilizer demand to attend this growth resulted not only from the expansion in cultivated area, but is also the consequence of larger investments in fertilizers to improve productivity. According to Brazilian estimates, the average amount of P used per hectare of sugarcane increased from 27 kg in the 1990s to 30 kg in the last decade. These rates are higher than the 16 kg P ha^{-1} that can be calculated from P consumption and planted area information given in Table 18.2 and IBGE (2007). An explanation may be that P fertilizer is often only applied in the first (planting) year of the sugarcane cycle and not for the full 7-year sugarcane cycle. Consumption statistics may therefore be more reliable than estimates based on fertilizer recommendations. During the same period, the annual average amount of P used per hectare of soybean grew from 16 to 26 kg as farmers invested more in fertility management than in new land. Soybean production required an extra input of 6.1 million t of P from 1990 to 2008 (IBGE 2007; ANDA 2010). The 1,200 million L of biodiesel (ANP 2009), consume 3% of the total amount of mineral P fertilizers added to soybean. By 2010, the amount of P fertilizers necessary to supply the projected national biodiesel demand of 1,800 million L will increase proportionally.

18.5 Land Degradation

Pastures and arable land, particularly in the tropics, normally are established by conversion of natural vegetation (13 million ha $year^{-1}$, FAO 2007) on P-deficient soils often with high extractable Al and Fe contents. Once the tight nutrient cycling of natural ecosystems is interrupted, sustained agriculture production has to be based on lime and P inputs. Of the 3,400 million ha classified as permanent meadows and pastures (Table 18.1), approximately 532 million ha have potential for expansion of agricultural production, 80% of which are in the southern hemisphere and close to 200 million ha in Brazil (WWF 2009). The report estimates that arable lands in Brazil extend over 70 million ha and that about 30% of its 200 million ha of pasture lands are degraded (WWF 2009). By recovering such degraded soils the country could double its arable land resulting in less intense deforestation in the Cerrado and Amazon (WWF 2009; Field et al. 2007).

However, if the 70 million ha of degraded pasture in Brazil (IBGE 2007) were converted into biofuel fields, assuming an average input of 28 kg P ha^{-1}, this would result in nearly 2 million t of additional P, which is of the same order of magnitude as the current fertilizer P consumption of Brazil (1.7 million t in 2007, Table 18.2). It is not clear how such an expansion in P demand could be satisfied and for how long. World P fertilizer consumption in 2007 amounted to 17.4 million t for a total agriculture area of 4,967 million ha (Table 18.1). The expected growth in world demand for phosphate fertilizers by 2012 is 2.2 million t P compared to 2006 figures (FAO 2008). Unless management policies are developed and environmental regulations implemented to induce P recycling at farm, municipality, and watershed scales, the environmental effects of such a dramatic increase in land conversion are grim.

These additional land pressures are in conflict with environmental goals of preserving ecosystem services, and are accompanied by an historic increase in the intensity and extension of soil degradation at a world scale (EEA 2003; Pimentel 2006; FAO 2007). Actual arable land is approximately 1,500 million ha worldwide (Table 18.1), which is about the same area as has been abandoned due to land degradation since farming began (Lal 1990). Because soil erosion removes surface soil, which normally has a considerably higher P content than the subsoil, this diminishes the potential for food production. Soil erosion rates depend on past and present land uses, soil type, climate, and land-surface forms (Pimentel 2006). The rate of loss of arable land is estimated at 10 million ha $year^{-1}$ (Faeth and Crosson 1994). Estimates obtained from analyzing global river transport of sediments (Beusen et al. 2005) or analyzing soil erosion with reference to land use and climate change (Yang et al. 2003) are coincident in reporting South East Asia as the region suffering the most serious erosion, followed by South America and Africa. Societal response to this situation is inadequate. Most consumers connect meat, fruits and vegetables with agricultural activities, but not with the soil and even less with P. In this context, education must assume a role in reinstating the link between food availability and resource conservation in the understanding of urban societies.

P carried by rivers to be buried in sediments of estuaries and near-shore marine environments of the continental margins (Ruttenberg and Berner 1993) will probably stay out of reach until land reserves of phosphate rock become so depleted that high phosphate ore prices could justify the investment of exploring such dilute marine deposits. However, at this point, P fertilizer prices will be much higher than nowadays and the current perception of an ever-growing food supply will have to be adapted to the new reality. Therefore, optimization of P resource management to minimize its gross transfer into the aquatic system should be a major issue when analyzing sustainable food production.

18.6 Concluding Remarks

Phosphate has been essential to the growth of human populations and their prosperity. But, P has been used in production systems in a way that leaked large amounts of P into down-stream ecosystems. The negative effects on these ecosystems, whose biological productivity are often P or N and P limited, are well known. Detrimental effects from P excess now occur at a global scale, although some regions, particularly in poorer countries, remain P-deficient. Agricultural production will always be P-dependent, and it is becoming increasingly clear that P is a finite resource that, by scarcity or price, will be a limiting factor in future production increases. How to feed growing populations has long been a preoccupation. In addition, there is now growing concern about agricultural expansion for biofuel production to offset fossil fuel use. Balancing arguments of CO_2 emissions versus P limitations on human wellbeing will not be an easy task. It is clear though that P use will have to be accompanied by greater efforts towards re-use, recycling, and strategic targeted applications.

References

Beaton JD, Roberts TL, Halstead EH, Cowell LE (1995) Global transfers of P in fertilizer materials and agricultural commodities. In: Tiessen H (ed) Phosphorus in the global environment. Wiley, New York, pp 7–26, SCOPE/ICSU/UNEP

ANDA (2010) Main fertilizer sector indicators. Associação Nacional para Difusão de Adubos, São Paulo, Brasil. Available at http://www.anda.org.br/estatisticas.aspx Last accessed 10 Aug 2010

ANP (2009) National Petroleum Agency (Agência Nacional de Petróleo, Gas Natural e Biocombustíveis). Resolution ANP no 7, 19.3.2008–DOU 20.3.2008

Beusen AHW, Dekkers ALM, Bouwman AF, Ludwig W, Harrison J (2005) Estimation of global river transport of sediments and associated particulate C, N, and P. Global Biogeochem Cycles 19: GB4S05, doi:10.1029/2005GB002453

Bilotta GS, Brazier RE, Haygarth PM, Macleod CJA, Butler P, Granger S, Krueger T, Freer J, Quinton J (2008) Rethinking the contribution of drained and undrained grasslands to sediment-related water quality problems. J Environ Qual 37:906–914

Chardon WJ, Aalderink GH, van der Salm C (2007) Phosphorus leaching from cow manure patches on soil columns. J Environ Qual 36:17–22

Chen M, Chen J, Sun F (2008) Agricultural phosphorus flow and its environmental impacts in China. Sci Total Environ 405:140–152

Cordell D, Drangert J-O, White S (2009) The story of phosphorus: global food security and food for thought. Global Environ Change 19:292–305

Craswell ET, Vlek PLG, Tiessen H (2010) Peak phosphorus – implications for soil productivity and global food security. In: Proceedings 19th World Congress of Soil Science: Soil solutions for a changing world 1–6 August 2010, Brisbane, Australia. Published on CDROM, ASSSI, Warragul, Australia

Delgado C, Rosegrant M, Steinfeld H, Shui S, Courbois C (1999) Livestock to 2020: the next food revolution. Food, agriculture and the environment discussion paper 28. IFPRI/FAO/ILRI. Available at http://www.ifpri.org/sites/default/files/publications/pubs_2020_dp_dp28.pdf Last accessed 10 Aug 2010

EEA (2003) Assessment and reporting on soil erosion: background and workshop report. European Environmental Agency technical report 94, EEA, Copenhagen

EEA (2005) Source apportionment of nitrogen and phosphorus inputs into the aquatic environment. European Environmental Agency Report 7. EEA, Copenhagen

Faeth P, Crosson P (1994) Building the case for sustainable agriculture. Environment 36(1):16–20

FAO (2007) The Agriculture–forest interface. Committee on Agriculture, 20th Session, Inf. 13. Food and Agriculture Organization of the United Nations, Rome

FAO (2008) Current world fertilizer trends and outlook to 2012. Food and Agriculture Organization of the United Nations, Rome

FAO (2009). FAOSTAT. Available at http://faostat.fao.org Last accessed 10 Aug 2010

Fargione J, Hill J, Tilman D, Polasky S, Hawthorne P (2008) Land clearing and the biofuel carbon debt. Science 319:1235–1237

Ferreira ES, Monteiro AO (1987) Efeitos da aplicação da vinhaça nas propriedades químicas, físicas e biológicas do solo. Bol Técnico Copersucar, São Paulo 36:3–7

Field CB, Campbell JE, Lobell DB (2007) Biomass energy: the scale of the potential resource. Trends Ecol Evol 23(2):65–72

Fixen PE (2009) World fertilizer nutrient reserves – a view to the future. Better Crops 93:8–11

Galvão SRS, Salcedo IH (2009) Soil phosphorus fractions in sandy soils amended with cattle manure for long periods. Braz J Soil Sci 33:613–622

Gerber P, Chilonda P, Franceschini G, Menzi H (2005) Geographical determinant and environmental implications of livestock production intensification in Asia. Biores Technol 96: 263–276

Hao X, Godlinski F, Chang C (2008) Distribution of phosphorus forms in soil following long-term continuous and discontinuous cattle manure applications. Soil Sci Soc Am J 71(1):90–97

Howarth RW, Jensen HS, Marino R, Postma H (1995) Transport to and processing of P in near-shore and oceanic waters. In: Tiessen H (ed) Phosphorus in the global environment. Wiley, New York, pp 323–346, SCOPE/ICSU/UNEP

IBGE (2007) Agrarian census. Brazilian Institute of Geography and Statistics. www.ibge.gov.br

IBGE (2010) Aggreagated Database (SIDRA). Brazilian Institute of Geography and Statistics. http://www.sidra.ibge.gov.br/

IFA (2009) Assessment of fertilizer use by crop at the global level, 2007/07–2007/08. Patrick Heffer, International Fertilizer Industry Association, Paris. Available at http://www.fertilizer.org/ifa/Home-Page/STATISTICS/FUBC Last accessed 10 Aug 2010

Jasinski SM (2004) Phosphate rock. In: US Geological Survey Minerals Yearbook 2004. USGS, Washington, DC, pp 56.1–56.10. Available at http://minerals.usgs.gov/minerals/pubs/commodity/phosphate_rock/phospmyb04.pdf Last accessed 10 Aug 2010

Kleinman PJA, Srinivasan MS, Dell CJ, Schmidt JP, Sharpley AN, Bryant RB (2006) J Environ Qual 35:1248–1259

Klink CA, Machado RB (2005) Conservation of the Brazilian cerrado. Conserv Biol 19(3): 707–713

Krauss UH, Saam HG, Schmidt HW (1984) International strategic minerals inventory summary report – phosphate. US Geological Survey Circular 930-C. Department of the Interior, Washington DC

Krueger T, Freer J, Quinton JN, Macleod CJA (2007) Processes affecting transfer of sediment and colloids, with associated phosphorus, from intensively farmed grasslands: a critical note on modeling of phosphorus transfers. Hydrol Proc 21:557–562

Letkeman LP, Tiessen H, Campbell CA (1996) Phosphorus transformation and redistribution during Pedogenesis of western Canadian soils. Geoderma 71:201–218

Lal R (1990) Soil erosion and land degradation: the global risks. In: Lal R, Stewart BA (eds) Soil degradation. Springer, New York, pp 129–172

Liu Y, Villalba G, Ayres RU, Schroder H (2008) Global phosphorus flows and environmental impacts from a consumption perspective. J Ind Ecol 12(2):229–247

McAlpine CA, Etter A, Fearnside PM, Seabrook L, Laurance WF (2009) Increasing world consumption of beef as a driver of regional and global change: a call for policy action based on evidence from Queensland (Australia), Colombia and Brazil. Global Environ Change 19:21–33

McGechan MB, Lewis DR, Hooda PS (2005) Modelling through-soil transport of phosphorus to surface waters from livestock agriculture at the field and catchment scale. Sci Total Environ 344:185–199

McMichael AJ, Powles JW, Butler CD, Uauy R (2007) Food, livestock production, energy, climate change, and health. Lancet 370:1253–1263

MEA (2005) Scenarios. Ecosystems and human well-being, vol 2. Millennium ecosystem assessment, Island Press, Washington DC

Meybeck M, Helmer R (1989) The quality of rivers: from pristine stage to global pollution. Paleogra Palaeoclimatol Palaeoecol 75:283–309

Neset TSS, Bader HP, Scheidegger R, Lohm U (2008) The flow of phosphorus in food production and consumption – Linköping, Sweden, 1870–2000. Sci Total Environ 396:111–120

OECD-FAO (2008) Agricultural outlook 2008–2017. Organisation for Economic Co-operation and Development–Food and Agriculture Organization of the United Nations. OECD, Paris. Available from http://www.oecdbookshop.org

Pickard WF (2008) Geochemical constraints on sustainable development: can an advanced global economy achieve long-term stability? Glob Planet Change 61:285–299

Pimentel D (2006) Soil erosion: a food and environmental threat. Environ Dev Sustain 8:119–137

Ruth L (2008) Bio or bust? The economic and ecological cost of biofuels. European Molecular Biology Organization. EMBO Rep 9(2):130–133

Ruttenberg KC, Berner RA (1993) Authigenic apatite formation and burial in sediments from non-upwelling continental margin environments. Geochim Cosmochim Acta 57:991–1007

Salcedo IH, Medeiros C (1995) Phosphorus transfers from tropical terrestrial to aquatic systems – Mangroves. In: Tiessen H (ed) Phosphorus in the global environment. Wiley, New York, pp 347–362, SCOPE/ICSU/UNEP

Schröder J (2005) Revisiting the agronomic benefits of manure: a correct assessment and exploitation of its fertilizer value spares de environment. Bioresour Technol 96:253–261

Seré C, Steinfeld H (1995) World livestock production systems: current status, issues and trends. FAO Animal Production and Health Paper 127. Food and Agriculture Organization of the United Nations, Rome

Sharpley AN, Hedley MJ, Sibbesen E, Hillbricht-Ilkowska A, House WA, Ryszkowski L (1995) Phosphorus transfers from terrestrial to aquatic ecosystems. In: Tiessen H (ed) Phosphorus in the global environment. Wiley, New York, pp 171–200, SCOPE/ICSU/UNEP

Smil V (2000) Phosphorus in the environment: natural flows and human interferences. Annu Rev Energy Environ 25:53–58

Somerville C (2006) The billion-ton biofuels vision. Science 312:1277

Soupir ML, Mostaghimi S, Yagow ER (2006) Transport from livestock manure applied to pastureland using phosphorus-based strategies. J Environ Qual 35:1269–1278

Steinfeld H, Gerber P, Wassenaar T, Castel V, Rosales M, de Haan C (2006) Livestock's long shadow: environmental issues and options. Food and Agriculture Organization of the United Nations, Rome

Steinfeld H, Wassenaar T (2007) The role of livestock production in carbon and nitrogen cycles. Annu Rev Environ Resour 32:271–294

Tamminga S (2003) Pollution due to nutrient losses and its control in European animal production. Livest Prod Sci 84:101–111

Tenkorang F, Lowenberg-DeBoer J (2009) Forecasting long-term global fertilizer demand. Nutr Cycl Agroecosyst 83:233–247

Tilman D, Hill J, Lehman C (2006) Carbon-negative biofuels from low-input high-diversity grassland biomass. Science 314:1594–1600

UN (2004) World population prospects: the 2004 revision. Population Division of the Department of Economic and Social Affairs of the United Nations Secretariat, Washington, DC

USEPA (1996) Environmental indicators of water quality in the United States. USEPA 841-R-96-002. US Environmental Protection Agency, Office of Water (4503F), US Government Printing Office, Washington DC

USGS (2007) Phosphate rock. Mineral commodity summaries. United States Geological Survey, Washington DC. Available from http://minerals.usgs.gov/minerals/pubs/commodity/phosphate_rock/ Last accessed 10 Aug 2010

Villalba G, Liu Y, Schroder H, Ayres RU (2008) Global phosphorus flows in the industrial economy from a production perspective. J Indust Ecol 12(4):557–569

Weikard HP, Seyhan D (2009) Distribution of phosphorus resources between rich and poor countries: the effect of recycling. Ecol Econ 68:1749–1755

White PJ, Hammond JP (2009) The sources of phosphorus in the waters of Great Britain. J Environ Qual 38:13–26

WWF (2009) O impacto do Mercado mundial de biocombustíveis na expansão da agricultura brasileira e suas consequências para as mudanças climáticas. Programa de Agricultura e Meio Ambiente, World-Wide Fund for Nature, Brasília, Brasil

Yang D, Kanae S, Oki T, Koike T, Musiake K (2003) Global potential soil erosion with reference to land use and climate changes. Hydrol Process 17:2913–2928

Appendix – General Conclusions

Upon inviting authors to contribute to this book we asked them to examine (1) the role of biological processes in soil phosphorus (P) cycling against the availability of inorganic P resulting from physicochemical processes and (2) the interactions between carbon, nitrogen, and P. In addition, authors were asked to point out gaps in methods and understanding.

When reviewing the chapters, we were once again impressed by the great variety of biological processes in soil P cycling, including for example enzymatic processes, microbial solubilization of inorganic P, and plant P acquisition. However, quantitative assessments of the contribution of these processes to soil P dynamics and plant P nutrition are rare. One noteworthy exception is the study by Achat et al. (mentioned in Chap. 3 by Frossard et al.) of a very low P-sorbing sandy forest soil with very low total P content, the majority of which was present in soil organic matter and microorganisms. In this situation, mineralization of microbial P largely determined the increase in phosphate concentration in the soil solution, although the release of P from soil organic matter and from mineral surfaces remained low. By contrast, Table 17.1 in Chap. 17 by Oberson et al. presents data for a cropped soil in which the daily rate of gross organic P mineralization was only 5–9% of the potential release of inorganic P by physicochemical processes. To reach quantitative conclusions, we need more assessments like the examples given above, and biological processes other than mineralization and immobilization should also be quantified, preferably at different timescales.

Although not always presented quantitatively, the book contains plenty of examples of how biological P cycling interacts with the physicochemical processes. Biological processes go "beyond organic P" and there is no clear separation between the dynamics of inorganic and organic P because biological activity and physicochemical processes are closely interrelated.

Information on the interaction of P dynamics with carbon can be found in nearly every chapter, whereas interactions with nitrogen are less frequently addressed. Examples for the role of carbon in P cycling include the net flux of carbon from the plant to the fungal partner in mycorrhizal symbioses and the exudation of lytic enzymes and organic acids by some of the fungi (Chap. 6 by Jansa et al.), the role of carbon metabolites in the regulation of plant response to changes in P supply

(Chap. 10 by George et al.), and the effects of carbon availability on microorganisms (Chap. 17 by Oberson et al.). For the interactions of nitrogen with P dynamics, Jouany et al. (Chap. 11) report some important findings for grasslands, and Reed et al. (Chap. 14) stress that in tropical forests, addition of N can increase soil phosphatase activity, presumably because it promotes the formation of N-rich proteins such as phosphatases. Obviously, it is very important to consider carbon and nitrogen dynamics when investigating biological P cycling. For a better, comprehensive understanding of nutrient cycling, combined studies for several elements should thus be continued.

Several chapters identified the same research topics and came to similar conclusions:

- Numerous authors claim that further progress will be made not only through improvement of single techniques (e.g. spectroscopic methods, Doolette and Smernik, Chap. 1), but in particular, through the combination of different methods. Examples are the combination of extraction of microbial cells from soil with molecular analysis and examination of single-cell ecophysiology (Bünemann et al., Chap. 2), measurements of soil P dynamics when conducting field trials with P-solubilizing microorganisms in order to distinguish P-related from other effects (Jones and Oburger, Chap. 7), and assessment of P fluxes and phosphatase activities in relation to the expression of associated genes (Nannipieri et al., Chap. 9). Likewise, a need for close collaboration between experimentalists and modelers has been identified (Schnepf et al., Chap. 5).
- Several chapters stress the importance of a better mechanistic understanding of the role of certain groups of organisms in order to manage them to the greatest benefit in agroecosystems and forests. Examples are the quantification of weathering by ectomycorrhiza (Jansa et al., Chap. 6), assessing the role of the macrofauna (Chapuis-Lardy et al., Chap. 8), measuring the utilization of organic P by mycorrhizae in forests (Fox et al., Chap. 13), and identifying the source of microbial P uptake (Oberson et al., Chap. 17). Thus, as Reed et al. (Chap. 14) put it, we need to know "which pools are accessible by which organisms on which timescales", and Oberson et al. (Chap. 17) add that we need to know how to render this P available to plants. We think that the combination of molecular tools with measurements of P fluxes could further advance our understanding of the action and actors. This is also in accordance with the word of caution by Wasaki and Maruyama (Chap. 4) not to forget conventional, culture-dependent methods when using molecular tools. Likewise, George et al. (Chap. 10) call for a systems approach in the management of plant P nutrition because of the complexity of gene-by-gene and gene-by-environment interactions.
- On the ecosystem level, several authors conclude that it is mandatory to better understand the effect of P limitation on net primary productivity and other ecosystem functions in order to predict the effects of climate change on ecosystems (Weintraub, Chap.12; Reed et al., Chap. 14; Belnap, Chap. 15). For agroecosystems, Oberson et al. (Chap. 17) suggest optimizing P-use efficiency through integrated P management (Fig. 17.2), while for the global P cycle,

increased reuse, recycling and strategic targeted applications are required (Tiessen et al., Chap. 18).

In this book we aimed at capturing the action, and have to conclude that we need to know more about the actors. In addition, a lot of action is urgently needed from society to improve the management of the limited P resource, which is today often present in excess.

Else K. Bünemann, Astrid Oberson and Emmanuel Frossard
Lindau, Switzerland

Index